Monographs on Endocrinology

Volume 4

Edited by

F. Gross, Heidelberg · A. Labhart, Zürich

T. Mann, Cambridge · L. T. Samuels, Salt Lake City

J. Zander, München

U. Westphal

Steroid-Protein Interactions

With 144 Figures

Springer-Verlag Berlin · Heidelberg · New York 1971

Ulrich Westphal

Professor of Biochemistry
University of Louisville, School of Medicine
Louisville, Kentucky 40202/USA

ISBN-13: 978-3-642-46264-1 e-ISBN-13: 978-3-642-46262-7
DOI: 10.1007/978-3-642-46262-7

Softcover reprint of the hardcover 1st edition 1971

To Ilse
in appreciation
of her understanding of
Husband — Manuscript Interactions

Preface

This is the story of steroid-protein interactions as one investigator sees it. Following the general concept of this monograph series, it emphasizes the results and interpretations obtained in the author's laboratory, and is therefore a subjective account. Recognizing, however, that the discussion of the subject would be incomplete if the material were limited to one scientist's work, the essential achievements of other investigators have been incorporated. An effort has thus been made to give a balanced presentation and to enable the reader to see in perspective the varied facets of the interactions between steroids and proteins.

Since this is the first comprehensive treatment of the topic, it seems appropriate to go to the roots, and try to find out how it all started. The first chapters, therefore, take the reader to the laboratories of those who very early conceived the significance of the attachment of dyes, drugs, and other conspicuous molecules to those colloids called proteins. The discovery of the steroid hormones set the stage for meaningful investigation of their interaction with proteins of various origins and functions — a process which is continuing today with increasing vigor.

The problems involved in steroid interactions with proteins relate to many fields — from physical chemistry to biochemistry, to endocrinology and medicine. An attempt has been made to introduce the readers to the subjects without expecting them to have any special knowledge in these separate disciplines, at the risk of giving explanations elementary to some. At the same time, full documentation was considered essential to account for the accepted results and their interpretation. Whenever space did not allow discussion of a particular point, reference was made to the bibliography for supplemental information. The author appreciates that editors and publisher have consented to this comprehensive presentation.

No author will ever be able to thank all those who have helped him. It has been my good fortune to witness as a young student the evolution of steroid chemistry in Adolf Windaus' laboratory. It was decisive to get involved in the early chemical elucidation of steroid hormones under the guidance of Adolf Butenandt, and to remain for many years in the unforgettable atmosphere of the *Arbeitskreis Butenandt*. Subsequent association with the medical department of Hermann Bennhold opened my eyes to the problems of protein binding and its biological and medical implications. In recent times, there were many who supported my work, from sending valuable compounds to making possible my participation in both national and international symposia and meetings, which proved to be most helpful in evaluating new developments. My indebtedness to all these men is obvious — and deeply felt.

Indispensable is the help from close associates. The contributions of my students and other co-workers are documented in the bibliography. Less apparent but of equal importance for a research laboratory, I am grateful for the loyal cooperation of Miss NANCY RUST and Mr. GEORGE B. HARDING who have given me highly competent assistance for many years. More than one secretary was called upon to help — special acknowledgement is given to Mrs. ELLEN FORD who kept order in the reprint files and typed the entire manuscript and bibliography. Most of the drawings originating from our laboratory were prepared by Mr. HARDING. Reading of the manuscript by Drs. L. T. SAMUELS and R. M. BURTON aided in clarifying the presentation.

Through the preparation of this monograph I have enjoyed, and acknowledge with gratitude, the patient and helpful cooperation of Dr. H. GÖTZE and his staff at the Spinger-Verlag.

Finally, I should like to express my appreciation to the US Public Health Service, National Institutes of Health, for a Research Career Award and continuing research support that has made — and is making — my work possible.

January, 1971 ULRICH WESTPHAL

Contents

I. Introduction

Corpora non agunt nisi liquida
(PARACELSUS, 1493—1541)
Corpora non agunt nisi fixata
(PAUL EHRLICH, 1854—1915)

These two statements seem to contradict each other. PARACELSUS postulates the "liquid", i. e., dissolved or dispersed state for a compound to act; PAUL EHRLICH sees the central requirement for activity in an attachment, a fixation of the substance. Between the two poles of these concepts lies much of the material that will be discussed in this book. The contradiction disappears when we interpret in context: for any activity, chemical, biochemical or biological, a compound must be in solution. The molecules, or ions, must not be bound to any great extent to themselves, in a crystalline or amorphous solid; they also must not be bonded to other structures, or other binding compounds, since the association may prevent them from reaching their targets or from exerting their specific effects. It is in this sense that PARACELSUS' claim must be understood: substances are not active unless they are in solution,—"liquid".

PARACELSUS' perception forms an important milestone in man's understanding of the chemical activity of any compound, including its effects on biological structures; it is as valid today as it was more than 400 years ago. However, it does not give us any information on how the biological effects are mediated. PAUL EHRLICH's postulate tries to fill this void: The substances do not exert their activity unless they become attached.

Impressed by his observations on binding of dyes to proteins, and by distinct differences in the binding affinities, EHRLICH had developed his side chain theory. Toxins or antigens or other substances become bound to specific receptors which are located at the cell and serve as a connecting link between the compounds and the reactive tissue. This attachment to the "chemoreceptor" was originally considered to be so firm as to be practically irreversible; later the concept was modified to include weaker binding forces between a receptor and the active substance, resulting in a dissociable complex.

These basic ideas, long ago conceived by PARACELSUS and by PAUL EHRLICH, form an essential part of any interpretation of drug action, of the biological activity of hormones and vitamins, of immunochemical processes, of enzyme reactions and other biochemical phenomena. Whatever the specific mechanism of the biological action of a biocatalyst may be, one of the first steps can be assumed to involve an interaction with a reactive tissue, consisting of protein or another macromolecular structure. This may be an interaction with a specific receptor protein located at the cell wall or inside the cell, at the confining membranes of subcellular particles, or within the mitochondria and the microsomes; it may be an interaction with an enzyme protein, resulting in an allosteric change of its conformation and a modification of its enzymatic activity. Or it may be an interaction with a gene-related structure in the nucleus, e. g., with a nuclear protein or a repressor molecule of the Jacob-Monod scheme, so that the gene expression is influenced in a positive or negative way.

The interaction of hormones or other biologically active compounds with target site proteins has become a field of most acute interest. Intensive studies in many laboratories are devoted to interactions between steroid hormones and cellular macromolecules involved in the mechanism of action. Receptor proteins have been recognized in cytoplasma and nucleus, and the relationships between them begin to emerge. However, these proteins have not yet been obtained in pure form, and their chemical properties await elucidation.

More information is available on ligand interactions with serum proteins. These natural macromolecules are readily accessible and can be obtained in pure form; their physicochemical properties have been studied extensively. Use of serum proteins for binding studies, especially those involving plasma albumin, have therefore provided much of our basic knowledge on ligand-protein interactions. Such studies have also brought into focus two problems intimately connected with the function of the blood proteins: that of transport of the ligands as protein-bound moieties in the circulating blood, and that of the effect of protein binding on the biological activity of the ligand.

Our main concern in interactions between biologically active compounds and proteins thus involves the three areas mentioned: transport, effect on biological function, and specific phases of the mechanism of action. It is in this order, historically, that the relationships between protein-binding phenomena and their biological consequences were conceived, and we may feel that it is the same order in which the general significance of ligand interaction with protein structures increases. The biochemist's interest certainly proceeded in this order—from the simple concept of a transport mechanism, over a mode of regulation of physiological function in which the binding protein plays the role of a reversible inhibitor, to a mechanism of action involving interaction with receptor proteins or with specific regulators of gene activity.

The scope of the present monograph has been essentially limited to steroid interactions with the various protein components of the blood serum, thus including possible transport function as well as regulatory mechanisms. The nature and biological significance of receptor proteins for steroid hormones will be the subject of a monograph by E. V. Jensen in this series (1970). In accordance, interactions of steroids with tissues and tissue proteins, with cells, subcellular fractions, and other particulate matter will not be discussed.

II. Early Studies on Interactions between Proteins and Small Molecules or Ions

The first observations on binding of ligands to proteins were made almost 70 years ago, long before the chemical structure of the steroids was elucidated. In 1903, the anatomist HEIDENHAIN (1903) described the dialysis of acidic azo dyes through parchment tubing and its inhibition by the addition of serum albumin. He interpreted this result correctly by assuming complex formation between dye and serum protein. This appears to be the first documented observation of ligand-protein interaction in solution and recognition of the consequences on the diffusibility of the ligand. HEIDENHAIN seems to have sensed and appreciated the biochemical significance and implications of his finding; however, he left it to others to further pursue the problem which lay beyond the field of anatomy.

A more comprehensive series of experiments was carried out a few years later by BUSCK (1906) who investigated the photodynamic effects of certain dyes ("photobiological sensibilizators"), such as fluorescein derivatives, on paramecia. Addition of serum reduced or abolished the toxicity toward the test organisms. At the same time, the presence of the serum decreased the diffusion rate of the dyes through various dialysis membranes, affected fluorescence and light absorption spectra, altered solubility and other properties. BUSCK interpreted these results as caused by the formation of complexes between albumin and the dyes. The dialysis method was found useful for this type of studies in several laboratories; for example, OSBORNE (1906) in the same year established that dissociable protein-cation complexes readily exchanged the bound cations during dialysis.

BECHHOLD (1907a, b) apparently was the first author to publish quantitative data on dye binding to proteins, and to give valid physicochemical interpretations of the phenomena. He worked in the institute of PAUL EHRLICH when he developed a novel technique which he called ultrafiltration. When BECHHOLD added horse serum to an aqueous solution of methylene blue, which by itself was readily filtered or dialyzed through the semipermeable membranes used, the ultrafiltration of the dye was greatly impaired; the ultrafiltrate had less than one third of the dye concentration present in the solution before filtration (Table II-1). Increasing dilution of the dye-serum mixture reversed the effect: the dye concentration in the ultrafiltrate approached more and more the concentration of the unfiltered solu-

Table II-1. *Ultrafiltration of a mixture of methylene blue and dialyzed horse serum (2.2% albumin)*[a]

Concentrations in mixture before filtration		Concentration of methylene blue in ultrafiltrate expressed as	
Methylene blue mg/100 ml	Albumin g/100 ml	mg/100 ml	% of concentration before filtration
125	1.1	36	29
63	0.55	31	50
31	0.28	20	66
16	0.14	11	70
8	0.07	7.5	94

[a] Calculated from experimental data of BECHHOLD (1907 a). Under the assumption, verified by BECHHOLD, that no dye adsorption to the ultrafilter occurs, the last column gives an approximate measure of the percent unbound methylene blue in the original mixture with serum.

tion. BECHHOLD interpreted these results as indicating that methylene blue was bound ("adsorbed") to the serum proteins and thus made incapable of passing through the semipermeable pores of the ultrafilter. He furthermore concluded that the "adsorption" was "split", i.e., that the dye-protein complex dissociated, with increasing dilution.

Similar conclusions were drawn from another simple experiment (BECHHOLD, 1907 a). A wool thread was dyed in an aqueous solution of 0.25% methylene blue. When the staining bath contained serum in addition to the same concentration of methylene blue, the wool dyed less intensively, indicating decreased availability of the dye by attachment to serum proteins. Dilution of both the serum-containing and the serum-free staining solutions eliminated the difference in color intensity of the stained wool.

Although these observations are essentially of a qualitative nature, they document for the first time the basic properties of ligand-protein interactions. BECHHOLD drew correct physicochemical conclusions which, however, were not recognized in their significance by contemporary researchers. BECHHOLD foresaw clearly the biological and medical implications of his findings beyond their importance for the understanding of colloidal systems. He noticed that the disinfecting action of tetrachloro-o-biphenol against diphtheria bacilli was greatly reduced when the pathogens were cultured in serum instead of bouillon. He then demonstrated by ultrafiltration that approximately 90% of the phenol was retained on the filter with the serum proteins, and concluded "daß in erster Linie chemische oder physikalisch-chemische Ursachen, nämlich die Bindung des Desinfizienz durch das Blutserum, die Herabsetzung der Desinfektionswirkung im Organismus bedingten" (BECHHOLD, 1907 a).

In 1909, SÖRENSEN published his classical studies on the significance of the hydrogen ion concentration on enzymatic processes. He noticed a change in color of certain indicator dyes on addition of protein when the measurements of electromotive force did not show a change in pH. The interpretation of this phenomenon led to the discovery of the "protein error" in pH determination by indicator dyes. SÖRENSEN recognized that binding of the dyes, e.g., those of the azo group, congo red, and others, to the genuine protein was responsible for the "Eiweißfehler"; degraded protein, or protein in the presence of active pepsin, was unable to produce consistent color change.

At the same time, FRIEDEMANN (1909) demonstrated dye binding to proteins in a more direct way. While engaged in studies on the aqueous humor of the eye, he found that the diffusion of fluorescein through semipermeable membranes was impaired by the addition of serum proteins. This important observation was ignored or forgotten until more than 10 years later when DE HAAN and VAN CREVELD (1921) investigated the mechanism of the permeation of fluorescein from the blood vessels into the anterior chamber of the eye—the classical system introduced by PAUL EHRLICH (1882) 40 years earlier. DE HAAN and VAN CREVELD clearly recognized that it was the binding of fluorescein to the serum proteins which prevented the free transit of the dye into the aqueous humor, resulting in the relatively low dye concentration in the anterior chamber which had puzzled earlier investigators. As supporting evidence, the Dutch authors performed ultrafiltration and pressure dialysis experiments which proved the binding of fluorescein to serum proteins and its dependence on the concentration ratio of dye to protein. Fig. II-1 demonstrates the decrease of relative binding with increasing total fluorescein concentration, and shows the similarity of the curve to that of an adsorption isotherm. On the basis of these *in-vitro* results, DE HAAN and VAN

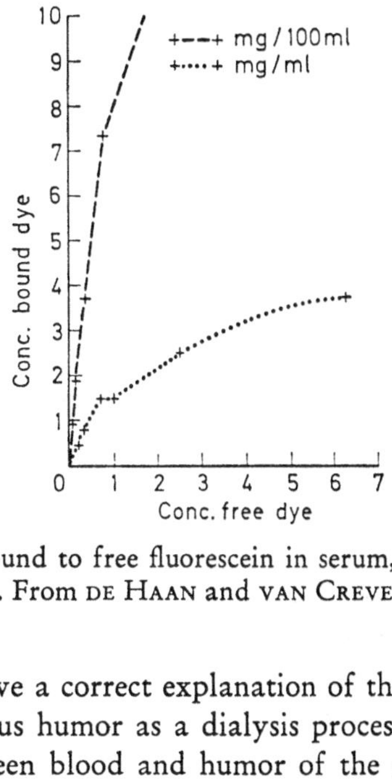

Fig. II-1. Ratio of bound to free fluorescein in serum, determined by ultrafiltration. From DE HAAN and VAN CREVELD (1921)

CREVELD (1921) gave a correct explanation of the passage of fluorescein into the aqueous humor as a dialysis process through the membrane barrier between blood and humor of the anterior chamber of the eye. The binding of the dye to the serum proteins, resulting in its decreased permeation, plays an essential role in this process.

A latent period of many years passed before the early observations, clearly documented and correctly interpreted, were recognized in their significance and accepted as evidence for a novel concept. This is the time when the question of "physical adsorption" versus "chemical union" was hotly debated—a dispute which may be seen already in PAUL EHRLICH's doctoral dissertation (1878) with respect to histological staining. How vigorous the argument became later, and how difficult it was for the contemporaries to accept the new concepts, is apparent from the cynical comments by which MATHEWS (1921) ridiculed observations on biological inactivation of atropine by serum proteins: "Colloids, those magic things, have magicked the atropine, have translated it so that it was for a time quite different from what it usually is. The authors explain the fact by calling it a case of adsorption".

The "authors" referred to by MATHEWS are STORM VAN LEEUWEN and his coworkers who had found that the biological activities of atropine (STORM VAN LEEUWEN and ZEIJDNER, 1921) and of pilocarpine (STORM VAN LEEUWEN, 1921) were decreased by "adsorption" or "binding" to rabbit serum. This inactivation was not caused by chemical destruction of the drugs which could be recovered unaltered by solvent extraction. It was recognized (STORM VAN LEEUWEN, 1924 b) "that these combinations are easily reversible", as indicated by the extractability, and by subsequent dialysis experiments. The reversible nature of the binding led to the assumption of displacement (STORM VAN LEEUWEN, 1924 b) of one ligand by another, a concept which could explain the loosening of the combination of atropine and rabbit serum by sodium citrate or peptones (STORM VAN LEEUWEN and ZEIJDNER, 1921), and other similar observations. Dialysis through parchment paper membranes has been extensively studied by BEUTNER (1925, 1926) who demonstrated reversible serum binding of alkaloids and their displacement. A similar displacement of serum protein-bound dyes by bile salts was observed by ROSENTHAL (1925) in ultrafiltration studies. Serum protein prevented rose bengal, bromo sulphalein, bilirubin and other dyes from permeating collodion membranes; after addition of taurocholate, the dyestuffs were liberated and appeared in the ultrafiltrate.

The ways of thinking and experimental approaches of the pharmacologist have greatly contributed to our understanding of the role played by proteins in the mode of action of drugs and other biologically active compounds (GOLDSTEIN, 1949); the similarity of the approaches to basically similar problems remains evident in the pharmacological literature (ARIENS, 1964; MAUTNER, 1967; MEYER and GUTTMAN, 1968; WAUD, 1968; EHRENPREIS et al., 1969). On the basis of the observations on protein binding described above, STORM VAN LEEUWEN developed a concept which in its essential points is valid today. Modifying and extending EHRLICH's side chain theory (EHRLICH, 1900; see also Chapter I), he postulated two types of receptors for a drug, the primary or dominant, and the secondary receptors (STORM VAN LEEUWEN, 1924 a). The dominant receptors are the loci at the reactive sites of target tissues where the first interaction takes place which ultimately results in the biological effect. Such primary receptor substances for steroid hormones have been recognized and characterized in recent years, as will be discussed by E. V. JENSEN in a monograph of this series (1970).

The secondary receptors (STORM VAN LEEUWEN, 1924 a) are potentially present anywhere in the organism but are not directly related to the primary action. The plasma proteins are typical secondary receptors. Other secondary receptors may be present in various organs or tissues where the compounds may be stored as, for example, the lipophilic progesterone in the adipose tissue. Other receptors of a secondary nature may be the enzymes present in liver and other organs which metabolize the drugs or hormones, often affecting their biological activity, or facilitating their excretion. The actual distribution of a compound in the organism depends on all these interactions, i.e., all the different receptors compete for the drug or hormone, in proportion to their concentration, or number of binding sites, and to the intrinsic binding affinity of the individual sites. GOLDSTEIN (1949) has given a schematic presentation of the

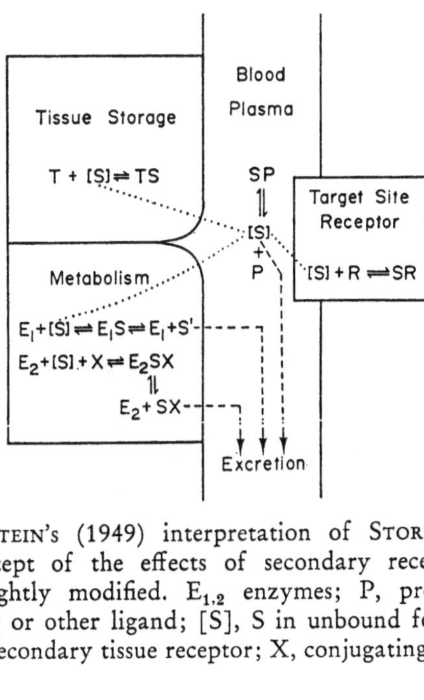

Fig. II-2. GOLDSTEIN's (1949) interpretation of STORM VAN LEEUWEN's (1924 a, b) concept of the effects of secondary receptors on primary drug action; slightly modified. $E_{1,2}$ enzymes; P, protein; R, primary receptor; S, drug or other ligand; [S], S in unbound form; S′, metabolite of S; T, secondary tissue receptor; X, conjugating compound

influence of these various receptors on the primary action of a drug, based essentially on STORM VAN LEEUWEN's concepts (1924 a, b), and incorporating some refinements. This scheme is given in Fig. II-2 with little modification. A modern version of the *in vivo* distribution and a critical evaluation of the role played by binding to various secondary receptors has been elaborated by TAIT and BURSTEIN (1964) and by BAIRD et al. (1969) for the steroids; additional considerations have been given by PETERSON (1959); YATES and URQUHART (1962); BEISEL et al. (1964 b); BRADLEY and WATERHOUSE (1966); and SAMUELS (1966).

III. First Observations on Reversible Binding of Steroids to Proteins

As early as 1913, the pharmacologist Ernst Oppenheimer, at the age of 25, published studies on the behavior of cardiac-active compounds of the digitalis series (1913). He observed a decrease of toxicity towards the isolated frog heart when serum was added to the drug solutions. After testing several possible interpretations experimentally, Oppenheimer suggested as the most likely explanation for the reduced toxicity a binding ("adsorption") of the cardiac-active steroid to the colloidal components of the serum. This binding was assumed to be easily dissociable so that the active compound is taken up by the heart cells which themselves have similar adsorption affinity; the result is a binding equilibrium ("Adsorptionsgleichgewicht") in which the steroid is distributed between the serum colloids and the heart cells. This interpretation clearly is a precursor of the concepts of Storm van Leeuwen; it was ignored or forgotten by the investigators of the time. Subsequent research, however, gives ample evidence of Ernst Oppenheimer's remarkable foresight.

It has been related above that Rosenthal (1925) observed displacement of dyes from their attachment to proteins by the addition of taurocholate, and attributed this effect to the great affinity of bile salts for proteins. A few years later, Lecomte du Noüy (1928) studied the influence of sodium oleate, glycocholate and taurocholate on surface tension of aqueous solutions. He had noted earlier (1922) that the effect of small amounts of the bile salts or of sodium oleate on surface tension was completely neutralized by serum, and described these experiments in detail for oleate. To gain insight into the quantitative relationships, he added decreasing amounts of the surface-active bile salts to the serum and determined the quantity which no longer showed an effect on surface tension. He considered this amount completely adsorbed to the serum protein to form an inactive complex. The quantity varied with different sera and ranged from 0.025 to 0.1 mg bile salt per ml serum.

This is the time when BENNHOLD (1924, 1926, 1938) began his classical studies on binding of small molecules to serum proteins. His work covered a great variety of acidic and basic dyes, natural pigments, drugs and other compounds. Binding was demonstrated by the rate of diffusion into gelatin, which was greater for the free, monodisperse dye than for the dye-protein complex. Another extensively used technique was a simple form of electrophoresis in a U-tube, which gave different migration rates for the unbound and the protein-bound compounds. BENNHOLD saw the significance of the binding to the serum proteins in the transport of the various substances in a protein-bound form (1932). This is in contradistinction to the earlier concepts of STORM VAN LEEUWEN and others, who emphasized the influence on availability of unbound, biologically active compounds by the binding to the "secondary receptors" in the blood serum. BENNHOLD elaborated his ideas of the serum proteins functioning as "vehicles" for the drugs and other substances by proposing a "directed transport" according to which the small molecules would be carried in a specific manner to certain organs as reactive tissues. Here the proteins would enter the cells, and the ligands would be detached from the binding protein by specific mechanisms (BENNHOLD, 1932, 1938).

Although these postulates were not generally accepted (ROTHLIN, 1945; GOLDSTEIN, 1949) the idea of the circulating proteins playing the role of a transport system proved attractive and had considerable influence on the interpretation of serum protein binding in subsequent years; it is still prevalent in the European literature (BENNHOLD and OTT, 1961; BENNHOLD, 1964, 1966; DESGREZ and DE TRAVERSE, 1966; ALY et al., 1968). This is illustrated in Fig. III-1 which shows the great number of different compounds which are bound to the proteins of the circulating blood as "Transportmilieu". It will become evident in the following chapters that for the circulating steroids, transport is only part of the significance of protein binding. An example for such transport function would be the solubilization of insoluble, e.g., lipidic, compounds by complex formation with serum proteins. In most other cases, the ligands, including steroid hormones, are water soluble at their normal concentrations. Therefore, they do not require a "vehicle" for transport; on the contrary, they would be able to permeate cell membranes more readily without association with the large protein molecules. Whether any of the steroid-binding serum proteins has a specific function in facilitating attachment of the hormones onto receptor sites or entrance into target cells is not known. A modern view on the influence of plasma proteins on the action of drugs has been given by THORP (1964), ARIENS (1964), and MAUTNER (1967).

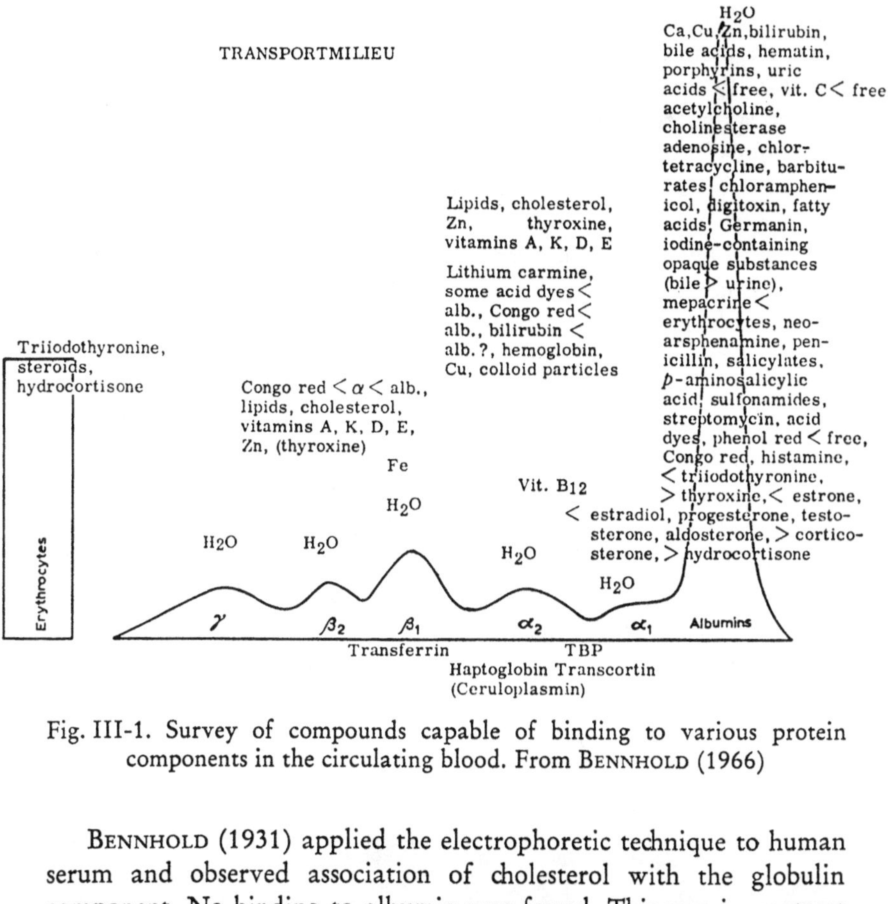

Fig. III-1. Survey of compounds capable of binding to various protein components in the circulating blood. From BENNHOLD (1966)

BENNHOLD (1931) applied the electrophoretic technique to human serum and observed association of cholesterol with the globulin component. No binding to albumin was found. This was in contrast to the behavior of the bile acids which migrated with the albumin fraction (BENNHOLD, 1938). These steroid derivatives are present as anions in the serum; their binding characteristics (see Chapter VI. 1), therefore, resembled those of acidic dyes and other anions which bind preferentially to serum albumin (BENNHOLD, 1938; GOLDSTEIN, 1949). A steroid of a different chemical nature, digitoxin, was also found to associate with the proteins of human serum, presumably the albumin component; this was concluded from the behavior of the steroid in gelatin-diffusion studies (HOEKSTRA, 1931). The results confirm by physicochemical techniques the earlier observations of OPPENHEIMER (1913) which have been reported above. More recent data on HSA* complexes with cardiac-active steroids will be given in Table VI-9.

It is apparent from these historical notes that the early observations on steroid-protein interactions were mainly concerned with biologically active steroids, and that the studies were stimulated by the endeavour to understand distribution and mode of action of the compounds in the living organism. In the late twenties and subsequent years a new class of steroid compounds of great physiological significance was discovered: the steroid hormones. BRUNELLI (1934) appears to have been the first one to study interaction of the estrogenic hormones with plasma proteins. Clearly under the influence of BENNHOLD's concept of a transport function of the serum proteins, BRUNELLI demonstrated in dialysis experiments using collodion membranes that estrone was bound to the proteins. This conclusion was believed to be corroborated by the observation of impaired ether extraction of the estrogen from its mixture with serum. The following chapters of the monograph will show to what measure that experiment of BRUNELLI's, performed more than 35 years ago, has grown: the study of steroid-protein interaction has become a field of its own, with numerous extensions into physicochemical, biological and medical areas. And the study of hormonal mechanisms of steroids, intimately involved with binding to biopolymers of various functions, has only begun.

* The abbreviations used in this monograph may be found in Appendix A.

IV. Basic Methods for Determination of Steroid-Protein Interactions*

1. General Principles

The basic requirement of any procedure for determination of protein binding is its capability of distinguishing the bound portion of a ligand from the free, i.e., unbound** fraction. This may be accomplished by actual physical separation of the two species each of which then can be analyzed; such a system is most lucid and leads to least ambiguous results. In many cases, physical separation is not carried out, and the distinction between the bound and unbound form is based on properties specific for one or the other. Examples are methods in which the biological activity of a compound is drastically changed by binding to a protein, as has been discussed for some of the early observations. Another instance is the measurement of surface tension in cases in which the effect of a surface-active compound is abolished by association with a protein. Changes of spectral properties by interaction with proteins are in the same category.

Most of the techniques used in the early work on protein binding provide valid qualitative information but do not lend themselves conveniently to quantitative interpretation. An important requirement for unambiguous determination of ligand-protein interaction is for the system to be in a state of thermodynamic equilibrium. The protein complexes which will be discussed here are held together by

---

* A review of the earlier work on protein-ligand interactions, including a critical evaluation of methods and presentation of binding data has been given by GOLDSTEIN (1949). More recent general treatments have been presented by KLOTZ (1953) as well as by EDSALL and WYMAN (1958). Methods have been discussed by HUGHES and KLOTZ (1956) and by ROSENBERG and KLOTZ (1960).

** Although the term "free" is still being used in protein-binding studies, the expression "unbound" is preferred since it may avoid confusion with the "free", i. e., unconjugated form of a steroid.

noncovalent bonds of a relatively weak, dissociable nature; their energy is of the order of that of hydrogen bonds, approximately one order of magnitude lower than the usual covalent linkages. As a consequence, the protein-ligand complexes are in a state of dynamic equilibrium; they dissociate readily and reversibly under ordinary conditions. This will not adversely affect the results of methods in which the bound and unbound portions of the small molecule are not separated, as long as the measuring process does not interfere with the equilibrium. With various other procedures, however, which are based on physical separation of bound and unbound species, this

Table IV-1. *Methods applied for determination of interaction between steroids and proteins*

No.	Method	Originally applied to	Reference
1.	Biological assay	Digitoxin and other cardiac-active steroids	OPPENHEIMER (1913)
2.	Tensiometry	Bile Salts	LECOMTE DU NOÜY (1922, 1928)
3.	Ultrafiltration	Taurocholate	ROSENTHAL (1925)
4.	Electrophoresis	Cholesterol	BENNHOLD (1931)
5.	Gelatin diffusion	Digitoxin	HOEKSTRA (1931)
6.	Dialysis	Estrogens	BRUNELLI (1934)
7.	Solvent extraction	Estrogens	BRUNELLI (1934)
8.	Equilibrium dialysis [a]	Digitoxin and other cardiac-active steroids	HAARMANN et al. (1940)
9.	Solubility determination	Estradiol, testosterone, progesterone	BISCHOFF and PILHORN (1948)
10.	Equilibrium dialysis	Testosterone	SCHELLMAN et al. (1954)
11.	Solvent partition	Testosterone	SCHELLMAN et al. (1954)
12.	Paper strip electrophoresis	Progesterone, cortisol	SZEGO and ROBERTS (1955) WESTPHAL et al. (1955)
13.	Ultracentrifugation	Progesterone, cortisol	WESTPHAL (1955)
14.	Boundary electrophoresis	Progesterone, cortisol	WESTPHAL (1955)
15.	Equilibrium paper electrophoresis	Corticosteroids	DAUGHADAY (1956 c)
16.	Paper chromatography	Progesterone, cortisol	WESTPHAL et al. (1956)
17.	Spectrophotometry	Δ^4-3-ketosteroids	WESTPHAL (1956), (1957 b)

Table IV-1 (continued)

No.	Method	Originally applied to	Reference
18.	Continuous flow electrophoresis	Progesterone, corticosteroids	DAUGHADAY (1958 c)
19.	Starch block electrophoresis	Cortisol	DOE et al. (1960)
20.	Multiple equilibrium dialysis	Progesterone, corticosteroids	WESTPHAL (1961), WESTPHAL et al. (1961)
21.	Equilibrium fractionation	Corticosteroids	WESTPHAL (1961), DEVENUTO et al. (1962)
22.	Gel filtration	Corticosteroids	DE MOOR et al. (1962)
23.	Overtaking electrophoresis	Estrogens	STRUCK (1963)
24.	Turbidimetry	Estrone	THERRIAULT and TAYLOR (1964)
25.	Gel equilibration	Testosterone	PEARLMAN and CRÉPY (1967 b)
26.	Competitive adsorption	Cortisol	HEYNS et al. (1967)
27.	Enzymatic assay	Cortisol	LEYBOLD et al. (1967)

[a] "Austauschdialyse" at 2 concentrations, without evaluation according to law of mass action.

equilibrium is constantly disturbed so that the process of separation is accompanied by a continuously developing series of new equilibria. The elements of irreversibility thus inherent in these methods prevent their use for a stringent thermodynamic evaluation. Examples for such techniques are ultrafiltration, "open-end" dialysis and gel filtration with small volume samples. In contrast, complete reversibility obtains in the method of equilibrium dialysis which therefore has been most widely adopted for determination of thermodynamically valid binding data.

Interactions between steroids and proteins have been investigated for more than half of a century. Many different techniques have been applied, based on diverse principles. Table IV-1 summarizes these methods in chronological order and gives names of the original authors involved. It may be noted that the methods applied during the earlier years lead essentially to qualitative results, illustrating the state of knowledge on protein-ligand interaction during that period.

2. Measurement of Biological Activity

Compared to the frequent application of biological activity as a criterion of drug-protein binding by pharmacologists, and by McLean and Hastings (1935) in their classical studies on serum binding of calcium, this quality has not been used extensively to assess protein interaction of steroids. Nevertheless, approximate quantitative results have been obtained for the binding of cardioactive steroid glycosides to whole serum and serum albumin of various species, using the isolated frog heart as test preparation (Fawaz and Farah, 1944; Farah, 1945); distinct differences in protein-binding affinity were observed (Suter, 1944; Rothlin, 1945) between various active glycosides.

In a more recent version of the bioassay principle, Blecher (1964) has proposed to determine binding of desoxycorticosterone to serum proteins in an *in-vitro* system. The inhibitory effect of this steroid on the net influx of α-aminoisobutyrate into rat thymic lymphocytes is eliminated by the presence of BSA, as a result of binding of desoxycorticosterone to the protein. The extent of reversal of the steroid effect serves to estimate the degree of interaction between steroid and protein. In another test system (Blecher, 1966), the glucocorticoid-mediated inhibition of the conversion of glucose to CO_2 and lipids (see also Matsui and Plager, 1966; Plager and Matsui, 1966) by free adipose cells is overcome by CBG, in direct relation to its concentration. These indirect assessments of steroid-protein interaction are carried out in highly complex biological systems, involving a variety of factors which are only partly understood. The conclusions concerning quantitative binding data, therefore, have to be taken with reservation.

Another application of a biological *in-vitro* system for measurement of steroid-protein interaction is the enzymatic procedure developed by Leybold et al. (1967). The plasma CBG binding sites are saturated with an excess of cortisol, and the system is subjected to the reductive action of Δ^4-3-ketosteroid hydrogenase. Since only non-protein bound cortisol is reduced to tetrahydrocortisol-5 α (3 α, 11 β, 17, 21-tetrahydroxy-5 α-pregnan-20-one), determination of the remaining cortisol gives a measure of CBG present. The results obtained by this method are comparable with those derived from other procedures.

The early results on steroid-protein interaction had been obtained before physicochemical concepts were applied to reversible binding of ligands to proteins. The similarity to the adsorption isotherm was then the closest approach to a quantitative treatment available. Although some of the methods introduced at the time have later been developed to useful quantitative procedures, others do not lend themselves readily for this purpose. For example, influence on surface tension (LECOMTE DU NOÜY, 1922, 1928) is exerted by only a limited number of steroid compounds; this technique has played no further role in the investigation of steroid-protein interaction.

3. Ultrafiltration

Applied early in studies on protein-ligand interactions, this method resembles dialysis in its principle: a semipermeable membrane is used, through which the low-molecular components of the system pass under a given pressure. Solvent and unbound steroid molecules thus form the ultrafiltrate, whereas the macromolecular compounds, including steroid-protein complexes, remain in the filtrans*. Both ultrafiltrate and filtrans can then be analyzed. The procedure can readily be performed under anaerobic conditions by applying pressure from a nitrogen cylinder. A detailed description of the use of ultrafiltration for protein-ligand interaction has been given by BOYER et al. (1947); equipment and techniques have been described by several authors (COOLIDGE, 1940; MALINOW and KORZON, 1947; FELDMAN et al., 1950; BENNETT and KIRBY, 1965). Ultrafiltration may be combined with separation of different proteins by the use of retarding membranes, possibly in a cascading series (ROSENFELD, 1965); this application resembles the method of differential dialysis (CRAIG, 1964). A rapid "diafiltration" method for binding studies has been described by BLATT et al. (1968).

Ultrafiltration (SHANNON, 1935; SMITH and SMITH, 1938) was used to show (BOETTIGER, 1946) that estriol was strongly bound to plasma proteins; the adsorption isotherm was followed in a very approximate way. LAVIETES (1937) developed a technique which has

* This term, unknown to WEBSTER, was used by BECHHOLD (1907 a) to designate the fluid to be filtered. Revival of the term seems desirable.

been applied to an analysis of steroid interaction with serum proteins (SANDBERG et al., 1957; UPTON and BONDY, 1958). The all-glass apparatus of TORIBARA (TORIBARA, 1953; TORIBARA et al., 1957), in which the filtrans is pressed through cellophane dialysis tubing by centrifugal force, has been used for studies on protein binding of steroid hormones (PETERSON et al., 1960; CHEN et al., 1961). A constant pH is maintained in the plasma by use of a 5% CO_2—95% O_2 gas mixture.

In the analysis of protein binding of steroid hormones or other neutral steroids by ultrafiltration, the assumption is made that the concentration of steroid in the ultrafiltrate equals the concentration of unbound steroid in the filtrans. This assumption is justified as long as the ligand under study is uncharged, i.e., as long as Donnan equilibria need not be taken into account. The bound portion of the steroid is then obtained as the difference between the total concentration of steroid in the filtrans and the concentration of unbound steroid in the ultrafiltrate.

A potential source of error in several methods is the adsorption of ligand or protein to the semipermeable membrane or to other surface material, or absorption by diffusion into component parts of the ultrafiltration equipment, such as rubber seals. These possibilities have to be checked by control experiments. Such tests should be made with any procedure involving small amounts of steroids, particularly trace amounts of radiolabeled compounds, in any container material applied. A good example for this complication is the observation of adsorption of estriol (BOETTIGER, 1946), estradiol, testosterone and progesterone (BISCHOFF and PILHORN, 1948) to collodion membranes in ultrafiltration; the adsorption for the latter two steroids from a saturated aqueous solution was virtually complete. As another example, it has been found (WESTPHAL, 1958) that progesterone and other steroids are avidly adsorbed by nitrocellulose (lusteroid) tubes used in ultracentrifugation. In this case, disappearance of the steroid from the solution appears to result from diffusion into the plastic; after initial formation of a monomolecular layer, the rate of the subsequent migration into the nitrocellulose was first order for the steroids. Passage of various steroids, including cortisol, progesterone, testosterone and estradiol, through silicone rubber (Silastic, Dow Corning) has also been observed (DZIUK and COOK, 1966). For these and other reasons, it is necessary to determine the steroid concentration in both the filtrate and the filtrans and calculate the bound portion from these data. Furthermore, a balance of total recovery of steroid will show any discrepancy due to adsorption or other loss.

Another possible inaccuracy is inherent in the ultrafiltration procedure. While the ultrafiltrate is formed the protein concentration in the filtrans increases. As a consequence, the binding equilibrium between steroid and protein is subject to continuous change. Reliable results can be expected only when the volume of the ultrafiltrate is small in relation to that of the filtrans, and when the rate of equilibration is slow in relation to filtration time; in this case and other errors excluded, the results of the ultrafiltration experiments would approach thermodynamic validity. Corrections should be made if the ultrafiltrate volumes are as high as 20—40% (SANDBERG et al., 1957) or 50% (PEARSON et al., 1967) of the total filtrans (PETERSON et al., 1960; MILLS et al., 1960; CHEN et al., 1961). In a study reported from YATES' laboratory (KELLER et al., 1966), ultrafiltration was performed by the TORIBARA technique, and the volume of ultrafiltrate was kept smaller than 10% of the filtrans. PLAGER et al. (1964) also filtered about 1/10 of the plasma volume. This appears to be a reasonable compromise to compensate for the concentration effect.

In spite of the drawbacks peculiar to this method, ultrafiltration is a practical technique which can be carried out in a relatively short time, and under conditions approximating a state of equilibrium. Use of a thermostated centrifuge appears desirable. One of the advantages of the ultrafiltration—in contrast to equilibrium dialysis—is the possibility of applying undiluted serum or plasma. In view of the complex nature of the multiple binding equilibria present in the blood serum, maintenance of the native concentration of all components may be advantageous in certain cases.

4. Electrophoresis

Migration in the electric field was the principal tool used by BENNHOLD (1932, 1938) who found that cholesterol was associated with the globulin fraction and not with albumin (BENNHOLD, 1931). Sterols are uncharged molecules under the usual experimental conditions and, therefore, do not show electrophoretic mobility in the unbound state. However, they will migrate in the electric field when bound to a plasma protein. The electrophoretic mobility of the protein remains essentially unchanged after association with a neutral steroid. If the steroid moiety is ionized (bile acid salt, steroid sulfate, glucosiduronate, amino steroid, etc.) the ionic charge will be added to the net charge of the protein, resulting in an altered electrophoretic mobility. This is a similar effect as observed with organic ions (BALLOU et al., 1945; KLOTZ and WALKER, 1947 b) and with other

Descending side						Ascending side and bottom fraction				
Progesterone experiment		Cortisol experiment					Progesterone experiment		Cortisol experiment	
% HSA	% Prog	% HSA	% Cort	Section		Section	% HSA	% Prog	% HSA	% Cort
0	0.16	0	3.8	DT		AT	2.7	2.8	7.1	0.6
0	0.27	0.1	10.3	DM		AM	22.9	22.5	29.4	17.6
20.2	18.7	13.9	14.8	DB		AB	29.2	26.5	21.2	19.3
						B	25.0	29.0	28.4	33.5

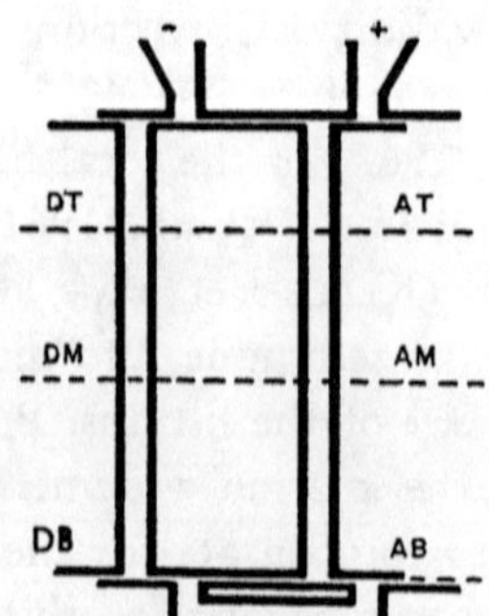

Fig. IV-1. Boundary electrophoresis of mixture of HSA and progesterone-4-^{14}C or cortisol-4-^{14}C, in Michaelis buffer, pH 8,6; $\mu = 0.1$; 2°; 4 hrs. Starting conditions: 1.2 μg progesterone-4-^{14}C per ml 1.3% HSA (molar ratio 1 progesterone: 50 HSA), or 1.7 μg cortisol-4-^{14}C per ml 1.2% HSA (1 cortisol : 36 HSA) in descending limb (DT, DM, DB) and bottom section (B). Figures denote percentage of total for HSA and steroids in various sections after completion of electrophoresis. Calculated from WESTPHAL (1955)

ions (LONGSWORTH and JACOBSEN, 1949), including higher fatty acids which form strong complexes with serum albumin (WESTPHAL and PRIEST, 1955).

The original apparatus used by BENNHOLD (1932) for what was then called cataphoresis, was a modification of the simple U-tube described by MICHAELIS (1909). Electrophoresis under these conditions is subject to disturbances by heat convection and other inadequacies, and the sampling for analysis of the different layers is cumbersome. A radical improvement came with the construction of the apparatus for boundary electrophoresis by TISELIUS (1957) which provided reproducible convection-free migration of the various protein components of a mixture. This equipment was used to demonstrate (WESTPHAL, 1955) binding of progesterone and of cortisol to human serum albumin (Fig. IV-1).

Electrophoretic procedures are well suited for a qualitative demonstration of steroid binding to proteins. Their principal merit lies in the electrophoretic characterization of the component to which the steroid is bound. However, the results do not lend themselves readily for quantification. In boundary electrophoresis, sampling for analysis is delicate and subject to error, due to instability of the liquid layers. The greater difficulty is a fundamental one and results from the dissociation of the steroid-protein complexes. Fig. IV-1 illustrates the situation for the electrophoretic migration of the HSA complexes of progesterone and of cortisol. After electrophoresis, the ratio of progesterone to albumin throughout the cell is essentially the same as in the starting solution, i. e., 1 : 1 in terms of percent of total applied. Progesterone is transported to the ascending side in firm association with HSA; only a very small percentage has dissociated and remains as unbound progesterone in the two albumin-free top sections on the descending side. In contrast, the ratio between cortisol and HSA changes from section to section, indicating the higher degree of dissociation. Less than 10% of the original cortisol: albumin ratio is found in the top ascending section, AT, after electrophoresis, whereas a relatively large portion stays as unbound cortisol in the two upper sections (DT, DM) of the descending limb from which the albumin has moved away. Clearly a quantification of these experimental results would be complex and inaccurate since they are determined by both dissociation constant and dissociation rate.

The ready dissociation of the steroid-protein complexes is even more troublesome in paper strip electrophoresis. This technique was introduced in 1955 for the study of serum protein interaction with estrogens (SZEGO and ROBERTS, 1955) as well as with progesterone and cortisol (WESTPHAL et al., 1955, 1956). Considerable trailing of steroid was observed, even for complexes of relatively high binding affinity, such as that between progesterone and HSA. The weak interaction between cortisol and HSA was demonstrable only by adjusting the analyzing procedure to the very small amounts of cortisol carried over the paper strip by the migrating HSA. Similar difficulties were experienced by other authors who studied serum protein binding of corticosteroids (UPTON and BONDY, 1958; REICH, 1960), progesterone (HASKINS and TAUBERT, 1963), and estrogens (ROSENBAUM et al., 1966 a) by paper-electrophoretic techniques.

DAUGHADAY (1956 c) has overcome the difficulties caused by the continuous dissociation of the steroid-protein complex during electrophoretic migration in an elegant technique termed equilibrium paper electrophoresis. The principle of the procedure is to perform the electrophoresis in an environment which contains the dissociable steroid at the concentration of the unbound portion. This is achieved by simply dialyzing the protein-steroid solution to be electrophoresed* against the electrophoretic buffer until equilibrium is reached. The temperature of this equilibrium dialysis should be as closely as possible that of the actual electrophoresis, and rise of temperature by the applied current should be avoided; this is particularly important in the study of complexes with high temperature coefficients. Since the dissociation of the steroid-protein complexes is reversible, the net result is apparent absence of dissociation. Actually, there is continuous dissociation and association as can be shown by electrophoresis of a radiolabeled steroid-serum mixture in a medium which has the equilibrium concentration of the unbound steroid in unlabeled form (unpublished observations). For application of the equilibrium paper strip electrophoresis in studies on CBG see Chapter VIII. 4 c.

An interesting variation of the paper strip electrophoresis has been described by STRUCK (1963); the technique is analogous to the retention electrophoresis (LANG, 1955; HAAN and LANG, 1957; LANG, 1959) for the detection of antigen-antibody interaction. The paper is impregnated with estrogen and the protein is allowed to migrate through the (presumably) homogeneously imbedded steroid (Überwanderungselektrophorese = overtaking electrophoresis). If binding occurs, the estrogen will migrate together with the protein, and the area behind the migrating complex becomes depleted of steroid. The extent of this depletion is utilized for an approximation of the quantity bound to the moving protein.

Dissociation of the steroid-protein complex in conventional paper strip electrophoresis is extensive because the limited amount of protein migrates through a large buffer layer where the dissociated

---

* Term used in recent publication (OSUGA and FEENEY, 1967).

steroid remains. Most of the difficulties would be eliminated if the flow of the protein solution were made continuous; this idea prompted the construction of a continuous paper-electrophoresis apparatus in our laboratory (SELDEN and WESTPHAL, 1957) before commercial models were readily available. Application of the principle of equilibrium electrophoresis in connection with the continuous-flow electrophoresis on a paper curtain by DAUGHADAY (1958 c) led to the demonstration of the CBG. For this and other uses of the continuous electrophoresis technique for various problems of steroid-protein interaction (WESTPHAL et al., 1961; WESTPHAL and DE VENUTO, 1966) see Chapter VIII. 4 c. Carrier-free continuous electrophoresis has been applied for studies on protein binding of ligands other than steroids (FRÖHLICH, 1967).

5. Equilibrium Dialysis

The first evidence of binding of a steroid hormone to serum protein was obtained by dialysis as discussed in Chapter III. A similar technique was used to demonstrate the binding of digitoxin and other cardiac-active steroids to serum proteins (HAARMANN et al., 1940); in this case, dialysis was continued to equilibrium so that a definite value for the bound and unbound portion of the steroid was obtained.

The procedure of HAARMANN et al. (1940) can be considered the forerunner of modern investigations of steroid-protein interactions by equilibrium dialysis, first reported by the group of SAMUELS (SCHELLMAN et al., 1954) and followed by extensive studies in other laboratories. Techniques for rapid dialysis have been developed (CABIB and ALGRANATI, 1960; CRAIG and KONIGSBERG, 1961; MCMENAMY, 1968; MCMENAMY et al., 1968); their applicability to small quantities of steroids has to be tested.

The principal merit of the equilibrium dialysis procedure (HUGHES and KLOTZ, 1956; ROSENBERG and KLOTZ, 1960; STEIN, 1965) for measuring a dissociable binding equilibrium is its thermodynamic validity. Since the unbound ligand can move freely through the semipermeable membrane, its equilibrium concentration, [S], is equal on both sides of the membrane. If the protein present inside the dialysis bag binds the ligand, S, the total ligand concentration inside the bag, $[S_t]$, will be

$$[S_t] = [S_{bd}] + [S]; \quad [S_{bd}] = [S_t] - [S] \tag{IV-1}$$

where $[S_{bd}]$ denotes the concentration of bound ligand. The quantity, $[S_{bd}]$, can be obtained as difference between the "outside" concentration, $[S]$, and the "inside" concentration, $[S_t]$. In the actual experiment, the equilibrium concentrations of both $[S]$ and $[S_t]$ should be determined by analysis and related to the starting concentrations. This gives the recovery value which should approximate 100%; a poor recovery may reveal technical inadequacies which would remain undetected if only one of the quantities, $[S]$ or $[S_t]$, were determined, and the other calculated by difference.

Evidently, the concentration of bound ligand depends on the concentration of protein, P. If $\bar{\nu}$ indicates the average number of molecules of S bound per molecule of P under the conditions of the experiment, and $[S_{bd}]$ and $[P_t]$ are the molar concentrations of bound ligand and of total protein, then

$$[S_{bd}] = \bar{\nu}[P_t]. \tag{IV-2}$$

For the characterization of a binding system (Edsall and Wyman, 1958), $\bar{\nu}$ should be determined at several values of $[P_t]$ at constant $[S]$; $\bar{\nu}$ should be a well-defined number, independent of $[P_t]$ within the limits of the concentrations applied. In most equilibrium dialysis studies on steroid-protein interactions that have been published, the protein concentration has been kept constant and the steroid level varied. Indications are that the information thus obtained is insufficient for a complete characterization of a binding system; additional data at different protein concentrations will be needed.

The evaluation of equilibrium dialysis data as outlined above assumes that the ligands which interact with the protein are uncharged. This assumption is valid for most steroids, which do not carry charges under normal experimental conditions. Even the phenolic estradiol in solution at pH 6.5—8.0 behaves essentially as a non-electrolyte; ionization would be less than 0.1% (Bischoff and Pilhorn, 1948). Charged steroid molecules, such as bile acids or acid steroid conjugates which are present as anions at neutral pH, would alter the net charge of the protein inside the dialysis bag. This would result, according to the laws of the Gibbs-Donnan equilibrium, in an unequal distribution of diffusible unbound ions inside and outside the semipermeable membrane, so that the concentration of bound steroid anions would not be adequately determined. This complication of steroid-protein interaction by steroid ions has not been thoroughly investigated. The interpretation would have to take into account that the total ring structure portion of the molecule, presumably in addition to the anionic group, participates in the interaction. This is evident from studies on binding of cholanic acids to human serum albumin (see Table VI-1) which show that the hydroxy groups in the steroid ring system strongly influence the binding affinity.

Another potential error is the adsorption of one or both of the interacting components to the membrane. This possibility can be checked for the ligand by an accurate recovery balance as outlined above. Control experiments with individual solutions of ligand and protein, in the absence of the other, also reveal adsorption to the membrane material and should be performed with compounds not previously tested. Similar complications by adsorption to wall surfaces have been discussed in detail for ultrafiltration experiments (Chapter IV. 3).

Experimental details for application of equilibrium dialysis techniques have been published (WESTPHAL, 1969). The multiple equilibrium dialysis procedure (WESTPHAL, 1961; WESTPHAL et al., 1961) was developed in order to measure binding affinities of different protein samples under strictly identical conditions of their solvent environment, including equal distribution of endogenous steroids among all binding proteins. It follows the same principle which was adopted by DAUGHADAY et al. (DAUGHADAY, 1959; DAUGHADAY et al., 1959, 1961) in their technique of double equilibrium dialysis. The equilibrium dialysis method can be readily adapted to the study of suspended material, e. g., of subcellular fractions (DE VENUTO et al., 1962; DAVIDSON et al., 1962); the interpretation and quantitative evaluation of such systems, however, requires careful consideration of the specific properties of the components involved.

6. Solvent Extraction

It appears theoretically possible to find conditions at which a protein-bound ligand would resist solvent extraction in comparison with the same ligand in protein-free solution. BRUNELLI (1934) has described an experiment in which extraction of estrone by ether was impaired by the presence of horse serum. However, this approach does not provide an adequate method for determination of protein binding, not even in a qualitative way.

7. Measurement of Solubility

When a neutral steroid in solution is in equilibrium with its pure crystalline phase, its thermodynamic activity is fixed. Its concentration is also fixed for a specific solvent, a given temperature and other conditions. Practically all steroids are sparingly soluble in water. If protein is added to an aqueous solution of steroid in the presence of solid steroid phase, and if the steroid is bound by the protein, the thermodynamic activity is changed from its equilibrium state, and additional solid phase will be dissolved to compensate for the bound quantity. At the new equilibrium, the total amount of dissolved steroid, i. e., the sum of the free solute and the protein-bound form, will be increased. For non-electrolytes such as neutral steroids, the increase in solubility may be equated to the bound portion (KLOTZ, 1953). For electrolytes, the thermodynamic activity of the unbound ions as well as the solubility product must be taken into account.

The phenomenon of solubility increase in the presence of serum was recognized more than 60 years ago to indicate interaction with protein components (MOORE and ROAF, 1904, 1905). In these first experiments, chloroform, ether, ethyl acetate, benzene and other volatile compounds were the solutes, and there is some question about the participation of the lipid phase of lipoproteins in the phenomenon of increased solubility (HAGGARD, 1923; GOLDSTEIN, 1949). Later studies dealt with the simple hydrocarbons ethylene and cyclopropane, which showed a definite increase of their solubilities in the presence of plasma (ORCUTT and SEEVERS, 1937). Modern investigations (WISHNIA, 1962; WISHNIA and PINDER, 1966; MOHAMMADZADEH-K et al., 1967) have applied the method of solubility increase to establish beyond doubt that simple alkanes such as ethane, propane, butane and pentane are bound to pure proteins, as shown for bovine serum albumin, hemoglobin, myoglobin, lysozyme and beta-lactoglobulin (WISHNIA, 1969 a, b). These recent publications demonstrate the refinements in physicochemical terms of which the solubility method is capable for studies on ligand-protein interactions. Number of binding sites, association constants, free energies, enthalpies and entropies were obtained, and certain conclusions concerning the protein structure in relation to binding sites were derived.

Studies based on solubility increase of steroid hormones in the presence of proteins, carefully conducted by F. BISCHOFF and his coworkers (BISCHOFF and PILHORN, 1948; BISCHOFF and KATHERMAN, 1948), mark the beginning of quantitative investigations in the field of steroid-protein interactions. This work, which will be discussed in Chapter VI. 5, resulted in molar ratios for the binding of estradiol, testosterone and progesterone to BSA, in general agreement with later findings. Similar magnitudes of binding values and approximate free energies of binding were obtained in several laboratories (BISCHOFF et al., 1954; EIK-NES et al., 1954; SANDBERG et al., 1957; AVIGAN, 1959).

The solubility method involves relatively high concentrations of steroids, and it is difficult to reach equilibrium (BISCHOFF and PILHORN, 1948). The procedure is time consuming; for example, it appears doubtful that shaking of crystalline corticosteroids for 1 hr at 25° in aqueous systems (MACEK et al., 1952) provides saturation values of solubility. The technique also suffers from lack of ready reproducibility. For these and other reasons, measurements of protein interaction by solubility increase have not been favored in recent years, especially since availability of radiolabeled steroids made the determination of minimal quantities practical. Nevertheless, the principle of solubility may be considered whenever concentrations close to the saturation of the binding sites appear desirable.

8. Turbidimetry

An elegant refinement of the solubility method has been described by THERRIAULT and TAYLOR (1964) who determined the insoluble portion of the steroid by turbidimetric analysis; the procedure was developed in a study of phosphatidylserine interaction with HSA. In the earlier work of BISCHOFF, the insoluble steroid phase was separated from the equilibrated steroid-protein solution by filtration, and the protein-bound steroid was determined in the clear filtrate by colorimetric procedures, following precipitation of the protein by ethanol. In the turbidimetric technique, the insoluble ligand portion was measured, and the protein-bound quantity was calculated as the difference between the light transmittance in the protein-free solution and in the presence of protein, at a given ligand concentration.

Several conditions must be met for this method to be applicable. The solubility of the ligand must be sufficiently low so that suspen-

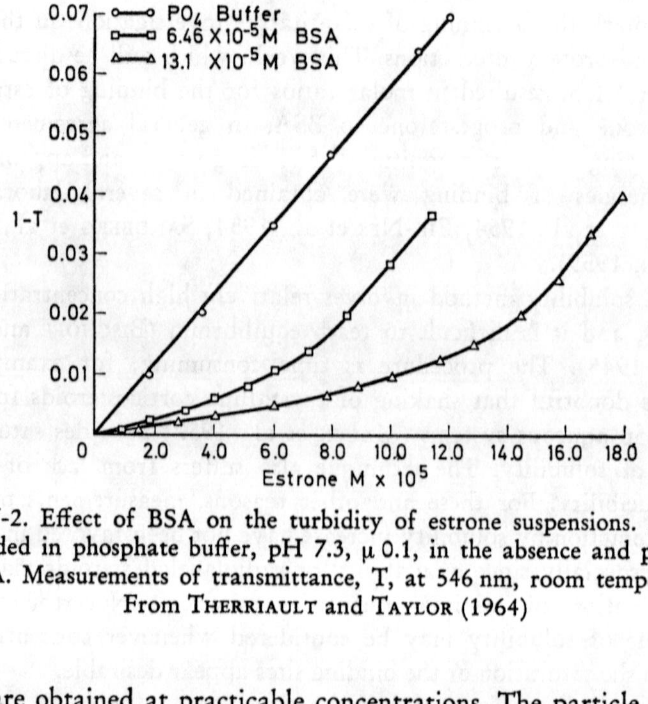

Fig. IV-2. Effect of BSA on the turbidity of estrone suspensions. Estrone suspended in phosphate buffer, pH 7.3, μ 0.1, in the absence and presence of BSA. Measurements of transmittance, T, at 546 nm, room temperature. From THERRIAULT and TAYLOR (1964)

sions are obtained at practicable concentrations. The particle size of the suspensions must be uniform, and the suspension must be stable without addition of stabilizers. A linear proportion must exist, within reasonable limits of concentration, between the concentration of the suspended ligand and the (non-specific) light absorption of the suspension. These characteristics of the suspended particles must not be altered by the presence of the protein.

The above conditions are fulfilled in the system estrone-BSA (THERRIAULT and TAYLOR, 1964). Fig. IV-2 shows the absorbance of estrone solutions of various concentrations, measured with a light scattering photometer at 546 nm, in protein-free buffer and in the presence of BSA. The increase of transmittance by the addition of BSA to the system is evident. Evaluation of the data according to the method of SCATCHARD (see Chapter V. 4) indicates (Fig. IV-3) that the binding follows the law of mass action and that unambiguous values for the apparent association constant and the number of binding sites can be obtained. In contrast, the equilibrium dialysis method is not capable of providing data at low $\bar{\nu}/[S]$ values, i. e., at high estrone concentrations, due to its low solubility. This has been pointed

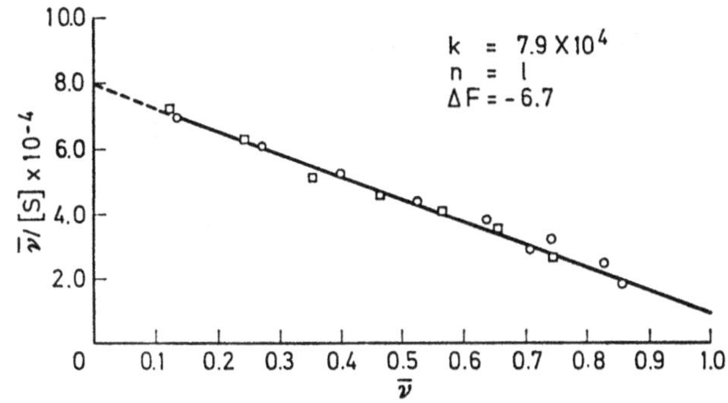

Fig. IV-3. Plot of $\bar{\nu}/[S]$ vs. $\bar{\nu}$, according to SCATCHARD, of data from Fig. IV-2. □ Results at 6.46×10^{-5} M BSA; ○ at 13.1×10^{-5} M BSA. From THERRIAULT and TAYLOR (1964)

out by SANDBERG et al. (1957) who studied estrone binding to HSA by equilibrium dialysis. This example illustrates the usefulness of the turbidimetric procedure for binding studies with ligands of limited solubility.

9. Solvent Partition

This technique, called partition analysis by KARUSH (1950), is based on the change of the distribution coefficient of a solute between an aqueous and an immiscible organic phase when a solute-binding protein is introduced into the aqueous phase. The partition coefficient, i. e., the ratio of equilibrium concentrations of solute in the aqueous and in the organic solvent, is a fixed quantity at given temperature, pH, and ionic strength. Values obtained for numerous steroids in various solvent systems have been published (ENGEL and CARTER,

1963). If to the aqueous phase a protein is added which forms complexes with the solute, the concentration of the unbound solute in the aqueous phase decreases, and solute moves in from the organic phase to compensate for the decrease. If the partition coefficient, which pertains to the unbound solute in the two phases, remains unaffected by the presence of the protein, a determination of the new equilibrium concentration of the solute in the organic solvent allows calculation of the corresponding concentration of unbound ligand in the aqueous protein phase, and thus by difference that of the bound portion.

The solvent partition method has been used originally in studies on protein interactions with mercurials (HUGHES, 1950) and with organic anions (KARUSH, 1951). Well controlled investigations on binding of testosterone to BSA have been conducted by SCHELLMAN et al. (1954) who also obtained some data on estradiol interaction with BSA.

The advantages of the partition analysis procedure are simplicity of execution and absence of membrane adsorption. However, there are some principal objections which call for considerable caution and critical interpretation of the results. A small but finite quantity of the organic solvent will become dissolved in the aqueous phase. Even assuming that this would not affect the binding properties of the ligand, any interaction of the protein with the organic molecule may alter the binding characteristics of the protein for the ligand-solute. Such alteration may be caused by direct competition, or by an indirect, e. g., allosteric effect of the lipophilic molecules of the organic solvent on the protein structure. Weak competition of toluene or hexanol with protein binding of dye anions has been observed (KARUSH, 1951). Although SCHELLMAN et al. (1954) assume competition effects to be smaller for n-heptane, it is now definitely established that n-alkanes, e. g., butane, pentane and others, form association complexes with several pure proteins (WISHNIA, 1962; WISHNIA and PINDER, 1966). Interactions between hydrocarbons and proteins have been studied by MOHAMMADZADEH-K et al. (1969 a, b) with reference to hydrocarbon structure and to protein structure. A conformational change of BSA by heptane interaction has been demonstrated by ALFSEN (1963). Careful control experiments would have to be made for each partition system in order to determine possible interaction of the organic solvent molecules with the protein, and its influence on higher protein structure and on the binding affinity toward the steroid under study. Possible surface denaturation effects at the liquid-liquid interphase may also interfere with the ligand-protein interactions.

10. Ultracentrifugation

If a solution containing a ligand and a binding protein is subjected to centrifugal fields of the magnitude applied in sedimentation velocity studies, the protein will sediment in association with part of the ligand whereas the unbound ligand does not sediment at the rate of the macromolecule. Since the small molecule also forms a concentration gradient in the ultracentrifuge tube, the concentration of unbound ligand is obtained by extrapolating to zero protein concentration a plot of total ligand concentration vs. protein concentration. The bound ligand can be calculated by difference of total and unbound ligand.

This procedure has been used by CHANUTIN et al. in their classical work on calcium binding to casein (CHANUTIN et al., 1942) and later by HAYES and VELICK (1954) in the study of coenzyme binding to alcohol dehydrogenase. LOKEN et al. (1960) and SCHOLTAN (1965) developed the technique further. Recent theoretical and experimental refinements have demonstrated the potential of ultracentrifugal methods for the analysis of interactions between macromolecules and small ligands (STEINBERG and SCHACHMAN, 1966). Sedimentation equilibrium is used to obtain the bound and unbound ligand concentrations at various protein concentrations; the results obtained for BSA-methyl orange interaction were in agreement with those found by equilibrium dialysis and sedimentation velocity.

A much less elaborate procedure has been used to demonstrate binding of steroid hormones to proteins by ultracentrifugation (WESTPHAL, 1955). A mixture of HSA and radiolabeled progesterone or cortisol in a preparative tube was subjected for 5 hrs to approximately 200,000 × g, and the content of the tube was analyzed in three fractions at different distances from the center of rotation. The percentage concentrations obtained for the free and bound portions of the steroids are in satisfactory agreement with values determined later with quantitative methods.

In more recent studies, ultracentrifugation methods have been applied (BEISEL et al., 1964 a) to assess the binding of cortisol to plasma proteins. Centrifugal forces of $115{,}000 \times g$ were employed for 6 hrs at 37° to plasma diluted with an equal volume of saline containing cortisol-4-^{14}C. Empirical correction was made to obtain values for undiluted plasma.

The method of density gradient centrifugation in sucrose solution has proved to be very useful for characterization of macromolecular high-affinity binders in various tissues. The technique will not be discussed here since it has been predominantly applied to receptor proteins of target tissues for steroid hormones such as estrogens (JENSEN et al., 1968; ROCHEFORT and BAULIEU, 1968; GORSKI et al., 1969), progesterone (MILGROM and BAULIEU, 1968), and aldosterone (EDELMAN and FIMOGNARI, 1968).

11. Paper Chromatography

Chromatographic demonstration of steroid-binding to proteins was obtained as a by-product of a study of the behavior of steroids on filter paper in an aqueous system (WESTPHAL et al., 1956). These experiments showed that it is possible, in principle, to give qualitative evidence for protein-steroid association by chromatography: progesterone migrated at the same rate as the protein, when an aqueous solution containing both steroid and serum or HSA was subjected to paper chromatography. Progesterone alone migrated in a narrow zone at a slower rate than the protein whereas in the presence of HSA the steroid was transported as far as the protein front. As would be expected dissociation of the complex occurred during the chromatography resulting in a progesterone concentration gradient which sloped toward the solvent front.

This technique at best gives qualitative results and has not been further applied to investigate steroid-protein interaction. Valid results would be obtainable if paper-, thin layer- or other chromatographic techniques were used in systems equilibrated with the concentration of the unbound ligand, in a similar way as discussed above for equilibrium paper electrophoresis. A combination of chromatographic separation of protein components and interaction of radiolabeled steroids with the proteins has been realized in studies on high-affinity binders (CHADER and WESTPHAL, 1968 b); details may be found in Chapter XI. 2.

12. Equilibrium Fractionation

Special techniques of centrifugation and ultracentrifugation are applied in the method of equilibrium fractionation (WESTPHAL, 1961; DE VENUTO et al., 1962). This technique is useful for studies on non-covalent binding of steroid compounds to subcellular fractions or other particles which can be separated by differential centrifugation. The fractions must be obtained in homogeneous form, free of cross

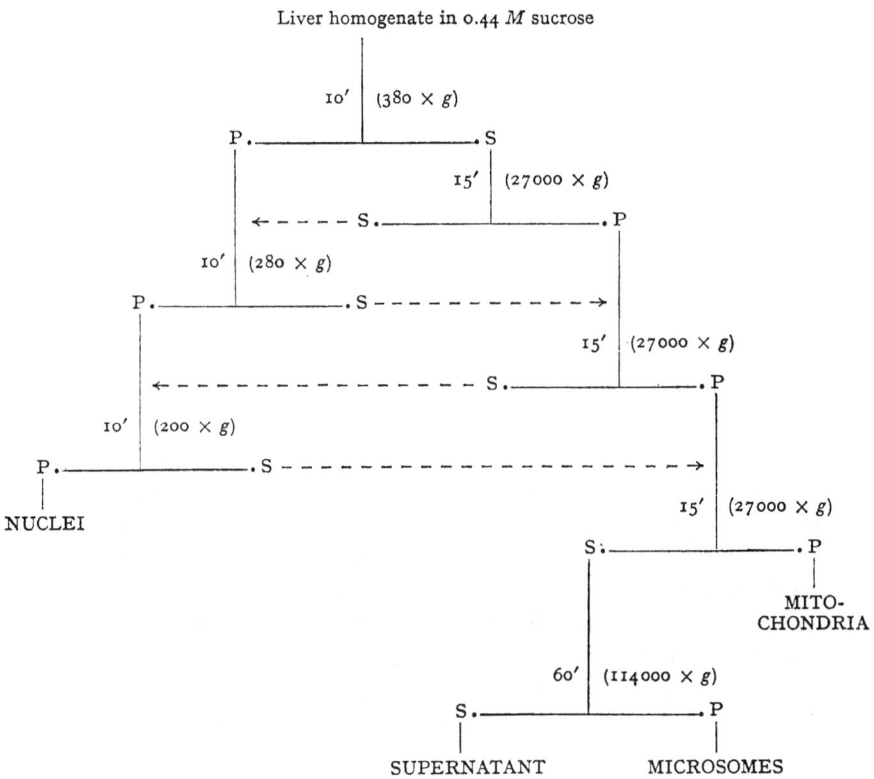

Fig. IV-4. Equilibrium fractionation of rat liver cells. P, precipitate; S, supernatant. From DE VENUTO et al. (1962)

contamination; this requires washing procedures which are ordinarily carried out by introducing fresh solutions at given stages into the fractionation system. Since the non-covalent interactions are regulated by equilibria of binding, a certain unbound percentage of the total steroid is removed by dissociation from its binding site with each addition of fresh wash medium. Loss of ligand in such procedure can be readily demonstrated (DE VENUTO et al., 1962).

The principle of the equilibrium fractionation technique is to maintain the binding equilibrium originally established when the cell material is homogenized prior to differential centrifugation. The initial sucrose solution used for homogenization of liver or muscle tissue (DOUNCE et al., 1955) is utilized throughout the entire fractionation procedure, and no fresh solvent medium is introduced. The original sucrose solution therefore remains in equilibrium with all fractions. Fig. IV-4 shows the scheme of the equilibrium fractionation technique for liver homogenate; experimental details have been described for homogenates of liver and muscle in sucrose solutions (DE VENUTO et al., 1962).

13. Spectrophotometry

A colored compound may undergo marked changes in its absorption spectrum when bound to a protein. This phenomenon has been early recognized in the "protein error" observed with indicator dyes used in pH determination in the presence of proteins (SØRENSEN, 1909) as discussed in Chapter II. Systematic investigations (KLOTZ, 1946 b; KLOTZ and LONGFELLOW, 1946; KLOTZ and WALKER, 1947 b; KLOTZ, 1953) on the interactions of BSA with azo dyes have shown that the differences in the absorption spectra for the bound and unbound dye can be used for quantitative determination of the concentration of the bound and unbound species.

Steroids of the usual chemical structures are not colored; many of them, however, show a high selective light absorption in the ultraviolet. This holds true for the natural steroid hormones in their original state; the responsible chromophoric group for most of them is the Δ^4-3-keto structure in ring A, except for the estrogens in which ring A is phenolic. It should be expected that an interaction of this chromophoric group with a protein would have a definite influence on its electronic configuration, resulting in a change of intensity and possibly a spectral shift of the ultraviolet absorption. It is recognized that even the formation of loose complexes in solution can be revealed by changes in the electronic spectra; such changes may be caused by very weak forces acting on valency electrons (BRAUDE, 1955).

It was found that the specific ultraviolet absorption of the Δ^4-3-keto grouping in the steroid hormones is markedly decreased when they are bound to proteins (WESTPHAL, 1957 b); the depression of the absorbance is proportional to the binding affinity. This finding is interpreted as indicating that the Δ^4-3-keto grouping is involved in the interaction of the steroid hormones with proteins. The absorption maximum of the α,β-unsaturated keto group in ring A, which is at about 248—249 nm in aqueous, protein-free solution, shows a shift of 2—3 nm to shorter wavelengths when the steroid interacts with a strongly binding protein. In many cases of weaker association, no displacement of the absorption maximum is observed although the absorbance is reduced (WESTPHAL and ASHLEY, 1958 a, 1959, 1962). Fig. IV-5 illustrates the depression of the absorbance of progesterone by interaction with HSA at a ratio of one molecule progesterone to two albumin molecules.

The reduction of light absorption of the Δ^4-3-ketosteroids resulting from interaction with the protein may be expressed as $\Delta\varepsilon$, the depression value of the molar extinction coefficient observed in the presence of the binding protein. It is calculated in percent of total steroid absorbance according to

$$\Delta\varepsilon = \frac{100\,(\varepsilon_w - \varepsilon_p)}{\varepsilon_w} \qquad \text{(IV-3)}$$

where ε_w is the molar extinction coefficient of the steroid in phosphate buffer, at the wavelength of its maximal absorption, $\lambda_{max}^{H_2O}$; and ε_p the molar extinction coefficient of the steroid in the same buffer in the presence of the protein, at the same $\lambda_{max}^{H_2O}$ of the protein-free solution. The depression values, $\Delta\varepsilon$, measure the lowering of ε as percentage of the values observed in the absence of protein. Fig. IV-5 indicates that they may be obtained directly from the absorbance values, without transforming them to molar extinction coefficients.

For purposes of comparison, $\Delta\varepsilon$ is a convenient quantity to express the approximate strength of binding of a series of Δ^4-3-ketosteroids to a protein (WESTPHAL and ASHLEY, 1958 a, 1959, 1962). Since $\Delta\varepsilon$ is determined as a relatively small difference in absorbance in the presence of a large background absorption by the protein, it is necessary that all experimental conditions are strictly identical, if the comparison of $\Delta\varepsilon$ obtained from different steroids is to be valid. Special precautions have to be taken to control concentrations and volumes of the solutions, as well as other experimental conditions (WESTPHAL, 1957b). The more recent development of split-compartment mixing cells by YANKEELOV (1963) facilitates the measurements.

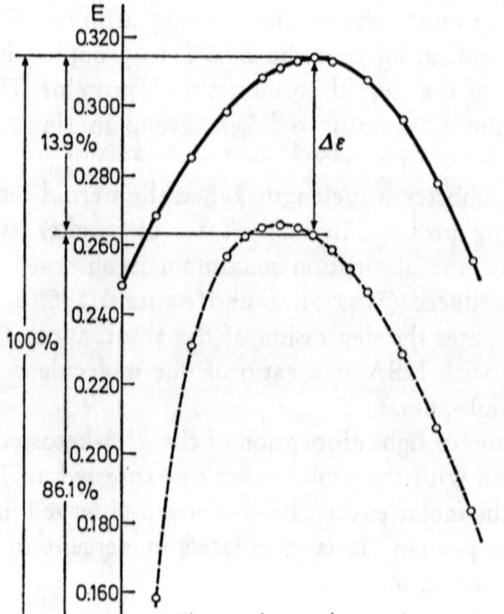

Fig. IV-5. Ultraviolet absorption of 2×10^{-5} M progesterone in phosphate buffer, pH 7.6, $\mu = 0.1$ (full curve), and of 2×10^{-5} M progesterone in the presence of 4×10^{-5} M HSA in the same buffer (broken curve). The method of determining the depression of the extinction coefficient by protein interaction, $\Delta\varepsilon$, as percentage of the extinction coefficient of unbound progesterone is indicated

In the range of concentrations and under the conditions generally employed in these spectrophotometric studies, the steroid solutions follow Beer's law: alterations in the extinction coefficients, therefore, cannot be attributed to displacements in dimer and polymer equilibria.

There is practically no change of pH on addition of the protein to the buffered solutions. Thus it is apparent that the spectral change results from the formation of an intermolecular complex between the steroid and the protein. The portions of bound and unbound steroid can be calculated (WESTPHAL et al., 1958) from the spectrophotometric data according to the procedure of KLOTZ (1946 b) in the following way.

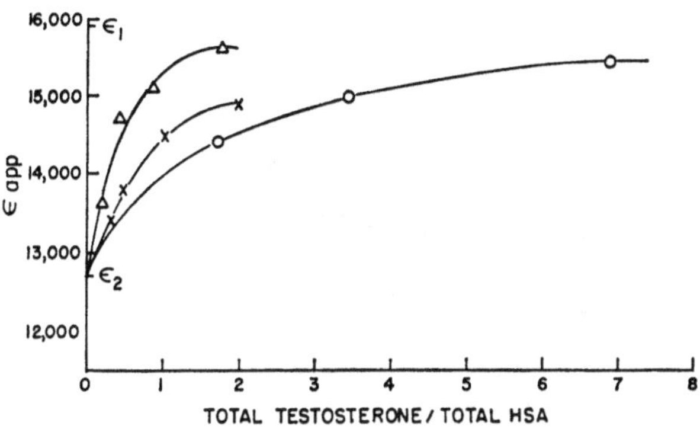

Fig. IV-6. Graphic determination of ε_2 for testosterone in the presence of human serum albumin (HSA). ε_{app} = molecular extinction coefficient of steroid, observed at various concentrations of steroid and/or HSA; ε_1 = molecular extinction coefficient of unbound steroid; ε_2 = molecular extinction coefficient of completely bound steroid. Concentrations of testosterone: Δ, 10^{-5} M; x, 2×10^{-5} M; O, 4×10^{-5} M. Phosphate buffer pH 7.6, $\mu = 0.1$; 25°. From WESTPHAL et al. (1958)

At a given steroid concentration, increase of the protein concentration will enhance the percentage of steroid bound to the protein. If ε_1 is the molecular extinction coefficient of the steroid in the absence of protein, i. e., at 0% binding, and ε_2 the molecular extinction coefficient at 100% binding, a numerical value for ε_2 can be obtained by graphic extrapolation to $x = 0$ of the observed molecular extinction coefficient (ε apparent) versus x, i.e., the ratio of total steroid/total protein. An example for this procedure is given in Fig. IV-6 which shows the extinction coefficients of testosterone solutions at several concentrations and at different testosterone/HSA ratios.

Wavelengths of maximal absorption, λ_{max}, and molecular extinction coefficients, ε, in phosphate buffer, pH 7.6, μ 0.1, have been measured (WESTPHAL, 1957 b; WESTPHAL and ASHLEY, 1958 a, 1959,

1962) for numerous Δ^4-3-ketosteroids (see Table VII-2). The extinction coefficients were found not to differ from those determined in alcohol or other solvents (DORFMAN, 1953; DUSZA et al., 1963) in agreement with the general experience of a negligible solvent effect on extinction coefficients (FIESER and FIESER, 1959). By use of the numerical values of the extinction coefficients for the unbound

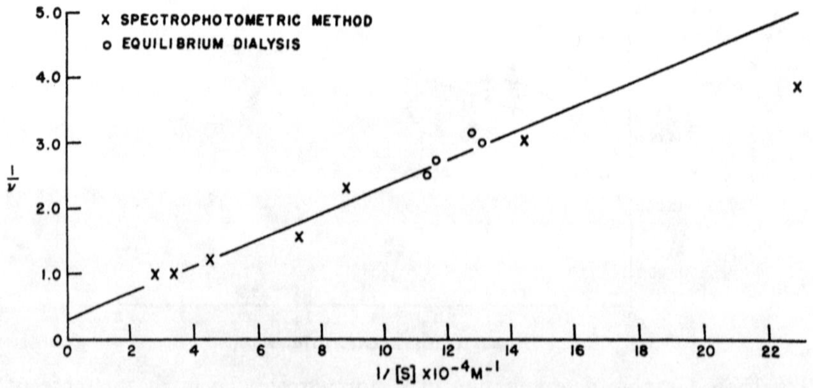

Fig. IV-7. Binding of testosterone to human serum albumin: $\bar{\nu}$ = moles steroid bound per mole total protein; [S] = concentration of unbound steroid in moles per liter; phosphate buffer pH 7.6, μ = 0.1; 25°. From WESTPHAL et al. (1958)

steroid, ε_1, and for the completely bound steroid ε_2, the fraction of the unbound steroid, α, can be obtained by interpolation according to

$$\alpha = \frac{\varepsilon_{app} - \varepsilon_2}{\varepsilon_1 - \varepsilon_2} \tag{IV-4}$$

where ε_{app} is the molecular extinction coefficient of the steroid observed at any given protein concentration (Fig. IV-6). The concentrations of unbound and bound steroid, [S] and [S_{bd}], respectively, are calculated according to

$$[S] = \alpha[S_t]; \tag{IV-5}$$
$$[S_{bd}] = (1 - \alpha)[S_t] \tag{IV-6}$$

where [S_t] is the total steroid concentration.

The binding parameters obtained by the spectrophotometric method are in agreement with those determined by other procedures, e.g., equilibrium dialysis, as shown by a comparison for several Δ^4-3-ketosteroids (WESTPHAL et al., 1958). The correspondence of the binding values resulting from the two methods is illustrated in Fig. IV-7 in which the data obtained for testosterone-HSA interaction are represented by the method of reciprocal plots; this procedure of evaluating binding parameters will be discussed in Chapter V. 4 c. Additional results from the spectrophotometric procedure will be given in Chapter VII. In a recent publication, RYAN (1968) has reexamined our spectrophotometric method with techniques and optical equipment reflecting modern developments over 11 years. The range of measurement was extended and the results were expressed as difference spectra, a mode of presentation which has been fruitfully applied to study environmental influences on chromophoric structures in proteins (SCHERAGA, 1961). The essential results of our studies were confirmed; no attempt was made to evaluate the difference spectra to obtain association constants or other binding parameters. An analysis of the spectral perturbations resulting from the steroid interactions with HSA has been reported recently (RYAN and GIBBS, 1970).

It should be mentioned in connection with the spectrophotometric methods that ATTALLAH and LATA (1968) have utilized fluorescence quenching titration for the study of steroid-protein interaction.

14. Gel Filtration

Equilibrium dialysis, ultrafiltration and other methods for the quantitative determination of ligand-protein interaction are based on the physical separation of the bound and unbound portion of the ligand. Another technique to accomplish such separation is chromatography—a principle used in various forms for chemical fractionation. The feasibility of chromatographic separation of unbound from protein-bound ligand was shown some years ago (WESTPHAL et al., 1950) with the azorubin-HSA complex formed by addition of an excess of the dye to HSA solutions or serum (WESTPHAL and GEDIGK, 1951). The unbound dye was adsorbed on anionotropic aluminum oxide in a narrow zone, and the albumin-bound azorubin passed over the column and was measured in the filtrate. The result obtained is a minimum value for the azorubin-binding capacity of the albumin.

The method is capable of demonstrating interference with the azorubin binding by the presence of competing anions, e.g., fatty acids (Westphal, 1953; Westphal et al., 1953).

Similar in several essential points to this chromatographic method for the determination of dye anion binding to albumin is the technique of gel filtration. In the simple application in a chromatographic column, the method has been introduced (de Moor et al., 1962) for the assay of CBG activity in blood serum or plasma (Doe et al., 1964). It is based on the principle of molecular sieving (Porath and Flodin, 1959; Porath, 1960) in which the unbound corticosteroid penetrates the network ("pores") of a gel suspension of Sephadex, a cross-linked dextran, and is thus retained and slowed down on the filtration column; the protein including the bound portion of the total corticosteroid is eluted without retention (molecular exclusion) and appears in the eluate immediately following the void volume. Unlabeled or radioactive corticosteroid is added to the test serum in excess of its binding capacity and equilibrated, so that the serum to be eluted is saturated with the steroid. The assay gives the binding capacity directly in μg steroid per 100 ml of serum or plasma. Details of this procedure will be given in Chapter VIII. 2 b.

The method of gel filtration in the simple form described (de Moor et al., 1962; Quincey and Gray, 1963; Doe et al., 1964; Westphal, 1969) is not an equilibrium procedure. The component parts of the corticosteroid-CBG complex are in a binding equilibrium subject to the law of mass action. As a consequence, the complex undergoes continuous dissociation as it moves down the Sephadex column, while the unbound molecules enter the internal volume (Gelotte, 1960) of the gel grains. Only ligand-protein complexes of high binding affinity such as those of corticosteroids with CBG will survive the gel filtration without complete dissociation; the result of the measurement will be a less-than-maximal binding capacity. Complexes characterized by smaller association constants, e.g., those between corticosteroids and serum albumin, will be completely dissociated before they reach the end of the gel filtration column; they are therefore not measurable by this procedure. This deficiency is turned to advantage in the determination by gel filtration of CBG capacity in serum or plasma (see Chapter VIII. 2 c); the interference with the CBG binding by the albumin present in relatively large quantities is thus eliminated.

The gel filtration procedure in the form described shares with other non-equilibrated binding systems the inability to yield thermodynamically valid results. At best, it may provide relative data on binding capacity and an approximate measure of binding affinity of the interacting macromolecules and thus facilitate a comparison of a number of test samples. Strict adherence to identical experimental conditions is necessary for such comparison to be valid. It should be noted that BURKE (1969 a) has recently described a steady-state gel filtration procedure which eliminates the problems of dissociation. This is achieved by greatly increasing the sample volume; this modification is therefore useful only when large amounts of test material are available. Application of the technique has been reported by BURKE (1969 c).

It is evident from these considerations that the use of gel filtration for the determination of steroid binding is limited to very specific cases, as for example the determination of CBG capacity; the results have to be interpreted with caution. In view of its simplicity and usefulness for comparative determination of CBG activity under strictly controlled conditions (SEAL and DOE, 1963; DOE et al., 1964; SEAL and DOE, 1965), additional details and critical discussion on the gel filtration method will be given in Chapter VIII. 2 c.

15. Gel Equilibration

Efforts have been made to combine the desirable features of gel filtration, especially rapidity and convenience, with those of an equilibrated system. In equilibrium dialysis, the macromolecule is separated from the ligand by the semipermeable membrane through which the ligand can diffuse. In its result, this process is very similar to the reactions occurring in a suspension of Sephadex gel of such dimension that the macromolecules are excluded from the internal volume, whereas the ligand can freely move through the internal and external volume of the gel suspension. One main difference in the two procedures is the rate of equilibration which is incomparably faster in the Sephadex system because of lack of a membrane. Application of gel filtration for desalting protein solutions demonstrates the similarity of the two methods.

HUMMEL and DREYER (1962) have described a method which is analogous in its basic principle to the equilibrium paper electrophoresis of DAUGHADAY (1956 c) and resembles even more the retention or overtaking electrophoresis of STRUCK (1963). The protein, P, is dissolved in a solution of the ligand, S, to which it binds. The formation of PS reduces the concentration of free ligand, [S], by an amount equivalent to [PS]. The solution is then placed on a Sephadex G-25 column which has been equilibrated with the same solution of S that was used to dissolve the protein. The column is eluted with the solution of S and the concentration of S in the eluate is measured. With the appearance of the protein immediately following the void volume, the concentration of total S in the eluate will rise above the original level by an amount equal to PS. Correspondingly, at some point after the protein peak the concentration of S in the eluate will fall below the base line level to form a trough. Excess of S found in the protein peak is equal to the amount represented by the trough. As the protein moves down the column it continues to remove S from the internal volume of the ligand solution until equilibrium is reached at the base-line concentration of S. Evidence of binding is given by the presence of excess S in the protein peak, and by the subsequent appearance of the trough indicating the missing amount of S; measurement of the two (equal) quantities yields the amount of S bound by P at the equilibrium concentration which is the base-line concentration of unbound S.

This method has been modified and refined recently (FAIRCLOUGH and FRUTON, 1966) and applied for the determination of binding parameters in the interaction between BSA and L-tryptophan or tryptophan derivatives. The results were in agreement with values obtained by equilibrium dialysis. The method lends itself to automation and offers advantages in speed, convenience and flexibility.

The equilibrium gel filtration (HUMMEL and DREYER, 1962; FAIRCLOUGH and FRUTON, 1966) has not yet been applied to study steroid-protein interactions; it has been discussed in view of its potential significance for this area of research. However, another method, also developed recently with the aim of utilizing the Sephadex gel in an equilibrated system, has served in the study of steroid binding. PEARLMAN and CRÉPY (1967) employed Sephadex G-25 in a batchwise fashion, thus resembling more closely a typical equilibrium dialysis procedure. In principle, the method utilizes a system of compartments one of which is inaccessible to the macromolecules by a semipermeable border. The protein, P, and the bound steroid, S_{bd}, are excluded by the cross-linked dextran gel and remain in the external phase; unbound steroid, S, is distributed freely between the external and internal phases of the system. Since the internal volume is inaccessible to the protein it is this internal phase that corresponds to the outside phase in equilibrium dialysis; *vice versa,* the external volume of the Sephadex gel suspension is the equivalent of the inside phase of an equilibrium dialysis system in which the protein is retained by the membrane.

In the batchwise gel equilibration procedure as applied to the study of testosterone interaction with human serum proteins (PEARLMAN and CRÉPY, 1967), radioactive steroid and protein are added to Sephadex G-25 in 0.155 M phosphate buffer of pH 7.4. The external volume is 3-times the internal volume. Equilibration is reached by shaking at 25° for one hour.

Determination of radioactivity in the external phase after equilibration provides S_x, the amount of total steroid, i.e., bound steroid, S_{bd}, plus unbound steroid, S, in the excluded volume, hence

$$S_{bd} = S_x - S. \qquad \text{(IV-7)}$$

The concentration of unbound steroid in the internal phase is equal to that in the external phase. Under the volume conditions described, and in the absence of protein, the distribution of steroid between external and internal phase would theoretically be 3 : 1 if there were no interference with the distribution by the gel material. The partition is given by (PEARLMAN and CRÉPY, 1967)

$$K' = \frac{S_x}{a - S_x} \qquad \text{(IV-8)}$$

where K′ is a partition factor for the given system, S_x the steroid in the external volume, and a the total steroid introduced into the system; $a - S_x$ is therefore the steroid in the internal phase.

In the absence of protein, S_x equals unbound steroid in the external phase, S; hence

$$S = K'(a - S_x). \qquad \text{(IV-9)}$$

Under the assumption that the partition of unbound steroid between external and internal phase is the same in the absence and presence of protein, S [Eq. (IV-9)] may be substituted in (IV-7) to give

$$S_{bd} = S_x - K'(a - S_x). \qquad \text{(IV-10)}$$

It is evident that the concentrations of S_{bd} and S can be calculated from the radioactivity values S_x and a and the partition factor K', using the values for the internal and external volumes.

The theoretical value for K' for the system described is 3, assuming undisturbed distribution between external and internal volume. However, the value obtained experimentally is much smaller; for testosterone in phosphate buffer, pH 7.4, a K' value of 1.13 was found. This indicates that the chemical activity of the steroid is decreased in the system, presumably by adsorption to the gel matrix of the Sephadex. Such deviation from the theoretical distribution between external and internal phase has been observed for other steroids; the K' values were dependent on their structure (PEARLMAN and CRÉPY, 1967). BURKE (1969 b) found adsorption to Sephadex G-25 and G-50 for testosterone, desoxycorticosterone, cortisol, and aldosterone, with decreasing affinity in the order given. Differences in adsorption affinity may be utilized for separation of chemically related steroids; this has been shown by VAN BAELEN et al. (1967 a, b) who separated estradiol, estrone and estriol by gel filtration using Sephadex G-25 and Sephadex LH-20. Similar divergence of K' from theory has been found with other compounds of various chemical classes when studied in their adsorption behavior toward cross-linked dextrans (GELOTTE, 1960; MARSDEN, 1965; JANSON, 1967).

PEARLMAN and CRÉPY (1967) have used the gel equilibration procedure to measure testosterone binding to BSA and HSA, and obtained association constants similar to those found in other laboratories. In application to serum, the value 1/P is determined at 50% steroid binding, where P equals total serum protein concentration. The serum is diluted 20 to over 350-fold, and steroid concentrations of the very low order of 10^{-9} M are applied. The value 1/P corresponds to the C-value (Chapter VIII. 2 a) when $S_{bd}/S = 1$, and equals n k when $P \gg S_{bd}$ according to

$$\frac{S_{bd}}{S} \cdot \frac{1}{P} = C \cong n\,k\,. \qquad \text{(IV-11)}$$

The experimental setup of this equilibrium method for assessing steroid-protein interaction is straightforward, and very little material is needed. The main advantage is the rapidity of equilibration, owing to the high ratio of surface area of the Sephadex gel to the volume of the internal phase.

A negative aspect of the method is the reversible adsorption of the steroid ligands which causes the deviation of the partition coefficient from its theoretical value. The magnitude of this discrepancy indicates serious interference of the measuring system with the measured quantities; the adsorptive forces of the gel surface constitute a second reversible binding system which competes with the binding sites on the protein for the ligand. Compensation is made (PEARLMAN and CRÉPY, 1967) by determining K′ experimentally in a protein-free system. Such correction would presume that K′ remains the same in the absence and presence of protein. The validity of this assumption has not been proven, and would be difficult to ascertain since the steroid concentration in the internal phase cannot be determined without upsetting the reversible adsorption system of the gel surface.

Whereas the gel equilibration procedure allows to calculate the association constant, an estimation of the number of binding sites was not attempted because too extensive an extrapolation would have been required (PEARLMAN and CRÉPY, 1967). This limitation has to be considered in the application of the original gel equilibration technique. GUÉRIGUIAN and CRÉPY (1969) have used the same principle to develop a new procedure to measure the concentration of the testosterone-binding β-globulin in serum.

In spite of the limitations inherent in this novel procedure, it may prove useful for the rapid assessment of protein binding to steroids, especially for comparative estimations under closely similar conditions. PEARLMAN (1970) has recently summarized the essential features of the technique, including calculations. Additional discussion of the method has been given by STONE and METZGER (1968).

16. Use of Radiolabeled Steroids and the Problem of Decomposition

With most methods discussed in the preceding sections, radiolabeled steroids are being used for the analysis of the interactions with proteins. The high sensitivity of quantitative determination makes their application a necessity whenever the low concentration of binding protein requires the use of minimal quantities of steroids. The investigator must be aware, however, of the particular properties of the radioactive steroids applied in order to avoid erroneous conclusions.

Whereas most steroids under usual experimental conditions are relatively stable substances, organic compounds labeled with radioisotopes have a general tendency to decompose on storage (BAYLY and EVANS, 1966, 1967). This is the result of self-irradiation which produces molecular alterations and, especially in the presence of oxygen, chemical reactions similar to those obtained with external ionizing radiation. The self-decomposition of carbon-14-labeled organic compounds (ROCHLIN, 1965) was recognized some time ago when autoradiolysis was observed during their storage in dry form (TOLBERT et al., 1953). The extent of decomposition ranged from 1 to over 60% for various compounds; cholesterol over a period of about 18 months showed approximately 40% destruction, based on the amount no longer precipitable with digitonin. Air oxygen was required for the decomposition reactions which resulted in oxidation at and near the 5,6-double bond of cholesterol; the two epimeric 7-hydroxy, the 7-oxo and the $5\alpha, 6\beta$-dihydroxy derivatives were formed (DAUBEN and PAYOT, 1956). Unlabeled cholesterol in air or ^{14}C-labeled cholesterol *in vacuo* were stable under otherwise similar conditions. Good stability was also observed (MANGOLD and SAND, 1968) for the purified methyl esters of ^{14}C-labeled oleic, linoleic and linolenic acid of low specific activity when stored for 10 yrs at —30° in the dark, oxygen excluded. The decomposition rate is increased by ultraviolet light so that the influence of light irradiation and of the ionizing radiation from the radioisotope may overlap. This has been shown with radiolabeled and unlabeled testosterone and other steroids which form decomposition products when exposed to ultraviolet light (SAVARD et al., 1953).

A few years later, self-decomposition of cortisol-4-^{14}C was observed in ethanol solution or in the solvent-free state (GREENAWAY and HOLNESS, 1959); auto-irradiation was assumed to be the cause. In these auto-radiolytic processes in polar solvents such as alcohols, water or others, H_2O_2 and the radicals OH and HO_2 have been suggested as the oxidizing agents; they are formed when O_2-saturated water is subjected to ionizing radiation (BARRON, 1954). On X-irradiation of cortisone solutions in water and in methanol, in the presence and absence of oxygen, androst-4-ene-3,11,17-trione is formed (WEISS, 1953; ALLINSON et al., 1955). The authors assume that OH radicals are responsible for the oxidative attack at C-17 since H_2O_2 in concentrations formed during radiolysis did not affect corticosteroids (ALLINSON et al., 1955). This result makes it more likely that radicals rather than H_2O_2 are involved in the decomposition of cortisol-4-^{14}C caused by ionizing radiation of external or internal origin.

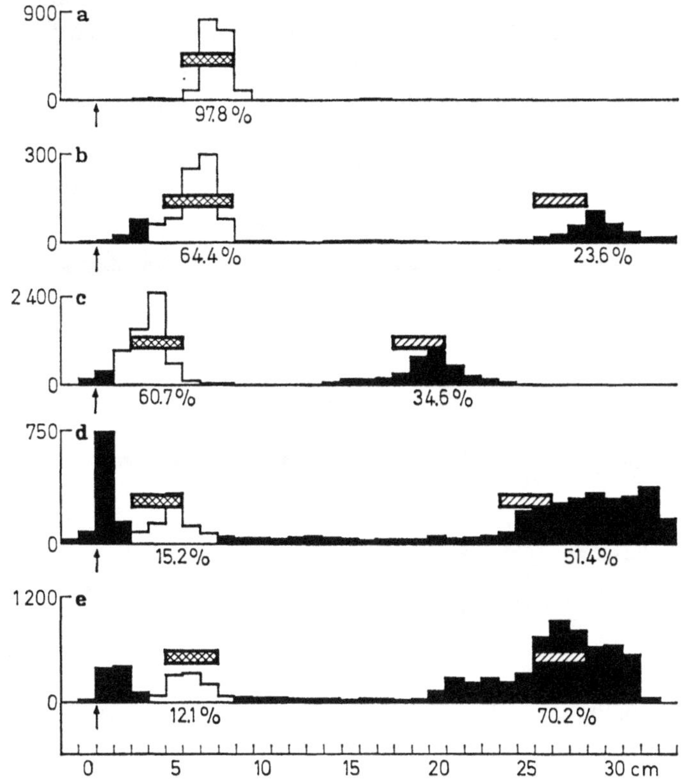

Fig. IV-8. Chromatographic analysis of samples of cortisol-4-^{14}C kept in 0.05 M phosphate, pH 7.4, 23°, light. Chloroform-formamide, 23°. (a) cortisol-4-^{14}C, starting material; (b), (c), (d), (e), after 2, 4, 16, 21 weeks, respectively. Clear area, cortisol-4-^{14}C; dark area, decomposition products; ▩ cortisol standard; ▨ corticosterone standard. Arrow indicates line of application. Ordinates, counts/min. From WESTPHAL et al. (1967)

Experiments in our laboratory (WESTPHAL et al., 1967) showed that radiochemically pure cortisol-4-^{14}C in aqueous solution at room temperature decomposed during several weeks to products of lower and higher polarity (Fig. IV-8). Light accelerated the breakdown (Fig. IV-9). Similar decomposition products were formed, although at a slower rate, when methanol was used as solvent, and when the solution was stored at —85° in the dark. The main product of the autoradiolysis was identified as 11β-hydroxyandrost-4-ene-3,17-dione by chromatographic methods, by CrO_3-oxidation to androst-4-ene-3,11,17-trione (adrenosterone), and by co-crystallization with the authentic 11β-hydroxy steroid to constant specific activity.

Spontaneous degradation of the corticosteroid structure, specifically of the dihydroxyacetone side chain, is not limited to radio-

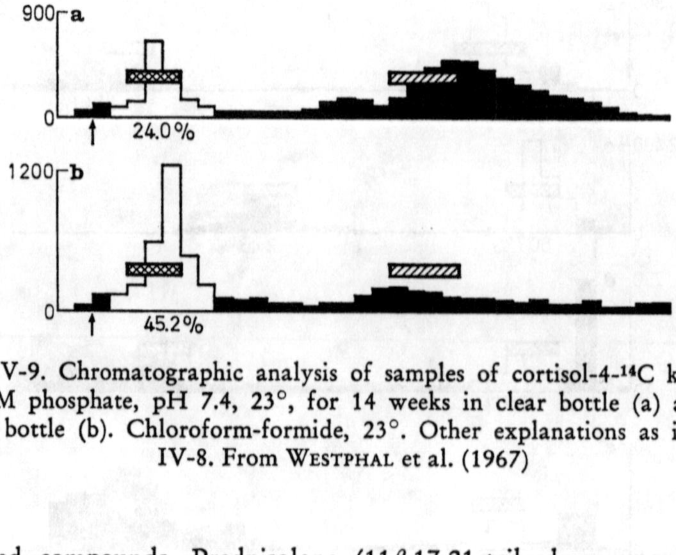

Fig. IV-9. Chromatographic analysis of samples of cortisol-4-^{14}C kept in 0.05 M phosphate, pH 7.4, 23°, for 14 weeks in clear bottle (a) and in dark bottle (b). Chloroform-formide, 23°. Other explanations as in Fig. IV-8. From Westphal et al. (1967)

labeled compounds. Prednisolone (11β,17,21-trihydroxypregna-1,4-diene-3,20-dione) has been observed (Chulski and Forist, 1958) to be converted to the corresponding 17β-etianic acid (11β,17α-dihydroxy-3-oxoandrosta-1,4-diene-17β-carboxylic acid) and to the 17-ketone (11β-hydroxyandrosta-1,4-diene-3,17-dione) when exposed to aqueous suspensions of magnesium oxide at pH 10.4. Similar decompositions were obtained (Kripalani and Sorby, 1967) when dilute solutions of cortisol in phosphate buffer of pH 7.4, μ 0.16, were incubated at 15°, 25° and 35°. These degradations occured under aerobic conditions, involving known oxidative cleavage reactions (Velluz et al., 1947; Caspi et al., 1961). Another type of spontaneous oxidation was the conversion of cortisol to 21-dehydrocortisol (11β,17-dihydroxy-21-oxopregn-4-ene-3,20-dione) in aqueous carbonate buffer of pH 9.1 (Monder, 1968). Incubation of 1,2-

^{3}H-cortisol for 48 hrs at 37° in phosphate buffer of pH 7.4 also resulted in the formation of 21-dehydrocortisol (OHTSUKA and KOIDE, 1968). Other 17α-hydroxycorticosteroids (cortisone, prednisolone) in the same buffer were autoxidized at about the same rate as cortisol. Corticosteroids lacking the 17-hydroxy group were considerably more stable; they were altered about 10 times more slowly than the 17α-hydroxysteroids (MONDER, 1968). In our autoradiolysis studies (WESTPHAL et al., 1967), corticosterone-4-^{14}C decomposed about as rapidly as cortisol-4-^{14}C, forming several products of higher as well as lower polarity. 21-Dehydrocortisol and 11β-hydroxyandrost-4-ene-3,17-dione have been identified as contaminants of tritiated cortisol preparations (MONDER and WALKER, 1970).

It is the experience of many investigators, and has been emphasized some years ago (EVANS and STANFORD, 1963; ROCHLIN, 1965; EVANS, 1966) that the self-irradiation decomposition of radiolabeled organic compounds is catalyzed or accelerated by contaminations. FRANKEL and NALBANDOV (1966) demonstrated in an extensive study that impurities in a variety of solvents are essential factors in the autoradiolytic reactions. GELLER and SILBERMAN (1967) observed strong effects of impurities on the self-decomposition of estrone-6,7-^{3}H and cortisone-4-^{14}C. We found a marked decrease in the spontaneous degradation of cortisol-4-^{14}C with increasing purification (see Table I in WESTPHAL et al., 1967). Certain proteins appear to inhibit the decomposition reactions possibly by adsorbing some of the impurities; the addition of EDTA also decreases the oxidation rate (MONDER, 1968).

The practical conclusions from these findings on self-irradiation decomposition of radiolabeled steroids are evident: Periodic checking of the radiochemical purity of the labeled steroids; scrupulous purification of all steroids and other reagents, elimination of trace contaminations from solvents, glassware and all materials that come in contact with the radiosteroids. Double-distilled deionized water should be used since heavy metals such as copper may be particularly effective in catalyzing the breakdown reactions. Air should be excluded from the systems whenever practicable. All these precautions are especially important for experiments of long duration, such as equilibrium dialysis, or in techniques in which the radiosteroids are exposed to air while forming a large surface, as for instance in paper chromatography. Experiments in our laboratory (WESTPHAL and

HARDING, 1968) have shown that pure cortisol is not affected by exposure to 37° at pH 7.4 in 0.05 M phosphate buffer, according to analysis by paper and thin layer chromatography. No degradation of cortisol was found either in the serum protein-containing inside solution of an equilibrium dialysis system after 48 hrs at 37°; the outside solution contained a few percent of less polar material.

V. Characteristics of Steroid-Protein Interactions

1. Molecular Characteristics of Steroids

If we consider, from the general viewpoint of ligand-protein interaction, which properties of a steroid molecule* have particular relevance to its interaction with a protein, three characteristics appear to be important:

1) electrostatic charge, if any;
2) "polarity", i.e., number of polar substituents;
3) stereochemical configuration.

Most of the natural steroid hormones and other products of steroid biosynthesis are uncharged, i.e., electroneutral molecules. This does not apply to a large group of basic plant steroids, alkamine derivatives, which belong to the class of the alkaloids. They contain primary, secondary or tertiary amino groups. No systematic studies have been published on their protein-binding qualities. A second group is that of the natural estrogenic hormones which possess a phenolic hydroxyl group; the pK values for estradiol-17β and estrone in aqueous solution have been reported to be 10.08 and 10.91, respectively (Kirdani and Burgett, 1967). At the neutral pH of the circulating blood, only a very small fraction of the estrogens would be present in ionic form, whereas the major portion would be uncharged. However, this normal equilibrium may be shifted if there is a high-affinity binding site for the estrogen phenolate anion.

---

* For chemistry of steroids see Klyne (1957); Shoppee (1958); Fieser and Fieser (1959); Lettré et al. (1959); Djerassi (1963); Engel (1963); Neudert and Röpke (1965). Comprehensive references for more than 21,000 steroid compounds are found in Jacques et al. (1965). For biological, biochemical and some of the applied aspects consult Dorfman and Shipley (1956); Clark and Grant (1960); Heftmann and Mosettig (1960); Soffer et al. (1961); Villee and Engel (1961); Young (1961 b); Applezweig (1962); von Eichler and Farah (1962); Zuckerman (1962); Martini and Pecile (1964); Dorfman and Ungar (1965); Pincus et al. (1966); Gray and Bacharach (1967); Heftmann (1970).

The numerous representatives of the bile acids form a third group which are carboxy anions at neutral pH. Other carboxyl containing natural steroid derivatives are the glucuronides (glucosiduronic acids), β-glucosidic conjugation products with D-glucuronic acid; at normal blood pH, they are present as glucosiduronate anions. Also occurring as anions in the blood are the sulfates, ester conjugates formed from

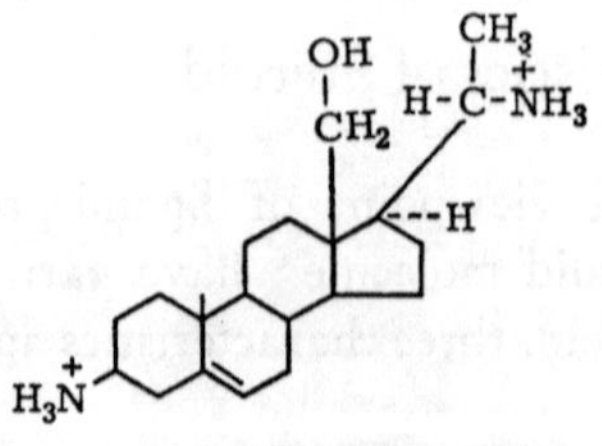

(1) Holarrhimine
Alkaloid

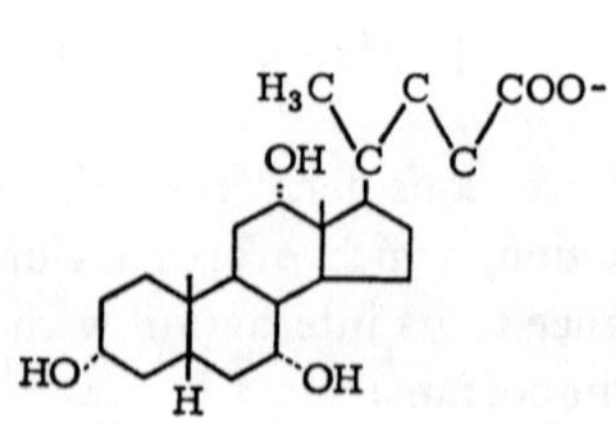

(2) Cholic acid anion
Bile acid

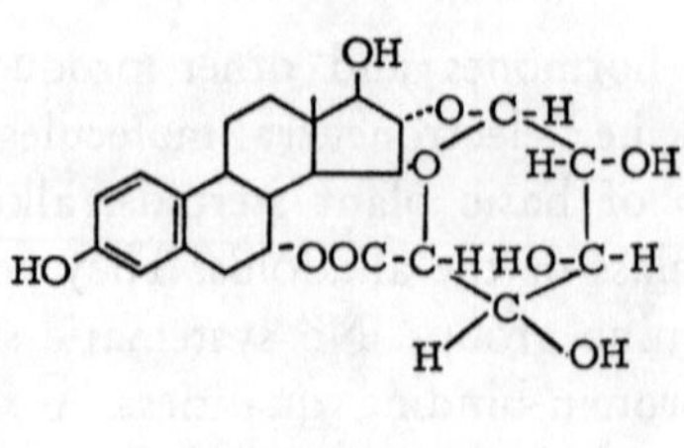

(3) Estriol glucosiduronic acid anion

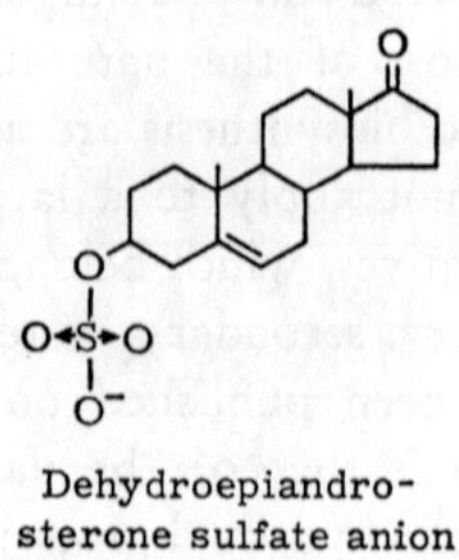

(4) Dehydroepiandrosterone sulfate anion

Fig. V-1. Types of steroids which carry electrostatic charges at blood pH

various steroid hormones. A bibliography including abstracts on steroid conjugates has been compiled by BERNSTEIN et al. (1966); the subject has also been treated by HADD and BLICKENSTAFF (1969). Examples for ion-forming steroids are given in Fig. V-1.

By far the major portion of the steroid hormones and of other naturally occurring steroid compounds are uncharged, neutral molecules. They vary in their possession of oxygen functions, present either as hydroxyl or oxo groups. The differences in the number and nature of these and possibly other electron-attracting, "polar" groups provide a classification which proved to be convenient for considerations of protein binding. According to the number of polar groups present in a steroid molecule, we may distinguish non-polar, mono-polar, duo-polar, etc. steroids. The basic hydrocarbon, e.g., cholestane, is then a non-polar compound, and we refer to an increase in "polarity" when the number of polar groups in the molecule increases. This usage of "polarity" is similar to that conventional in chromatography.

Consideration of protein interaction with steroids under the viewpoint of the number of their polar groups is useful since the thermodynamic stability of many steroid-protein complexes is governed by the "polarity rule". According to this rule, the binding affinity between a protein and a steroid decreases with an increasing number of polar groups. This behavior, characteristic of serum albumin and other proteins, was first observed by EIK-NES et al. (1954); it was later established (WESTPHAL and ASHLEY, 1962) with a large number of Δ^4-3-ketosteroids (see Chapter VII). It is noteworthy that most natural steroids of distinct biological activity are at least "duo-polar". Fig. V-2 gives the formulas of the major types of steroids discussed with respect to their protein-binding properties.

For an understanding of the interaction between the steroid and the large protein molecule, at the specific binding site, knowledge of the stereochemical structure of the ligand is necessary. Whereas the conventional formulas of the steroids (Fig. V-1, V-2) give us schematic indication of the spatial arrangement of the constituent atoms, more information on the conformational structure can be derived from simple models, designed to represent standard valency angles (FIESER and FIESER, 1959). These models are similar to the perspective formulas shown in Fig. V-3 and V-4; they clearly indicate the position of a substituent in the β- and α-position, i. e., on the upper or front side of the molecule, and on the lower, or rear side, respectively. Equatorial and axial bonds can be distinguished, and the various intramolecular interactions, which influence the stability of the molecule, can be recognized and approximately assessed. TURNER (1954) has discussed in detail the conformational and other molecular properties of steroids in their possible significance for protein interaction.

In a more recent publication, KIER (1968) has assessed the preferred conformations of 20-oxopregnane hormones by molecular orbital calculations. He developed models representing progesterone, corticosterone and cortisol in the D-ring conformation. These results, obtained on purely theoretical arguments, have been challenged by

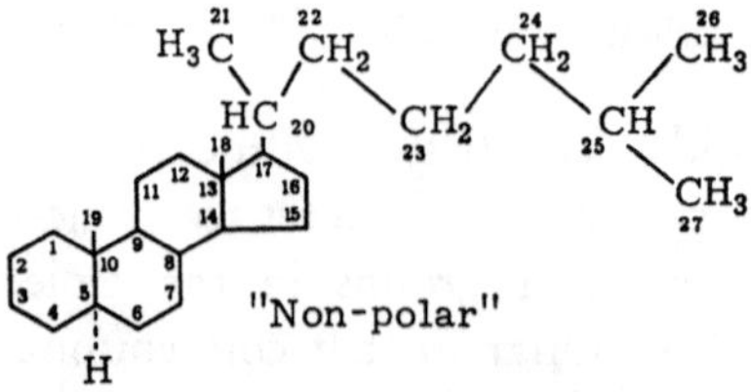

(5) Cholestane

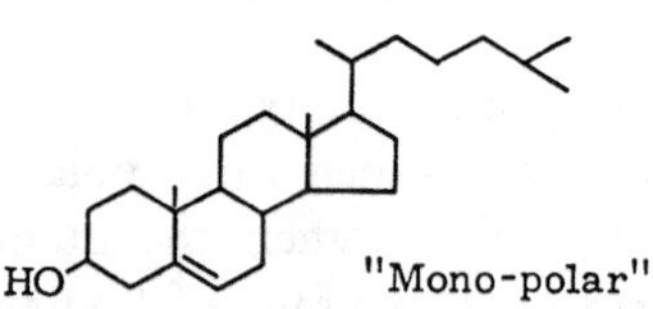

(6) Cholesterol

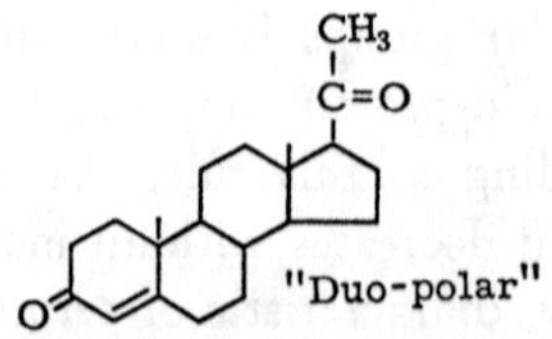

(7) Progesterone

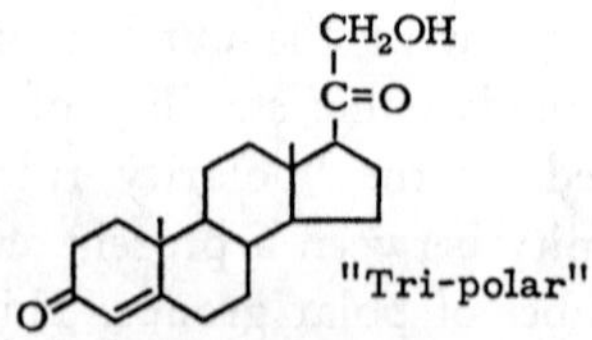

(8) Cortexone=Desoxy-corticosterone

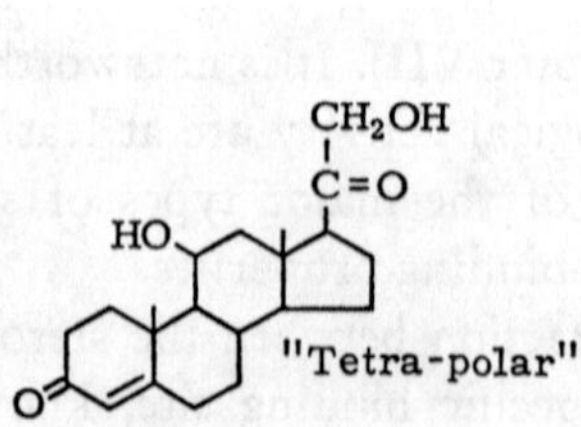

(9) Corticosterone

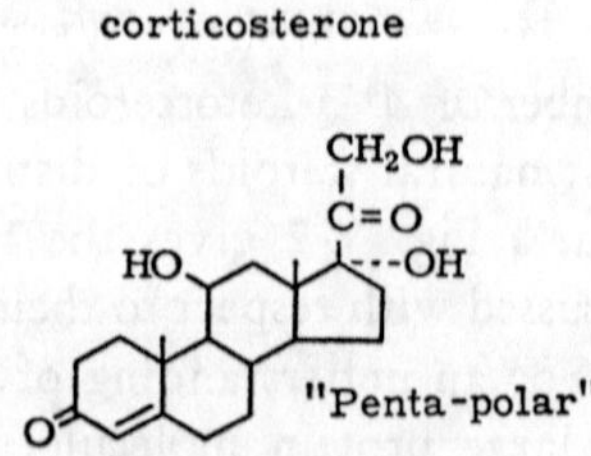

(10) Cortisol

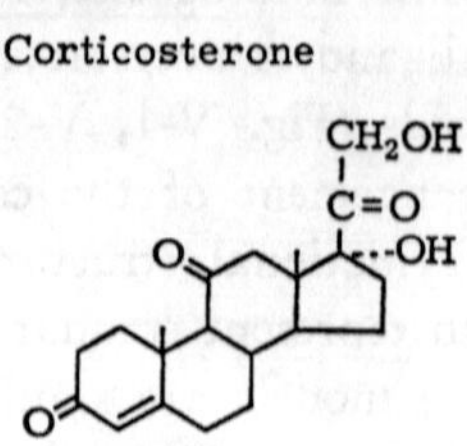

(11) Cortisone

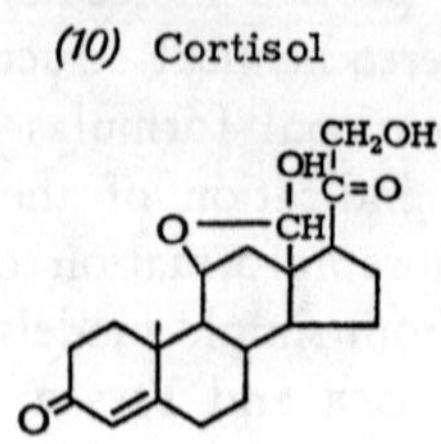

(12) Aldosterone

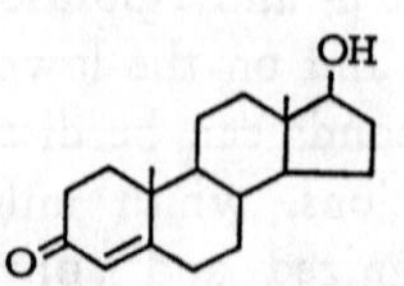

(13) Testosterone

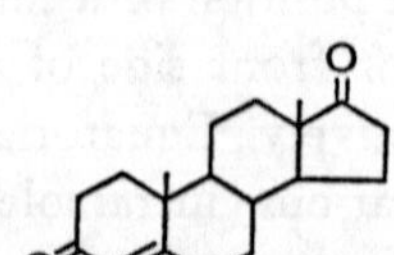

(14) Androstenedione

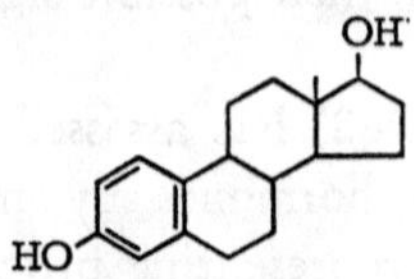

(15) Estradiol

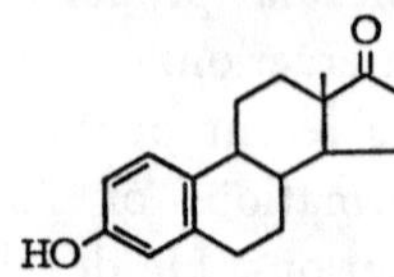

(16) Estrone

Fig. V-2. Types of steroids of different "polarity"

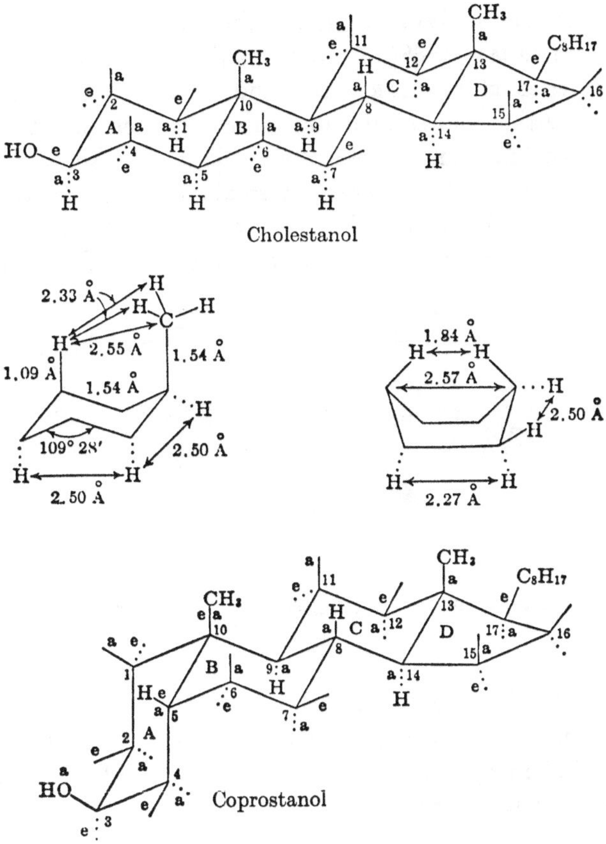

Fig. V-3. Perspective formulas of cholestanol and coprostanol, and dimensions of saturated six-membered rings. Substituents in α and β position indicated by broken and solid lines, respectively. From FIESER and FIESER (1959)

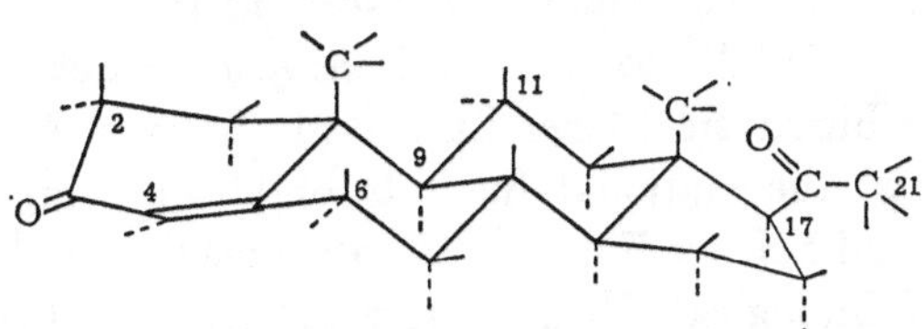

Fig. V-4. Perspective formula of progesterone. Substituents in α and β position indicated by broken and solid lines, respectively. From WESTPHAL (1961)

crystal structure analysis of the 4-chloroderivative of cortisone (COOPER and DUAX, 1969). A conformation deviating only slightly from that reported by KIER was found, in agreement with optical rotatory dispersion and circular dichroism measurements. Interatomic bond distances and angles were obtained. Similar studies with 6β-bromoprogesterone (GOPALAKRISHNA et al., 1969) revealed a highly distorted A-ring, due to the Δ^4-double bond and the ketone oxygen at C-3. The rings B and C had the chair form, ring D was found to

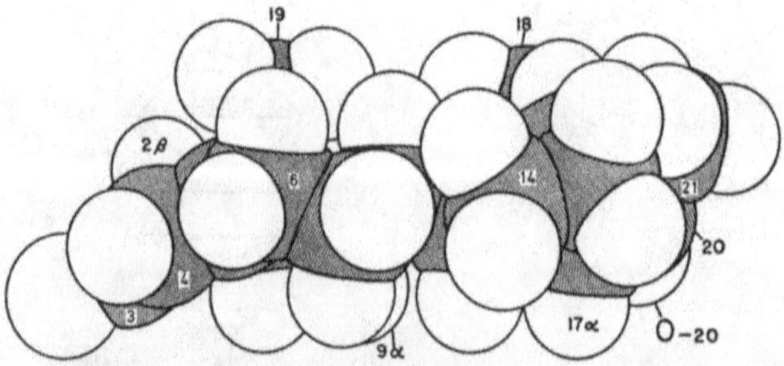

Fig. V-5. Stuart-type model of progesterone, side view. From WESTPHAL and ASHLEY (1959)

be a distorted half-chair. Similar measurements were made for testosterone (COOPER et al., 1968) and estriol (COOPER et al., 1969). These results should be important for consideration of binding sites in high-affinity binding serum proteins such as CBG and in nuclear receptor proteins.

The most lucid information on the spatial arrangement of the steroid, and on the conformational structure to fit that of the binding site in the protein molecule, is provided by the Stuart-type models (Fig. V-5). They approximate the actual dimensions of the various atoms, bonded at true relative distances. A recent development is the improved version of the COREY-PAULING models with connectors by KOLTUN. These "CPK" atomic models have been developed by the Biophysics and Biophysical Chemistry Study Section, Atomic Models Subcommittee of the National Institute of Health, implemented by the U.S. National Science Foundation in connection with the American Society of Biological Chemists (see HARTE and RUPLEY, 1968). Use of these spatial models for the steroids in connection with similar models for the polypeptide chain of the protein may provide valuable insight into the molecular basis of the interaction.

2. Molecular Aspects of the Interaction

One of the essential properties of protein molecules is their ability to interact specifically with small molecules of various chemical structures to form protein-ligand complexes (KLOTZ, 1953; EDSALL and WYMAN, 1958; TANFORD, 1961; STEINHARDT and BEYCHOK, 1964; WEBER, 1965; STEINHARDT and REYNOLDS, 1970). These interactions are regulated by the law of mass action, and the mathematical formulations developed for their characterization are based on this law.

The ligand-protein complexes are held together by noncovalent bonds. Three types of such bonds are generally considered to prevail in interactions with proteins: (1) hydrogen bonds; (2) ionic bonds; and (3) hydrophobic bonds. The first two groups may sometimes be treated together as interactions between polar groups; for it is not always possible to specify the electrostatic contribution to the hydrogen bond precisely, and therefore not profitable to distinguish between hydrogen bonds and ionic bonds (SCHERAGA, 1963). Such interactions may be designated summarily as "hydrogen bonds"; for a comprehensive description see SCHERAGA (1963). Hydrogen bonding has been more recently discussed by DONOHUE (1968), who gives additional literature references. The association of lipids with proteins and hydrophobic bonding have been reviewed by GURD (1960 a).

Numerous possibilities exist for the formation of hydrogen bonds between steroids and proteins; aliphatic, alicyclic and aromatic hydroxy groups of the steroids as well as hydroxy, amino and peptide structures of the proteins can donate hydrogen, while oxo groups, peptide carbonyl and other groupings can function as hydrogen acceptors. Hydrophobic bonds are formed between large areas of the steroid molecule, which is basically a nonpolar structure, and aliphatic and aromatic side chains of the proteins, including sulfide groupings. LOW (1953) has given a lucid account of the configurational properties of the structural elements in proteins, and has illustrated the side chains of the amino acid residues as van der Waals packing models (Fig. 1 in LOW, 1953), ordered according to their "polarity". These are the molecular structures, incorporated in the polypeptide chains in accordance with the given protein conformation, with which the steroids interact.

Concerning noncovalent bonding of the -SH group (CECIL, 1963), hydrogen bonds have been suggested, but little evidence has been obtained to support this view. It has been rather assumed that the cysteine side chain is sufficiently apolar to enter hydrophobic interactions. NEMETHY and SCHERAGA (1962) consider the sulfur atom in cysteine and methionine as equivalent to a CH_2 group. These authors have calculated the thermodynamic constants of the hydrophobic bonds that are possible between isolated pairs of non-polar side chains of alanine, valine, leucine, isoleucine, methionine, cysteine, proline and phenylalanine.

One particular class of molecular interactions is that between proteins and ions. If we consider an ion of fundamental characteristics, the proton, we have an example of acid-base equilibria in protein solution, reactions of great chemical and biological significance for which theory is well developed. Large protein molecules may contain hundreds of basic sites with which protons can associate. Other small ions may interact with anionic or cationic sites in the protein, or possibly with a dipole structure. Such binding may occur with steroid compounds such as bile acids or other charged steroid derivatives such as sulfates or glucosiduronates. It must be assumed in these cases, however, that the interaction is not mediated by electrostatic bonding alone; there is evidence that the neutral portion of the steroid molecule participates in the binding process. Both polar and apolar interaction may be realized in the same complex, depending on the structure of the steroid and the protein. Importance of both electrostatic attraction and hydrophobic interaction has been generally accepted for binding of long-chain fatty acids to serum albumin (SPECTOR et al., 1969).

Proteins are also capable of interacting with metal cations, but the degree of binding varies with the chemical nature of both components, and with the net charge of the protein. Generally, dissociable ionic binding prevails, but with some elements of the transition group including copper, zinc, iron, manganese, the association constants are very high and exchange is very slow. Examples for these are the binding of iron in hemoglobin, and that of zinc in carbonic anhydrase. The relevance of metal-protein binding to steroid-protein interaction will become evident in Chapter XIII. 3 where metal ion inhibition of the association between progesterone and AAG is discussed.

Among the neutral small molecules that interact with proteins, water is of unique importance (SINANOĞLU and ABDULNUR, 1965). Interactions of proteins and other biological and synthetic polymers with water molecules play an essential role in establishing macromolecular conformation (KLOTZ, 1965); this has been shown by results obtained from a very diverse group of experimental techniques. Details of the molecular nature of the influence of water on macromolecules have not been elucidated; a strong correlation between protein hydration and the sum of the polar residues (minus the amides) has been demonstrated by BULL and BREESE (1968 a, 1968 b). Ion-dipole interactions provide the binding to polar amino acid side chains; hydrogen bonding to peptide groups appears to be responsible for a major portion of water uptake in protein hydration. There appears to be an almost unlimited number of accessible binding sites for water, even after alteration of the native structure of a protein.

Characteristic of electrostatic interaction and of hydration is the great number of binding sites on the protein that are capable of attaching the small molecule (KLOTZ, 1953). The association of a protein with hydrogen ions or with water may thus be considered devoid of the influence of a structural specificity of the ions. This is in contrast to many other ions, including organic anions and cations: very high selectivity has been observed in the binding affinity of anionic haptens to antibody proteins (LANDSTEINER, 1945). Interaction of closely related organic anions to serum albumin also shows highly specific differences, for instance between positional isomers, or with various stereoisomers (KLOTZ, 1953).

A high selectivity of interaction is generally observed with small neutral molecules such as steroids. The number and position of electron attracting and electron repelling groups in the steroid molecule influences the binding affinity; the steric configuration is another determining factor. With some proteins, no binding of steroids is detectable; other proteins form complexes of association constants greater than 10^9 liters/mole. The number of binding sites appears limited. In many instances, only one principal binding site has been demonstrated per protein molecule; it has to be realized, however, that low ligand solubility may prevent detection of additional types of binding sites of lower affinity. Common to all known steroid-protein complexes is the reversible nature of the association, regulated by a binding equilibrium according to the law of mass action. The binding affinity generally decreases with increasing temperature, and may become difficult to detect at a temperature which does not yet inactivate the protein irreversibly. Dependency of the association constant on pH has been observed, with an optimum close to neutral in several cases. The protein concentration appears to influence the binding affinity of the steroid-protein complex in a manner which is not fully understood.

To arrive at a meaningful interpretation (EDSALL and WYMAN, 1958) of the interaction between a protein and a ligand such as a steroid, we consider a macromolecule, P, which contains n binding sites each of which is capable of attaching a ligand, S. The sites on P which may associate with S are not necessarily unoccupied when S is absent. A water molecule may be attached at the binding site for the ligand S; an example for such substitution is given by the observation of HAUROWITZ (1951) that the binding site for oxygen at the iron atom of hemoglobin is occupied by a water molecule in oxygen-free solution. Binding of water molecules to sites specific for the interaction with other ligands has been demonstrated in other cases (EDSALL and WYMAN, 1958). The ligand-protein interactions then involve a displacement of water from the binding sites by S, not a simple addition of S. The amount of bound water is determined by its activity relative to the activity of S. Since the activity of water is virtually constant in dilute aqueous solutions, the measured association constants for the interaction between P and S implicitly include the activity of water.

Occupation of a steroid-binding site by a water molecule may not be as likely to occur as in the other cases mentioned, because of the difference in chemical structure and polarity between steroids and the ligands which have been found to be replaceable by water. Attachment of a lipid molecule would appear to have a greater probability, and this is indeed suggested by results on inhibition of progesterone binding to HSA (WESTPHAL, 1966) and to AAG (GANGULY et al., 1967) by the presence of lipidic substances (for details see Chapters VI. 5 b and XIII. 2). Unlike in the case of water binding, where the constant water activity in all systems studied in dilute aqueous solution automatically includes the effect of water on the association constant, the lipid present in varying amounts would have an unpredictable influence on the binding affinity. This explains the necessity in studies on steroid-protein interactions to completely remove any lipidic material from the protein molecules or from the solvent systems used. Once the steroid-binding sites are unoccupied, except for the presence of water molecules, the ligand S can enter a true binding equilibrium according to its activity.

Thermodynamically valid, true equilibrium constants are expressed in terms of the activities of the combining species. The experimentally determined constants, however, are obtained from concentration terms rather than from activities, they depend therefore on the solvent, salt concentrations, etc. They can be transformed to the true thermodynamic constants by extrapolation to infinite dilution of all solutes involved, a condition under which activities by definition become equal to concentrations. Since this refinement is rarely justified by the accuracy of the experimental data it is usual practice not to determine the true thermodynamic constants (TANFORD, 1961).

With increasing concentration of S in the solution, the average number of sites on P which are occupied by S will increase, varying from zero when [S] is zero to a limiting value of n when [S] becomes infinite. This average number of occupied sites per P molecule at a particular concentration of S is denoted by the symbol $\bar{\nu}$; it may also be expressed as the average number of molecules of S bound by one molecule of P, under given experimental conditions. The value of $\bar{\nu}$ constitutes a statistical average; in general, even if all the binding sites are equivalent, a statistical distribution of the bound S molecules among all available sites will take place. For any value of $\bar{\nu}$ determined experimentally, there exist at any given moment some P molecules to which more than $\bar{\nu}$ molecules of S are bound, while other P molecules have less. The ratio $\bar{\nu}/n$ gives the fraction of all the available binding sites which are occupied by molecules of S at any moment.

3. Some Pertinent Questions

For an understanding of the physicochemical and biological meaning of steroid hormone binding to a protein, several questions (SCATCHARD, 1949; SCATCHARD et al., 1954) should be considered. First, we would like to know how many ligand molecules, $\bar{\nu}$, are bound per protein molecule under given conditions, and what the theoretical maximum number, n, of the binding sites is. The value of n for a given ligand is a unique property of a protein, and has some similarity with a valence or coordination number (SCATCHARD et al., 1954). It provides the protein a definitive binding capacity.

The second question concerns the binding affinity; it cannot be answered independently of the first one. If the protein has only one binding site for the ligand, the affinity or association constant, k, completely describes the strength of interaction. The value of k generally is dependent on temperature, pH and other conditions. If there is more than one binding site, each site has its intrinsic association constant, $k_1, k_2 \ldots k_n$. If the binding sites are equivalent and independent of each other, $k_1 = k_2 \ldots = k_n$, then a single constant, k, is sufficient to characterize all the n binding sites. This ideal case is rarely encountered, and may be improbable; increasing refinement of our measuring techniques can be expected to bear that out. We may also see that the attachment of one molecule of S to P either supports or hinders the binding of a second S molecule at another site; we speak of positive or negative interactions, respectively. The latter case is observed more frequently; for instance, all interactions in acid-base equilibria are negative. A well-known example for positive interaction is the binding of oxygen to hemoglobin; attachment of an oxygen molecule to one heme group increases the tendency for its neighbors to pick up more oxygen. In the field of steroid-protein interactions, no complete analysis of this problem has as yet been made. For several proteins, only one binding site has been found for the steroids examined; low steroid solubility may be a limiting factor for full analysis. In other cases, it is conceivable that the steroid molecule occupying a primary binding site influences the binding affinity of a second binding site by steric hindrance, by allosteric effects or possibly other mechanisms.

Whereas these questions can be elucidated by the physical chemist, a third one, that of the chemical nature of the binding site, needs additional help from the protein chemist. Nature and sequence of the amino acid residues of the site and its environment, the secondary and tertiary structure of the binding site area, its general location in the protein molecule—these are problems to be explored. A fourth question concerns the reasons for the complex formation; it is intimately involved with the third one. Whereas this question can be answered to some extent for many electrostatic and dipole interactions, we are far from understanding the molecular mechanism underlying the binding of steroids to proteins. An analysis of the hydrophobicity of the binding site and adjacent areas may assist in the elucidation of the problem.

Further questions are of concern to the biological chemist and to the endocrinologist: what is the physiological significance of the interaction? What effect does the protein binding of a steroid hormone have on its biological activity? Do the binding proteins form integral parts of the mechanism of action? Clearly, a thorough understanding of these and other basic questions is required for a meaningful interpretation of the hormone-protein interactions in relation to biological function.

In the circulating blood there may be a second steroid hormone, S_2, which binds to the serum protein at the same site as the first one, S_1. In that case, the amount of S_1 bound, at a given concentration (activity) of free S_1 in solution, is also dependent on the concentration of S_2; S_1 and S_2 compete for the same binding site on P. The binding ratio will be regulated by the concentrations of S_1 and S_2 and their binding affinities. Examples for this type of competitive binding are known for the interaction of corticosteroid hormones and progesterone with human CBG which has one binding site for these different types of steroid hormones (see Chapter XI). Similar competition has been found for the progesterone-binding site of AAG by corticosteroid hormones and testosterone (see Chapter XIII).

Inhibition of binding by a second ligand does not necessarily indicate that the two ligands are attached to the same binding site. For example, binding of S_2 at one of the sites may have an allosteric effect resulting in an altered affinity of a site on P which is specific for S_1. Methods will be discussed which allow some delineation between competitive and non-competitive binding.

Competition phenomena in steroid binding to proteins may play an important role in the manifestation of biological activity of the steroid hormones. Increased secretion of a hormone, e.g., progesterone, may lead to the displacement of another hormone, e. g., cortisol, from a common binding site and thus may be responsible for a higher concentration of the unbound, i. e., biologically active form of cortisol. The implications of competitive steroid binding have not yet been fully investigated.

4. Presentation of Binding Data

a) General Considerations

The simplest form of quantitative expression for steroid binding to a protein is the determination of that portion of total steroid which is associated with the macromolecule. The unbound or free portion of the steroid, S, is often given as fraction of the total steroid, S_t,

$$\alpha_1 = \frac{S}{S_t} . \tag{V-1}$$

The bound fraction then becomes

$$\frac{S_{bd}}{S_t} = 1 - \alpha_1 , \tag{V-2}$$

and 100 $(1 - \alpha_1)$ is a measure of the percent steroid bound to the protein. This latter value is a convenient expression for the attachment of steroids to the individual components of a protein mixture, e. g., to assess the participation of the different serum proteins in the interaction. Its significance is limited to the particular conditions of the experiment, and its application is useful strictly for the comparison of binding parameters in a given system.

Expressing the binding properties of a steroid as percent of total bound does not provide information on involvement of the binding protein. Whenever it is of interest to relate the binding to the particular protein under study, or the mixture of proteins present, it is practical to define a term of specific binding, such as μg steroid bound per g of protein. This expression is useful, for instance, in the presentation of data for steroid binding to unknown mixtures of tissue proteins or similar problems. If the protein is well defined, the value $\bar{\nu}$, indicating the average number of steroid molecules bound per molecule of protein, may be calculated according to

$$\bar{\nu} = \frac{[S_{bd}]}{[P_t]} \tag{V-3}$$

where the concentrations of S_{bd} and P_t are expressed in moles per liter. It should be kept in mind, however, that these expressions for specific binding do not have general validity but are limited to the particular conditions of a given experiment. They are useful since they can be readily obtained and serve well for comparison of binding systems.

The simple expressions for steroid binding given by equations (V-1), (V-2), and (V-3) make no reference to the conditions of steroid concentration under which they were obtained. Since the formation of the steroid-protein complex is regulated by an equilibrium subject to the law of mass action, any change in the concentration of unbound steroid would alter the percentage of the bound portion. A convenient expression, combining affinity, C, has been introduced (DAUGHADAY, 1958 b) to account for steroid concentration as well as protein concentration; it is defined as

$$C = \frac{[S_{bd}]}{[S][P_t]} \text{ g}^{-1} \text{ liter} \tag{V-4}$$

where the concentrations of bound and unbound steroid may be expressed in any convenient term, and the protein concentration is given in g/liter. The use of (V-4) is appropriate whenever the protein, P, is ill defined as for instance in the case of a serum protein mixture or of an undefined fraction of tissue proteins. Its general applicability is somewhat limited by the fact that the C-value is obtained by entering P_t, the total protein concentration. If instead of total protein, $[P_t]$, the concentration of unbound protein, [P], would be used in equation (V-4), the term would provide the association constant, k, as outlined below. For the C-value to be valid for practical purposes, $[P_t]$ must be large in relation to $[S_{bd}]$ so that $[P_t]$ becomes nearly identical with [P]. Additional comments concerning the use of (V-4) as an expression of binding data will be given in Chapter VIII. 2 a.

The presentation of binding data by the terms (V-1) to (V-4) is not satisfactory thermodynamically, even if we ignore the difference between activities and concentrations. An adequate analysis of ligand interaction with a macromolecule clearly must follow the law of mass action. Let us consider the idealized case that all the n combining sites on P are equivalent and independent, and set the activities of the components equal to their concentrations. In the special case when $n = 1$, the value $\bar{\nu}$, i. e., the average number of molecules of S bound by one molecule of P, varies only between 0 and 1. The only reaction taking place is

$$S + P \rightleftarrows SP, \tag{V-5}$$

and the association constant, k, is given by

$$k = \frac{[SP]}{[S][P]} = \frac{1}{K}. \tag{V-6}$$

Eq. (V-6) indicates further that the dissociation constant, K, is equal to the reciprocal of k.

If, in Eq. (V-6), $[P] \gg [SP]$, [P] becomes $\cong [P_t]$. Inserting Eq. (V-3), we obtain

$$k \cong \frac{[SP]}{[S][P_t]} = \frac{\bar{\nu}}{[S]} \tag{V-7}$$

for the special case when $n=1$. If we have one set of n equivalent and independent binding sites,

$$\bar{\nu}/[S] = n\,k\,. \tag{V-8}$$

This expression (V-8) is often useful when binding affinities of different complexes are determined under strictly identical conditions for comparison. Note that $\bar{\nu}/[S]$ in Eq. (V-7) differs from C in Eq. (V-4) only by expressing $[P_t]$ in moles/liter and g/liter, respectively. The C-value obtained with a pure protein, in great excess over [SP], can therefore be used for an approximate calculation of the affinity constant according to

$$\bar{\nu}/[S] = C \times \text{mol. wt.} = n\,k\,. \tag{V-9}$$

If α denotes the fraction of binding sites on P which are unoccupied, of all the molecules of P and SP present in solution, then

$$\alpha = \frac{[P]}{[SP]+[P]} = \frac{K}{[S]+K} = \frac{1}{1+k[S]}\,. \tag{V-10}$$

Accordingly, the average number of moles of S bound, for all the molecules of P and SP present, is given by

$$\bar{\nu} = 1-\alpha = \frac{[SP]}{[SP]+[P]} = \frac{[S]}{[S]+K} = \frac{k[S]}{1+k[S]}\,. \tag{V-11}$$

If the protein has n binding sites instead of one as assumed for Eqs. (V-10)—(V-11), and if these n binding sites are independent of each other, and are equivalent in their binding affinity for S, then the probability that S is bound to the available binding sites ranging from $i=1$ to n, becomes

$$\bar{\nu} = \sum_{i=1}^{n} \bar{\nu}_i = n\,\bar{\nu}_i\,, \tag{V-12}$$

which indicates the total average number of occupied sites per molecule of P. Then the probability that any site, chosen at random from any molecule present, is occupied by S becomes

$$1-\alpha=\frac{\bar{\nu}}{n}=\frac{k[S]}{1+k[S]}, \text{ or} \tag{V-13}$$

$$\bar{\nu}=\frac{n\,k[S]}{1+k[S]} \text{ or} \tag{V-14}$$

$$\frac{\bar{\nu}}{n-\bar{\nu}}=k[S]. \tag{V-15}$$

The simplifying assumption of one single set of equivalent and independent binding sites is not realistic for the general case. Protein molecules are likely to contain several sets of binding sites, and we assume again that within one of m different sets the binding groups are independent and equivalent, i.e., each one has the same intrinsic association constant. The first set then has n_1 binding sites, each with association constant k_1, the second set n_2 such groups, each with association constant k_2, and so forth. Starting with the highest binding affinity for the first set of sites, we order the k values as $k_1 > k_2 > k_3 > \ldots k_{m-1} > k_m$. The total number of binding sites in the protein, n, is given as

$$n=n_1+n_2+n_3+\ldots\ldots\ldots+n_m=\sum n_i, \tag{V-16}$$

and the average number of sites occupied by S molecules becomes

$$\bar{\nu}=\sum_{i=1}^{m}\frac{n_i\,k_i[S]}{1+k_i[S]}. \tag{V-17}$$

As mentioned before, interactions between the sites are not considered in Eqs. (V-14) and (V-17). It is obvious from Eq. (V-17) that the mathematical evaluation of a binding equilibrium may become very complex if it involves a protein with a number of different binding groups. In cases actually analyzed, only very few sets of binding sites have been derived from the experimental data, with one intrinsic association constant assessed for all the binding sites within each set. For example, if we assume three sets of equivalent and independent binding sites, Eq. (V-17) would be resolved to

$$\bar{\nu}=\frac{n_1\,k_1[S]}{1+k_1[S]}+\frac{n_2\,k_2[S]}{1+k_2[S]}+\frac{n_3\,k_3[S]}{1+k_3[S]}. \tag{V-18}$$

For the quantitative characterization of a ligand-protein binding system it is necessary to obtain information on the number(s) of binding sites, n, and the association constant(s), k. The values determined are usually unambiguous if there is only one class of a limited number of binding groups. However, in the more general case of several types of binding sites with different affinity constants, the more complex problem of determining $n_1, n_2 \ldots n_m$ and the corresponding constants $k_1, k_2 \ldots k_m$ has to be solved. The experimental data provide paired values for $\bar{\nu}$ and [S] by a number of equilibrium methods discussed in Chapter IV. Several procedures are available for analysis of these data so that values for n and k can be obtained and their validity tested. We will consider those methods which have been applied in the field of steroid-protein interactions and have led to reasonable interpretations of the binding systems. It is evident that the use of computers in the analysis of multiple binding equilibria will be very helpful.

b) The Logarithmic Plot

This procedure has found little application in studies on steroid binding but is mentioned here because it provides some basic information in a comparatively simple way. The values of $\bar{\nu}$ are plotted against log[S] or against —log[S], which may be interpreted as "pS", in analogy to pH in the case of proton concentrations. This form of presentation has been called the "formation function" by BJERRUM; it corresponds to the usual titration curve in acid-base equilibria, in which bound protons are plotted as ordinate against pH as abscissa. Transforming equation (V-15) by use of logarithms, we obtain for a single set of equivalent groups:

$$\log k + \log[S] = \log \frac{\bar{\nu}}{n - \bar{\nu}} . \qquad \text{(V-19)}$$

Figure 4 on p. 616 of EDSALL and WYMAN (1958) gives an example for the application of the procedure for systems with one and two sets of binding groups.

The presentation of binding data by the method of logarithmic plot has the advantage of covering a wide range of concentrations conveniently. Inspection of the data will show whether there is a similarity to the ideal curve and may then allow at least an approximate estimation of n, the number of binding sites. If n is known, the value of log[S] at which $\bar{\nu} = n/2$ gives —log k according to Eq. (V-19). In the more common case of greater complexity of the binding sites, the curve obtained will differ distinctly from the ideal curve;

the deviations may indicate marked interactions between the binding groups, or may show whether the binding sites fall into one or more different classes with different k values as described by Eq. (V-17). In these more complex cases, the plot of the formation function is often inadequate for determination of n values. An example is given in the application of the logarithmic plot to evaluate the interaction of cortisol with HSA and BSA (BRUNKHORST and HESS, 1965).

c) The Reciprocal Plot

Representation of binding data by the method of reciprocal plots has been used extensively in the fundamental studies of KLOTZ and his coworkers (KLOTZ et al., 1946; KLOTZ, 1950, 1953) on protein interactions with anions and other ligands. Both sides of Eq. (V-14) are inverted to give

$$\frac{1}{\bar{\nu}} = \frac{1}{\mathrm{n\,k[S]}} + \frac{1}{\mathrm{n}}. \qquad \text{(V-20)}$$

If equation (V-20) holds, a plot of $1/\bar{\nu}$ against $1/[S]$ gives a straight line; the intersection with the ordinate axis yields $1/n$, and the slope of the line is $1/n\,k$. As an example, Fig. V-6 shows the evaluation of binding data obtained by equilibrium dialysis in the interaction between progesterone and delipidated HSA in the absence and presence of myristic acid. Within the limits of the solubility of progesterone in the aqueous system, a single set of independent and equivalent binding sites is assumed; the straight lines, calculated by the method of least squares, are in reasonable agreement with the assumption. It follows from the intersects with the ordinate, according to (V-20), that the number of binding sites for the progesterone-HSA complex in the absence of myristic acid is approximately 4. Addition of a total of 5 moles myristic acid per mole HSA present in the system reduces this value to $n = 2.0$. The $n\,k$ values, calculated from the slope of the straight lines, are $1.9 \times 10^5\,M^{-1}$ and $1.1 \times 10^5\,M^{-1}$ in the absence and presence of myristic acid, respectively. Evidently, the effect of myristic acid consists of a reduction of the number of binding sites without affecting the apparent intrinsic association constant of about $5 \times 10^4\,M^{-1}$.

Whereas this procedure gives adequate results for simple systems with few binding sites which are equivalent and independent, it may lead to erroneous evaluation in cases of greater complexity. Binding values obtained at high ligand concentrations in search for secondary or higher types of binding sites will concentrate close to the ordinate. Unambiguous extrapolation of the plotted values to the ordinate axis may then be very difficult or impossible to achieve—just at those concentrations which are required to reveal additional binding sites of lower affinity.

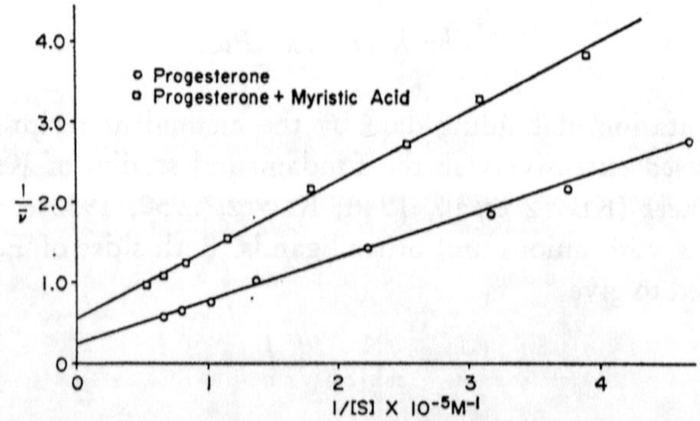

Fig. V-6. Interaction of progesterone with delipidated HSA, evaluated by reciprocal plots. Lower curve in absence, upper curve in presence of 5 moles of myristic acid per mole HSA. Equilibrium dialysis, phosphate buffer of pH 7.4, 4°. Lower curve: $n = 4.2$; $nk = 1.9 \times 10^5\ M^{-1}$. Upper curve: $n = 2.0$; $nk = 1.1 \times 10^5\ M^{-1}$. (Unpublished studies with G. B. HARDING)

The method of reciprocal plots has been applied extensively in the study of steroid-protein interactions after it was first used in such systems by SCHELLMAN et al. (1954) for the investigation of testosterone binding to BSA. Adequate results were obtained in many cases, primarily due to the fact that many of the steroid-protein systems studied gave low values of $n = 1$ or 2 under the conditions employed. This is in contrast to the binding of various small anions, cations and other charged ligands to proteins which involve a greater number of often non-equivalent binding sites. In many cases, however, the low number of steroid-binding sites may be apparent only, resulting from the low water solubility of the steroid which limits the analysis to low ratios of steroid to protein. A few examples for the evaluation of steroid-protein interactions by reciprocal plots will be presented in subsequent chapters of this monograph.

The limitations of Eq. (V-20) in describing binding parameters obtained in studies on steroid-protein complexes were recognized in the initial report from SAMUEL's laboratory (SCHELLMAN et al., 1954), and amendments were made. Validity of Eq. (V-20) is not a unique consequence of identical sites but will usually be observed at the very low ligand concentrations often employed in steroid binding studies; the values obtained with the use of (V-20) must be considered only as

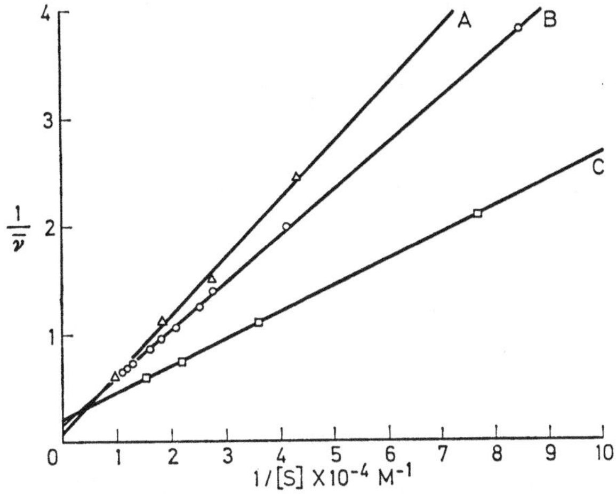

Fig. V-7. Binding of testosterone at 25° to three preparations of BSA of increasing purity dissolved in phosphate buffer, pH 7.4 ± 0.1. Preparation A was studied by equilibrium dialysis; B and C were studied in partition experiments. Data plotted according to Eq. (V-21). From SCHELLMAN et al. (1954)

relative binding parameters. To account for non-identical binding sites, the original equation (VON MURALT, 1930; KLOTZ, 1946 a) governing multiple association between ligands and proteins was transformed to

$$\frac{1}{\bar{\nu}} = \frac{1}{n\langle k\rangle} \cdot \frac{1}{[S]} + \frac{1}{n}\left(1 + \frac{\mu^2}{\langle k\rangle^2}\right) \quad \text{(V-21)}$$

where

$\langle k\rangle = \frac{\sum_{i=1}^{n} k_i}{n}$; $\mu^2 = \langle k^2\rangle - \langle k\rangle^2$, and

$\langle k^2\rangle = \frac{\sum_{i=1}^{n} k_i^2}{n}$; k_i is the association constant for the ith site. In the absence of information on the heterogeneity of the binding sites, only the value for average distribution, $n\langle k\rangle$, has theoretical significance. Figure V-7 shows the reciprocal plots for data on testosterone binding to BSA, and Table V-1 gives the corresponding binding parameters, calculated according to Eqs. (V-20) and (V-21).

Eq. (V-20) which describes a ligand-protein binding system in the form of reciprocal plots is the exact analog of the Lineweaver-Burk Eq. (V-22) used widely in expressing the results of enzyme kinetic studies:

$$\frac{1}{v} = \frac{K_m}{V_{max}[S']} + \frac{1}{V_{max}} \qquad \text{(V-22)}$$

where v is the initial velocity of the enzyme-catalyzed reaction, V_{max} the maximum velocity at very high substrate concentration, [S'], and K_m the Michaelis-Menten constant.

Eq. (V-22) is frequently applied to recognize competitive inhibition of enzyme reactions. A plot of $1/v$ against $1/[S']$ gives a straight line (analogous to the lower curve in Fig. V-6) in which the ordinate intercept equals $1/V_{max}$ and the slope is K_m/V_{max}. If a competitive inhibitor is added to the system, the intersect with the ordinate axis remains the same but the slope of the line is increased. In the case of non-competitive inhibition, both the slope and the intercept are increased (compare upper curve in Fig. V-6).

In an analogous manner, Eqs. (V-20) or (V-21) may be used to analyze whether an inhibitor of steroid binding to a protein is com-

Table V-1. *Binding of testosterone to three BSA samples at 25°, pH 7.4 ± 0.1* [a]

Protein	$n\langle k\rangle$ $\times 10^{-4}\ M^{-1}$	n (apparent)	k (apparent) $\times 10^{-3}\ M^{-1}$
A	1.9	10	1.8
B	2.3	6	3.8
C	4.0	5	7.9

[a] Data of Fig. V-7 used to calculate n k by use of Eq. (V-21), and n (apparent) and k (apparent) with Eq. (V-20). From SCHELLMAN et al. (1954).

petitive or non-competitive. It should be noted, however, that the interpretation of such data may not be unequivocal. Seemingly competitive plots may be given by systems which are inhibited as a result of allosteric interactions, or of other conformational interference. Examples will be found in metal ion inhibition of the progesterone-AAG association (Chapter XIII).

d) The Scatchard Plot

Some of the difficulties in the evaluation of the reciprocal plots are eliminated in a somewhat different presentation of binding data which was developed by SCATCHARD (1949). The same Eq. (V-14) which served as the basis for the method of reciprocal plots is used in the form (V-15); both sides are multiplied by $(n-\bar{\nu})/[S]$ to obtain

$$\frac{\bar{\nu}}{[S]} = k\,(n-\bar{\nu})\,. \tag{V-23}$$

A plot of $\bar{\nu}/[S]$ as ordinate against $\bar{\nu}$ as abscissa gives a straight line if the binding sites are equivalent and independent. Figure V-8 shows this type of graph for the same binding data that were represented in Fig. V-6. It is evident from Eq. (V-23) that the intercept on the abscissa, when $\bar{\nu}/[S]=0$, gives n, the maximum number of interacting groups; the intersection with the ordinate gives n k. The graph indicates that $\bar{\nu}/[S]$ approaches its maximum limiting value as [S] approaches zero, which occurs when $\bar{\nu}$ approaches zero; at high [S] values, even though $\bar{\nu}$ is increasing, $\bar{\nu}/[S]$ approaches zero as $\bar{\nu}$ is approaching n. This method of representing data gives more even relative weight to the different points on the curve than the reciprocal plot which stresses more the values obtained at very low [S].

The Scatchard plot has proved to be particularly useful for the recognition of different classes of binding sites. Whereas in the evaluation by reciprocal plots, the extrapolation of the experimental points close to the ordinate, that is at the higher values of [S], is very difficult because of crowding of the reciprocal values, this problem is easier to solve for the curves obtained by the Scatchard procedure which provides a more even distribution of the binding data. In the general case of a binding system described by (V-17), $\bar{\nu}/[S]$ approaches an upper limit of $\Sigma\, n_i k_i$ when $\bar{\nu}$ is approaching zero. On the other hand, when $\bar{\nu}/[S]$ nears zero, which occurs as [S] becomes very large, it is evident from (V-17) that $\bar{\nu}$ approaches a limiting value of $n = \Sigma n_i$. Extrapolation of Scatchard curves may not always be unambiguous, but it is at least possible with this method to obtain approximate estimates of the various values of n and k for different types of binding groups. Computer analysis of the experimental data to obtain a best fit for at least two members of Eq. (V-18) has proven very helpful in our laboratory when more than one set of binding sites is involved. Examples will be given in later chapters, especially in Chapter XIII.

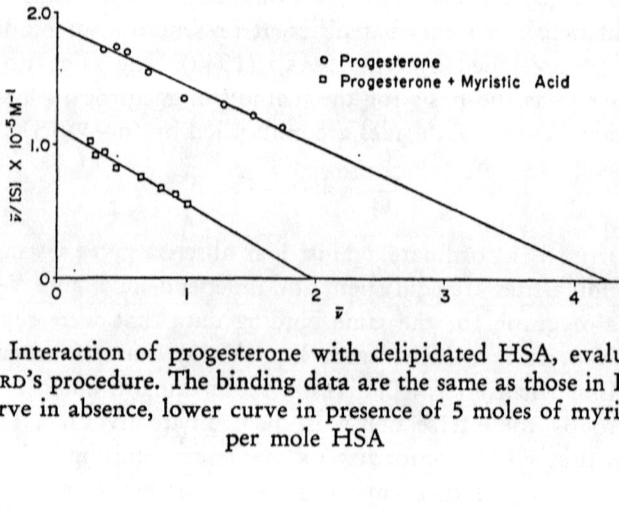

Fig. V-8. Interaction of progesterone with delipidated HSA, evaluated by SCATCHARD's procedure. The binding data are the same as those in Fig. V-6; upper curve in absence, lower curve in presence of 5 moles of myristic acid per mole HSA

e) Approximate Free Energy, Enthalpy and Entropy

For a complete investigation of a ligand-macromolecule complex, the effect of temperature should be determined, and the changes in free energy, heat content and entropy of the reaction calculated by application of the Gibbs-Helmholtz equation (TANFORD, 1961). The intrinsic association constants are determined at several temperatures, and the corresponding thermodynamic properties are evaluated according to

$$\Delta F^{\circ} = -RT \ln k = -2.30\, RT \log k \qquad \text{(V-24)}$$

where ΔF° is the change of Gibbs free energy in calories/mole; R the molar gas constant = 1.987 calories per degree per mole; T the absolute temperature in °K. Numerical ΔF° values for steroid-protein interactions will be given below in kcal/mole.

Heat content or enthalpy is calculated according to the van't Hoff equation written as

$$\Delta H^\circ = -R \frac{d \ln k}{d(1/T)} \quad \text{(V-25)}$$

where ΔH° is the change of enthalpy. If ΔH° can be assumed constant over a temperature range from T_1 to T_2, Eq. (V-25) may be rearranged and integrated directly to give

$$\ln\left(\frac{k_2}{k_1}\right) = \frac{\Delta H^\circ}{R}\left(\frac{1}{T_1} - \frac{1}{T_2}\right). \quad \text{(V-26)}$$

Conversion to denary logarithms and expressing ΔH° in calories per mole results in

$$\log\left(\frac{k_2}{k_1}\right) = \frac{\Delta H^\circ}{4.576}\left(\frac{1}{T_1} - \frac{1}{T_2}\right) \quad \text{(V-27)}$$

from which the heat of reaction can be calculated if we know the association constant at two or more different temperatures. If 4.576 log k is plotted against 1/T, the resulting curve will be a straight line if ΔH° is independent of temperature; the slope of the line gives ΔH° directly. If the curve is not linear, the slope of the tangent to it, at any given value of 1/T, provides ΔH° for the corresponding temperature, T. If ΔH° varies considerably with temperature, the integration of (V-25) requires the formulation of ΔH° as a function of T and the integration of the resulting expression.

Once the free energy and enthalpy of the interaction have been determined, the entropy may be calculated in cal mole^{-1} deg^{-1} according to

$$\Delta S^\circ = \frac{\Delta H^\circ - \Delta F^\circ}{T} \quad \text{(V-28)}$$

where ΔS° is the change of entropy. Values for ΔH° and ΔS° in steroid-protein interactions have not been determined extensively. A few cases will be discussed in Chapter VI. 5 e, and other examples will be given in Chapters XI. 4 and XIII. Information on thermodynamic parameters of steroid-protein interactions will be valuable in future studies for an interpretation of the reactions involved.

f) Competition between Different Steroids for the Same Binding Site

For an understanding of the binding equilibria in natural protein systems containing several steroids, it is necessary to determine whether the interaction of one steroid takes place at the binding site of another steroid, or at a different site. Such knowledge is needed in the interpretation of the binding in the blood serum of steroid hormones such as cortisol and progesterone to CBG, of testosterone and estradiol to the steroid binding β-globulin, or of various steroids to AAG.

If two steroids, S_1 and S_2, are bound to the same binding sites, n, of a protein or other macromolecule, and if the n sites are equivalent and independent, the binding of S_1 in the presence of S_2 is given (EDSALL and WYMAN, 1958) by the equation

$$\bar{\nu}_{S1} = \frac{n\,k_{S_1}[S_1]}{1 + k_{S_1}[S_1] + k_{S_2}[S_2]}\,. \tag{V-29}$$

Since an analogous relationship would hold for the binding of S_2 in the presence of S_1, it follows that the relation of bound S_1 to bound S_2 is

$$\frac{\bar{\nu}_{S_1}}{\bar{\nu}_{S_2}} = \frac{k_{S_1}[S_1]}{k_{S_2}[S_2]}\,. \tag{V-30}$$

These relations may be extended to binding by proteins containing several different sets of binding sites. In the actual testing of the question of competitive binding of two steroids to the same site of a protein, it is convenient to use radiocarbon-labeled S_1 and tritium-labeled S_2, and determine $\bar{\nu}_{S1}$ in the presence of S_2. If the experimental $\bar{\nu}_{S1}$ agrees with the value calculated by Eq. (V-29), the binding is competitive. A similar test can be performed with the aid of Eq. (V-30); if the equilibrium dialysis experiment is performed at varying concentrations of S_1 and a constant, sufficiently high concentration of S_2, any discrepancy from competitive binding at the same site becomes conspicuous. Examples are given in Chapters VI. 5 g and XIII. 4.

g) Calculation of Distribution of a Steroid between Two Binding Proteins

Knowledge of the association constants and the numbers of binding sites is a prerequisite for assessing the distribution of a steroid between two different proteins, e.g., in the mixture of serum proteins. According to Eq. (V-6),

$$[SP] = k[S][P] \tag{V-31}$$

where [SP] indicates the concentration of the steroid-protein complex; [S] and [P] the concentrations of unbound steroid and unbound protein, respectively; and k the association constant characteristic for the steroid and protein involved, and valid for a given temperature, pH, solvent, and other conditions. All concentrations are expressed in moles/liter. The Eq. (V-31) holds for any binding protein; if we designate the two proteins considered in this calculation as P_A and P_B, we obtain

$$[SP_A] = k_A[S][P_A] \text{ and} \tag{V-32}$$

$$[SP_B] = k_B[S][P_B] \tag{V-33}$$

where k_A and k_B are the association constants for the complexes of S with P_A and P_B, respectively. Dividing Eq. (V-32) by (V-33) results in

$$\frac{[SP_A]}{[SP_B]} = \frac{k_A[P_A]}{k_B[P_B]}. \tag{V-34}$$

If we substitute convenient terms and rearrange (V-34), as shown in detail by YATES and URQUHART (1962), a quadratic equation is obtained

$$\left(1 - \frac{k_A}{k_B}\right)[SP_B]^2 + \left([S] - [\Sigma S] - [\Sigma P_B] - \frac{k_A}{k_B}[\Sigma P_A] + \frac{k_A}{k_B}[\Sigma S] - \frac{k_A}{k_B}[S]\right)[SP_B] + [\Sigma P_B]([\Sigma S] - [S]) = 0 \tag{V-35}$$

where $[\Sigma S]$ equals total concentration of the steroid; k_A and k_B are the association constants (for each binding site) of the steroid complexes with P_A and P_B, respectively. The symbols $[\Sigma P_A]$ and $[\Sigma P_B]$ indicate the total molar concentrations of binding sites on P_A and P_B, respectively, calculated from the total molar concentration of the protein and its number of binding sites (n_A, n_B) for the steroid under study:

$$[\Sigma P_A] = n_A[P_{A_{total}}], \tag{V-36}$$

$$[\Sigma P_B] = n_B[P_{B_{total}}]. \tag{V-37}$$

Eq. (V-35) is solved for $[SP_B]$, the concentration of steroid bound to P_B, by use of the quadratic formula. The concentration of the P_A-

bound steroid may then be obtained by deducting $[SP_B]$ from the total concentration of protein-bound steroid:

$$[SP_A] = [\Sigma S] - [S] - [SP_B]. \qquad \text{(V-38)}$$

Eq. (V-35) and related equations used in the calculation of $[SP_A]$ and $[SP_B]$ are derived under the assumption that the binding of S to P_A is independent of its binding to P_B, i.e., that association constants and numbers of binding sites are not affected by the presence of the second protein. Absence of protein-protein interaction between P_A and P_B, at least in a manner which would affect the binding affinity to S, is the main condition for this assumption. In the actual application of Eq. (V-35) for the determination of proportional binding of a steroid to two proteins, as present for instance in a blood serum, the values for k_A, k_B, n_A, n_B are obtained from independent studies, mostly from data reported in the literature. The total concentrations of P_A and P_B, as well as $[\Sigma S]$ are determined for the particular system under study, e.g., a blood serum. Chemical, electrophoretic, and ultracentrifugal methods are available (COOPER, 1960) to measure the concentrations of the major serum protein components, including the steroid-binding albumin. Quantitative determination of other proteins involved in steroid interactions may be more difficult because of the much smaller amounts in which they occur; these proteins will be discussed in subsequent chapters. Reliable methods are available for the quantification of most steroid hormones (DORFMAN, 1962b; DORFMAN, 1969); more recent techniques based on competitive protein binding (DICZFALUSY, 1970) will be discussed below (Chapters VIII. 2b and XII. 3). The concentration of unbound steroid, [S], may be obtained experimentally by equilibrium dialysis. Examples for the application of (V-35) in calculating proportional binding of cortisol, corticosterone, or progesterone to HSA and to CBG have been given (YATES and URQUHART, 1962; WESTPHAL, 1966); see also Chapter VIII. 5b. β.

The equations used above have been derived under the assumption that only one steroid be present in the system containing the two binding proteins, i.e., HSA and CBG. In reality, however, as for example in a blood serum, several steroid hormones are present which compete for the same binding sites; in the case of a pregnancy serum there are at least progesterone, cortisol, and corticosterone. Formulas were therefore developed (WESTPHAL, 1966) to assess the simultaneous binding of two steroids to the two proteins. They permit an approximate calculation of the distribution of the three steroid hormones mentioned since the binding constants for cortisol and corticosterone to transcortin are similar at 37°; the same two steroids also bind with comparable affinity to serum albumin. The condition for validity of the calculations is that the steroids compete for the same binding sites on the protein. The equations are based on formulas derived for the competition between different ligands for the same binding sites in a protein (EDSALL and WYMAN, 1958), which state that the molar ratio of ligand S_1 to ligand S_2, both bound to the binding protein, equals the ratio of the products of association constant and concentration for the two ligands:

$$\frac{\bar{\nu}_{S_1}}{\bar{\nu}_{S_2}} = \frac{k_{S_1}[S_1]}{k_{S_2}[S_2]} . \qquad \text{(V-39)}$$

The premise of our approach is that the total concentration of binding sites of a given protein, $[\Sigma\,P]$, is distributed between two ligands, S_1 and S_2, according to their competitive binding to the protein, given by their concentrations and association constants:

$$[\Sigma\,P] = [\Sigma\,P_{S_1}] + [\Sigma\,P_{S_2}]; \qquad \text{(V-40)}$$

$$\frac{[\Sigma\,P_{S_1}]}{[\Sigma\,P_{S_2}]} = \frac{[S_1\,P]}{[S_2\,P]} = \frac{k_{S_1}[\Sigma\,S_1]}{k_{S_2}[\Sigma\,S_2]} . \qquad \text{(V-41)}$$

Rearrangement of this equation gives the proportional concentration of the binding sites which "belongs" to S_1:

$$[\Sigma\,P_{S_1}] = \frac{k_{S_1}[\Sigma\,S_1]\cdot[\Sigma\,P]}{k_{S_1}[\Sigma\,S_1] + k_{S_2}[\Sigma\,S_2]} . \qquad \text{(V-42)}$$

$[\Sigma\,P_{S_2}]$ is then found from equation (V-40). These "proportional concentrations" of binding sites, i.e., $[\Sigma\,P_{S_1}]$ and $[\Sigma\,P_{S_2}]$, are calculated separately for S_1 (e.g., progesterone) and S_2 (e.g., cortisol) for both protein P_A, $[\Sigma\,P_{A\,S_{1,2}}]$, and P_B, $[\Sigma\,P_{B\,S_{1,2}}]$. Examples for the distribution of progesterone, cortisol and corticosterone between CBG and HSA have been given (WESTPHAL, 1966; see also Chapter VIII. 5 b. β).

It should be noted that the apportionment of CBG and HSA to the two steroids (progesterone and cortisol) involves an approximation. They are apportioned to the two steroids according to total steroid concentrations and association constants. It is not taken into account that this apportionment of one protein, e.g., CBG, to the two steroids is also depending on the relative binding of the two steroids to the other binding protein, in this case HSA. This approximation is based on the relatively constant great excess of binding sites in HSA. More complex calculations are required when in addition to HSA and CBG other high-affinity binding proteins, such as the steroid-binding β-globulin, have to be included in the binding distribution.

In a recent publication, LEBEAU et al. (1969) have described procedures to determine binding parameters for steroids interacting with several binding proteins. The bound and unbound fractions of a steroid are determined by equilibrium dialysis at 4°, and apparent affinity constants and concentration of binding sites are evaluated by computer with the aid of formulas of multiple binding equilibria. Systems containing up to two high-affinity binding proteins and one protein of low affinity are considered. Graphical evaluation in the form of a proportion graph has been found advantageous for the analysis of the binding data which may be obtained by any technique that determines the bound and unbound portions of the steroid (BAULIEU and RAYNAUD, 1969, 1970).

An interesting approach to the problem of distinguishing between binding proteins of different affinities has been worked out by MILGROM and BAULIEU (1970). The method utilizes differential dissociation of the ligand. It is based on the observation that the apparent dissociation rate of a high-affinity steroid-protein complex is low, particularly at low temperature, in comparison with that of a steroid complex of low affinity such as given by HSA. The unbound steroid is removed instantaneously by the addition of an adsorbent such as dextran-coated charcoal. Analyzing a multiple binding system at given time intervals and extrapolating the apparent 'decay' curves to time zero allows distinction between low-affinity ("nonspecific") and high-affinity ("specific") binding, and permits assessment of concentration of binding sites and other parameters. A valuable discussion of the various problems encountered in measuring steroid-protein binding systems, and of practical methods, has been given by BAULIEU et al. (1970).

It seems likely that these newer methods will be used in the more complex biological binding systems that await elucidation. It appears equally possible that additional techniques will be developed with the specific purpose of adapting the measuring tools to the extremely small quantities of binding macromolecules involved in cellular and subcellular interactions. The more classical procedures described in this chapter will remain important for accurate determination of various physicochemical parameters which characterize the binding proteins, especially after chemical purification or final isolation has been achieved. The problems involved in the measurement of biological binding systems will be discussed once more in Chapter VIII. 2 in connection with CBG.

VI. Steroid Interactions with Serum Proteins Other than Specific Glycoproteins

1. Interactions of Bile Acids

Binding of bile acids to serum proteins was demonstrated early (ROSENTHAL, 1925) when it was observed that sodium taurocholate at neutral pH displaces acidic dyes (rose bengal, bromsulphalein and others) from their association with blood proteins. Similar competition for anionic binding sites occurs on HSA, as shown by displacement of the anionic dye azorubin by cholic and desoxycholic acid at pH 7.7 (WESTPHAL et al., 1953). No displacement of bromsulphalein from HSA by bile acids was noted at pH 9.1 (KUCEROVA et al., 1966). Various other observations indicated binding of bile acids to serum proteins (LECOMTE DU NOUY, 1922, 1928; RUDMAN and KENDALL, 1957 a; BENNHOLD and OTT, 1961) and to purified HSA (COHN, 1948). Serum protein binding suppressed the hemolytic activity of desoxycholic acid on red cell membranes (CATANESE et al., 1969).

Quantitative data were obtained by RUDMAN and KENDALL (1957 b) who investigated the interaction of a series of cholanic acids, including conjugated forms, with HSA (COHN fraction V). The binding was measured by equilibrium dialysis at pH 7.6, 5°. Table VI-1 shows the $\bar{\nu}/[S]$ values for cholanic acids with 1, 2 or 3 oxygen functions. These values, calculated from the data of RUDMAN and KENDALL (1957 b), give a very approximate assessment of the binding affinity according to Eqs. (V-7) and (V-8); no indications for the presence of more than one type of binding sites were obtained. Combining (V-7) and (V-8) to

$$\bar{\nu}/[S] = n\,k = k', \qquad \text{(VI-1)}$$

the approximate free energy of association can be calculated according to (V-22),

$$\Delta F^\circ = -R\,T \ln k'. \qquad \text{(VI-2)}$$

These values have been incorporated in Table VI-1.

Table VI-1. *Interaction of cholanic acids and HSA. Equilibrium dialysis; 5°; pH 7.6;* $\bar{\nu}/[S]$ *and* ΔF° *values calculated from* Rudman and Kendall *(1957 b) according to Eqs. (V-7) and (VI-2), respectively*

Bile acid	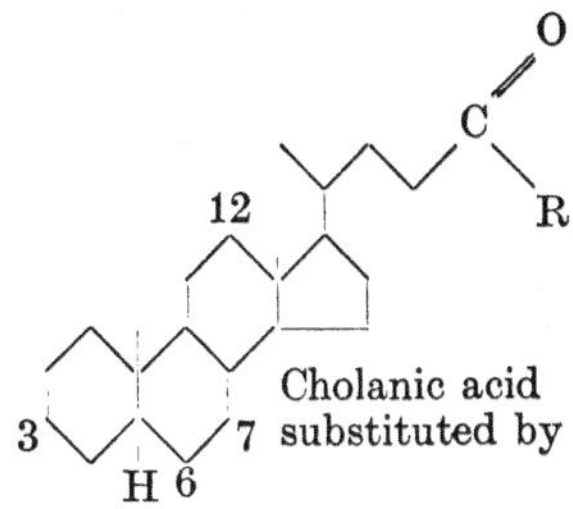 Cholanic acid substituted by	$\bar{\nu}/[S]$ ($\times 10^{-4}$ M^{-1})	ΔF° (kcal/mole)
Lithocholic	3α-OH; R=OH	9.9	—6.3
7α-Hydroxycholanic	7α-OH; R=OH	13.2	—6.5
Desoxycholic	3α, 12α-di-OH; R=OH	1.3	—5.2
Hyodesoxycholic	3α, 6α-di-OH; R=OH	2.1	—5.5
Chenodesoxycholic	3α, 7α-di-OH; R=OH	1.9	—5.4
Glycodesoxycholic	3α, 12α-di-OH; R=glycine	1.3	—5.2
Taurodesoxycholic	3α, 12α-di-OH; R=taurine	1.3	—5.2
Diformyldesoxycholic	3α, 12α-di-O-C(=O)H; R=OH	2.1	—5.5
Hydroxyketocholenic	3α-OH, 12=0; R=OH; Δ^{9-11}	0.84	—5.0
Cholic	3α, 7α, 12α-tri-OH; R=OH	0.45	—4.6
Glycocholic	3α, 7α, 12α-tri-OH; R=glycine	0.44	—4.6
Taurocholic	3α, 7α, 12α-tri-OH; R=taurine	0.27	—4.4
Triformylcholic	3α, 7α, 12α-tri-O-C(=O)H; R=OH	0.59	—4.8

For desoxycholic acid and cholic acid, the authors (Rudman and Kendall, 1957 b) have determined by the method of reciprocal plots the number of binding sites, n, and the association constant, k, for each site. Table VI-2 shows that the values for n are relatively high, *i.e.*, higher than those found for neutral steroids (see below); they are lower than the number of binding sites observed in HSA for many other anions. The apparent association constants, k′ = n k, and ΔF° calculated for these k′ values are in fair agreement with $\bar{\nu}/[S]$ and ΔF° of Table VI-1 if one considers that the basic condition, i.e., $[SP] \ll [P]$ in (V-7) and (V-8), was not fulfilled in these experiments.

Table VI-2. *Interaction of cholanic acids with HSA. Equilibrium dialysis; NaCl-phosphate; pH 7.6; 5°. Method of reciprocal plots.* From RUDMAN and KENDALL (1957 b)

Binding Parameter	Equation	Desoxycholic Acid	Cholic Acid
k	(V-18)	1.35×10^3 M^{-1}	1.54×10^3 M^{-1}
n	(V-18)	12	4
$k' = n\,k$	(VI-1)	1.62×10^4 M^{-1}	0.62×10^4 M^{-1}
$\Delta F°$	(VI-2)	—5.4 kcal/mole	—4.8 kcal/mole

The hydrogen ion concentration was without influence on the binding affinity between desoxycholic acid and HSA from pH 5.6 to about 9; it then decreased rapidly to almost zero at pH 11 (RUDMAN and KENDALL, 1957 b). Denaturation was excluded when reversibility of this pH effect was found.

The competition between bile acids and anionic dyes for binding sites on serum albumin, mentioned above, suggests that the hydroxycholanic acids interact with anion-binding sites on the albumin molecule. However, in a similar way as the paraffin chain of fatty acid anions determines the affinity of binding to serum albumin, presumably by hydrophobic bonding, the steroid molecule connected with the carboxyl at C-23 influences the strength of interaction. Table VI-1 shows that the polarity rule is followed: the binding affinity, $\bar{\nu}/[S]$, and consequently $-\Delta F°$, decrease with increasing number of hydroxy groups. Conjugation of the carboxy group with glycine or taurine has no additional effect on the binding affinity. This indicates that replacing the C-24 carboxyl group by a peptide-bonded glycine molecule and thus increasing the distance between sterane nucleus and anionic site apparently does not influence the binding affinity. Similarly, further substitution of the —COOH group by a $-CH_2-SO_3H$ group in taurocholic acid hardly affects the association with HSA. Cationic centers may be assumed to be the binding sites for the cholanate anions, as postulated for other organic anions (KLOTZ, 1953); additional interaction of the steroid ring structure with adjacent groups in the albumin molecule is suggested by the binding behavior according to the polarity rule.

2. Interactions of Steroid Conjugates

Another group of steroid derivatives which are present as anions at the pH of the circulating blood are the conjugated forms, e.g., the sulfates and the glucosiduronates. The significance of the conjugated forms, especially the sulfates, for secretion and biotransformation of steroid hormones may be greater than had been previously realized (BAULIEU et al., 1965); their binding to plasma proteins therefore is of considerable biological importance. The significance of protein binding for hepatic clearance, biological activity and other hormone-related processes has been discussed by TAIT and BURSTEIN (1964) and by BAIRD et al. (1969).

Early studies had suggested that the conjugated estrogens are bound to serum proteins (RAKOFF et al., 1943). DAUGHADAY (DAUGHADAY and BREMER, 1955; DAUGHADAY, 1956 b) found in equilibrium dialysis experiments after oral administration of cortisone acetate weak binding of 17-hydroxycorticosteroid glucosiduronates to plasma proteins; the binding affinity was many times smaller than that of the free hydroxycorticosteroids (65% vs. 94% bound). Further experiments indicated interaction with serum albumin: administration of estrone sulfate resulted in the appearance of estrogenic activity in the albumin fraction (ANTONIADES et al., 1957 a). SANDBERG et al. (1957) injected radiolabeled steroids intravenously into human subjects and demonstrated albumin binding of the conjugated forms of estrogens and cortisol. The conjugated metabolites of estradiol, estrone, progesterone and testosterone became so avidly attached to HSA or BSA, that SLAUNWHITE and SANDBERG (1958) utilized this interaction for a method of extracting urinary steroid conjugates by dialyzing the urine against 5% BSA at pH 3—5. The conjugated steroids were recovered from the albumin by addition of ethanol. The interaction of the corticosteroid conjugates with serum albumin was weaker than that of the conjugates of less polar steroids. Binding of estradiol glucosiduronate to plasma proteins was found to be in competition with binding of unconjugated estradiol (GIORGI and CROSIGNANI, 1966); similar competition was observed between estrone and estrone sulfate (GIORGI and CROSIGNANI, 1969). Studies with pure conjugates of estriol demonstrated binding of estriol glucosiduronate and estriol sulfate to HSA (TOUCHSTONE and CHANG, 1963). The sulfate esters of dehydroepiandrosterone, testosterone, 17-acetoxypregn-5-en-3β-ol-20-one, and pregn-5-en-3β-ol-20-one were bound to rabbit and rat plasma, in addition to human plasma (WANG and BULBROOK, 1967).

In similar injection experiments (SANDBERG et al., 1957; ANTONIADES et al., 1957 b), the association of the radioactive steroid hormones and their conjugated metabolites was studied by the low temperature-ethanol fractionation procedure of COHN (COHN et al., 1946, 1950). A major portion of both forms of the steroids was found attached to the albumin fraction; other fractions, particularly IV-1 in case of the corticosteroids, also contained the labeled steroids and their conjugated metabolites (ANTONIADES et al., 1960). However, this distribution must be interpreted with caution since it is not certain that the presence of a steroid conjugate in a given protein fraction is a true indication of its original binding to the protein precipitate in this fraction (SAMUELS, 1957; RALL, 1957). This becomes particularly obvious in the "soluble" fraction VI, obtained at the highest ethanol concentration, which may contain more steroid than was originally present in unbound form.

It is of interest in this connection that SANDBERG and SLAUNWHITE (1959) did not find binding of the urinary conjugated metabolites of cortisol to transcortin (CBG). The sulfate of the unmetabolized corticosterone forms a CBG complex of approximately half the affinity of the CBG-corticosterone complex (LEBEAU et al., 1969; LEBEAU and BAULIEU, 1970).

To avoid the uncertainty about the original binding to a particular plasma protein as concluded from protein fractionation studies, OERTEL et al. (1962, 1965) explored the interaction of steroid conjugates with plasma proteins by zone electrophoresis and gel filtration. The authors administered ammonium dehydroepiandrosterone-7α-^{3}H-sulfate orally to a normal subject and found on separation of the plasma proteins that the sulfated conjugates were clearly associated with the albumin fraction. Binding to high-density and low-density lipoproteins, prepared by flotation, was negligible. The authors assume on the basis of extensive studies on the chromatographic properties of steroid conjugates (OERTEL, 1961, 1962; OERTEL and KAISER, 1962) that the dehydroepiandrosterone sulfate occurs partially or totally in the form of nonpolar complexes, sulfatides, which are relatively unstable and decompose readily to form steroid sulfate and diglyceride (OERTEL, 1964). More recent electrophoretic studies

Table VI-3. *Interaction of steroid sulfates with serum albumin. Equilibrium dialysis; pH 7.4*

Steroid	Protein	Temp.	Solvent [a]	Method	n_1	k_1 M^{-1}	n_2	k_2 M^{-1}
Dehydroepiandrosterone sulfate [c]	BSA	5—6°	K-R	Reciprocal plots	20	1.3×10^5	—	—
Dehydroepiandrosterone sulfate [c]	BSA	5—6°	P	Reciprocal plots	18 [b]	2.0×10^5	—	—
Pregnenolone sulfate [c]	BSA	5—6°	P	Reciprocal plots	19	1.4×10^5	—	—
Androsterone sulfate [d, e]	HSA	4°	S	Scatchard	1	3.6×10^5	16	1.9×10^3
Dehydroepiandrosterone sulfate [e]	HSA	4°	S	Scatchard	2	6.3×10^4	18	1.5×10^3

[a] K-R, Krebs-Ringer bicarbonate; P, phosphate buffer; S, 0.9% saline.
[b] At unbound steroid sulfate concentrations $[S] > 2\times10^{-6}$ M, a second set of binding sites was observed.
[c] Puche and Nes (1962).
[d] Almost identical values were obtained for etiocholanolone sulfate.
[e] Plager (1965).

by Oertel et al. (1969) showed that the sulfo conjugates, produced *in vivo* after injection of tritiated dehydroepiandrosterone into human subjects, were associated with the albumin fraction, whereas the glucosiduronates were bound to the α_1-globulins.

There is general accord among various laboratories that albumin is the main protein responsible for the binding of steroid conjugates to plasma proteins. Spectrophotometric studies have shown (Westphal and Ashley, 1962) that the glucosiduronate of the androsta-3,5-dien-3-ol-17-one interacts strongly with HSA and β-lactoglobulin. Puche and Nes (1962) reported strong binding to BSA for dehydroepiandrosterone sulfate and pregnenolone sulfate (Table VI-3). The authors assume possible interaction with the cationic sites of arginine, in a similar way as suggested by Klotz et al. (1946) for methyl orange which also carries an anionic sulfate group. It is of interest that increasing concentrations of sulfate ions displaced the dehydroepiandrosterone sulfate from the BSA binding sites (Puche and Nes, 1962); addition of 1.3 and 2.5 M Na_2SO_4 to the outside solution raised the unbound portion of the steroid sulfate from the original value of 11% to 68% and 83%, respectively. Wang and Bulbrook (1967) have reported displacement of dehydroepiandrosterone sulfate from binding sites in human plasma by non-steroidal organic sulfates.

Table VI-3 also contains data on binding of androgen sulfates to HSA as reported by Plager (1965). This author concludes from his studies that albumin alone accounts for the binding of steroid sulfates; no evidence for involvement of other plasma proteins was obtained. The percentage binding to HSA was identical for the sulfates of androsterone and etiocholanolone (Fig. VI-1); almost the same binding values were found for dehydroepiandrosterone sulfate. Analysis of the equilibrium dialysis data according to Scatchard's procedure indicated at least two types of binding sites of different affinity for androsterone sulfate (Table VI-3); the values for etiocholanolone sulfate were almost identical. Two sets of binding groups were also found by the Scatchard procedure for dehydroepiandrosterone sulfate (Table VI-3). The first type (n_1) of this steroid conjugate had a lower binding affinity than that of the other two sulfates studied; the second set of binding sites was similar for all three conjugates. Plager assumes that both androsterone sulfate and etiocholanolone sulfate are bound to the same sites of the HSA molecule; he found that the two sulfates displace each other in a similar manner from the albumin-binding sites. The binding capacity of albumin for the steroid conjugates exceeds greatly their concentration under normal physiological conditions.

Table VI-3 seems to indicate a principal difference between steroid sulfate binding to BSA and to HSA, concerning the values

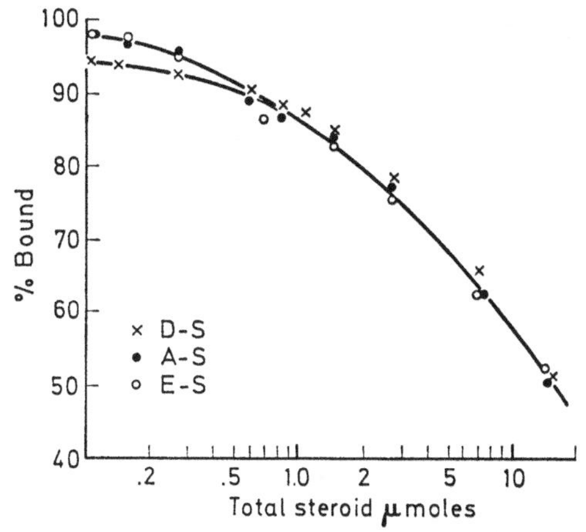

Fig. VI-1. Binding of dehydroepiandrosterone sulfate (D-S), androsterone sulfate (A-S) and etiocholanolone sulfate (E-S) to HSA. Equilibrium dialysis, 0.9% saline, pH 7.4, 4°. Abscissa shows total amount of steroid sulfate added to dialysis system consisting of 2 ml 0.8% HSA and 6 ml saline. From Plager (1965)

of n_1. The reason for this is not known. The albumins used were a crystalline BSA preparation (95% electrophoretic purity) and Cohn fraction V from human plasma (97% electrophoretic purity); neither preparation was delipitated. Both albumin preparations may therefore be assumed to contain an unknown quantity of endogenous lipid which may interfere with the binding process (see Chapter VI. 5 b). The HSA and BSA solutions used in the equilibrium dialysis experiments had concentrations of 0.8% and 5.0%, respectively.

Sandberg et al. (1966) studied the binding of steroid conjugates to human plasma and HSA. The glucosiduronates of estriol and estrone were biosynthesized *in vitro* by the livers of various species

(SLAUNWHITE et al., 1964). Equilibrium dialysis experiments showed that they associated with HSA to about the same extent as with total plasma proteins, indicating that albumin is primarily responsible for their binding in plasma. The percentage association of the glucosiduronates of estriol and of estrone with 1% HSA was slightly higher than that with 1 : 5 diluted human plasma.

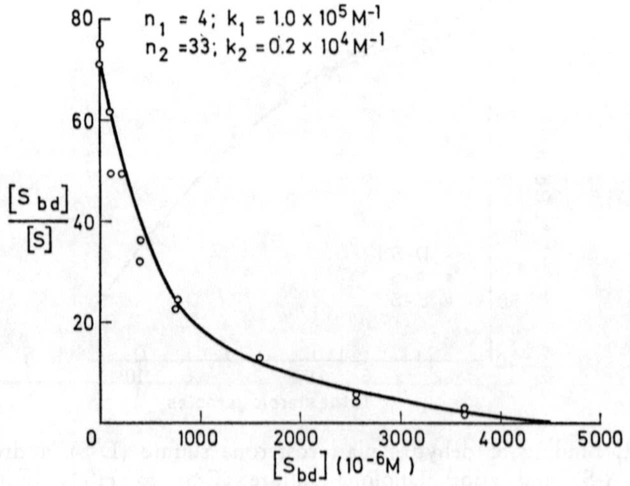

Fig. VI-2. Interaction of estrone sulfate with HSA. Equilibrium dialysis; pH 7; 5°. Adapted from SANDBERG et al. (1966)

The interaction of estrone sulfate with HSA was reported (PURDY et al., 1961) to have a binding capacity of 50 moles per mole albumin as determined by equilibrium dialysis; dialysis data were later evaluated according to Scatchard's procedure (SANDBERG et al., 1966). Figure VI-2 shows the graphic analysis of the binding data; two sets of binding sites, $n_1 = 4$ and $n_2 = 33$ were found. The association constants were of the same order as those of the androgen sulfates (Table VI-3). Another similarity with the androgen sulfates was observed in displacement studies: dehydroepiandrosterone sulfate and etiocholanolone sulfate displaced radiolabeled estrone sulfate from the binding sites on HSA almost as effectively as equal concentrations of estrone sulfate did. These observations corroborate the interpretation of steroid sulfate binding as anion interaction. The significance of the anionic sulfate group for the complex formation follows also from the fact that unconjugated steroids cannot easily displace each other from HSA (SANDBERG et al., 1966).

3. Interaction of Monopolar Neutral Steroids; Lipoproteins

Protein binding of nonpolar steroids, e. g., saturated steranes, has not been investigated. Although interaction of these hydrocarbons with proteins by hydrophobic bonding may be expected, such binding is presently of no particular interest from a biological viewpoint. This is in contrast to interaction of unsaturated and aromatic hydrocarbon derivatives of steroids whose covalent binding to tissue proteins, e. g., to soluble and particulate proteins of the skin (MILLER, 1951) is of interest for the role of polycyclic aromatic hydrocarbons in carcinogenesis (HEIDELBERGER, 1959; MILLER and MILLER, 1966; HUGGINS et al., 1967). The problem of carcinogenic hydrocarbons from steroids has been recently reviewed by BISCHOFF (1969). Binding parameters and thermodynamic data have been reported for hydrophobic interaction between HSA and polycyclic aromatic hydrocarbons, including carcinogenic compounds (FRANKE, 1968, 1969); *in-vitro* binding of 3,4-benzpyrene to BSA has been studied by ANGHILERI (1968). SLUYSER (1968) measured binding of carcinogens to histones by gel filtration. Interaction of various proteins with saturated hydrocarbons has been mentioned in Chapters IV. 7 and IV. 9.

By far the most important representative of the "monopolar" steroids is cholesterol. The highly saturated four-ring system and the aliphatic side chain accord this compound the properties of a lipid, including very low solubility in water and high solubility in organic solvents. State of solution and transport in the circulating blood is made possible by association with proteins; attachment of cholesterol to the euglobulins has been recognized early (BANG, 1918; HANDOVSKY, 1925; GARDNER and GAINSBOROUGH, 1927; MACHEBOEUF, 1929 a, 1929 b) although mere presence of sterol in the precipitated globulin (TROENSEGAARD and KOUDAHL, 1926) cannot be considered adequate proof for protein binding under physiological conditions.

Nearly all the lipids transported and metabolized in higher mammals, particularly man, are present in the blood as constituents of lipoproteins (ONCLEY, 1958; GURD, 1960 b; LINDGREN and NICHOLS, 1960; VANDENHEUVEL, 1962; SCANU, 1965; POLONOVSKI, 1966; CORNWELL, 1967; FREDRICKSON et al., 1967; FURMAN et al., 1967; NICHOLS, 1967; SHAPIRO, 1967; SCHUMAKER and ADAMS, 1969; TRIA and SCANU, 1969). The principal lipids are nonesterified cholesterol, cholesteryl esters, glycerides, phosphatides, and nonesterified fatty acids (STOFFEL, 1967). The term lipoprotein in its broadest definition would include all known lipid-protein structures, the largest chylomicra as well as fatty acid-albumin complexes; the molecular weights would range from 70,000 to about 10^{11}. However, by far the major portion of the serum lipoproteins are in a molecular weight range from approximately 200,000 to 10 million. Cholesterol and cholesteryl esters constitute up to 50% or more of the lipoproteins; they are present in numerous molecular entities forming this complex and heterogeneous class of proteins. Free cholesterol can be rapidly exchanged between the various transport forms and many tissues (FREDRICKSON et al., 1967; GRAHAM and GREEN, 1967), most swiftly between plasma, liver and red blood cells. However, it is assumed that the plasma cholesterol serves primarily as a structural component of the lipoproteins which provide transport for other lipids. The liver is the site of lipoprotein synthesis (RADDING et al., 1958; RADDING and STEINBERG, 1960; HAFT et al., 1962).

A discussion of the lipoproteins is beyond the scope of this monograph. The bibliography contains additional references on separation (DELALLA and GOFMAN, 1954; KUNKEL and TRAUTMAN, 1956; ONCLEY et al., 1957). Structural properties of the lipoproteins are discussed by FISHER and GURIN (1964); LEVY and FREDRICKSON (1965); LEVY et al. (1967); SKIPSKI et al. (1967); DAY and LEVY (1969). For recent chemical studies on the lipid-free protein moieties or apoproteins see RUDMAN et al. (1968); GOTTO et al. (1969); LINDGREN et al. (1969); SCANU et al. (1969); SHORE and SHORE (1969).

FRANCIS et al. (1965) have described an α-globulin, "transcholesterin", which binds specifically cholesterol and dihydrocholesterol. It was demonstrated in normal and lipoprotein-depleted human serum by immunoelectrophoretic techniques. No immunological relation to the lipoproteins was observed (LING and KRASTEFF, 1968); its chemical nature and physiological significance await clarification. A similar transcholesterin in the serum of rats (LING et al., 1968) has been reported to increase after different high-fat or atherogenic diets. The mobilization of this cholesterol-binding alpha-globulin is assumed to represent a homeostatic mechanism to buffer the high cholesterol concentrations in the serum. Further work on this protein, especially the question of its relationship to atherosclerotic processes, can be anticipated with considerable interest.

Table VI-4. *Uptake in vitro of steroids by HSA and human low-density lipoprotein (S_f 3—8). Calculated from* AVIGAN *(1959). Method: Solubility; 37°; pH 7.45. Concentration of either protein, 1 g/100 ml*

Steroid	moles Steroid per mole HSA	Lipoprotein binds x-times more x
Cholesterol	0.6	8.7
Cholest-4-en-3-one	0.3	9.8
Sitosterol	1.1	3.0
Progesterone	2.8	2.6
Cortexone	3.5	2.9
Corticosterone	9.2	1.2
Testosterone	5.6	0.9

In studies on interaction of cholesterol and other steroids with serum proteins, AVIGAN (1959) devised a useful method of incorporating the lipids from a dispersion on celite. The lipoproteins, especially the low-density fraction (S_f 3—8) took up the greatest part of the radiolabeled sterols, without reaching saturation even after 48 hrs at 37°; serum albumin bound less steroid and became saturated after a few hours. These lipoproteins, loaded *in vitro* with radiolabeled cholesterol, showed a similar metabolic behavior as natural lipoproteins. Table VI-4 demonstrates the higher affinity with which the "monopolar" steroids are attached to the low-density lipoproteins as compared to albumin. The results also indicate greater incorporation of the more lipophilic monopolar sterols than of the duo- and higher-polar steroid hormones; this might be expected if one considers the ratio of hydrophobic to hydrophilic portions in the various molecules. Complex formation of 7-dehydrocholesterol and

7-hydroxycholesterol with serum albumin and other proteins has been reported (VENDT, 1957) to protect the steroids against air oxidation. Observations with rodent urine (FINLAYSON and BAUMANN, 1957) showed that cholesterol and 5α-cholest-7-en-3β-ol (Δ^7-cholestenol) are present in association with albumin and, to a higher degree for cholesterol, with globulin. Cholesterol and Δ^7-cholestenol were unesterified to 74% and 24%, respectively.

In contrast to the more familiar ligand-protein interactions which are characterized by a definite number of binding sites specifically adapted for the association with molecules of certain chemical structures, the incorporation of lipids in the lipoproteins appears to be governed by rules less specific for individual lipid molecules. In this respect, early experiments of MACHEBOEUF, TAYEAU and POLONOVSKI (for references, see GURD, 1960 b) should be remembered in which displacement of the endogenous lipids from lipoproteins was achieved by a number of different compounds possessing hydrocarbon chains, e. g., higher fatty acids. Such displacement was also effected by steroid compounds such as saponin and cholic acid. Studies with the lipid-free apoproteins, made possible by recent accomplishments in several laboratories, are likely to elucidate the nature of the bonding mechanisms involved.

4. Interactions of Duo- and Polypolar Neutral Steroids. Qualitative Observations with Steroid Hormones

One of the early questions in steroid binding studies was that of the particular serum protein components with which the steroid compounds associated. It was recognized that the serum albumin molecule enters reversible binding equilibria with all duo- and polypolar steroid compounds investigated, in accordance with the general property of this protein to interact with numerous substances of varied chemical nature. Binding of steroids to serum albumin will be discussed in detail in Section 5 of this chapter. Special sections will also be devoted to steroid interactions with serum glycoproteins of globulin nature, i. e., CBG (Chapters VIII—XI), the steroid-binding β-globulin (Chapter XII), and AAG (Chapter XIII). It should be remembered that in the early work of BRUNELLI (1934), as well as in the studies of SZEGO and ROBERTS (1946), and ROBERTS and SZEGO (1946), indications were obtained for participation of globulins in estrogen binding; globulins were also suspected to interact with corticosteroids (SAVARD et al., 1952).

In the studies of FORBES and HOOKER (1949), and HOOKER and FORBES (1949), on free and bound progesterone in plasma, the criterion of solubility in organic solvents was used to indicate the free state; this definition of binding is different from that measured by equilibrium procedures which characterize reversible, non-covalent interaction between steroids and proteins. Attempts to assess binding of 17-ketosteroids (GARDNER, 1954) and of a number of steroid hormones (ANTONIADES et al., 1957 a; ANTONIADES et al., 1957 b; SANDBERG et al., 1957) to the different serum protein components by application of COHN's cold ethanol fractionation procedure have to be interpreted with caution as discussed above (Chapter VI. 2). However, generally valid results have been obtained in equilibrium dialysis or ultrafiltration studies in which the steroids interacted *in vitro* with the various fractions (LEVEDAHL, 1954; DAUGHADAY, 1956 b; SANDBERG et al., 1957; DAUGHADAY, 1958 b). Especially the extensive studies of SANDBERG et al. gave a comprehensive picture of the binding potentials of the various COHN fractions, corroborating the predominant ability of albumin to associate with the major types of steroids tested. As far as the exact quantitative nature of these data is concerned, it must be remembered that the various protein fractions may have undergone changes from the native state, accompanied by a decrease in binding activity. Such alterations presumably are least pronounced for Fraction V (albumin); they are considerable for other fractions, e. g., IV-4 which contains CBG.

In an effort to maintain the serum proteins in their native state, electrophoretic methods were applied to separate the serum protein components and simultaneously follow their association with steroids. WESTPHAL et al. (1956) dissolved progesterone-4-^{14}C and cortisol-4-^{14}C in human or rat serum and analyzed the electrophoretically separated protein components for bound radiosteroids. These experiments were done at unphysiologically high steroid concentrations, to compensate for the low specific activities then available in commercial radiolabeled steroids; they demonstrated nevertheless that up to 90% of progesterone was bound to the migrating proteins of human serum, whereas the bound portion of cortisol appeared much smaller. In similar qualitative observations, association with serum proteins was shown for progesterone and cortisol by paper chromatographic techniques (WESTPHAL et al., 1956).

These experiments were inadequate in that the high steroid concentrations used, i. e., more than 100 times physiological, tend to mask protein components of low binding capacity. Furthermore, continuous dissociation of the steroid-protein complexes during migration in the electric field invalidates any conclusion of binding to a certain protein fraction on the basis of simultaneous presence of the steroid. Only in the migrating front would such conclusion be valid in a qualitative sense.

The first problem has been overcome by the availability of ^{14}C- and particularly ^{3}H-labeled steroids of increasingly higher specific activities. The second difficulty was solved in an elegant manner by DAUGHADAY (1956 c) who performed the electrophoretic separation in a medium which contained the equilibrium concentration of the unbound steroid so that virtually no dissociation was apparent (equilibrium paper electrophoresis, see Chapter IV. 4). Although this technique gave clear evidence of albumin binding for cortisol and corticosterone, it did at first not allow detection of other binding proteins of low capacity such as CBG, presumably because of oversaturation of the CBG binding sites with the radiolabeled steroids of relatively low specific activity. The usefulness of the equilibrium paper electrophoresis procedure for demonstration of CBG in animal sera was shown ten years later (WESTPHAL and DE VENUTO, 1966) when radiosteroids of higher activity were available (see Chapter VIII. 4 c).

Continuous flow electrophoresis (SELDEN and WESTPHAL, 1957) was used in our laboratory (WESTPHAL et al., 1961) to separate human serum into the four major components which were then assayed with radiolabeled progesterone and corticoids in a multiple equilibrium dialysis system. The original unfractionated serum was included in the multiple binding equilibrium system. The results are shown in Table VI-5. The binding distribution is distorted by the use of the unphysiologically high steroid concentrations which conceal the bind-

Table VI-5. *Binding of progesterone and corticosterone to human serum and its electrophoretic protein components. Multiple equilibrium dialysis; 25°; pH 7.6*

Steroid	Steroid bound to					Steroid unbound	Steroid bound to original serum	Steroid unbound in original serum
	Albu. 4.6%	α-Glob. 0.7%	β-Glob. 0.8%	γ-Glob. 0.9%	All components (calc'd)			
Progesterone								
μg/ml [a]	6.12	0.13	0.21	0.07	6.53	0.19	5.73	0.19
% of total	91.0	2.0	3.1	1.0	97.1	2.9	96.8	3.2
% of total [b]	88.9	1.4	6.7	3.0	100.0	—	—	—
Corticosterone								
μg/ml [a]	4.92	0.13	0.08	0.04	5.17	0.66	3.99	0.66
% of total	84.4	2.2	1.4	0.7	88.7	11.3	85.8	14.2

[a] Average values of quadruplicate determinations.

[b] For comparison, values obtained from separation by continuous electrophoresis (pH 7.6; 4°) of human serum containing progesterone-4-^{14}C. From Westphal et al. (1961).

ing to CBG. However, the data demonstrate that the totaled interaction with the individual fractions, present in approximately normal serum concentrations, nearly equals the association with the original serum. Table VI-5 also shows that direct separation by continuous electrophoresis of human serum to which progesterone-4-^{14}C had been added results in approximately the same binding distribution to the electrophoretic protein components. In making a quantitative comparison it must be remembered that the relative concentrations of the protein components may not be the same as those applied in the multiple equilibrium dialysis experiment.

The binding affinity for progesterone and cortisol of several purified proteins from human plasma were determined (WESTPHAL et al., 1961) by equilibrium dialysis (Table VI-6). The apparent association

Table VI-6. *Binding affinity of progesterone and cortisol to human plasma protein components. Equilibrium dialysis; 4°; pH 7.6.* From WESTPHAL et al. (1961)

Protein	Conc. in plasma g/l	Progesterone		Cortisol	
		n k [a] M^{-1}	$\Delta F°$ kcal/mole	n k [a] M^{-1}	$\Delta F°$ kcal/mole
Prealbumin	0.4	0.51×10^4	—4.7	0.15×10^4	—4.0
Fibrinogen	3.0	1.46×10^4	—5.3	0.92×10^4	—5.0
Globin, modified	—	1.59×10^4	—5.3	0.04×10^4	—3.3

[a] See Eq. (V-8).

constants and the corresponding free-energies of binding are small in comparison with those for HSA, AAG or CBG as will be seen below. The polarity rule is followed for all these protein-steroid complexes, in accordance with the behavior of most other serum proteins. Approximate association constants and free energies of interaction have been reported by SANDBERG et al. (1957) for the binding equilibria of a number of steroids with the plasma protein fractions obtained by COHN's cold ethanol procedure. Since most of these fractions are protein mixtures, the molecular constants had to be calculated by assuming certain molecular weights.

5. Steroid Interaction with Serum Albumin

a) Experimental Data

Reversible binding of steroids to protein has been studied more widely with serum albumin than with any other protein. Easy availability in relatively pure form, good stability, considerable information on binding affinity for many different types of compounds, and at least a general knowledge of its chemical structure may be reasons for this preference. From the viewpoint of biological significance, the low specificity and comparatively low affinity of steroid interaction with albumin would make these complexes less significant than the associations with serum proteins of higher affinity.

Before the advent of radiolabeled steroids for routine use in biochemical studies, binding was measured by methods which required relatively large amounts, in accordance with the lower sensitivity of biometric, colorimetric or spectrophotometric assay procedures. The solubility method was used in the laboratories of BISCHOFF (BISCHOFF and KATHERMAN, 1948; BISCHOFF and PILHORN, 1948; BISCHOFF and STAUFFER, 1954; BISCHOFF et al., 1954), SAMUELS (EIK-NES et al., 1954), SANDBERG and SLAUNWHITE (SANDBERG et al., 1957; SLAUNWHITE et al., 1963), and AVIGAN (1959), and provided fundamental information. As a measure of protein binding, the "dispersing power", r, of the protein for a given steroid is determined from the difference of solubility in the protein solution and the protein-free solvent; within certain limits, r is proportional to the protein concentration. It can be expressed as moles steroid bound per mole albumin, r, according to

$$r = \frac{[S_{bd}]}{[P_t]} = \frac{[S]_{Alb} - [S]_{Buf}}{[P_t]} \qquad \text{(VI-3)}$$

where $[S]_{Alb}$ and $[S]_{Buf}$ are the solubilities of the steroid in the albumin solution and in the buffer solvent, respectively; $[P_t]$ is the concentration of albumin. All concentrations are given in moles/liter. Values of r on the basis of solubility determinations are shown in Table VI-7 for various steroids.

It might seem that the r-values were a valid measure for the number of binding sites, n, in a protein. However, it is easy to see that they will be affected by the solubility of the steroid in the

Table VI-7. *Dispersion (r) of neutral steroids by binding to serum albumin. Method: Solubility; r in moles of steroid per mole albumin, calculated from data given in references*

No.	Steroid	Protein	Temp. °C	pH	r	Reference
1	Estrone	BSA	37.5	5.3	0.11	BISCHOFF and KATHERMAN (1952)
2	Estrone	BSA	37.5	8.5	0.33	BISCHOFF and KATHERMAN (1952)
3	Estrone	HSA	5	7.0	0.15	SANDBERG et al. (1957)
4	Estradiol	BSA	37.5	5.5	0.42	BISCHOFF and PILHORN (1948)
5	Estradiol	BSA	37.5	8.5	1.0	BISCHOFF and KATHERMAN (1952)
6	Estradiol	HSA	5	7.0	0.69	SANDBERG et al. (1957)
7	Estriol	HSA	5	7.0	0.33	SLAUNWHITE et al. (1963)
8	Androst-4-ene-3, 17-dione	HSA	5	7.0	0.24	SLAUNWHITE et al. (1963)
9	Testosterone	BSA	25	5.5	2.0	BISCHOFF and PILHORN (1948)
10	Testosterone	BSA	25	7.2	2.5	BISCHOFF and PILHORN (1948)
11	Testosterone	BSA	37.5	5.3	2.3	BISCHOFF and PILHORN (1948)
12	Testosterone	BSA	37.5	7.4	2.7	BISCHOFF and PILHORN (1948)
13	Testosterone	BSA	37.5	8.0	3.1	BISCHOFF and PILHORN (1948)
14	Testosterone	HSA	5	7.0	0.56	SANDBERG et al. (1957)
15	Testosterone	HSA	37	7.5	5.6	AVIGAN (1959)
16	Testosterone propionate	BSA	37.5	5.3	0.87	BISCHOFF et al. (1952)
17	Testosterone propionate	BSA	37.5	8.3	1.4	BISCHOFF et al. (1952)
18	Androst-4-ene-3, 11,17-trione	HSA	5	7.0	0.07	SLAUNWHITE et al. (1963)
19	Androst-4-en-11β-ol-3,17-dione	HSA	5	7.0	1.5	SLAUNWHITE et al. (1963)
20	Progesterone	BSA	25	5.5	0.46	BISCHOFF and PILHORN (1948)
21	Progesterone	BSA	25	7.1	0.61	BISCHOFF and PILHORN (1948)
22	Progesterone	BSA	37.5	5.3	0.61	BISCHOFF and PILHORN (1948)

Table VI-7 (continued)

No.	Steroid	Protein	Temp. °C	pH	r	Reference
23	Progesterone	BSA	37.5	7.4	0.81	BISCHOFF and PILHORN (1948)
24	Progesterone	BSA	37.5	8.1	1.3	BISCHOFF and PILHORN (1948)
25	Progesterone	HSA	5	7.0	0.24	SANDBERG et al. (1957)
26	Progesterone	HSA	37	7.5	2.8	Avigan (1959)
27	Pregnane-3,11,20-trione	HSA	5	7.0	0.31	SLAUNWHITE et al. (1963)
28	17-Hydroxy-pregn-4-ene-3,20-dione	HSA	5	7.0	0.06	SLAUNWHITE et al. (1963)
29	17-Acetoxy-6α-methylpregn-4-ene-3,20-dione	HSA	5	7.0	0.02	SLAUNWHITE et al. (1963)
30	Cortexone	HSA	5	7.0	2.0	SLAUNWHITE et al. (1963)
31	Cortexone	HSA	37	7.5	3.5	AVIGAN (1959)
32	Corticosterone	HSA	5	7.0	2.6	SANDBERG et al. (1957)
33	Corticosterone	HSA	37	7.5	9.2	AVIGAN (1959)
34	Cortisone	HSA	5	7.0	0.71	SANDBERG et al. (1957)
35	Cortisol	HSA	5	7.0	0.43	SANDBERG et al. (1957)
36	3α,17,21-Trihydroxypregnane-11,20-dione	HSA	5	7.0	0.28	SLAUNWHITE et al. (1963)
37	11β,17,21-Trihydroxy-6α-methylpregna-1,4-diene-3,20-dione	HSA	5	7.0	0.35	SLAUNWHITE et al. (1963)
38	Cholest-4-en-3-one	HSA	37	7.5	0.3	AVIGAN (1959)
39	Cholesterol	HSA	37	7.5	0.6	AVIGAN (1959)
40	Sitosterol	HSA	37	7.5	1.1	AVIGAN (1959)

aqueous solvent system. If the association constant were infinite or very high, the dispersion r would directly indicate n. Since the steroid-albumin complexes listed in Table VI-7 are markedly dissociated under the conditions of the experiments, the concentration of maximal solubility may be reached when the binding to the protein is still below its total binding capacity. The dispersion, r, obtained from solubility studies therefore provides only minimum values for the number of binding sites.

It is evident from Table VI-7 that the r-values in many cases are below unity. Since they have been obtained in the presence of an excess of undissolved steroid, the solubility of the steroid in the aqueous system can be recognized as a limiting factor. For the same reason, the dispersion value, r, is mostly considerably lower at 5° than at the higher temperatures. Values for the heat of dispersion in BSA solution have been reported by BISCHOFF and PILHORN (1948) for estradiol, progesterone and testosterone to be 13.7, 5.5 and 1.9 kcal/mole, respectively.

Table VI-8. *Reduction of dispersing power (r) of BSA for estradiol by the presence of other steroids. Solubility; 37.5°; pH 5.5; r in moles steroid per mole BSA.* Calculated from BISCHOFF and KATHERMAN (1948) and BISCHOFF and PILHORN (1948)

Steroid	r	Solution saturated with Progesterone r	Solution saturated with Testosterone r
Estradiol	0.42	0.28	0.20
Progesterone	0.65	—	—
Testosterone	2.3	—	—

Table VI-8 shows the use of solubility data to demonstrate inhibition of estradiol binding to BSA by progesterone and testosterone. Whether the reduced dispersion of estradiol in the presence of the two other steroid hormones is caused by competitive displacement from the binding site(s), by an influence on the activity of estradiol in the aqueous solution, by an allosteric effect on albumin, or by still another mechanism is not evident from the experimental results.

More complete data on interaction of steroids with BSA and HSA have been determined by a number of other methods. Table VI-9 contains the results of these studies performed over the years in various laboratories. In this table, the number of binding sites, n, is given as estimated by standard procedures; in a few cases, dispersion values, r, as calculated from solubility data according to eq. (VI-3), are presented instead. Whenever possible, intrinsic association constants, k, have been listed; however, in most cases, information on number and quality of binding sites is missing, and n k values have been given. Free (Gibbs) energy of the interaction between steroid and protein is computed either for the intrinsic association constant, k, or for n k.

The data in Table VI-9 show that the binding affinity between steroid and serum albumin decreases with increasing number of polar groups in the steroid molecule, as an indication of the general validity of the polarity rule (see Chapter V. 1). This is evident not only for the steroid hormones but also for the cardiac-active steroid compounds (No. 21 to 27). In addition, a comparison between the cardiac glycosides and the corresponding aglycones shows that the presence of the sugar components lowers the association constant. Removal of the three digitoxose molecules from digitoxin increases the n k value more than 3-fold (No. 21-2, 22-1); elimination of one molecule of rhamnose from strophanthin has a similar effect although to a lesser degree (No. 26, 27). It should be noted that the binding parameters of the cardiac-active steroids were determined in a 1% HSA solution which contained 3% ethanol. The presence of the ethanol reduced the affinity constants somewhat (SCHOLTAN et al., 1966). Studies on binding of g-strophantin, digoxin, digitoxin and digitoxigenin to human plasma, HSA and BSA were recently reported by KUSHINSKY (1969). The method of gel filtration over Sephadex G-50 was used at the very low steroid glycoside concentrations of 2.5×10^{-7} g/ml in 4% albumin. The affinity constants obtained differed from those given in Table VI-9.

LUKAS and DE MARTINO (1969) have recently studied binding of digitoxin and other cardiac steroid (glycoside)s to HSA. In contrast to the general experience with other steroid compounds, they observed a marked increase of the association constant with rising temperature (No. 22-2 to 22-4 in Table VI-9). Calculation of thermodynamic parameters gave for ΔF° at 1°, 20°, and 37°, —6.0, —6.6 and —7.2 kcal/mole, respectively. A positive enthalpy change $\Delta H^\circ =$ 2.8 kcal/mole, and positive entropy change, $\Delta S^\circ = 32$ cal mole^{-1} deg^{-1} were obtained. As observed with other steroid-protein interactions, the reaction is spontaneous (negative ΔF°); it is associated with a large gain of entropy. The positive enthalpy, indicating endothermic reaction, is unusual; positive ΔH° values have not been found in other steroid-protein complexes investigated so far.

b) Quality of Albumin

It is immediately apparent in Table VI-9 that the affinity constants, n k, of a given steroid complex with one of the two serum albumins, determined in different laboratories or at different times under similar conditions of temperature and pH, show considerable variation (e.g., No. 2-1, 2-2 and 2-3; the testosterone-BSA complexes at various temperatures; progesterone-HSA complexes and others). Since the results listed in Table VI-9 were obtained by accepted methods in the laboratories of experienced investigators, the main reason for the differences is seen in the quality of the albumin preparations applied. A good example is the analysis of testosterone-BSA interaction in SAMUELS' laboratory (No. 10-6, 10-7, 10-8) which resulted in different n k values for three different BSA preparations (see also Fig. V-8 and Table V-1). In this case, the sample of the highest electrophoretic purity gave the highest n k value. Another example is the comparison of two progesterone-HSA complexes at 4° (No. 13-2 and 13-3). The complex, No. 13-2, was obtained with a 4-times crystallized HSA preparation of high purity which was not further treated. The progesterone complex, No. 13-3, was formed with the same HSA preparation after extraction with chloroform-methanol (3 : 1, v/v) at 4°. It is evident that the delipidation of HSA with the organic solvents increases the association constant k, as well as n k, approximately 2.5 times. The number of binding sites is $n = 2$ for both albumin preparations. The analysis of the binding parameters for the two complexes by the method of reciprocal plots is given in Fig. VI-3. A similar phenomenon has been observed with the progesterone-AAG complex whose association constant is multiplied several fold after delipidation of the pure AAG by organic solvents (see Chapter XIII).

The relatively high n k value for the cortisol-HSA complex at 4° (No. 19-5) and several analogous cases may also be explained by the delipidated state of the HSA preparation used. A similar interpretation presumably holds for the high affinity constant of the estrone-BSA complex (No. 1-1), in comparison with the estrone HSA interaction (No. 1-2), which is contrary to the general experience gained in our and other laboratories that BSA binds steroids and other ligands with lower affinity than HSA does. The BSA preparation used by THERRIAULT and TAYLOR (1964) had been carefully freed of lipid (THERRIAULT and TAYLOR, 1960).

The reasons for the higher binding affinity of HSA for progesterone after removal of lipids by organic solvents are not known in detail; the nature of the extracted material has not been analyzed. Crystalline HSA is known to contain small amounts of fatty acid which cannot be removed by repeated crystallization (KENDALL, 1941; COHN et al., 1947); analyses of fatty acid contents of 26 serum albumin preparations from various commercial sources, representing different species, have been published by CHEN (1967). The most prominent among the contaminating acids are palmitic, stearic, oleic and linoleic acids (SAIFER and GOLDMAN, 1961), lipids which form stable complexes with serum albumin; the association is mediated by the anionic site and by hydrophobic interaction of the aliphatic chain with non-polar regions in the albumin molecule. CHEN (1967) uses charcoal to remove the fatty acids from albumin. This method seems to be somewhat more effective than the extraction procedure of GOODMAN (1957) and the acid separation technique of WILLIAMS and

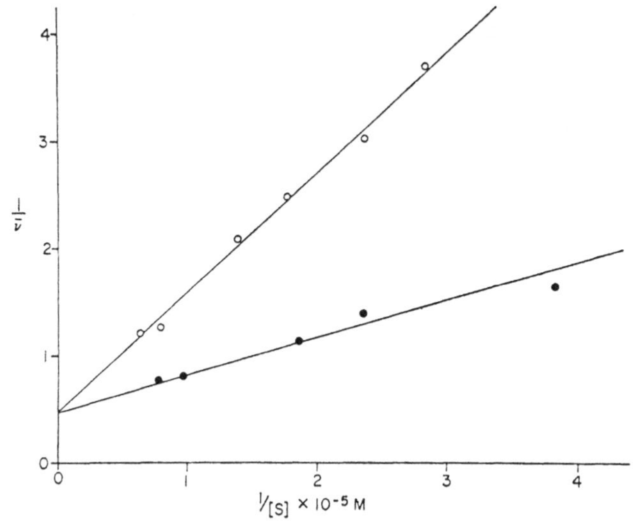

Fig. VI-3. Binding of progesterone to an untreated (open circles) and delipidated (black circles) HSA preparation, in 0.05 M phosphate, pH 7.4; 4°; $n=2.17$, $k=0.41\times10^5\ M^{-1}$ for untreated HSA; $n=2.13$, $k=1.0\times10^5\ M^{-1}$ for delipidated HSA. See Table VI-9, No. 13-2 and 13-3. From WESTPHAL (1966)

Table VI-9. *Interaction of neutral steroids with serum albumin. n, Number of binding sites; r, dispersion in moles/mole albumin; k, apparent association constant; $\Delta F°$, standard free energy change*

No.	Steroid	Protein [r]	Method [m]	Temp. °C	pH	n or r [a]	k $M^{-1} \times 10^{-4}$	n k [b] $M^{-1} \times 10^{-4}$	$\Delta F°$ [c] kcal/mole	Reference [d]
1–1	Estrone	BSA [j]	T	~23	7.3	1	7.9	7.9	—6.7	Therriault and Taylor (1964)
1–2	Estrone	HSA	ED	5	7.0	1.1	4.0	4.4	—5.9 [c]	Sandberg et al. (1957)
2–1	Estradiol	BSA	S	25	7.3	0.43 [a]	—	>72 [b]	<(—8.0) [c]	Eik-Nes et al. (1954)
2–2	Estradiol	BSA	PA	25	7.4	—	—	30 [b]	—7.5 [c]	Schellman et al. (1954)
2–3	Estradiol	BSA	GE	25	7.4	—	—	9.1	—6.8 [c]	Pearlman et al. (1969)
2–4	Estradiol	HSA	ED	5	7.0	—	—	16 [b]	—6.7 [c]	Sandberg et al. (1957)
3	Androstan-17-one	BSA	S	37	7.3	0.19 [a]	—	>97 [b]	<(—8.2) [c]	Eik-Nes et al. (1954)
4	Androst-4-en-3-one	BSA	S	25	7.3	0.68 [a]	—	>68 [b]	<(—8.0) [c]	Eik-Nes et al. (1954)
5	Androstane-3,17-dione	BSA	S	37	7.3	4.1 [a]	—	1.9 [b]	—5.9 [c]	Eik-Nes et al. (1954)
6	Androsterone	BSA	S	25	7.3	1.9 [a]	—	4.3 [b]	—6.6 [c]	Eik-Nes et al. (1954)
7	Etiocholanolone	BSA	S	25	7.3	1.1 [a]	—	1.6 [b]	—5.8 [c]	Eik-Nes et al. (1954)
8	Androst-4-ene-3,17-dione	BSA	S	37	7.3	3.2	—	1.6 [b]	—5.8 [c]	Eik-Nes et al. (1954)
9	Dehydroepiandrosterone	BSA	S	37	7.3	3.8	—	3.8 [b]	—6.5 [c]	Eik-Nes et al. (1954)
10–1	Testosterone	BSA	PA	3	7.4	13	0.14	1.8	—5.4 [n]	Alfsen (1963)
10–2	Testosterone	BSA	ED	3	7.4	20	0.09	1.8	—5.4 [n]	Alfsen (1963)
10–3	Testosterone	BSA	ED	4	7.6	—	—	2.9 [b, t]	—5.7 [c]	Westphal (1961)
10–4	Testosterone	BMA	ED	4	7.6	—	—	2.5 [b, u]	—5.5 [c, u]	Westphal (1961)

10–5	Testosterone	BSA [e]	PA	8	7.4	7	—	1.7 [b]	—5.5 [c]	SCHELLMAN et al. (1954)
10–6	Testosterone	BSA [f]	ED	25	7.4	10	0.18	1.9	{—5.8 [c] —4.4	SCHELLMAN et al. (1954)
10–7	Testosterone	BSA [e]	PA	25	7.4	6	0.38	2.3	{—6.0 [c] —4.9	SCHELLMAN et al. (1954)
10–8	Testosterone	BSA [g]	PA	25	7.4	5	0.79	4.0	{—6.3 [c] —5.3	SCHELLMAN et al. (1954)
10–9	Testosterone	BSA	PA	25	7.4	5.0	—	3.0	{—6.1 [c] —5.8	LEVEDAHL (1955)
10–10	Testosterone	BSA [h]	PA	25	7.5	4.0	—	2.8	—6.1 [c]	LEVEDAHL and PERLMUTTER (1956)
10–11	Testosterone	BSA	PA	25	7.4	—	—	2.2	—5.9 [c]	OYAKAWA and LEVEDAHL (1958)
10–12	Testosterone	BSA	PA	25	9.2	—	—	3.2	—6.1 [c]	OYAKAWA and LEVEDAHL (1958)
10–13	Testosterone	BSA	PA	25	7.4	7	0.28	2.0	—5.9 [p]	ALFSEN (1963)
10–14	Testosterone	BSA	ED	25	7.4	4	0.49	2.0	—5.9 [p]	ALFSEN (1963)
10–15	Testosterone	BSA	GE	25	7.4	—	—	2.4	—6.0 [c]	PEARLMAN and CRÉPY (1967 b)
10–16	Testosterone	BSA	S	37	7.3	2.8 [a]	—	2.3 [b]	—6.2 [c]	EIK-NES et al. (1954)
10–17	Testosterone	BSA	ED; PA	37	7.4	9.3	—	2.0	—6.1 [c]	LEVEDAHL and BERNSTEIN (1954)
10–18	Testosterone	BSA [i]	PA	37	7.7	6.2	—	2.3	—6.2 [c]	LEVEDAHL and PERLMUTTER (1956)
10–19	Testosterone	BSA	PA	37	7.4	—	—	1.5	—5.9	OYAKAWA and LEVEDAHL (1958)
10–20	Testosterone	BSA	ED	37.5	5.3	3 [a]	—	1.4 [b]	—5.9 [c]	BISCHOFF and STAUFFER (1954)

Footnotes see p. 114

Table VI-9 (continued)

No.	Steroid	Protein [r]	Method [m]	Temp. °C	pH	n or r [a]	k $M^{-1} \times 10^{-4}$	n k [b] $M^{-1} \times 10^{-4}$	$\Delta F°$ [c] kcal/mole	Reference [d]
10–21	Testosterone	BSA	ED	37.5	7.4	4 [a]	—	1.8 [b]	—6.0 [c]	BISCHOFF and STAUFFER (1954)
10–22	Testosterone	BSA [e]	PA	37.5	7.4	4	—	2.0	—6.1 [c]	SCHELLMAN et al. (1954)
10–23	Testosterone	HSA	ED	5	7.0	2.1	1.2	2.5	—5.6 [c]; —5.2	SANDBERG et al. (1957)
10–24	Testosterone	HSA	ED	20	7.4	—	—	2.5	—5.9 [c]	VERMEULEN and VERDONCK (1968 a)
10–25	Testosterone	HSA	SP; ED	25	7.6	2.3	2.4	5.4	—6.4 [c]; —6.0	WESTPHAL et al. [d] (1958)
10–26	Testosterone	HSA	ED	25	7.6	—	—	4.1 [b, s]	—6.3 [c]	WESTPHAL (1961)
10–27	Testosterone	HSA	GE	25	7.4	—	—	4.2	—6.3 [c]	PEARLMAN and CRÉPY (1967 b)
10–28	Testosterone	HSA	ED	37	7.4	—	—	2.0	—6.1 [c]	VERMEULEN and VERDONCK (1968 a)
11	17-Methyltestosterone	BSA	S	37	7.3	2.3 [a]	—	1.9 [b]	—6.1 [c]	EIK-NES et al. (1954)
12	11β-Hydroxyandrost-4-ene-3,17-dione	HSA [j]	ED	4	7.4	—	—	0.8 [b]	—5.0 [c]	WESTPHAL and HARDING (unpubl.)
13–1	Progesterone	HSA	ED	4	7.4	—	—	11 [b]	—6.4 [c]	WESTPHAL (1964)
13–2	Progesterone	HSA	ED	4	7.4	2.2	4.1	8.9	—6.3 [c]; —5.8	WESTPHAL (1966)

13–3	Progesterone	HSA [j]	ED	4	7.4	2.1	10.0	21.3	—6.7 [c] —6.3	Westphal (1966)
13–4	Progesterone	HSA [j]	ED	4	7.4	3.0	—	41	—7.1	Westphal and Harding (unpubl.)
13–5	Progesterone	HSA	ED	5	7.0	0.9	—	5.7	—6.0 [c]	Sandberg et al. (1957)
13–6	Progesterone	HSA	UC	20	6.0	—	—	2.6 [b]	—6.0 [c]	Scholtan et al. (1968)
13–7	Progesterone	HSA	ED	24	7.4	—	—	8.5 [b]	—6.7 [c]	Westphal (1964)
13–8	Progesterone	HSA	ED	25	7.6	—	—	8.9 [b]	—6.7 [c]	Westphal et al. (1958)
13–9	Progesterone	HSA	SP	25	7.6	2.6	2.2	5.8	—6.5 c —5.9	Westphal et al. [d] (1958)
13–10	Progesterone	HSA	ED	37	7.4	—	—	6.1 [b]	—6.8 [c]	Westphal (1964)
13–11	Progesterone	HSA [j]	ED	37	7.4	2.6	4.6	12.0	—7.2 [c] —6.6	Westphal and Harding (unpubl.)
13–12	Progesterone	HSA	ED	45	7.4	—	—	4.3 [b]	—6.7 [c]	Westphal (1964)
13–13	Progesterone	RtSA	ED	4	7.4	—	—	14.4 [b]	—6.5 [c]	Westphal and Harding (unpubl.)
13–14	Progesterone	RtSA	ED	37	7.4	—	—	5.4 [b]	—6.7 [c]	Westphal and Harding (unpubl.)
13–15	Progesterone	RbSA	ED	4	7.4	—	—	11.7 [b]	—6.4 [c]	Westphal and Harding (unpubl.)
13–16	Progesterone	RbSA	ED	37	7.4	—	—	3.8 [b]	—6.5 [c]	Westphal and Harding (unpubl.)
14	Δ^{16}-Dehydroprogesterone (Pregna-4,16-diene-3,20-dione)	HSA	ED	25	7.6	—	—	11.0 [b]	—6.9 [c]	Westphal and Ashley (1962)
15–1	Cortexone	BSA	S	37	7.3	1.2 [a]	—	0.65	—5.8 [c]	Eik-Nes et al. (1954)
15–2	Cortexone	HSA [j]	ED	4	7.4	3.3	5.6	19	—6.7 [c] —6.0	Westphal and Harding (unpubl.)

Footnotes see p. 114

Table VI-9 (continued)

No.	Steroid	Protein [r]	Method [m]	Temp. °C	pH	n or r [a]	k $M^{-1} \times 10^{-4}$	n k [b] $M^{-1} \times 10^{-4}$	$\Delta F°$ [e] kcal/mole	Reference [d]
15–3	Cortexone	HSA	ED	25	7.6	—	—	4.7 [b]	—6.4 [e]	Westphal et al. [d] (1958)
15–4	Cortexone	HSA	SP	25	7.6	2.0	1.8	3.6	—6.2 [e] —5.8	Westphal et al. [d] (1958)
16–1	Corticosterone	HSA	ED	5	7.0	2.1	0.43	0.9	—5.0 [e]	Sandberg et al. (1957)
16–2	Corticosterone	HSA	ED	5	7.0	1.0 [k] 20 [v]	0.36 [k] 0.012 [v]	0.6	—4.5 [k] —2.6 [v]	Slaunwhite et al. (1963)
17–1	Cortisone	BSA	ED	34	7.5	—	—	0.5 [b]	—5.2 [e]	Schellman et al. (1954)
17–2	Cortisone	BSA	S	37	7.3	1.1 [a]	—	0.18 [b]	—4.7 [e]	Eik-Nes et al. (1954)
17–3	Cortisone	HSA	ED	5	7.0	—	—	0.5 [b]	—4.7 [e]	Sandberg et al. (1957)
17–4	Cortisone	HSA	UC	20	6.0	—	—	0.25 [b]	—4.6 [e]	Scholtan et al. (1968)
18	Prednisone	HSA	UC	20	6.0	—	—	0.46 [b]	—4.9 [e]	Scholtan et al. (1968)
19–1	Cortisol	BSA	ED	4	7.4	—	—	0.24 [b]	—4.3 [e]	Daughaday (1956 b)
19–2	Cortisol	BSA	ED	25	7—8	~1	—	0.34 [b]	—4.8 [e]	Brunkhorst and Hess (1965)
19–3	Cortisol	BSA	ED	25	7—8	10—20	—	0.25 [b]	—4.6 [e]	Brunkhorst and Hess (1965)
19–4	Cortisol	HSA	ED	4	7.4	—	—	0.6 [b]	—4.8 [e]	Daughaday (1956 b)
19–5	Cortisol	HSA [j]	ED	4	7.4	—	—	1.0 [b]	—5.1 [e]	Westphal and Harding (unpubl.)
19–6	Cortisol	HSA	ED	5	7.0	—	—	0.5 [b]	—4.7 [e]	Sandberg et al. (1957)
19–7	Cortisol	HSA	UC	20	6.0	—	—	0.22 [b]	—4.5 [e]	Scholtan et al. (1968)

19–8	Cortisol	HSA	ED	25	7.6	~10	—	1.2 [b]	—5.5 [c]	WESTPHAL et al. (1958)
19–9	Cortisol	HSA	SP	25	7.6	—	—	1.2 [b]	—5.5 [c]	WESTPHAL et al. (1958)
19–10	Cortisol	HSA	ED	25	7—8	~1	—	0.49 [b]	—5.0 [c]	BRUNKHORST and HESS (1965)
19–11	Cortisol	HSA	ED	25	7—8	10—20	—	0.44 [b]	—5.0 [c]	BRUNKHORST and HESS (1965)
19–12	Cortisol	HSA	ED	25	7.4	5.3	—	0.41	—4.9 [c]	KRIPALANI and SORBY (1967)
19–13	Cortisol	HSA	ED	37	7.4	—	—	0.3	—4.9 [c]	WESTPHAL (1966)
20–1	Aldosterone	HSA	ED	4	7.6	—	—	0.26 [b]	—4.3 [c]	DAVIDSON et al. (1962)
20–2	Aldosterone	HSA	ED	37	7.6	—	—	0.17 [b]	—4.6 [c]	DAVIDSON et al. (1962)
21–1	Digitoxigenin	HSA	ED	1	7.4	1	3.9	4.6	—5.8	LUKAS and DE MARTINO (1969)
21–2	Digitoxigenin	HSA	UC	20	7.4	—	—	3.4 [b]	—6.4 [c]	SCHOLTAN et al. (1966)
21–3	Digitoxigenin	HSA	ED	37	7.4	1 [w]	3.4 [w]	4.1	—6.5	LUKAS and DE MARTINO (1969)
22–1	Digitoxin	HSA	UC	20	7.4	—	—	1.0 [b]	—5.7 [c]	SCHOLTAN et al. (1966)
22–2	Digitoxin	HSA	ED	1	7.4	1.1	6.2	6.7	—6.0	LUKAS and DE MARTINO (1969)
22–3	Digitoxin	HSA	ED	20	7.4	1.0	8.7	8.7	—6.6	LUKAS and DE MARTINO (1969)
22–4	Digitoxin	HSA	ED	37	7.4	0.9	11.4	10.7	—7.2	LUKAS and DE MARTINO (1969)
23	Gitaloxin	HSA	UC	20	7.4	—	—	0.8 [b]	—5.6 [c]	SCHOLTAN et al. (1966)
24	Digoxin	HSA	UC	20	7.4	—	—	0.7 [b]	—5.4 [c]	SCHOLTAN et al. (1966)
25	Convallatoxin	HSA	UC	20	7.4	—	—	0.25 [b]	—4.8 [c]	SCHOLTAN et al. (1966)
26	g-Strophanthidin	HSA	UC	20	7.4	—	—	0.14 [b]	—4.4 [c]	SCHOLTAN et al. (1966)
27	g-Strophanthin	HSA	UC	20	7.4	—	—	0.11 [b]	—4.3 [c]	SCHOLTAN et al. (1966)

Footnotes see p. 114

FOSTER (1959); comparative details on the latter and a modified charcoal method have been reported by SOGAMI and FOSTER (1968).

The evaluation of the progesterone-binding data by the method of reciprocal plots (Fig. VI-3) is in agreement with the assumption,

---

(Footnotes to Table VI-9)

[a] Figures marked [a] represent r values from solubility.
[b] Figures marked [b] represent $\bar{\nu}/[S]$ values. See Eq. (V-8).
[c] Calculated from k for each binding site, n, or for n k ($\bar{\nu}/[S]$) when marked [c].
[d] References marked [d] contain data used for calculation.
[e] BSA "Prep. B".
[f] BSA "Prep. A".
[g] BSA "Prep. C".
[h] BSA, Armour Lot 370295 A.
[i] BSA, Armour Lot 212112.
[j] Albumin preparation delipidated.
[k] n_1; k_1.
[m] Abbreviations for methods applied:

ED — equilibrium dialysis
GE — gel equilibration
PA — partition analysis
PE — paper strip electrophoresis
S — solubility
SP — spectrophotometry
T — turbidimetry
UC — ultracentrifugation
UF — ultrafiltration.

[n] Calculated for n k. $\Delta F°$ values for No. 10-1 and 10-2 given by ALFSEN (1963) as 6.1 and 6.4 kcal/mole, respectively.
[p] Calculated for n k. $\Delta F°$ values for No. 10-13 and 10-14 given by ALFSEN (1963) as 6.2 and 5.8 kcal/mole, respectively.
[r] BMA, bovine mercaptalbumin.
BSA, bovine serum albumin.
HSA, human serum albumin.
RbSA, rabbit serum albumin.
RtSA, rat serum albumin.
[s] $\bar{\nu}/[S] = 4.05 \pm 0.47 \times 10^4\ M^{-1}$, average value from 12 determinations.
[t] $\bar{\nu}/[S] = 2.92 \pm 0.09 \times 10^4\ M^{-1}$, average value from 6 determinations.
[u] $\bar{\nu}[S] = 2.45 \pm 0.18 \times 10^4\ M^{-1}$, average value from 8 determinations, calculated for half the MW of the mercury dimer.
[v] n_2; k_2.
[w] A second set of sites was observed, $n_2 = 2$, each having $k_2 = 3.5 \times 10^3\ M^{-1}$.

but does not prove, that competitive binding occurs between progesterone and the solvent-removable lipid. The difference in the binding curves of Fig. VI-3 may therefore be best interpreted as caused by lipid binding in the untreated HSA either to the progesterone binding site, or to a hydrophobic area adjacent to the binding site in such a way that the association with progesterone is sterically hindered.

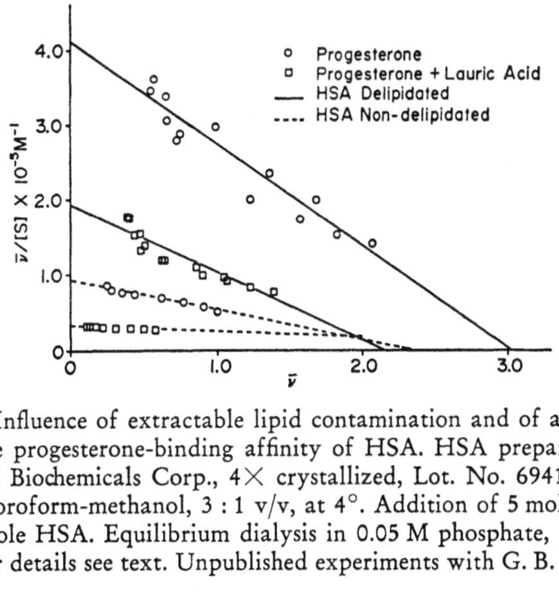

Fig. VI-4. Influence of extractable lipid contamination and of added lauric acid on the progesterone-binding affinity of HSA. HSA preparation from Nutritional Biochemicals Corp., 4× crystallized, Lot. No. 6941. Delipidation by chloroform-methanol, 3 : 1 v/v, at 4°. Addition of 5 moles of lauric acid per mole HSA. Equilibrium dialysis in 0.05 M phosphate, pH 7.4, 4°. For other details see text. Unpublished experiments with G. B. Harding

This interpretation was corroborated by recent studies (with G. B. Harding, unpublished) with the progesterone complexes of a 4-times crystallized HSA preparation. Fig. VI-4 shows that the complex formed with the untreated HSA (broken line, circles) has an n k value of $9 \times 10^4 M^{-1}$, and n is slightly over 2, in good agreement with the data in Table VI-9, No. 13-2. After removal of lipids by extraction with chloroform-methanol (3 : 1, v/v) at 4°, the HSA preparation interacted with progesterone at $n k = 4.1 \times 10^5 M^{-1}$, and n approximately 3 (solid line, circles). The HSA concentration was 1 mg/ml in all cases.

Addition of 5 moles of lauric acid per mole albumin to the binding system resulted in a decrease of the n k value of the progesterone-

HSA complex (solid line, squares) to about $2 \times 10^5\ M^{-1}$, accompanied by a reduction of n to the original value of about 2. Combination of the untreated HSA with the same amount of lauric acid yielded a progesterone complex (broken line, squares) of further reduced affinity, so that an n k value of about $3 \times 10^4\ M^{-1}$ was obtained. No change of n could be determined although the extrapolation over a relatively long distance makes this point uncertain. It should be added that the evaluation of these binding data for only one type of binding sites is a simplification; the slightly curved trend of the experimental points for the delipidated HSA may indicate this. However, the low solubility of progesterone in the aqueous system does not justify an analytical interpretation of the binding data for secondary binding sites. Similar results were obtained with myristic acid, which inhibited the affinity of the progesterone-HSA complex to about the same extent as lauric acid (see Figs. V-6 and V-8). The effect of palmitic acid was qualitatively similar, but not as marked. The interaction of these and other long-chain fatty acids with HSA has been studied by GOODMAN (1958).

PAGNINI et al. (1968) observed that the interaction of digitoxin and other cardiac-active steroids with BSA was inhibited by oleic acid and palmitic acid (GENAZZANI et al., 1968). Another good illustration of the inhibition of steroid-albumin interaction by long-chain fatty acids may be seen in the results of KOCH et al. (1969) on association of corticosterone with albumin in rat plasma. The authors found decreased albumin binding of corticosterone or progesterone in late pregnancy plasma in which the concentration of non-esterified fatty acids is increased. Raising the fatty acid level by lipase action inhibited corticosterone association with albumin.

These various observations on displacement of albumin-bound steroid hormone by fatty acids raise the question of drug interference with hormone-protein interaction in the blood (BRODIE, 1965). STENLAKE et al. (1968) have reported that acetylsalicylic acid, administered in recommended therapeutic doses for periods of one week, has no significant effect on plasma protein-binding of 11-hydroxycorticosteroids. However, when the corticosteroids were added *in vitro* to human plasma at levels exceeding the CBG capacity, acetylsalicylic acid produced an increase of the unbound 11-hydroxycorticosteroid fraction (STENLAKE et al., 1969). Phenylbutazone, ibufenac and indomethacin had no significant effect.

In-vivo studies in the rat by MAICKEL et al. (1969) showed that phenylbutazone and a number of other anti-inflammatory drugs increased the percentage of unbound corticosterone as measured by labeling of the endogenous steroid pool with tritiated corticosterone. Structural analogues of phenylbutazone and imidazolones, lacking anti-inflammatory properties, did not affect the ratio of unbound to bound corticosterone. An increase of the unbound fraction was also seen *in vitro* with phenylbutazone and salicylic acid, but only at higher concentrations of the drugs. The general problem of drug interaction with plasma proteins, including displacement effects on steroid hormones, has been discussed by KRIEGLSTEIN (1969).

c) Binding Parameters

A single binding site has been found for the interaction of estrone with BSA and with HSA (Table VI-9, No. 1). In contrast, the n values observed for the testosterone-BSA complex vary greatly from about 4 to 7, with values up to 20 reported. Since the affinity constants, n k, range over a much smaller limit, it is likely that one reason for the differences in the n values is the difficulty of interpreting the binding data; it has been pointed out that arbitrariness cannot be avoided in the extrapolation of the curves obtained by different methods. This has been emphasized by BRUNKHORST and HESS (1965). Approximately 2 binding sites have been observed for the HSA complex of testosterone (Table VI-9, No. 10-23, 10-25). An approximate value of $n = 2$ has been observed in our and in other laboratories for the HSA complexes of progesterone, cortexone and also in earlier studies for corticosterone; more recent work has been interpreted to indicate two sets of binding sites in HSA for corticosterone, i.e., $n_1 = 1$; $n_2 = 20$ (Table VI-9, No. 16-2). As pointed out in the preceding section, delipidation of HSA under mild conditions results in an increase of binding sites, as shown for progesterone, to about $n = 3$ (see Fig. VI-4).

Fairly wide ranges have been observed in the affinity constants and the number of binding sites for the albumin complexes with cortisol. A delipidated HSA preparation again gave a relatively high $\bar{\nu}/[S]$ value (Table VI-9, No. 19-5). BRUNKHORST and HESS (1965) have obtained important results on concentration dependency of the binding parameters. At constant protein concentration, the n k values decreased with increasing cortisol concentrations from 5.3 to 4.4×10^3 M^{-1} for the cortisol-HSA complex, and from 3.7 to $2.5 \times 10^3 M^{-1}$ for the cortisol-BSA complex (see also Table VI-9, No. 19-2, 19-3, 19-10, 19-11). The lowest total cortisol: albumin ratios inside the dialysis bags were less than 1 : 1000 in these experiments, the highest several thousand times greater. The Scatchard plot of these data (Fig. 1 in BRUNKHORST and HESS, 1965) shows a sharp increase of the affinity constant at the lowest cortisol concentrations indicating heterogeneity of binding sites. Any extrapolation of this end of the curve to the abscissa yields values of n far below unity, indicating that this heterogeneity of binding sites does not reside within the major portion of the albumin molecules. The curves for cortisol complexes of HSA and BSA have similar shapes indicating a similar type of heterogeneity for the albumins of both species; the $\bar{\nu}/[S]$ values for the interaction with HSA are higher than those for BSA.

One explanation of this phenomenon may be the presence of a macromolecular contamination in the albumin preparations with a higher binding affinity for cortisol. Another possible interpretation would be some type of microheterogeneity of the plasma albumins which is known in various forms (SCHULTZE and HEREMANS, 1966; MOORE and FOSTER, 1968). Such heterogeneity may be caused by polymerization (SAIFER and PALO, 1969); it is conceivable that polymeric forms have different binding affinities for steroids. Unpublished studies with G. B. HARDING show a consistently lower progesterone affinity for a HSA polymer of approximately tetrameric size. Heterogeneity may also be based on genuine differences in the primary structure of albumin molecules; this also could lead to variations in binding affinity. A convincing demonstration of true microheterogeneity in the primary structure of HSA has been given years ago by HUGHES (1950) who showed that only two thirds of the normal albumin can be transformed to the mercaptalbumin dimer. In agreement with this finding, two thirds of a sulfhydryl group has been determined as statistical average per molecule HSA or BSA (BENESCH and BENESCH, 1956).

WONG and FOSTER (1969 a, 1969 b) have used the membrane equilibrium salting-out method to investigate the microheterogeneity of BSA. By establishing solubility—ammonium sulfate concentration profiles, they excluded lipophilic contaminants and denaturation effects (differences in three-dimensional folding *per se)* as causing the microheterogeneity; variations in monomer—dimer content were not responsible either. Subfractions of different solubility had essentially the same sulfhydryl content (about 0.60 mole/mole). The authors conclude that differences in the microheterogeneous population are based on variations in the covalent structure.

It should be noted in this connection that the N and F forms which result from microheterogeneity of serum albumin (FOSTER et al., 1965; PETERSEN and FOSTER, 1965) differ significantly in their ability to bind hydrocarbons (WISHNIA and PINDER, 1964); the F conformation of BSA, which is produced around pH 4, binds only one-fourth as much butane and one-fifth as much pentane as native BSA. It is conceivable, therefore, that similar differences may occur among the components of microheterogeneous albumin systems for the binding of steroids. It will be pointed out below (Chapter XIII. 4) that in studies on steroid binding to AAG at low steroid concentrations observations have been made which resemble closely those observed with serum albumins.

d) Influence of Albumin Concentration

Most determinations of numbers of binding sites and association constants of steroid-protein complexes, including those described in the preceding paragraphs, have been done by varying the steroid level at a constant protein concentration. BRUNKHORST and HESS (1965) made the puzzling observation that determination of binding parameters resulted in different affinity constants when different protein concentrations were applied: analysis of cortisol-HSA complexes by the method of reciprocal plots gave n k values of 5.2, 4.6 and $3.7 \times 10^3\,M^{-1}$ at HSA concentrations of 0.3, 0.5 and 1.0%, respectively. These results cannot be explained by microheterogeneity and appear unrelated to the above observations of decreasing binding affinity with increasing concentration of the steroid. Additional studies in which the steroid concentration was held constant and the albumin concentration was varied showed similarly that the $\bar{\nu}/[S]$ values decreased with increasing concentration of BSA or HSA from 0.5 to 4.0%. Plotted according to Scatchard's method, parallel straight lines with positive slopes were obtained for the cortisol interaction with HSA and BSA; again, the general level of binding affinity was lower for BSA.

In view of the significance of these findings of BRUNKHORST and HESS, we have measured the binding affinity of cortisol to delipidated HSA at constant steroid concentration over a 100-fold range of protein concentration (KERKAY and WESTPHAL, 1969). Our experiments are in agreement with the above results. In addition, we have included studies in which protein concentration and steroid concentration were varied in parallel, so that the initial molar ratio of steroid: HSA was kept constant at 1 : 100 through the whole range of protein concentrations. The results (Table VI-10, columns 1-3) show that the $\bar{\nu}/[S]$ values at the lowest protein concentration of 0.1 mg/ml are more than 3 times those found at levels of 5.0 mg HSA per ml or higher. In contrast, the $\bar{\nu}/[S]$ values of the HSA complex with progesterone are practically not affected by the same change of HSA concentration under similar experimental conditions (Table VI-10, column 4).

No obvious explanation of this dependency of steroid-binding affinity on protein concentration is available. BRUNKHORST and HESS have suggested an interpretation on the basis of the changing ratios of steroid to protein; presence of competitive inhibitors in the protein might have explained the results. However, our observation that the same concentration dependency obtains when the ratio of steroid to HSA is kept constant, would eliminate that interpretation.

We will see below (Chapter XIII. 4) that the binding affinity between progesterone and AAG is also influenced by the protein concentration. However, in that case, the effect is the opposite; the $\bar{\nu}/[S]$ values become greater with increasing concentration of AAG, and the explanation may be given by the presence of inhibiting compounds. Table VI-10 shows further that the effect of protein concentration on steroid interaction with HSA is not the same for different steroids: it is strong for the pentapolar cortisol, but was not observed for the duopolar progesterone. It may be considered that increased protein-protein interaction between HSA molecules in the more concentrated solutions decreases binding of the hydrophilic cortisol but does not affect that of the more hydrophobic molecule of progesterone.

Table VI-10. *Effect of Protein Concentration on Apparent Affinity of Cortisol and Progesterone Complexes of HSA, measured by Equilibrium Dialysis at 4° in 0.05 M Phosphate, pH 7.4. The molar ratios (column 2) were calculated on the basis of total steroid and total HSA present in the dialysis system.* From KERKAY and WESTPHAL (1969)

HSA conc. mg/ml	Steroid : HSA	Cortisol $\bar{\nu}/[S]$ $M^{-1} \times 10^{-4}$	Progesterone $\bar{\nu}/[S]$ $M^{-1} \times 10^{-5}$
0.1	1 : 10	2.9	1.9
	1 : 100	1.9	1.9
0.5	1 : 50	1.4	2.1
	1 : 100	1.4	2.0
1.0	1 : 100	1.1	2.0
	1 : 100	1.1	2.0
2.0	1 : 200	0.8	2.0
	1 : 100	0.9	2.1
5.0	1 : 500	0.7	2.2
	1 : 100	0.6	2.0
10.0	1 : 1000	0.6	1.8
	1 : 100	0.5	1.8

e) Influence of Temperature

Although Table VI-9 contains data on albumin binding at different temperatures for several steroid hormones, the effect of temperature on the interactions has not been investigated extensively. In view of the inherently different binding properties of albumin preparations of varying purity and lipid content as used in different laboratories, it is not possible to use the data obtained by different investigators for calculation of temperature dependency. SLAUNWHITE et al. (1963) determined steroid binding to HSA by equilibrium dialysis at 5° and 25° and found a lower percentage of all steroids bound at the higher temperature. The difference was not significant for progesterone and testosterone, and only slight for estrone and estradiol; the binding affinity was greatly reduced at 25° for cortisol, and even more so for corticosterone. Only a few studies are available for testosterone and progesterone which can be used for calculation of thermodynamic parameters; these are compiled in Table VI-11.

It has been observed that the binding affinity between serum albumin and various ligands generally decreases with increasing tem-

Table VI-11. *Influence of temperature on steroid-albumin complexes*

Complex	No. in Table VI-9	Temp. °C	n k $M^{-1} \times 10^{-4}$	$\Delta F°$ kcal/mole	Temp. range for $\Delta H°$ and $\Delta S°$ °C	$\Delta H°$ kcal/mole	$\Delta S°$ cal $mole^{-1}$ deg^{-1}
Testosterone-BSA	10-5	8	1.7	—5.5	8—25	+3.0	+30
	10-7	25	2.3	—6.0	25—37.5	—2.1	+13
	10-22	37.5	2.0	—6.1	8—37.5	+1.0	+23
Testosterone-BSA	10-11	25	2.2	—5.9	25—37	—5.8	+ 0.2
	10-19	37	1.5	—5.9			
Testosterone-BSA	10-1	3	1.8	—5.4	3—25	+0.8	+22
	10-13	25	2.0	—5.9			
Progesterone-HSA	13-1	4	11.0	—6.4	4—24	—2.1	+15
	13-7	24	8.5	—6.7	24—37	—4.7	+ 7
	13-10	37	6.1	—6.8	37—45	—8.6	— 6
	13-12	45	4.3	—6.8	4—45	—4.0	+ 9
Progesterone-HSA [a]	13-3	4	10.0 [a]	—6.3 [a]	4—37	—4.0 [a]	+ 9 [a]
	13-11	37	4.6 [a]	—6.6 [a]			

[a] Delipidated HSA used; data calculated for one of the two binding sites present.

perature although the effect is small (KLOTZ and URQUHART, 1949). A similar relationship between temperature and apparent association constant has been established for the steroid complexes of CBG and AAG; with these two serum proteins, the increase of dissociation with rising temperature is very marked (see Chapters XI and XIII). In contrast, a slightly higher n k value has been determined for the testosterone-BSA complex at 25° as compared to 8° or 3° (Table VI-11). A small positive enthalpy change is obtained for this system, again in contradistinction to all other steroid-protein interactions analyzed so far, and also to most anion-protein associations studied. The negative Gibbs energy, indicating spontaneous formation of the testosterone-BSA complex, is thus essentially given by a relatively large positive entropy change, since

$$\Delta F^\circ = \Delta H^\circ - T\Delta S^\circ. \tag{VI-3}$$

A positive entropy change has also been found for the HSA-progesterone complex (Table VI-11) when the integration for the enthalpy change, ΔH°, is made over the whole temperature range studied; the enthalpy change is negative. The thermodynamic parameters of this complex are thus quite similar to those of anion-albumin complexes (KLOTZ and URQUHART, 1949). Enthalpy and entropy changes of the progesterone-HSA complex show considerable variation with temperature. The negative enthalpy change of the complex is highest at the highest temperature range, i.e., when the data are integrated between 37° and 45°. Similarly, highest negative enthalpy changes were found for the testosterone-BSA complex at the higher temperature range studied, i.e., 25—37° (Table VI-11). It has been ascertained that the reduced association constant at 45° (Table VI-11, No. 13-12) is not the result of denaturation but that the effect is reversible; full restoration of the binding affinity was observed when the albumin solution was brought back to 4°. These results, although approximate in the precision of the individual values, show that the interaction itself is favored at the higher temperatures so that the heat content of the system decreases. This may be an indication of the conformational adaptability of the albumin molecule to the interacting steroid.

The positive entropy changes observed in albumin binding of testosterone and progesterone require explanation. A simple association between the components,

$$S + P \rightarrow SP \qquad \text{(VI-4)}$$

should reduce the degrees of freedom of the system and thus result in a negative entropy change. KLOTZ in studies on anion binding to serum albumin (KLOTZ, 1950) has encountered positive entropies; they are explained by displacement of the structured water molecules bound to the interacting ions, and of the hydration water associated with the protein molecule. Liberation of water molecules from a highly ordered or ice-like state (FRANK, 1965) would greatly increase their entropy since the absolute entropy of ice and liquid water is 9.8 and 15 cal/mole/degree, respectively (KLOTZ, 1967).

In analogy to KLOTZ' interpretation of the positive entropy changes in dye anion-albumin interaction (KLOTZ and URQUHART, 1949), it may be assumed that hydration water is displaced from the albumin as well as from the steroid, so that (VI-4) becomes

$$S\,(H_2O)_{a+x} + P\,(H_2O)_{b+y} \rightarrow (H_2O)_a\,S - P\,(H_2O)_b + (x+y)\,H_2O\,. \qquad \text{(VI-5)}$$

Part of the water molecules are thus removed from their ordered state to form free solvent molecules, thus increasing the degrees of freedom of the system. This mechanism may in part explain the positive ΔS° values for the steroid-albumin complexes listed in Table VI-11. The water molecules involved in the randomization may originate from protein side chains (KLOTZ, 1965) which participate in the steroid interaction, and may include part of the polypeptide backbone structure (WARNER, 1965). It may also be assumed that the steroids, hydrated in aqueous solution, will lose part of this hydration water by the interaction with the protein. On the basis of molecular models, WARNER (1965) has suggested a highly ordered hexagonal structure for the water molecules hydrating a steroid such as cortisol (Fig. VI-5). Positive entropy change may be expected from the randomization of these various types of structured water molecules during formation of the steroid-albumin complex.

Since the interpretation of the positive entropy changes according to KLOTZ has greater validity for ionic interactions than for binding of uncharged molecules, an additional explanation should be considered (SCHELLMAN et al., 1954). KARUSH has emphasized the unique property of serum albumin to exist in many molecular configurations of approximately equal energy (KARUSH, 1950). This conformational adaptability permits complex formation with molecules of very different structure. It implies a large conformational entropy for the dissolved protein. Very large positive entropy changes occur in proteins on interaction with small molecules such as alcohol or phenobarbital (MCELROY, 1943). The entropy changes produced in steroid binding may result from a very local relaxation between coiled or otherwise ordered polypeptide sequences; for example, it has been suggested that the steroid molecules break or bridge ammonium-

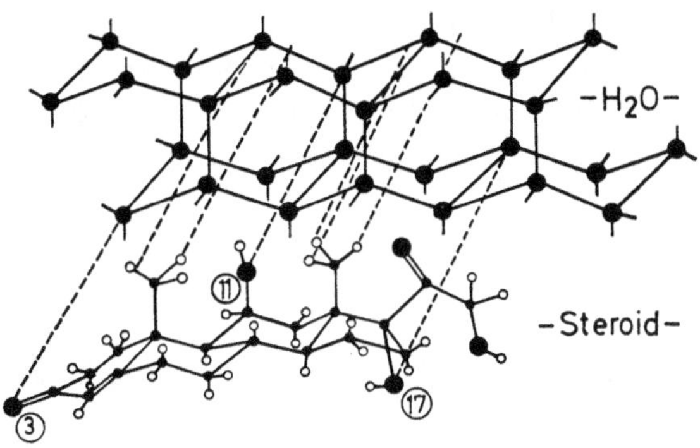

Fig. VI-5. Structured water molecules hydrating cortisol. From WARNER (1965)

phenolic cross links to reduce local strain and thus increase entropy with very small enthalpy change (SCHELLMAN et al., 1954). Conformational changes in the BSA molecule by interaction with testosterone have been demonstrated by ALFSEN (1963).

The temperature dependency of the negative enthalpy change of the progesterone-HSA complex (Table VI-11) indicates a greater decrease in heat content at higher temperatures than at 4°. Integration between 37 and 45° leads to $\Delta H^\circ = -8.6$ kcal/mole, a value higher than the free energy $-\Delta F^\circ$. Apparently, the stabilization of the system by complex formation is greater at this temperature, possibly due to an increased conformational adaptability of HSA. According to Eq. (VI-3), the entropy change would thus become negative as additional indication of increased order by the complex formation.

Another reason for the entropy change being less positive at the higher temperature is seen in the effect on the water of hydration.

It is known that increased temperature results in displacement and subsequent randomization of the water molecules present as hydration water in an ordered structure (KLOTZ, 1965). It follows as a qualitative explanation that at the higher temperature a smaller number of ordered water molecules are present on the protein and steroid surfaces which can be removed and randomized; their contribution to a positive entropy change is therefore smaller than at 4°. A similar interpretation can be given for the thermodynamic data of the testosterone-BSA complex (Table VI-11, No. 10-5, 10-7 and 10-22) which are qualitatively similar to those of the progesterone-HSA complex. The thermodynamic data calculated from experiments in other laboratories for the testosterone-BSA complex (Table VI-11, No. 10-11 and 10-19; 10-1 and 10-13) are in qualitative agreement with the above interpretation.

f) Influence of pH and Ionic Strength

A careful study of the pH dependence of anion binding to serum albumin has been carried out by KARUSH (1951). A slight increase was observed in the association of methyl orange to BSA between pH 6.4 and 7.6 although the opposite might have been expected from the increase in negative charges on the protein. Correspondingly, the decrease in binding affinity above pH 7.6 was less than would be estimated as a result of the increase in net negative charge of BSA. KLOTZ (1953) has made similar observations with BSA and HSA over the pH region of 6 to 9 and has interpreted the increased binding to result from conformational changes which expose new or auxiliary binding sites (KLOTZ et al., 1952) as shown by spectrophotometric studies (KLOTZ and WALKER, 1947 a). These observations are not limited to anion binding but have been made also with neutral organic molecules such as aminoazobenzene for which binding to BSA and HSA was found to increase between pH 7 and 9 (KLOTZ and AYERS, 1952). The postulated unfolding of the albumin molecule could effectively change both n and k of the binding system, and thus explain the pH dependence observed.

In comparison with the extensive studies on the pH dependence of anion binding to albumin little work has been devoted to the influence of pH changes on steroid-protein interactions. Early studies of BI-

SCHOFF and co-workers (BISCHOFF and PILHORN, 1948) have shown that the dispersing power of BSA solutions for testosterone and progesterone increases when the pH is raised from 5.3 to 8.0; the dispersion (r) of testosterone enhanced from 3 to 4 moles/mole BSA when the pH was changed from 5.2 to 7.4 (BISCHOFF and STAUFFER, 1954). A very similar increase of the solubility of testosterone in 1% BSA

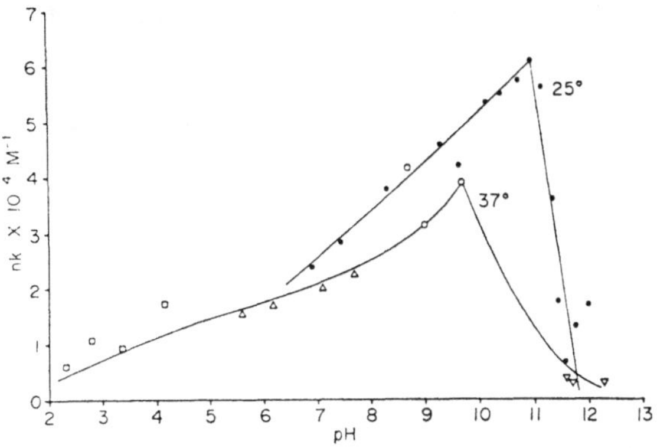

Fig. VI-6. Apparent association constants (n k) of testosterone-BSA complexes at different pH values. Curve 25°, BSA, Lot No. 370295 A; unbuffered system. Curve 37°, BSA, Lot No. 212112; ⊡ Sörensen phosphate; △ protein titrated with HCl; ⊙ phosphate; ▽ protein titrated with NaOH. Each point obtained with 5—6 steroid concentrations by method of reciprocal plots. Graph prepared from values in LEVEDAHL and PERLMUTTER (1956)

solution was observed between pH 6.5 and 8.5 (SCHELLMAN et al., 1954). Within this pH range, the hydrogen ion concentration did not appreciably influence the solubility of the two hormones in aqueous protein-free solution (BISCHOFF and PILHORN, 1948). LEVEDAHL and PERLMUTTER (1956) found a progressive increase of the apparent association constant of the testosterone-BSA complex between pH 2.3 and 11 (Fig. VI-6). The number of binding sites, n, was in general agreement with that observed in other laboratories; the logarithm of the dissociation constant was a linear function of pH. The binding decreased abruptly above pH 11 which may be attributed to denaturation at pH 11.5 and above. SANDBERG et al. (1957) observed a similar strengthening of the association of estrone and progesterone with

HSA between pH 4 and 9; the binding of estrone seemed to be somewhat reduced at pH 11.

LEVEDAHL and PERLMUTTER (1956) have suggested on the basis of the pH dependency of testosterone-BSA interaction that tyrosine groups may be involved in the binding mechanism. Previous results from the same laboratory have shown (LEVEDAHL and BERNSTEIN, 1954) that acetylation of 48% or 91% of the free amino groups of BSA did not affect the binding affinity to testosterone. This led to the assumption that the ε-amino groups of lysine are not involved in the interaction.

The influence of ionic strength on steroid-albumin interaction has received but little attention. A decrease of steroid solubility has been observed in aqueous protein-free buffer solution with increasing NaCl or Na_2SO_4 concentrations for estradiol, progesterone and testosterone at 25° and 37.5° (BISCHOFF and PILHORN, 1948), and with increasing ionic strength (phosphate) for testosterone at 25° (SCHELLMAN et al., 1954). Doubling the ionic strength of the buffer resulted in decreased binding of estrone to HSA (SANDBERG et al., 1957). Recent observations (GANGULY and WESTPHAL, 1968 a) indicate an increase of the apparent association constant of the progesterone-HSA complex from $1.0 \times 10^5\,M^{-1}$ to $2.1 \times 10^5\,M^{-1}$ when the NaCl concentration of the equilibrium dialysis system is changed from 0 to 4 M. A lesser enhancement of the binding affinity in the presence of sodium chloride solution has been found (unpublished experiments) for the cortisol-HSA complex at 4°; the $\bar{\nu}/[S]$ value increases from $0.9 \times 10^4\,M^{-1}$ in the absence of NaCl to $1.2 \times 10^4\,M^{-1}$ at 4 M NaCl concentration. These effects of neutral salts on HSA-steroid interaction are much less marked than those observed for the complexes of AAG with progesterone and other steroids. This phenomenon will be discussed in Chapter XIII. 5.

g) Competition between Steroids in HSA Interaction

The question of whether the binding of a steroid to HSA is affected by the presence of another steroid has been studied by equilibrium dialysis in the laboratory of SANDBERG and SLAUNWHITE (SANDBERG et al., 1957). Such an influence might be apparent by competitive binding to the same binding sites, by non-competitive inhibition of the interaction mediated at allosteric sites, or by other effects on the conformational structure of HSA which would lead to changes in the binding affinity for the original steroid. SANDBERG et al. (1957) determined the reciprocal binding affinity of several radiolabeled steroid hormones (progesterone, testosterone, corticosterone, estrone) in the absence and in the presence of saturating concentrations of a second steroid. No interference was observed between the steroid hormones studied. Later experiments (SLAUNWHITE et al., 1963) included estradiol and estriol which also failed to interfere with the binding of testosterone and progesterone. However, steroid solubility may have been a limiting factor in these experiments; the more soluble cortisol produced about 65% inhibition of corticosterone binding, and *vice versa*. Although these results on simultaneous interaction of more than one steroid with HSA make it unlikely that competition for the same binding sites takes place between the steroids tested, such mechanism would not be excluded for HSA association with cortisol and corticosterone.

The problem of competition for the same binding sites was recently investigated for the HSA complexes of progesterone and cortexone (HARDING and WESTPHAL, unpublished). This system was chosen because of the close structural relationship between these two steroids which differ only by the presence of a hydroxyl group or hydrogen at C-21. The experiment was performed by equilibrium dialysis at 4° with delipidated crystalline HSA, progesterone-4-^{14}C and cortexone-1,2-^{3}H. The bound and unbound fractions of both steroids were determined over a range of progesterone concentrations; the cortexone concentration was kept constant, as was the concentration of HSA. The results are plotted in Fig. VI-7 as bound against unbound steroid concentration. Curve A shows the progesterone-HSA interaction in the absence of a second steroid. Curve B indicates progesterone binding to HSA for the same set of progesterone concentrations as used for curve A, but with cortexone present. Clearly, the presence of this 21-hydroxy progesterone inhibits the binding of progesterone to HSA. The cluster of points, D, shows that the interaction of cortexone with HSA in the presence of progesterone is inversely related to the total concentration of progesterone; the numbers 1—9 indicate the increasing progesterone concentrations. This would imply increasing amounts of unbound cortexone with rising progesterone levels as apparent from curve C which gives the unbound cortexone

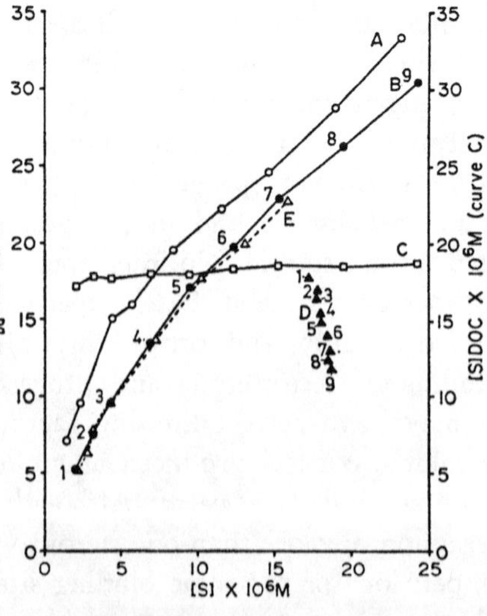

Fig. VI-7. Inhibition of progesterone-HSA interaction by constant amount of cortexone (DOC). Phosphate buffer of pH 7.4; 4°. HSA, 1.01 mg/ml (1.46×10^{-5} M) inside bag; 4-^{14}C-progesterone varied from 1.33 μg/ml (4.24×10^{-6} M) to 12 μg/ml (3.82×10^{-5} M); 1,2-^{3}H-cortexone, 8 μg/ml (2.42×10^{-5} M). Steroid concentrations calculated for total volume of dialysis system at the start of the experiment. Abscissa: [S] for progesterone (curves A, B, C), for DOC (curve E, cluster D). Left Ordinate: [S_{bd}] for progesterone (curves A, B); for DOC (curve E, cluster D). Right Ordinate: [S] unbound for DOC (curve C). The numbers given for cluster D correlate the values with the corresponding concentrations of bound and unbound progesterone, indicated by identical numbers in curve B. ○ A Progesterone alone; ● B Progesterone in presence of DOC; □ C DOC unbound in presence of progesterone; ▲ D DOC bound and unbound in presence of progesterone; △ E DOC alone

in the presence of the different concentrations of unbound progesterone shown on the abscissa. For comparison, curve E illustrates the binding of cortexone alone to the same HSA preparation; the HSA concentration and the different cortexone concentrations are equal to those used for progesterone in the experiments summarized by curves A and B.

For a quantitative evaluation of the multiple binding system, the experimental data of Fig. VI-7 were plotted according to SCATCHARD as seen in Fig. VI-8. This method permits one to obtain n and k values as given in the legend. The number of primary binding sites for progesterone and cortexone is approximately n = 3. This follows from a simplified evaluation, assuming only the first member of Eq. (V-18), i.e., one set of binding sites. The rationale for this is the same as that given for Fig. VI-4. It is evident that the presence of cortexone reduces n k to about half its original value; conversely, the interaction between cortexone and HSA decreases with increasing levels of progesterone in the system (cluster D).

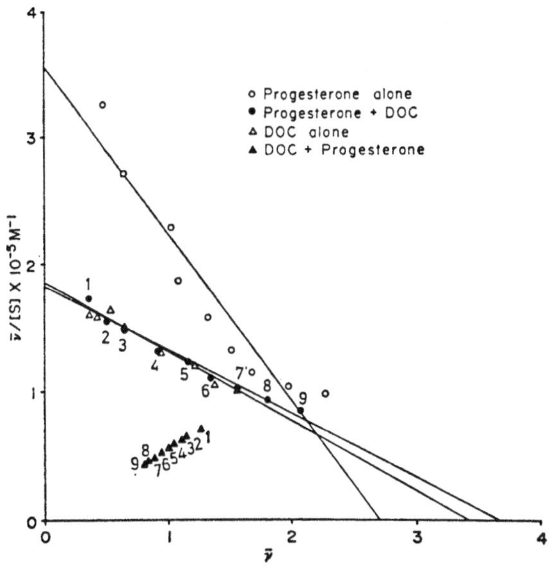

Fig. VI-8. Inhibition of progesterone-HSA interaction by constant amount of cortexone (DOC). Scatchard plots of binding data described in Fig. VI-7. ○ Progesterone alone; n = 2.7; n k = 3.6×10^5 M^{-1}; ● Progesterone in presence of DOC; n = 3.7; n k = 1.8×10^5 M^{-1}; ▲ DOC, in presence of varying progesterone concentrations identified in curve ● by identical numbers; △ DOC alone; n = 3.4; n k = 1.8×10^5 M^{-1}

In an effort to characterize the inhibition of HSA binding of progesterone by cortexone, the data were analyzed for competitive interaction by the Eqs. (V-29) and (V-30). Table VI-12 shows a comparison between the observed and calculated values of $\bar{\nu}$ for progesterone, and of the ratio $\bar{\nu}$(progesterone)/$\bar{\nu}$(cortexone). There is reasonable agreement between the experimental values and those calculated according to the formulas which were derived for com-

Table VI-12. *Competition of cortexone (DOC) for the progesterone-binding sites of HSA. Calculated from data given in Fig. VI-7. The first column shows the number of experiment in accordance with Fig. VI-7*

Expt. No.	$\bar{\nu}$ for progesterone		$\bar{\nu}_{prog}/\bar{\nu}_{DOC}$	
	observed	calculated by Eq. (V-29)	observed	calculated by Eq. (V-30)
1	0.36	0.33	0.30	0.24
2	0.51	0.48	0.44	0.37
3	0.66	0.61	0.59	0.50
4	0.92	0.86	0.88	0.78
5	1.17	1.06	1.16	1.06
6	1.36	1.24	1.42	1.35
7	1.58	1.39	1.78	1.64
8	1.81	1.58	2.13	2.11
9	2.09	1.74	2.61	2.59

petitive binding. This result would indicate that progesterone and the structurally closely related cortexone interact at the same binding sites in the HSA molecule.

Considering the almost omnipotent ability of serum albumin to interact with small ligands, studies of steroid competition for HSA binding sites may not contribute significantly to our knowledge on steroid behavior in biological binding systems. Somewhat more specific steroid association takes place with AAG; observations on competitive binding of steroids to specific binding sites in that glycoprotein will be discussed in Chapter XIII. 4. Nevertheless, the albumin molecule has served as a valuable matrix in the study of many parameters characteristic for the interaction between steroids and proteins. Such investigations may gain new impetus after future decisive progress in our knowledge of the chemical structure of the albumin molecule; exploration of the binding of a steroid to a defined polypeptide chain, and attempts to enhance the binding affinity by chemical modification remain challenging possibilities.

VII. Influence of Molecular Structure on Steroid Interaction with HSA and Other Proteins

1. Steroid Structure and Binding to Serum Albumin; Polarity Rule

One of the early results of studies on steroid-albumin interaction was the observation that the binding strength was inversely related to the number of polar groups in the steroid molecule (EIK-NES et al., 1954). This finding is the basis of the "polarity rule" (see Chapter V. 1) which has since been confirmed in many instances (WESTPHAL, 1961); the subsequent pages provide additional evidence for the general validity of this concept which results from the hydrophobic nature of the bond. The polarity rule is not unique for the interaction between steroids and albumin; a similar relationship has been observed for example for the binding affinity of various anions to albumin (WESTPHAL et al., 1953). More recent confirmation has been given by SCHOLTAN (1968).

Validity of the polarity rule has been previously seen (Table VI-1) in the interaction of bile acids with HSA. Table VII-1 shows the decrease of the free energy of binding with entrance of hydroxy groups into the progesterone molecule. Testosterone is included in this group of steroids since it may be considered derived from progesterone by replacement of the acetyl side chain at C-17 with a hydroxy group. The table also indicates that the results obtained with three different experimental methods are in agreement. The last column shows the values for depression of ultraviolet absorbance, $-\Delta\varepsilon$, in the presence of HSA.

For a more extensive study of the relationship between the chemical structure of a steroid and its binding affinity to a given protein, a method had to be applied which was more convenient than equilibrium dialysis for handling larger numbers of different steroids, and which could be used with unlabeled steroids (WESTPHAL and ASHLEY, 1958 a, 1959, 1962). The spectrophotometric procedure (Chapter

IV. 13) met these conditions; the $\Delta\varepsilon$ values had been found to be proportional to the binding affinity of the steroid-protein complex and a comparison of the spectrometric method with the equilibrium dialysis technique had shown its general validity. The limitation of the spectrophotometric procedure is given by the requirement of the presence of the Δ^4-3-keto grouping as a strong chromophor. Fortunately, this group is present in all natural hormones with the excep-

Table VII-1. *Approximate free energy of binding to HSA for steroids of different polarities, and depression of ultraviolet absorbance, $\Delta\varepsilon$, in presence of HSA. $\Delta F°$ in kcal per mole; pH 7.0—7.5. Figures in parentheses indicate references given as footnotes*

Method	Equilibrium Dialysis	Ultrafiltration	Equilibrium Dialysis	Spectrophoto-metry	—$\Delta\varepsilon$ %
Steroid	4°	5°	25°	25°	25°
Progesterone	—6.3 (1)	—6.1 (2)	—6.7 (3)	—6.5 (4)	13.9
Testosterone	—5.8 (2)	—5.8 (2)	—6.4 (4)	—6.4 (4)	13.2
Cortexone	—	—	—6.4 (4)	—6.2 (4)	12.6
Corticosterone	—5.2 (2)	—5.2 (2)	—	—	5.7
Cortisol	—4.8 (5)	—4.8 (2)	—5.5 (4)	—5.5 (4)	5.7

(1) Westphal (1966). (2) Sandberg et al. (1957). (3) Westphal (1964). (4) Westphal et al. (1958). (5) Daughaday (1956 b).

tion of the estrogens; many steroid derivatives are known which have the same α,β-unsaturated keto grouping in ring A.

Our first objective was to investigate the influence of polar, i.e., electron attracting, substituents. Fig. VII-1 illustrates the principle of these studies and their evaluation. Introduction of an 11β-OH-group into the three Δ^4-3-ketosteroids listed results in a decrease of the absolute value of $\Delta\varepsilon$, for the interaction with HSA as well as that with BLG. This is evident from a horizontal comparison. A comparison of the $\Delta\varepsilon$ values in vertical direction shows similar depressions by entrance of a 21-OH and a 17α-OH group. The change of $\Delta\varepsilon$ as a result of the structural alteration may be calculated, for instance for progesterone and 11β-hydroxyprogesterone in Fig. VII-1, as —13.9—(—6.6) = —7.3; the negative sign of the change indicates weakening of the binding affinity by the entrance of the hydroxy group.

Some sixty Δ^4-3-ketosteroids have been subjected to the spectrophotometric procedure to determine the strength of interaction with HSA and, in most cases, with BLG. The results and pertinent optical parameters have been compiled in Table VII-2; formulas are given in Fig. VII-2. The absolute $\Delta\varepsilon$ values for interaction of the steroids

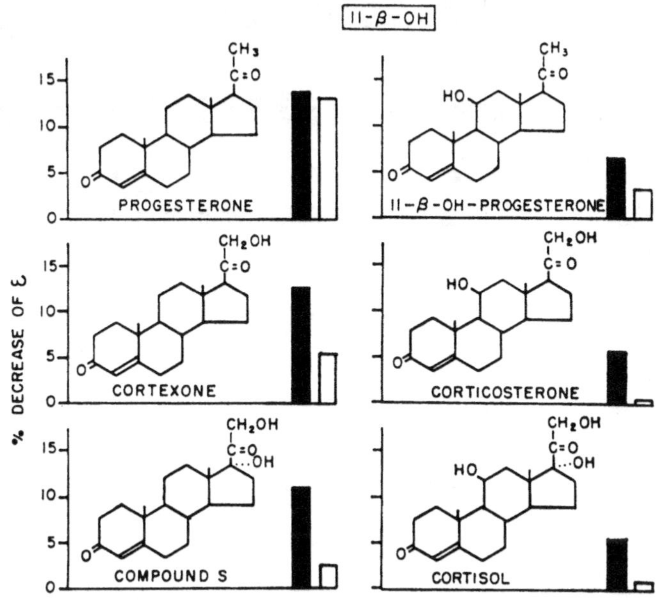

Fig. VII-1. Change of $\Delta\varepsilon$ of steroid-HSA interaction by introduction of hydroxy groups. Steroid concentration, 2×10^{-5} M; HSA concentration, 4×10^{-5} M $=0.276\%$; BLG concentration, 0.276%. Phosphate buffer pH 7.6, $\mu=0.1$; 25°. Full bars, HSA; open bars, BLG. From Westphal (1961)

with HSA are, with only a few exceptions, greater than those for binding to BLG. Two examples are given (No. 67, 68 in Table VII-2) to show that interaction with these proteins, and its measurement by the spectrophotometric procedure, is not limited to steroid structures; this is illustrated in Fig. VII-3 for the two unrelated ketones containing a conjugated double bond.

By pairing the steroids listed in Table VII-2 in such a way that the two partners of a pair are identical except for one substituent or functional grouping, the effect of this substituent or grouping on

Table VII-2. *Depression of ultraviolet absorbance, -$\Delta\varepsilon$, of Δ^4-3-ketosteroids by interaction with human serum albumin (HSA) and β-lactoglobulin (BLG)*[a]. From WESTPHAL and ASHLEY (1958 a, 1959, 1962)

No.	Steroid	Formula (Fig. VII-2)	R_1	R_2	R_3	R_4	$\lambda_{max}^{H_2O}$ nm	ε_W	-$\Delta\varepsilon$ in presence of HSA %	BLG %
1	Progesterone	I	H	H	H	H	249	16,700	13.9	13.3
2	11α-Hydroxyprogesterone	I	α-OH	H	H	H	248	15,600	2.9	0.0
3	11β-Hydroxyprogesterone	I	β-OH	H	H	H	248	16,000	6.6	3.3
4	17α-Hydroxyprogesterone	I	H	OH	H	H	248	15,700	12.8	4.2
5	11-Deoxycorticosterone = cortexone	I	H	H	OH	H	249	16,300	12.6	5.4
6	Cortexone acetate	I	H	H	OAc	H	248	16,400	11.6	6.3
7	2-Methylcortexone	I	H	H	OH	CH_3	248	15,200	17.6	8.9
8	Corticosterone	I	β-OH	H	OH	H	248	15,800	5.7	0.3
9	Epicorticosterone	I	α-OH	H	OH	H	248	15,400	4.4	3.3
10	11-Dehydrocorticosterone acetate	I	=O	H	OAc	H	244	16,000	11.7	8.6
11	17α-Hydroxycortexone	I	H	OH	OH	H	248	16,600	11.0	2.6
12	17α-Hydroxycortexone acetate	I	H	OH	OAc	H	248	15,200	15.1	11.4
13	1-Dehydro-17α-hydroxycortexone	II	H	H	H	OH	250	16,100	4.5	2.3
14	17α,20β,21-Trihydroxypregn-4-en-3-one	III	—	—	—	—	249	16,000	9.7	4.0
15	17β,20,21-Trihydroxypregn-4-en-3-one	III	—	—	—	—	249	15,800	7.5	3.0
16	Cortisone	I	=O	OH	OH	H	244	16,000	7.3	2.9
17	Cortisone acetate	I	=O	OH	OAc	H	245	16,200	14.8	4.8
18	Δ^1-Dehydrocortisone = prednisone	II	=O	H	H	OH	244	15,800	7.9	3.9

19	2-Methylcortisone	I	=O	OH	OH	CH_3	244	14,700	13.5	2.3
20	Cortisol	I	β-OH	OH	OH	H	248	15,900	5.7	0.9
21	Cortisol-21-acetate	I	β-OH	OH	OAc	H	248	15,800	6.7	0.5
22	Cortisol-21-propionate	I	β-OH	OH	OPr	H	248	16,500	6.3	1.8
23	Δ^1-Dehydrocortisol = prednisolone	II	β-OH	H	H	OH	248	15,000	3.5	0.5
24	2-Methylcortisol	I	β-OH	OH	OH	CH_3	248	15,300	9.2	3.1
25	Δ^1-Dehydro-6-methylcortisol = Medrol	II	β-OH	CH_3	H	OH	248	14,600	4.2	2.3
26	9α-Fluorocortisol-21-acetate	I [b]	β-OH	OH	OAc	H	244	16,900	4.1	2.7
27	2-Methyl-9α-fluorocortisol	I [b]	β-OH	OH	OH	CH_3	243	15,700	11.6	13.7
28	Δ^4-Androstenedione-3,17	IV	H	=O	—	H	248	16,300	12.7	10.1
29	Testosterone	IV	H	OH	H	H	249 [c]	15,900 [c]	13.2	3.5
30	2-Methyltestosterone	IV	H	OH	H	CH_3	248	15,000	16.5	6.5
31	17-Methyltestosterone	IV	H	OH	CH_3	H	249	16,000	12.5	1.5
32	11α-Hydroxy-17-methyltestosterone	IV	α-OH	OH	CH_3	H	248	15,000	4.7	1.4
33	11β-Hydroxy-17-methyltestosterone	IV	β-OH	OH	CH_3	H	248	15,600	5.5	0.5
34	Adrenosterone	IV	=O	=O	—	H	244	14,800	9.2	6.5
35	Androsta-1,4-diene-3,11,17-trione	VI	—	—	—	—	243	15,100	7.1	2.2
36	11β-Hydroxyandrosta-1,4-diene-3,17-dione	VI	—	—	—	—	246	14,500	1.4	3.7
37	4-Methyltestosterone	V	H	CH_3	CH_3	H	259	15,300	16.0	6.7
38	6α-Methyltestosterone	V	H	CH_3	H	α-CH_3	249	14,900	13.4	7.5
39	6β-Methyltestosterone	V	H	CH_3	H	β-CH_3	250	15,500	12.7	4.6
40	6α,17-Dimethyltestosterone	V	CH_3	CH_3	H	α-CH_3	250	15,400	15.4	6.0
41	6β,17-Dimethyltestosterone	V	CH_3	CH_3	H	β-CH_3	250	16,400	12.4	5.2
42	19-Nortestosterone	V	H	H	H	H	248	16,200	13.6	3.6
43	4-Methyl-19-nortestosterone	V	H	H	CH_3	H	259	15,900	18.2	5.0
44	6α-Methylandrost-4-ene-3,17-dione	VI	—	—	—	—	248	15,800	9.4	2.3

Footnotes see p. 139

Table VII-2 (continued)

No.	Steroid	Formula (Fig. VII-2)	R_1	R_2	R_3	R_4	$\lambda_{max}^{H_2O}$ nm	ε_W	$-\Delta\varepsilon$ in presence of HSA %	BLG %
45	6β-Methylandrost-4-ene-3,17-dione	VI	—	—	—	—	249	16,200	7.2	2.1
46	9α-Chlorocortisol-21-acetate	I [b]	β-OH	OH	OAc	H	247	17,100	4.7	0.0
47	9α-Bromocortisol-21-acetate	I [b]	β-OH	OH	OAc	H	248	15,300	5.2	0.0
48	3-(3-Oxo-17β-hydroxyandrost-4-en-17α-yl) propionic acid γ-lactone (SC-5233) [d]	VI	—	—	—	—	248	16,300	11.8	4.1
49	3-(3-Oxo-17β-hydroxy-19-nor-androst-4-en-17α-yl)-propionic acid γ-lactone (SC-8109) [d]	VI	—	—	—	—	248	17,100	13.9	4.6
50	6,17α-Dimethyl-17β-hydroxy-androsta-4,6-dien-3-one	VI	—	—	—	—	299	22,900	6.0	—
51	6-Dehydroprogesterone (pregna-4,6-diene-3,20-dione)	III	—	—	—	—	293	25,100 [e]	11.2	—
52	9(11)-Dehydroprogesterone (pregna-4,9(11)-diene-3,20-dione)	III	—	—	—	—	246	17,700	16.1	—
53	16-Dehydroprogesterone (pregna-4,16-diene-3,20-dione)	III	—	—	—	—	248	24,100	0.3	2.3
54	6α-Methyl-17-hydroxypregn-4-ene-3,20-dione	III	—	—	—	—	248	15,500	10.8	—
55	6-Methyl-17-hydroxypregna-4,6-diene-3,20-dione	III	—	—	—	—	298	22,300	3.7	—

56	6-Methyl-17-acetoxypregna-4,6-diene-3,20-dione	III	—	—	—	—	298	21,800	3.5	—
57	16α-Methylcortisone	I	=O	OH	OH	H	244	15,400 [e]	14.8	—
58	1-Dehydro-16α-methylcortisone (=16α-methylprednisone)	II	=O	H	α-CH_3	OH	244	15,400 [e]	9.8	—
59	1-Dehydro-16β-methylcortisone 21-acetate (=16β-methylprednisone 21-acetate)	II	=O	H	β-CH_3	OAc	243	14,200 [c]	9.9	—
60	16α-Methylcortisol	I	β-OH	OH	OH	H	247	15,700	6.9	—
61	16α-Methylcortisol 21-acetate	I	β-OH	OH	OAc	H	247	16,900 [e]	5.2	—
62	3-(3-Oxo-9α-fluoro-11β,17β-dihydroxyandrost-4-en-17α-yl) proprionic acid γ-lactone (SC-9837) [d]	VI	—	—	—	—	243	16,100 [e]	12.2	—
63	3-(3-Oxo-7α-acetylthio-17β-hydroxyandrost-4-en-17α-yl) propionic acid γ-lactone (SC-9420) [d] (=aldactone)	VI	—	—	—	—	243	20,200 [e]	11.7	—
64	4-Methyl-19-nortestosterone 17-acetate	V	H	H	CH_3	H	258	16,300	13.5	14.7
65	Androsta-3,5-dien-3β-ol-17-one glucosiduronate	VI	—	—	—	—	239	17,400	20.9	25.0
66	1-Dehydro-6α-fluoro-16α-methylcortisol 21-acetate	II	β-OH	α-F	α-CH_3	OAc	246	14,500 [e]	12.1	—
67	Mesityloxide	f	—	—	—	—	243	9,100	3.7	6.1
68	Isophorone	f	—	—	—	—	243	12,300	8.4	7.1

[a] Concentration of steroid, 2×10^{-5} M; HSA, 4×10^{-5} M = 0.276%; BLG, 0.276%. Phosphate buffer pH 7.6; μ 0.1; 25°. —$\Delta\varepsilon$ calculated by Eq. (IV-3).

[b] Halogen at C-9.

[c] In agreement with published values (Bischoff and Stauffer, 1954; Schellman et al., 1954).

[d] Code number of Searle and Co.

[e] Value reported obtained in alcoholic solution; for references see Westphal and Ashley (1962). It has been demonstrated (Westphal and Ashley, 1958 a) that there are no differences between the maximal extinction coefficients of Δ^4-3-ketosteroids in methanol or ethanol and in water. [f] See Figure VII-3.

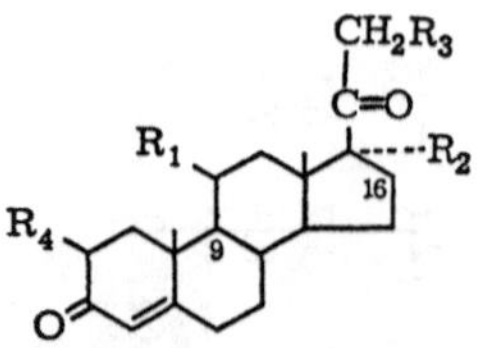

I

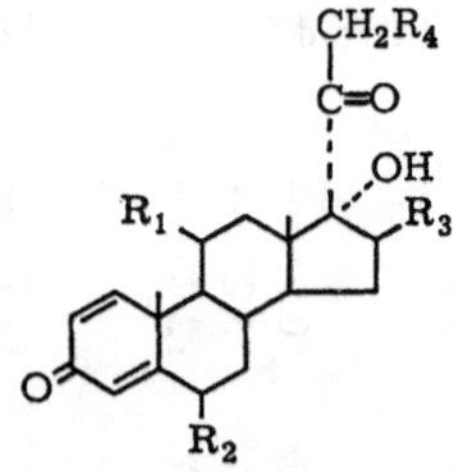

II

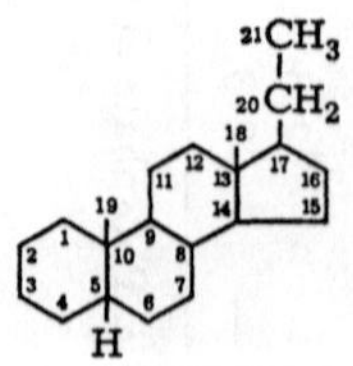

III Pregnane

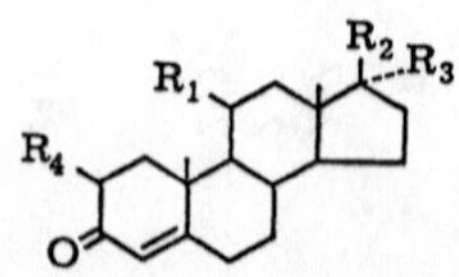

IV

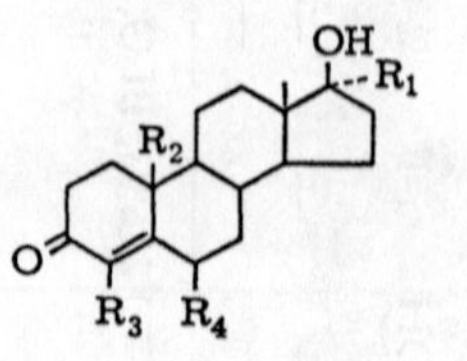

V

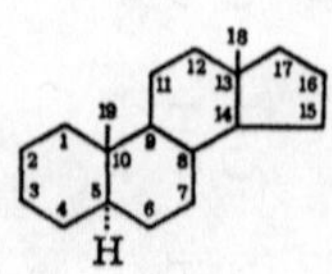

VI Androstane

Fig. VII-2. Formulas to illustrate the structures of steroids used in Table VII-2

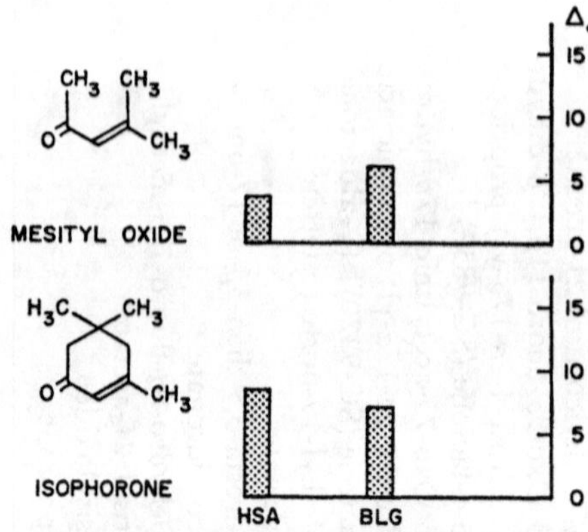

Fig. VII-3. $\Delta\varepsilon$ Values for interaction of α,β-unsaturated ketones with human serum albumin (HSA) and β-lactoglobulin (BLG). See No. 67 and 68 in Table VII-2. From WESTPHAL (1961)

the $\Delta\varepsilon$ value can be appraised. More than 60 such comparisons by pairing have been made and the average effects of a given structural change have been calculated. The results are given in Table VII-3. They show that introduction of electron-attracting substituents weakens the interactions with HSA without exception, those with BLG in most cases (Section 1). The effect of a hydroxy group is stronger than that of an oxo group (Section 2). Dehydrogenation also weakens the binding affinity if the new double bond is conjugated to the keto group in ring A (Section 5). Shift of a methyl group at C-6 from the α-position to the intermediate position at C-6 plus introduction of a double bond at Δ^6 leads to a marked weakening of the complex. Replacement of an α-hydrogen (rear side of steroid molecule) by an electron-repelling methyl group results in a positive change of $\Delta\varepsilon$, indicating a strengthening of the interaction with HSA (Section 4). The effects of electron-attracting and electron-repelling substituents can also be interpreted by a general increase of the binding affinity to albumin with increasing hydrophobicity of the steroid molecule. This would characterize the bond as predominantly hydrophobic.

It may be seen for several cases in Table VII-3 that the effect of a given substituent is greater when located near the Δ^4-3-keto grouping. For instance, a hydroxy group at C-11 destabilizes the HSA-steroid complex more than one at C-17 or C-21. Apparently, the distance from the chromophoric system in ring A has a modifying effect on the ultraviolet absorption; substituents close to the unsaturated keto group have a stronger influence than those in a more remote position. This is to be expected if we consider the influence of the protein interaction on the ultraviolet absorption caused by an effect on the electronic configuration of the chromophoric group. It should rather be noted as remarkable that the inductive effects traverse considerable distances in the steroid molecule, as indicated by an influence of a C-21-hydroxy group on the absorption of the chromophor in ring A. An inductive effect through the whole length of the steroid molecule has been recognized early by BUTENANDT and WESTPHAL (1934) when the larger dissociation constant of the phenolic hydroxy group at C-3 of estriol was interpreted as the result of a greater electrostatic induction of the diol structure at C-16 and C-17 in comparison with that of the oxo group at C-17 in estrone. Another example for an inductive effect from C-17 to C-3

Table VII-3. *Influence of functional groups in Δ^4-3-ketosteroids on interaction with human serum albumin (HSA) and β-lactoglobulin (BLG). Change of $\Delta\varepsilon$ values*[a] *by structural changes in the steroid molecules; pH 7.6, 25°.* Compiled from WESTPHAL and ASHLEY (1958 a, 1959, 1962)

Section	Structural change		Number of pairs compared	Average change of $\Delta\varepsilon$ for HSA	BLG
1	Introduction of	11β-OH	5	−5.5	−3.9
		11α-OH	3	−8.3	−5.2
		11-=O	3	−1.3	−0.6
		17α-OH	3	−0.9	−3.8
		21-OH	2	−1.6	−4.8
		9α-Br	1	−1.5	−0.5
		9α-Cl	1	−2.0	−0.5
		9α-F	1	−2.6	+2.2
		11-=O plus 17α-OH	2	−4.7	−4.6
2	Reduction of	11-=O to 11β-OH	4	−4.0	−0.8
		20-=O to 20β-OH	1	−1.3	+1.4
3	Acetylation of	17α-OH (a) = axial	1	−0.2	—
		17β-OH (e) = equatorial	1	−4.7	—
		21-OH	4	+2.9	+2.8
4	Introduction of	2α-CH_3	4	+4.5	+2.0
		4-CH_3	2	+3.7	+2.3
		6α-CH_3	3	−0.1	+0.2
		6β-CH_3	3	−2.0	−1.1
		10β-CH_3	3	−1.8	+0.4
		16α-CH_3	4	+2.0	—
		17α-CH_3	3	+0.7	−1.0
5	Introduction of	conjugated Δ^1	5	−3.1	−1.0
		conjugated Δ^6	1	−2.7	—
		conjugated Δ^6 when 6α-CH_3 (e) becomes 6-CH_3	2	−8.0	—
		unconjugated $\Delta^{9(11)}$	1	+2.2	—
6	Steric alteration	11β-OH → 11α-OH	3	−1.9	—
		17β-CH(OH)-CH_2OH → 17a-position	1	−2.2	—
		6β-CH_3 → 6α-CH_3	3	+2.0	—

[a] Change of $\Delta\varepsilon$ is defined as $\Delta\varepsilon$ for the original steroid minus $\Delta\varepsilon$ for the structurally changed steroid, so that a negative value for the change indicates weakening, and a positive value strengthening of the binding affinity. Note that all $\Delta\varepsilon$ values in Table VII-2 have a negative sign.

in the androstane series has been observed more recently (PETERSON and CHEVLI, 1967) in studies on rates of solvolysis.

The presentation in Table VII-3 of structural influences on protein interaction in the form of "average" changes of $\Delta\varepsilon$, calculated from $\Delta\varepsilon$ changes observed with individual pairs of steroids differing in one substituent, clearly is an oversimplification. The original data given in Table VII-2 as well as the individual calculations (WESTPHAL and ASHLEY, 1958 a, 1959, 1962) show differences in the spectral effects of a given substituent, depending on its location, and on the structure of the remainder of the steroid molecule. A detailed interpretation of these differences is not possible today. However, there is sufficient agreement between $\Delta\varepsilon$ changes with analogous pairs to indicate the order of magnitude for a given structural group's influence, and to distinguish an affinity-weakening from a strengthening effect.

Recent investigations of SCHOLTAN et al. (1968) have confirmed by ultracentrifugation the general results of the spectrophotometric studies which established the polarity rule. The authors selected bisguanylhydrazone derivatives of numerous steroids which differed in the number of hydroxyl or methyl groups in their molecules. The guanylhydrazones themselves have a higher binding affinity to HSA than the free ketones. Introduction of methyl into a steroid guanylhydrazone increased, and substitution of H by OH decreased the affinity constants. Oxo groups were less effective than hydroxyls in reducing the binding affinities. This is shown in Table VII-4 which gives the affinity constants and the free energies; the formulas of the steroid derivatives studied may be seen in Fig. VII-4. The apparent association constants were obtained as the intersections of Scatchard plots with the ordinate and therefore constitute $\bar{\nu}/[S]$ or n k values. Table VII-4 shows that the binding strength increases by incorporation of aliphatic carbon into steroid No. 1 to form steroids No. 2 and No. 3 which is progesterone (see formulas I—III in Fig. VII-4). Entrance of oxygen reduces the binding strength continuously from No. 3 down until the pentapolar cortisol is reached with steroid No. 14. Other studies showed that introduction of CN, NO and COOH groups also decreases the affinity of interaction with HSA.

A comparison of the apparent association constants, and their decline with increasing polarity of the steroids shows a relatively greater effect when fewer hydroxy groups are present in the mole-

Table VII-4. *Apparent association constants, $n k = \bar{\nu}/[S]$, and free energy, $\Delta F°$, of HSA complexes with steroid-bisguanylhydrazones, at 20°*. Data from SCHOLTAN et al. (1968)

No.	Bisguanylhydrazone of steroid	Formula (Fig. VII-4)	including OH at C-atom	$n k = \bar{\nu}/[S]$ $M^{-1} \times 10^{-4}$	$\Delta F°$ kcal/mole
1	Estr-4-ene-3,17-dione	I	—	0.8	—5.3
2	21-Norpregn-4-ene-3,20-dione	II	—	3.1	—6.0
3	Progesterone	III	—	5.3	—6.3
4	11-Oxopregn-4-ene-3,20-dione	III	11 [a]	2.0	—5.8
5	Deoxycorticosterone	III	21	1.6	—5.6
6	Deoxycorticosterone-21-glucoside	IV	2′,3′,4′,6′	1.6	—5.6
7	17β-Hydroxy-17α-pregn-4-ene-3,20-dione	V	17β	1.2	—5.5
8	11α-Hydroxypregn-4-ene-3,20-dione	III	11α	0.9	—5.3
9	17β,21-Dihydroxy-17α-pregn-4-ene-3,20-dione	V	17β,21	0.7	—5.2
10	11α,17-Dihydroxypregn-4-ene-3,20-dione	III	11α,17α	0.6	—5.0
11	16β,17β-Dihydroxy-17α-pregn-4-ene-3,20-dione	V	16β,17β	0.5	—5.0
12	6β,11α-Dihydroxypregn-4-ene-3,20-dione	III	6β,11α	0.3	—4.7
13	7α,14α-Dihydroxypregn-4-ene-3,20-dione	III	7α,14α	0.2	—4.5
14	Cortisol	III	11β,17α,21	0.2	—4.4

[a] Oxo group at C-11.

cule. The reduction of $\Delta F°$ per OH group thus decreases from about 1,000 cal/mole for progesterone (Table VII-4, No. 3 and 8) to a few hundred cal/mole in oxygen-rich steroids. This is shown in Fig. VII-5 which includes carbohydrate-free steroids and steroid glycosides. The effect of the carbohydrate hydroxyls is much weaker than that of OH groups located at steroid carbons. This is evident from a comparison of the binding parameters of progesterone (Table VII-4, No. 3) with those of cortisol (No. 14) and of the deoxycortico-

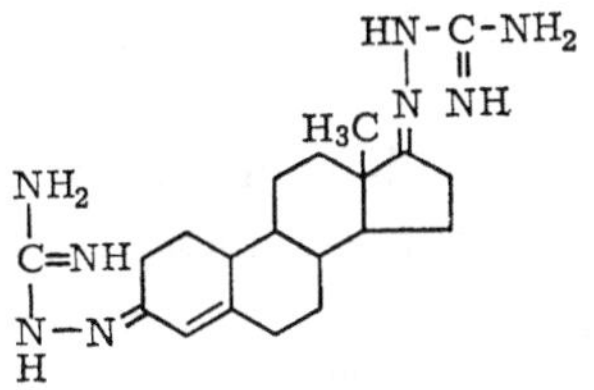

I

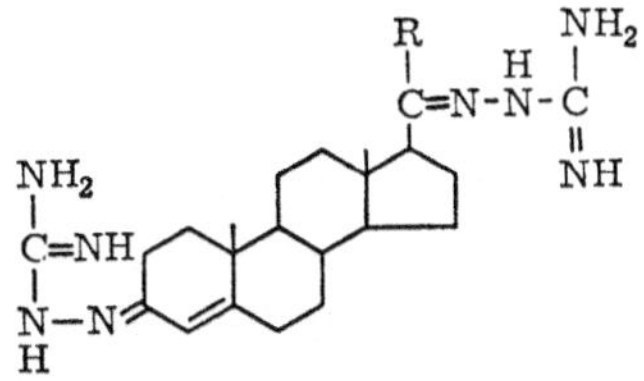

II R = H

III R = CH_3

IV R = 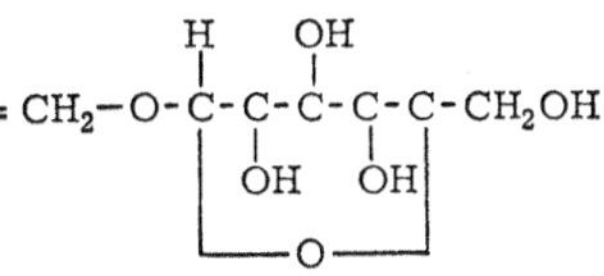

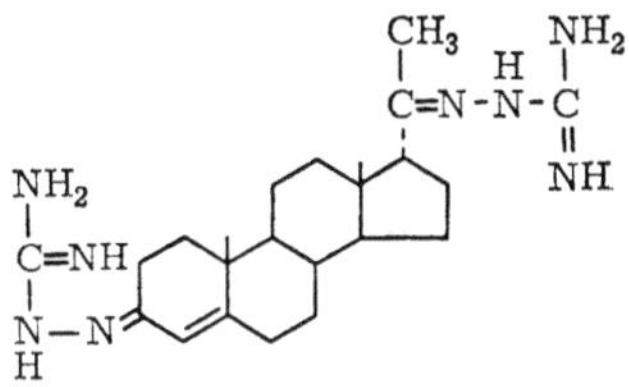

V

Fig. VII-4. Formulas of steroids listed in Table VII-4

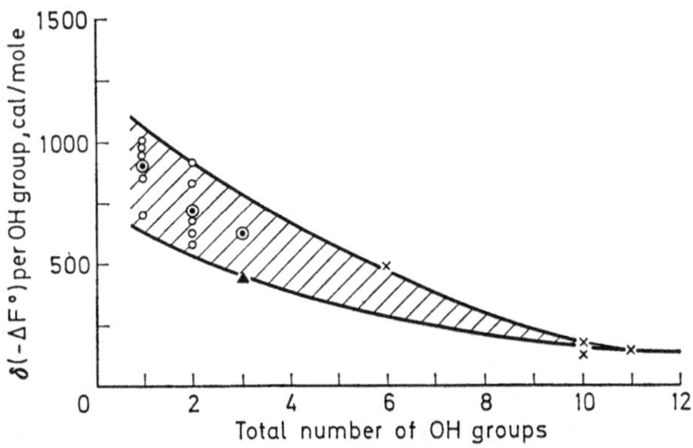

Fig. VII-5. Influence on HSA interaction of introduction of hydroxy groups into steroid guanylhydrazones, steroid hormones and digitalis glycosides. Ordinate, decrease, δ, of $-\Delta F°$ (cal/mole) of steroid-HSA complex per OH group introduced; abscissa, total number of OH groups in steroid molecule. From SCHOLTAN et al. (1968). ○ and ⊙ Steroid guanylhydrazones, individual and average values, respectively; x digitalis glycosides; ▲ steroid hormones

sterone-glycoside (No. 6). Introduction of three OH groups into the pregnene nucleus reduces $-\Delta F^{\circ}$ by far more than the attachment of four hydroxyls at the sugar component. This result is in accordance with the interpretation that the steroid-albumin complex is held together mainly by hydrophobic bonds, operating between the four-ring system of the steroid and given hydrophobic sites or areas on the HSA molecule. Hydroxyl groups of a sugar attached by a glycosidic bond would thus be removed from close contact with the protein, and their influence would be relatively small.

The results of the equilibrium dialysis and spectrophotometric studies (Tables VII-1, VII-2) as well as those of the ultracentrifugation experiments (Table VII-4) show that the binding affinity to HSA increases with decreasing polarity or with increasing hydrophobicity. The general hydrophobicity of a molecule can be assessed by the partition between an organic and an aqueous solvent. If the hydrophobic nature of the steroids is the main determinant in their interaction with albumin, a positive correlation should be demonstrable between binding affinity and partition coefficient.

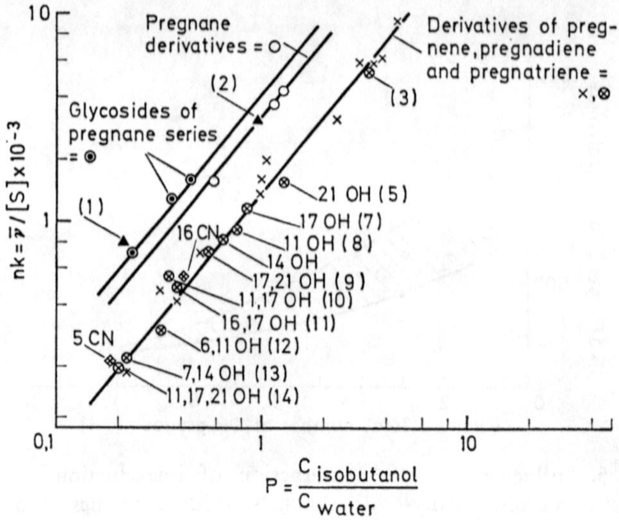

Fig. VII-6. Relationship between apparent association constant, $n\,k = \bar{\nu}/[S]$, and the partition coefficient, P, for the isobutanol-water system. The numbers in parentheses refer to the serial numbers of the steroid derivatives in Table VII-4. From SCHOLTAN et al. (1968)

Scholtan et al. (1968) measured the partition coefficients, P, between isobutanol and water for numerous steroids and steroid derivatives which had been studied for binding affinity to HSA. Fig. VII-6 shows the relationship of the apparent association constants, n k, to the partition coefficient P; both ordinates have logarithmic scales. It is evident that a linear relationship exists between log k and log P. The greater the relative solubility in the organic phase the higher the affinity to HSA. With increasing number of OH groups in the steroid molecule, both P and k decrease. The two parameters for pregnane derivatives with one, two or three double bonds are located on a nearly straight line, whereas different straight lines are occupied by the P and k values for saturated pregnane compounds and for glycosidic derivatives, respectively. The three linear arrangements for these different types of steroids show that for a given binding affinity, the hydrophobicity decreases from the unsaturated to the saturated steroids; this also means that at a given degree of hydrophobic character the binding affinity is higher for the saturated molecules. This effect is more pronounced for the glycosidic compounds, in accordance with the relatively low influence of the sugar hydroxyls as discussed above.

2. Steric Considerations

The relationship between $\Delta\varepsilon$ and strength of protein binding may be utilized to obtain additional information on the mechanism of steroid-protein interaction. In Table VII-2 a number of steroids are listed which differ only in the steric configuration of one substituent, attached to the steroid molecule either in α or in β-position (cf. Fig. V-3, V-4). Comparison of the $\Delta\varepsilon$ values of such pairs of epimeric steroids can be expected to reveal whether the influence of a given substituent on binding affinity is stronger when that group is located in the α-position, i. e., on the rear side of the steroid molecule, or when it is attached to the front side.

The evaluation of the steric effect is illustrated in Fig. VII-7 for the hydroxyl group at C-11 as substituent in three Δ^4-3-ketosteroids. It is evident that the affinity-decreasing effect of the electron-attracting group in all cases is stronger in the α-OH derivative. Section 6 of Table VII-3 shows a similar result for the ethylene glycol side chain

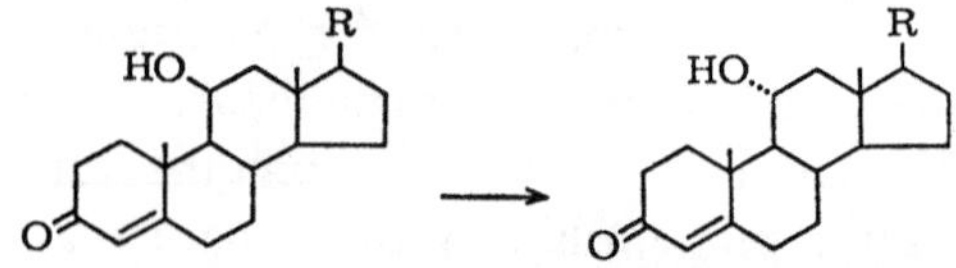

11β-OH ⟶ 11α-OH

	Difference of Δε:
-Progesterone	-3.7
-Cortexone	-1.3
-17-Methyltestosterone	-0.8

11α-OH Weakens interaction with HSA more than 11β -OH

Fig. VII-7. Influence of steric position of C-11 hydroxy group on steroid-HSA interaction. From WESTPHAL (1961)

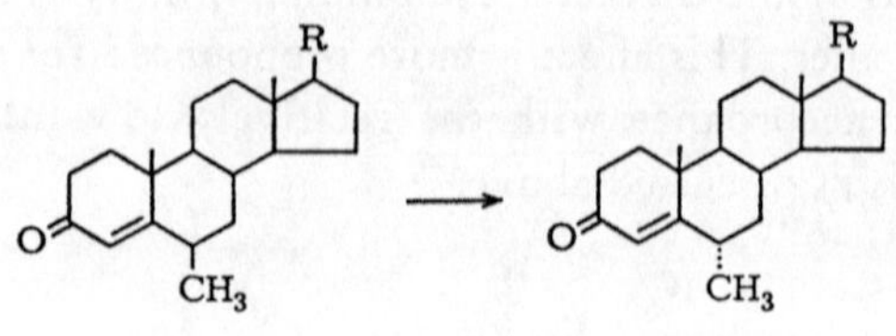

6β-CH_3 ⟶ 6α-CH_3

	Difference of Δε:
-Testosterone	+0.7
-Androstenedione	+2.2
-17-Methyltestosterone	+3.0

6α-CH_3 Causes stronger interaction with HSA than 6β-CH_3

Fig. VII-8. Influence of steric position of C-6 methyl group on steroid-HSA interaction. From WESTPHAL (1961)

at C-17; the steroid-HSA complex is weakened to a greater extent when the 2-carbon chain with two electron-attracting hydroxyl groups is in the α-position. Fig. VII-8 depicts analogous cases in which the influence of a methyl group at C-6 is analyzed. The electron-repelling alkyl group generally increases the binding affinity;

the effect is markedly stronger with the 6α- than with the 6β-methyl compounds.

It is concluded from these studies with epimeric Δ^4-3-keto-steroids that the interaction with HSA takes place at the rear side of the steroid molecule, seen as the lower side in Fig. V-4. The space-filling molecular model of progesterone depicted in Fig. V-5 reveals that the lower or alpha-side presents a fairly plane surface, in contrast to the upper or beta-side which is convex, "turtle-like"

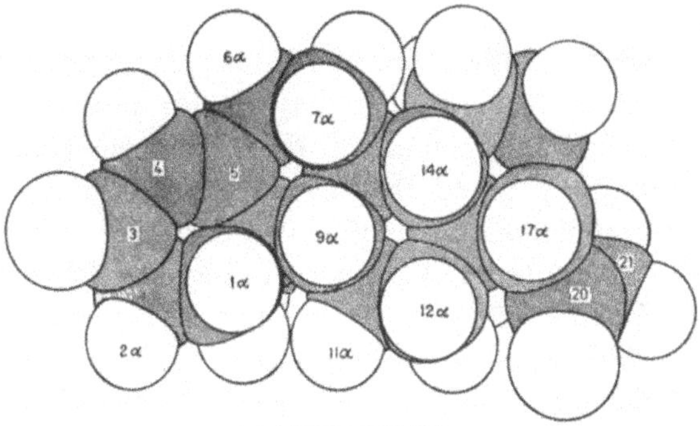

Fig. VII-9. Stuart-type model of progesterone, seen from the rear side. From Westphal and Ashley (1959)

(Westphal and Ashley, 1959). The configurational structure of the lower side appears to be less sterically hindered and thus more capable of interaction. The "rear attack" on the steroid molecule by various reagents, predominant in chemical reactions, may in part result from this steric configuration. Fig. VII-9 shows the details of this structural arrangement for progesterone as seen from the rear side; the numbered substituents may form the points of contact with the binding groups of an interacting protein.

This concept gains in validity when one considers the nature of the binding forces between steroid and protein. The association between neutral steroids and proteins is generally assumed to be mediated by hydrogen bonds and hydrophobic interactions, including van der Waals forces. Common to both of these binding mechanisms is the comparatively low energy of the bond, of the order of 5 to 10 kcal per mole, which explains the ready dissociation of the complexes. Furthermore, the forces are highly dependent upon distance; the van der Waals forces for spherical atoms are inversely proportional to the seventh power of the distance between the atomic centers. For efficient interaction, therefore, a close fit of the steroid and protein surfaces would be required. It may be seen in Fig. V-5 that the α substituents form essentially one plane which is shown in detail in Fig. VII-9, and which includes the oxygen function at C-3. The geometry of the rear side of the steroid molecule thus appears to be better suited for a close approach of the protein than the curved front side.

It is of interest that MARCUS and TALALAY (MARCUS and TALALAY, 1955; TALALAY, 1957) arrived at very similar conclusions in their studies on structural specificity of steroid-enzyme complexes (see Chapter XVI). They found that introduction of hydroxyl groups into the testosterone molecule resulted in a decreased affinity between steroid and β-hydroxysteroid dehydrogenase. This effect was somewhat greater for the 11α-hydroxyl than for the 11β-hydroxyl groups (cf. MARCUS and TALALAY, 1955, Table I) thus resembling the differential influence of an 11α- and an 11β-hydroxyl group on the interaction of progesterone with HSA. TALALAY considers it likely that the oxidizing enzymes attach to the rear side of the steroid molecule rather than to the front. MUNCK, SCOTT and ENGEL (MUNCK et al., 1957; SCOTT and ENGEL, 1957) compared the equilibrium constants of complexes between α- and β-hydroxy epimers of 11-hydroxyprogesterone or testosterone and adenosine, and demonstrated a greater interference of the substituent at the α position. They concluded that the binding involves the rear side of the steroid molecule. In studies on the corticosteroid-binding globulin of human plasma, DAUGHADAY (1958 b) observed a lower complexing affinity for epi-(11α-hydroxy) cortisol than for cortisol (11β-hydroxy). This behavior of the two epimers resembles that towards HSA. Similarly, epi-cortisol (11α-OH) was found (ZIMMERING et al., 1967) to interact with a specific immunoglobulin, i. e., the antibody against the conjugate of testosterone-17-hemisuccinate and BSA, with lower affinity ($k = 1.4 \times 10^4\ M^{-1}$) than the 11$\beta$-OH containing cortisol ($k = 3.6 \times 10^4\ M^{-1}$).

As apparent from Figs. V-4, V-5 and VII-9, the substituent in the 9α position has axial conformation and points straight out of the rear surface of the steroid molecule. According to the above concept, electronegative substituents in this position should weaken the interaction with HSA. This is the case as Table VII-3, Section 1, shows; the influence on the $\Delta\varepsilon$ values has indeed been found to be proportional to the electron-attracting capacity of the halogens, increasing from Br (—1.5), to Cl (—2.0), to F (—2.6). Since the 9α-halogen group occupies a fairly central location on the rear side of the steroid molecule (Fig. VII-9), the influence on the interaction with HSA can be interpreted best if one assumes an attachment at the rear side of the steroid.

It becomes apparent from these considerations that the rear surface of the steroid molecule plays a dominant role in interactions with protein structures. The α substituents forming the rear side plane (see Fig. VII-9) seem to act as points of contact with the surface of the protein. It is also apparent from Figs. V-5 and VII-9 that the oxygen located at C-3 is included in this planar arrangement. The great influence on physiological activity of certain substituents at these "contact points", e. g., at 17α or 9α, is in agreement with this concept.

It follows from Table VII-3, Section 6, that the steric inversion of a 6-methyl group from the equatorial 6α position to the axial 6β position (cf. Fig. VII-8) results in a weakening of the interaction with HSA. In this case, an electron-releasing substituent is moved from an equatorial rear side to an axial front side position. It was of interest to measure the interaction when the 6-methyl group is shifted from the α position to a horizontal location, as given in the half-chair conformation of a Δ^6-compound (see FIESER and FIESER, 1959, p. 15). The spectrometric data (Table VII-3, Section 5) show that a considerable weakening of the interaction is the consequence of this alteration. The effect is interpreted as the combined influence of the Δ^6 double bond and the steric shift of the methyl group. It has been pointed out before that increased electron density on the rear side of the steroid molecule weakens the interaction with HSA. As shown by the example in Section 5, Table VII-3, introduction of a Δ^6 double bond alone results in a lowering of the $\Delta\varepsilon$ value. The electron-releasing alkyl group apparently increases further the electron density in the area of interaction with the protein molecule, and consequently the $\Delta\varepsilon$ values show an additional reduction.

In reference to the decrease of the $\Delta\varepsilon$ values as a result of the inversion of the 6α-methyl to the 6β-methyl group, the above interpretation is in accordance with the concept of RINGOLD and BOWERS (1961) who conclude from geometrical considerations that the inductive effect of a 6β-axial substituent (RINGOLD et al., 1957) will be greater than that of the corresponding 6α-substituent. They predict in this connection that C-6 substitution of a Δ^4-3-ketone by a strong electron-releasing group would result in a lowering of the excitation energy and a consequent bathochromic shift in the position of maximal absorption (K band) in the ultraviolet spectrum. This bathochromic shift has indeed been observed in 6β-methyl-substituted Δ^4-3-ketosteroids, in aqueous as well as in alcoholic solution (for details see Table III in paper by WESTPHAL and ASHLEY, 1962).

Table VII-2 lists observations on four spirolactones (compounds No. 48, 49, 62 and 63), steroid derivatives that block the physiological action of aldosterone. The $-\Delta\varepsilon$ values in the presence of HSA are similar for all four of these blocking agents, in spite of structural differences. This would indicate that the typical spirolactone grouping, in which the two methylene groups are located toward the rear side of the steroid molecule in a comparatively rigid five-ring structure, has a determining influence on the interaction with HSA. Since the same spirolactone ring is common to all of these aldosterone-blocking steroids (RINGOLD et al., 1957) of otherwise different structures (BROWN et al., 1960), this particular grouping may be responsible for the interaction with protein structures of the receptor sites and thus for the competition with aldosterone-like steroids in the target tissues, as postulated by KAGAWA (1960).

The highest $-\Delta\varepsilon$ values that we ever observed are those for the enol glucosiduronate of androst-4-ene-3,17-dione (Compound No. 65, Table VII-2; see also Fig. VII-10). Although the accuracy of these values is less than that of the data for the other steroids (for details, see WESTPHAL and ASHLEY, 1962) the high $-\Delta\varepsilon$ values appear to give qualitative indication of a strong interaction with HSA and BLG. Such behavior might be expected from a steroid that carries an ionized carboxyl group under the conditions of the experiments, and therefore would show the comparatively strong albumin interaction of anions, in addition to that of the steroid structure.

Special mention should be made of the spectrometric data obtained for Δ^{16}-dehydroprogesterone (Compound No. 53 in Table VII-2). This compound possesses two α,β-unsaturated keto groupings, Δ^4-3-one and Δ^{16}-20-one, located at distant sites of the molecule. If the extinction coefficients of these two chromophores were strictly additive, the ε value for Δ^{16}-dehydroprogesterone would be 26,500, i.e., the sum of that of progesterone (16,700; see Table VII-2, No. 1) and that of Δ^{16}-pregnene-3,20-dione (9,800; see DORFMAN, 1953). The actual value of 24,100 for Δ^{16}-dehydroprogesterone is indicative of an interaction between the chromophoric systems. ALLEN and BERNSTEIN (1955) have given a number of examples in which the extinction coefficients of $\Delta^{4,16}$-pregnadiene-3,20-diones are smaller than calculated as additive contributions from the individual chromophoric systems.

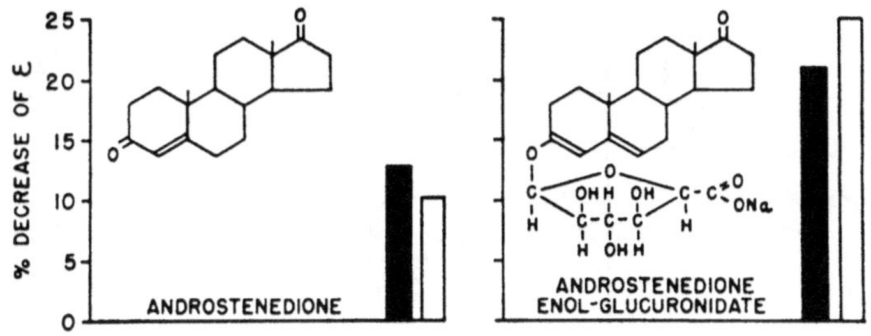

Fig. VII-10. $\Delta\varepsilon$ values of androst-4-ene-3,17-dione and its enol glucosiduronate in the presence of HSA (solid bars) and BLG (light bars)

The low $-\Delta\varepsilon$ value of Δ^{16}-dehydroprogesterone in the presence of HSA (Compound No. 53, Table VII-2) seemingly indicates lack of interaction with this protein. Since this was unlikely for a Δ^4-3-ketosteroid of this structure, the interaction was determined by equilibrium dialysis. The average binding-affinity values obtained, $\bar{\nu}/[S]$, were 9.1, 10.4, 12.1, and 12.3×10^4 as compared with 8.9×10^4 for progesterone (Table VI-9). The binding of Δ^{16}-dehydroprogesterone to HSA, therefore, was even stronger than that of progesterone.

To explain this discrepancy between the spectrometric and equilibrium dialysis procedures, it is assumed that the interaction of one chromophoric system with HSA diminishes or eliminates the interaction of this chromophore with the other unsaturated keto grouping. For example, if the Δ^4-3-keto group of Δ^{16}-dehydroprogesterone interacts with HSA more strongly than progesterone does, its extinction coefficient might be reduced by $\Delta\varepsilon \cong 15\%$, i.e., from 16,700 to 14,200.

If the extinction of Δ^{16}-pregnene-3,20-dione is now added in full amount (9,800), a total ε value of 24,000 would be obtained. This is the magnitude of ε that has been actually observed in the presence of HSA. Fig. VII-11 shows these relationships. This interpretation would explain why steroids that contain two isolated chromophoric systems instead of one may yield abnormally low negative $\Delta\varepsilon$ values in spite of strong interaction with a protein. Evidently, a similar calculation

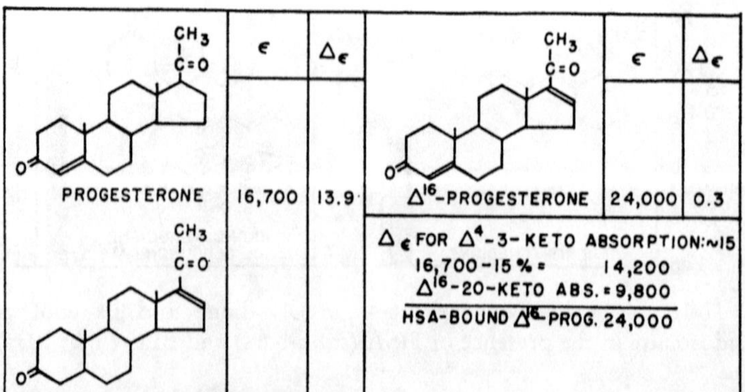

Fig. VII-11. Interaction of Δ^{16}-dehydroprogesterone with HSA

can be made for the other chromophoric group and for any partial participation of the two chromophores in the interaction with HSA. As a practical consequence, the spectrophotometric procedure is not suitable for the measurement of protein interaction with steroids (or other ligands) with two or more chromophoric groupings that interact with each other.

Although the interpretation of the influence of steroid substituents on interaction with HSA on the basis of electron distribution appears attractive, there are several instances that seem to contradict such a generalization. For example, the introduction of 6α-methyl into Δ^4-3-ketosteroids decreases the interaction slightly (Table VII-3, Section 4), whereas a strengthened binding should be expected according to the above concept. The one example of introduction of a $\Delta^{9(11)}$ double bond (Table VII-3, Section 5) would also be in contrast to the general hypothesis. Furthermore, a strict application of inductive influences would perhaps lead to the prediction of a stronger interaction for the 11β-hydroxy derivatives than has actually been observed.

These "exceptions" to the general concept may indicate a weakness of this simple interpretation that explains the observed binding phenomena essentially on the basis of electron densities, which by themselves are not fully understood. For instance, the assumption that the electron movement in an 11β-hydroxysteroid is restricted to the bond between C-11 and oxygen may be questioned; it is conceivable that it involves larger portions of the molecule, resulting in an electron shift from distant areas towards the β-hydroxyl group at C-11 and thereby imparting electronegativity also to the α-side at C-11. The electron density at C-11, rear side, would thus increase from

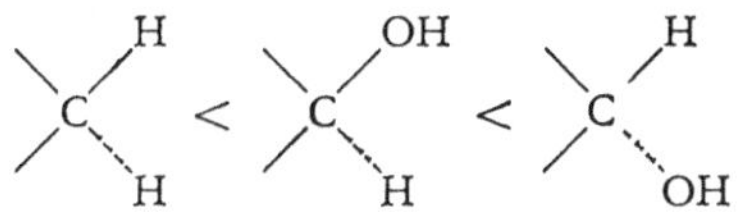

in accordance with the spectral observations of decreasing interaction with HSA in the same sequence.

Another limitation of the concept may be that it does not take into account possible molecular deformations occurring at various points of the steroid molecule and resulting in unexpected association or lack of it. Perhaps more important is the possible influence of a given substituted steroid on the conformation of the interacting protein molecule resulting in a closer or more remote association than expected from the experience with the unsubstituted steroid. Any specific binding affinities of the steroid to individual side chains or particular conformational structures of the protein are not taken into account. The interpretation of steroid binding to proteins in electronic terms thus appears as an unduly simplified explanation of a highly complex phenomenon. However, this concept is considered useful as describing comprehensively the known influences of the steroid structure on interaction with such proteins as HSA.

It may not be superfluous to emphasize that the comparative strength of interaction of physiologically active steroids with certain proteins does not by itself reveal an obvious relationship to biological activity. Even binding affinity to protein structures of the target site would not automatically indicate the degree of physiological effectiveness, unless we understand the complete mechanism by which the physiological action is mediated. Although the study of protein interaction of physiologically active compounds does not provide an immediate key to the relation between chemical structure and biological activity, an intimate knowledge of the chemical basis of such interactions is considered important for a complete understanding of the mechanism of hormone action.

3. Protein Structure in Relation to Steroid Interaction

The preceding discussion was concerned with the influence of steroid structure on protein binding, especially interaction with HSA. Most of the results were obtained by the spectrophotometric procedure. The same method has been used to compare the effect of different protein structures on binding to steroids. Clearly, our knowledge of primary and conformational structure of proteins lags far behind that of steroid chemistry; answers to our questions will therefore be more limited than in the case of the steroids.

With the aid of the spectrophotometric method, steroid interaction has been studied with a number of proteins from various sources. The steroids selected were progesterone, cortexone and cortisol so that comparative values for molecules of increasing polarity could be obtained. Fig. VII-12 shows the $\Delta\varepsilon$ values for the three pregnane derivatives in the presence of various proteins at equal weight/volume concentrations, i.e., 2.76 g/l; this is the concentration which corresponds to 4×10^{-5} M for HSA. All steroid concentrations were 2×10^{-5} M. It is evident that the proteins of animal origin follow the polarity rule in the interactions with the steroids; trypsin inhibitor from soybean does not obey the rule. No interaction with gelatin could be demonstrated. The $\Delta\varepsilon$ values for BSA are smaller than those for HSA, in agreement with numerous results from various laboratories (Table VI-9). The decrease of interaction with entrance of hydroxyl groups into the progesterone molecule is particularly marked for BLG; this is in general accordance with the experience with other steroids (Table VII-3).

The interactions of steroids with most of the proteins discussed so far, notably HSA, BSA, BLG and others, follow the polarity rule; hydrophobic binding is favored over hydrogen bonding. It may be assumed then that these proteins are characterized by a predominance of accessible hydrophobic side chains or other structural features which favor nonpolar binding in their interactions with steroids. In view of the general importance of these hydrophobic interactions it may be of interest to consider an assessment of different proteins according to the chemical nature of their side chains. Three parameters have been proposed to account for the hydrophobic or hydrophilic nature of proteins on the basis of their primary structure: a) the non-

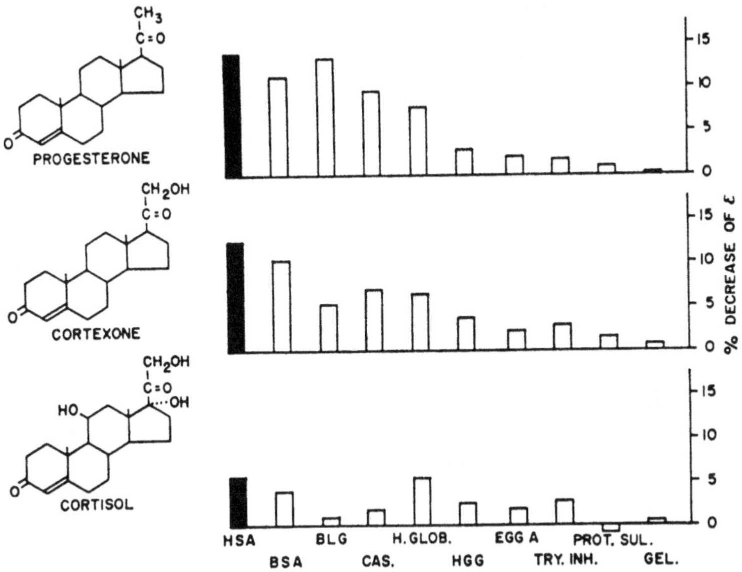

Fig. VII-12. $\Delta\varepsilon$ values of steroids (2×10^{-5} M) in the presence of various proteins (2.76 g/l) in phosphate buffer, pH 7.6, μ 0.1; 25°. CAS., casein; H.GLOB., "modified" human globin; HGG, human γ-globulin (COHN fraction II-1,2; 98% electrophoretic purity); EGG A, egg albumin; TRY.INH., trypsin inhibitor from soybean; PROT.SUL., protamine sulfate; GEL., gelatin. From WESTPHAL (1961)

polar side chain frequency, NPS, introduced by WAUGH (1954); b) the ratio, p, of the volume occupied by the polar residues to that occupied by the nonpolar residues, according to FISHER (1964); c) the average hydrophobicity, $H\Phi_{avg}$, developed by BIGELOW (1967), which is based on TANFORD's (1962) free energies of transfer of amino acid side chains from an organic environment to an aqueous medium. BIGELOW has determined average hydrophobicities (for cal-

Table VII-5. *Average hydrophobicities, $H\Phi_{avg}$, and nonpolar side-chain frequencies, NPS, of steroid-binding proteins in comparison with proteins of limited steroid-binding ability. $H\Phi_{avg}$ computed according to* BIGELOW *(1967); NPS calculated according to* WAUGH *(1954); see Appendix B for calculations*

Protein	$H\Phi_{avg}$ cal/residue	NPS	k (cortisol)/k (progesterone) 4°	37°
CBG-Hu[a]	1302	0.42	0.86	0.33
BLG	1224	0.38	—	0.07[b]
Casein	1198	0.38	—	0.19[b]
AAG, human[a]	1139	0.35	0.03	0.05
BSA	1134	0.33	—	0.35[b]
HSA	1128	0.32	0.03	0.10
CBG-Rb[a]	1095	0.36	2.5	—
CBG-Rt[a]	1062	0.33	1.0	—
Trypsin	954	0.29	—	1.0[b, c]
Lysozyme[d]	949	0.25	—	—

[a] $H\Phi$ and NPS calculations made for the polypeptide moiety only.
[b] $\Delta\varepsilon$ (cortisol)/$\Delta\varepsilon$ (progesterone) at 25°; from WESTPHAL (1957 b).
[c] See also RYAN (1968).
[d] Steroid binding to lysozyme and chymotrypsin (WESTPHAL and ASHLEY, 1958 a) questionable; see RYAN (1968).

culation see Appendix B) for some 150 proteins of widely divergent properties and found good correspondence between $H\Phi_{avg}$ and NPS, whereas FISHER's polarity index did not correlate well with the other two measures.

The original objective in the development of these different parameters for assessing the "nonpolar" and "polar" qualities of a protein had been to find an explanation for the stability of globular proteins; KAUZMANN's description of hydrophobic bonding (1959) had led to a picture of globular protein molecules with most of the nonpolar residues inside the molecule, most of the charged groups outside, and the polar but uncharged groups both inside and out. In the present context, the indices of hydrophobic nature may be of interest for general information; the average hydrophobicities and the NPS values of some proteins with steroid-binding properties are given in Table VII-5. The NPS frequencies of various types of proteins are in general accordance with the HΦ values although the "all or nothing" procedure of calculating NPS (see Appendix B) would seem to be a rather crude way of describing the hydrophobic nature of a protein.

The assessment of steroid-protein interaction by such general criteria as the average hydrophobicity, or NPS frequency, is clearly an oversimplification. The large protein molecule binds one, or a few, relatively small steroid molecules; the binding site, therefore, should be assumed to have the approximate size of the steroid molecule. The spatial relationship can be approximated from the dimension of the hemoglobin molecule, mol. wt. 67,000, which has been found by X-ray analysis to be $64 \times 55 \times 50$ Å (DICKERSON, 1964). Serum albumin, mol. wt. 69,000, is a somewhat more elongated molecule with an axial ratio of about 5 or 6. The dimensions of the progesterone molecule are approximately $13 \times 7 \times 5$ Å (FIESER and FIESER, 1949; WESTPHAL, 1958). Even assuming the participation of several amino acid residues neighboring the steroid-binding site, the relative portion of the protein molecule involved in the interaction remains small. RYAN and GIBBS (1970) conclude on the basis of spectral perturbation studies that interaction of testosterone with HSA involves the tryptophan residue. Studies on the influence of chemical blocking of the specific residues in BSA and HSA on testosterone binding suggested to OYAKAWA and LEVEDAHL (1958) that tyrosine was involved in the interaction.

4. Spectral Observations on Steroid Interaction with Compounds Other than Proteins

The spectrophotometric method was applied to examine possible interaction between Δ^4-3-ketosteroids and individual amino acids and amides. The three steroids, progesterone, cortexone, and cortisol were used in order to cover a wide range of polarity; their spectral behavior was studied at a concentration of 3×10^{-5} M. The influence of the amino acid compounds was tested at a concentration of 12×10^{-5} M, a fourfold molar ratio. Table VII-6 shows that of the acyclic nitrogenous compounds tested only tryptophan gives clear indication of interaction, more strongly for the duopolar progesterone than for the pentapolar cortisol. This is in accordance with a solu-

Table VII-6. *Influence of small nitrogenous molecules on ultraviolet absorption of Δ^4-3-ketosteroids. Depression of absorbance, —$\Delta\varepsilon$, expressed as percentage reduction of ε of the steroid solution (3×10^{-5} M) by the presence of nitrogenous compounds (12×10^{-5} M). Phosphate buffer, pH 7.6, μ 0.1; 25°* (WESTPHAL, 1957 a).

Compound	—$\Delta\varepsilon$ for		
	Pro-gesterone	Cor-texone	Cor-tisol
Formamide	0.0	0.7	1.4
Acetamide	1.0	0.0	0.0
Succinimide	0.4	1.0	0.0
l-Histidine	0.0	0.9	0.2
l-Tryptophan	3.2	1.4	1.8
l-Tyrosine	0.0	0.0	0.3
l-Leucinamide	0.0	0.1	0.6
Glutathione (red)	0.5	0.3	0.8
Glutathione (oxid)	0.0	0.0	0.0
Pyridine	0.9	1.7	2.0
Nicotinamide	0.8	0.3	1.3
Pyrimidine	0.0	0.5	0.6
Uracil	1.7	1.8	2.8
Uridine	1.7	1.6	0.4
Uridine monophosphate	0.6	1.2	1.2
Uridine diphosphate	0.4	1.5	1.4
Uridine triphosphate	0.5	0.7	1.7
6-Methyluracil	1.8	0.4	4.8
Cytosine	2.1	0.8	1.7
Cytidine	3.6	4.6	3.8
5-Methylcytosine	0.5	2.2	1.2
Thymine	0.1	0.7	0.8
Thymidine	0.0	0.0	0.5
Purine	0.1	0.0	0.7
Hypoxanthine	1.2	2.6	5.2
Adenine	9.4	9.3	8.3
Adenosine	5.4	6.1	6.0
Adenosine monophosphate	4.2	4.5	5.7
Adenosine diphosphate	2.7	4.0	4.9
Adenosine triphosphate	4.1	4.9	5.3
Guanosine	4.1	5.1	7.4
Diphosphopyridine nucleotide	5.3	6.0	8.6
Triphosphopyridine nucleotide	3.6	3.7	6.2
Diphosphopyridine nucleotide, reduced	0.3	0.9	3.8
Ribonucleic acid (yeast; 60 μg/ml)	2.8	4.3	4.6

bility increase of testosterone in aqueous solutions of amino acids which was found greatest with l-tryptophan (ABELSON et al., 1960). It also agrees with the aforementioned spectral observation by RYAN and GIBBS (1970) on perturbation of the tryptophan residue in HSA by interaction with testosterone. No significant spectral change was effected by tyrosine, histidine, or glutathione; leucinamide and other amides did not show interaction either (Table VII-6).

The same technique was applied to explore possible interaction of the three steroids with pyridines and pyrimidines (Table VII-6). With the exception of cytidine, the effects were negligible. Marked depressions of the steroid absorbance, however, was obtained with some purine derivatives. Whereas the unsubstituted purine nucleus does not influence the steroid spectra, introduction of an amino group results in a compound (adenine) which decreases ε by almost 10%. The interaction with progesterone appears to be stronger than that with cortisol. This relationship is reversed by replacement of the amino group of adenine by a hydroxy group: hypoxanthine is found to interact more efficiently with cortisol than with cortexone, and the interaction with progesterone is least pronounced. This behavior, contrary to the polarity rule, is shown by all those purine derivatives tested which contain additional hydroxy groups. Nucleotides do not appear to be significantly different from the nucleosides in their interaction with the three steroid hormones; no particular effect of the phosphate ester groups can be recognized. Diphosphopyridine nucleotide (NAD) and triphosphopyridine nucleotide (NADP) exert their influence on ε conversely to the polarity rule; this is interpreted to be in accordance with their hydrophilic nature. The reduced form, DPNH ($NADH_2$), has practically lost the capacity to interact with progesterone and cortexone, but still depresses the cortisol absorption.

It is evident from these experiments that the monocyclic nitrogenous heterocycles and their derivatives in general have little capability of interacting with the Δ^4-3-ketosteroids. A condensed system of at least two rings seems to be required; the positive effect of tryptophan is in accordance with this assumption. Unsubstituted purine is inactive; entrance of functional groups gives rise to interacting compounds. Adenine exhibits the strongest interaction and follows the polarity rule; adenosine and guanosine influence the steroid spectra in contradiction to the polarity rule. The binding forces may be assumed to be hydrophobic linkages and hydrogen bonds.

Scott and Engel (1957) concluded from solubility studies that adenine, adenosine and adenosine monophosphate formed complexes with progesterone, cortexone and testosterone. Additional investigations from the same laboratory (Munck et al., 1957) with the method of solvent partition extended these results and demonstrated interactions of these purine compounds with a number of steroids. Testosterone complexes were observed with several coenzyme components. Molinari and Lata (1962) studied interaction of steroids with pyrimidines and purines by solvent partition; Lata and Dac (1965) observed increased water solubility of testosterone by addition of urea and ureides. The reversible reaction between cortisol and diphosphopyridine nucleotide (NAD) at alkaline pH (Monder, 1963) apparently follows a different binding mechanism. Fluorescence studies by Fazekas and Webb (1967) led to the conclusion that cortisol and other Δ^4-3-ketosteroids interact with nicotinamide adenine dinucleotide to form complexes with characteristic chromatographic and spectral properties.

Some preliminary experiments are listed in Table VII-6 which indicate interaction between the Δ^4-3-ketosteroids and ribonucleic acid from yeast. These observations were made many years ago; since then, equilibrium dialysis studies have demonstrated formation of weak complexes between testosterone or estradiol and thymus DNA, especially after heat denaturation (Ts'o and Lu, 1964). The method of gel filtration also gave indication of extremely weak interaction between DNA and several steroid hormones (Sunaga and Koide, 1967). However, a study of the melting curves (Parr and Koide, 1969) showed no influence of cortisol in concentrations of 1.4 to 140 μM on DNA from human placenta, calf thymus or rat liver. Cohen and Kidson (1969) found in equilibrium dialysis studies that progesterone, estradiol, testosterone, and corticosterone were bound by denatured but not by native DNA. Analysis of steroid interaction with synthetic polynucleotides revealed a specific requirement for guanine residues. In accordance with the polarity rule, polyhydroxylated steroids (cortisol, ecdysone) bound poorly or not at all. Further studies (Cohen et al., 1969) indicated that two functional groups were required in the steroid molecule for interaction with the guanine residues in polyguanylic acid, i.e., the 3 and 20 (or 17) oxygen functions in the C_{21} steroids (or the androgenic and estrogenic structures); the authors assume the formation of two hydrogen bonds between steroid and purine.

SCHOLTAN et al. (1968) used ultracentrifugation to study the interaction of steroids and their bisguanylhydrazones with ribonucleic acid. The general binding properties of the RNA complexes were similar to those of the complexes with HSA although the binding affinity was somewhat lower. The polarity rule was obeyed as demonstrated by introduction of OH and CH_3 groups into the steroid compounds (SCHOLTAN, 1968). Hydrophobic bonding is assumed to be mainly responsible for the interaction (SCHOLTAN, 1967).

It should be mentioned in this context that MAHLER et al. (1966) observed interactions between a number of simple steroidal diamines and DNA and interpreted them by several possible mechanisms. Subsequent studies (MAHLER and BAYLOR, 1967) revealed that two of them, i.e., $3\beta,20\alpha$-diaminopregn-5-ene (irehdiamine A) and 5α-pregnan-$3\beta,20\alpha$-ylenebis(trimethylammoniumiodide) (malouetine) are effective inhibitors of phage replication, presumably by interference with DNA duplication. Spectral methods have been applied to explore the interaction of irehdiamine A with several synthetic homopolyribonucleotides (LEFRESNE and JACQUEMIN-SABLON, 1969) and with transfer RNA (PAOLETTI and PAOLETTI, 1969).

The examples given in this chapter for interaction between steroids and proteins or other (macro)molecules, and the attempted interpretations show clearly that our knowledge of the underlying chemistry is scant and essentially limited to the influence of the steroid structure. The spectral method applied many years ago in the author's laboratory is a relatively crude device to assess interaction between steroid and protein; results with refined techniques (RYAN, 1968) can be expected with interest. Additional findings on protein structure in relation to steroid interaction will be discussed in Chapter XIII. 6 with reference to AAG. In future work, extensive use should be made of molecular models for a better understanding of the steric and spatial relationships between the components of the complexes. However, this will require further elucidation of the primary and higher molecular structures of the binding proteins.

VIII. Corticosteroid-Binding Globulin (CBG) of Man and Other Species

1. Fundamental Observations

In 1956, DAUGHADAY (1956 a) and BUSH (1957) independently demonstrated by equilibrium dialysis techniques the presence in human plasma of a specific protein capable of binding corticosteroids with very high affinity but low capacity. These qualities clearly distinguished the interacting protein from albumin; it was termed the corticosteroid-binding globulin (CBG). At the same time, SANDBERG, SLAUNWHITE and ANTONIADES (1957) reported steroid binding data obtained with human plasma and plasma protein fractions which could not be accounted for by albumin binding; they were interpreted (SANDBERG and SLAUNWHITE, 1958; SLAUNWHITE and SANDBERG, 1959) as indicating the existence of a special protein of high binding affinity for corticosteroid hormones which was named transcortin. The background to this demonstration of transcortin has been recounted by SANDBERG et al. (1966). UPTON and BONDY (1958) confirmed in ultrafiltration studies the strong binding of small amounts of cortisol to a plasma protein other than albumin. DESGREZ (1967) has summarized some of the observations that led to the recognition of CBG.

The important finding by DAUGHADAY of at least two macromolecular binding systems for cortisol, corticosterone and other steroid hormones (DAUGHADAY, 1958 b) is shown in the original Fig. VIII-1. The apparent binding affinity, expressed as C-value (see Eq. (V-4) decreases with increasing amounts of steroid added to the plasma. Recalculation of the data to obtain the percentage of bound and unbound cortisol resulted in Fig. VIII-2 which shows that at 4° binding is practically complete up to a cortisol concentration of approximately 20 μg per 100 ml of plasma. The high-affinity binding sites then become saturated, and an increasing portion of the total cortisol remains in an albumin-bound or unbound state. BUSH (1957)

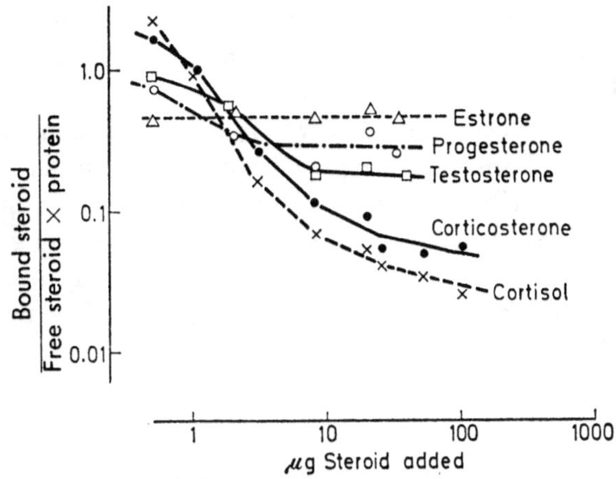

Fig. VIII-1. Binding affinity of normal human plasma for cortisol-4-C^{14}, corticosterone-4-C^{14}, progesterone-4-C^{14}, testosterone-4-C^{14}, and estrone-16-C^{14} in the presence of increasing amounts of total steroid. Equilibrium dialysis system containing 10 ml of plasma and 80 ml of Krebs phospho-saline buffer, pH 7.4; 4°. From DAUGHADAY (1958 b)

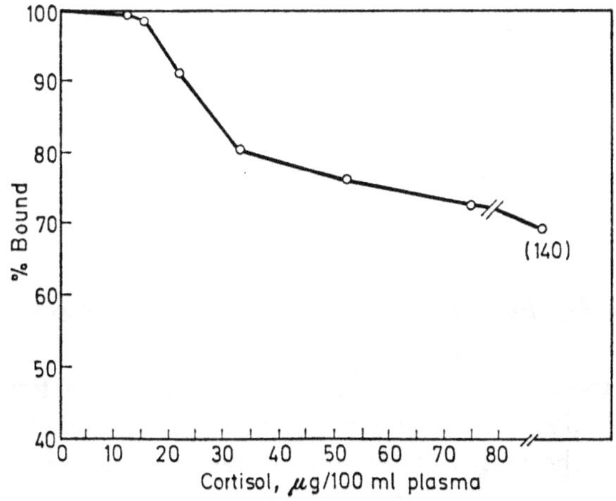

Fig. VIII-2. Percentage of cortisol binding by human plasma plotted against the total cortisol concentration of plasma, calculated from data summarized in Fig. VIII-1. From DAUGHADAY (1958 b)

drew similar conclusions from equilibrium dialysis studies at 37° where the dissociation of the cortisol-protein complex is considerably higher than at 4°.

Figure VIII-1 documents furthermore that more than one binding system exists in plasma also for progesterone and testosterone. It will be pointed out below (VIII. 4. c) that electrophoretic studies by DAUGHADAY (1958 c) gave evidence of progesterone binding to human CBG—a conclusion which is generally accepted today (see Chapter VIII. 5. a). High-affinity binding of testosterone will be discussed in Chapter XII.

2. Methods of Measurement

The quantitative determination of CBG activity in serum or plasma has inherent difficulties. These are caused by the very low concentration of this protein in the blood, by spontaneous dissociation of the complex of CBG with the radiolabeled corticosteroid which is generally used for the assay, by possible interference from other binding proteins, and by competitive or non-competitive binding inhibition by other steroid hormones which may or may not be known as to structure or quantity. Efforts to apply an immunochemical technique for quantitative assay of CBG in human serum (BEIER et al., 1963) have resulted in levels more than 20 times those obtained by other procedures—the discrepancy awaits explanation. Zinc sulfate precipitation (KNIGGE and HOAR, 1963) is not considered to provide valid measurement of the CBG-bound and unbound portion of the corticosteroids as present in the plasma.

a) Equilibrium Dialysis and Related Techniques

The methods most commonly used for determination of CBG activity in a given serum or plasma are equilibrium dialysis and gel filtration. In the dialysis technique, a trace amount of radiolabeled corticosteroid is dialyzed to equilibrium against a diluted serum sample enclosed in a bag of regenerated cellulose. The outside solution then represents the concentration of unbound steroid; the bound steroid is obtained by subtracting the unbound from the total steroid inside the dialysis bag. A condition for valid results is complete equilibration not only for the unbound steroid inside and outside the membrane but also for the added radiolabeled and endogenous corticosteroid hormone. The method measures the bound and unbound portion of the total, i.e., endogenous plus added corticosteroid; the result may be expressed in percent of total, or, after determination of the endogenous corticoid concentration, as μg per 100 ml bound and unbound.

To refer the ratio of bound: unbound corticoid to the protein concentration, the expression

$$C = \frac{[S_{bd}]}{[S] \times [P_t]} \, g^{-1} \, \text{liter} \qquad (V\text{-}4)$$

is used as explained in Chapter V. 4. The C-value is determined by equilibrium dialysis, often in the form of multiple equilibrium dialysis (Chapter IV. 5).

As discussed in Chapter V. 4, the C-value differs from the association constant, k, mainly by the use of total protein concentration, $[P_t]$, expressed in g per liter, instead of concentration of unbound protein, $[P]$, expressed in moles per liter. It therefore gives an accurate measurement of binding affinity only if the total protein concentration is large in relation to the concentration of bound protein, so that $[P] \cong [P_t]$. In the practical determination of the C-value as an indicator of the CBG binding activity of a serum, a compromise is made between measuring binding affinity and binding capacity; the concentration of the endogenous plus added radiolabeled corticosteroid is small enough to be well below saturation of the binding sites, and high enough to reveal differences in concentration of binding sites in different sera. This necessary compromise is the main reason why the use of a single C-value for a given serum is not considered completely satisfactory as a measure of binding activity. This is also the reason why the application of the multiple equilibrium dialysis technique (WESTPHAL, 1961; WESTPHAL et al., 1961) is preferred for a critical comparison of different sera, e.g., in animal experimentation. The results obtained with these methods over many years have provided ample evidence for their validity.

An important advantage of the use of the C-values obtained by multiple equilibrium dialysis at 37° is independence from inhibition by other steroids present in the individual sera. This is a consequence of the different sera being measured under conditions which equalize any effects on binding by competing or otherwise inhibitory steroids among all sera present in the same multiple equilibrium dialysis system. It is the unique and significant feature of the multiple equilibrium dialysis technique at 37° that the steroids are being equally distributed (see WARREN and SALHANICK, 1960) throughout all dialysis bags containing the different sera. Differences in binding activity (C-values) found between different sera are therefore the result of differences of binding affinity and binding capacity in the macromolecular components (binding proteins) and not of the concentrations of potentially competing or inhibitory steroids which may have been originally present in different quantities in the different sera.

Even if competition for the macromolecular binding sites occurs, for example, between cortisol and progesterone, such competition would equally affect all serum samples. It would therefore not be the cause of an artifactual decrease of binding activity for the test steroid in those serum samples in which the original concentration of competing steroid was particularly high. This does not exclude the possibility that the general level of the C-values of all samples of the multiple dialysis system may be somewhat lower or higher, depending on the presence of greater or smaller total amounts of competing steroids; however, the C-values measured for the individual serum samples will correctly represent the binding activity of the given macromolecular systems. Experience over the years has shown that the general level of C-values is closely comparable when large numbers of different serum samples are used in a multiple equilibrium dialysis system. This is being monitored in our laboratory by inclusion of the same normal control sample in the different multiple systems. The validity of the results obtained by the multiple equilibrium dialysis method at 37° has been confirmed in numerous cases by other techniques.

The equilibrium dialysis procedure, in the author's opinion, gives the most satisfactory answer to the question of what percentage of a given steroid hormone in a serum is bound and what portion is unbound. It has been ascertained in determinations of cortisol-binding affinity with sera of man, rabbit and guinea pig, that the C-values obtained with the 10-fold diluted sera are essentially the same as those of the undiluted sera (Table VIII-1). The sera used in these analyses had been stripped of endogenous steroids by gel filtration at

45°; the ratio of radiocortisol to total protein was the same in the experiments with diluted and undiluted serum.

It follows from the results given in Table VIII-1 that the C-values obtained by the standard procedure in 1 : 10 diluted serum may be applied to a calculation of the bound and unbound portion of the corticosteroid in the undiluted serum. When x equals the percent unbound steroid and 100 — x the percentage bound then

$$C = \frac{100 - x}{x \cdot [P_t]}, \qquad \text{(VIII-1)}$$

and x is obtained by using the protein concentration (g per liter) of the undiluted serum. The concentrations of bound and unbound corticosteroid are obtained in μg/100 ml by multiplying (100 — x) and x, respectively, by the total corticoid concentration in μg/100 ml, divided by 100.

Table VIII-1. *Influence of serum dilution on cortisol-binding affinity, determined*[a] *as average C-values of duplicate comparative equilibrium dialyses with several sera for each species, stripped of endogenous steroids by gel filtration at 45°; 0.05 M phosphate, pH 7.4, 37°*

Species	Serum Diluted 1 : 10	Serum Undiluted [b]
Human	0.29	0.28
Rabbit	0.24	0.22
Guinea Pig	0.06	0.06

[a] Unpublished experiments with G. B. HARDING and R. G. ALLEN. From WESTPHAL (1969).

[b] The concentration of added radiocortisol was 10-times that used with the diluted serum.

One condition for the validity of the C-values as a predominant expression of CBG activity is that the binding capacity, given by the concentration of high-affinity binding sites for corticosteroids, is not exceeded by the addition of the radioactive test corticosteroid. When increasing amounts of radiocorticosteroid are added to a given serum in an equilibrium dialysis experiment (Fig. VIII-1), the C-values decline gradually until the CBG binding sites for corticoids are occupied, and the level characteristic of binding to serum albumin is reached (DAUGHADAY, 1958 b). In contrast to the high-affinity CBG, albumin has a low binding affinity for corticosteroids, but the high albumin concentration provides a practically unlimited capacity. It follows from these considerations that, in determination of the C-value as a measure of CBG activity, the sum of endogenous and added corticosteroid must be well within the limit of the concentration of high-affinity (CBG) binding sites. Even then, the binding value obtained includes a contribution from binding protein other than CBG, i. e., mainly albumin. Unless the CBG concentration is extremely low, this contribution is usually of the order of 8—12% of the total binding value (WESTPHAL, 1966; GALA and WESTPHAL, 1966 b; DIAMOND et al., 1969).

Another way of eliminating interference with the CBG assay by endogenous steroids is the complete removal of the endogenous steroids from the serum. This can be done by gel filtration at 45° as will be described in Chapter VIII. 3. This technique is considered the method of choice for determination of accurate binding parameters in special cases as, for example, of the association constants and concentration of CBG binding sites in mammalian sera. However, the procedure is not practical in the routine assay of CBG activity or in animal experimentation involving many samples.

The interference by endogenous steroids is not limited to corticosteroid hormones. In confirmation of earlier suggestions (DAUGHADAY, 1958 c), progesterone has been reported to form a strong complex with CBG (DE MOOR and STEENO, 1963; SEAL and DOE, 1966; SANDBERG et al., 1966); the association constant at 37° is three times that of the CBG-cortisol complex in human serum (see Table VIII-12). The association apparently occurs at the same binding site at which cortisol and corticosterone interact. The presence of progesterone has to be considered, therefore, in the interpretation of experimental data on CBG activity, especially in pregnancy sera (SANDBERG et al., 1968). Testosterone also interacts with CBG (DOE and SEAL, 1963; GUÉRIGUIAN and PEARLMAN, 1968); see Table VIII-12. LEBEAU et al. (1969) reported binding of testosterone and estradiol to CBG.

The equilibrium dialysis procedure has been used in our laboratory for the determination of CBG concentration and association constants of the steroid-CBG complexes in unfractionated serum (WESTPHAL, 1967). Pooled normal serum of various species, including the human, were cleared of endogenous steroids by gel filtration at 45°. Bound and unbound portions of the steroid (cortisol, corticosterone, or progesterone) were measured in 1 : 10 diluted serum by equilibrium dialysis at 4° and 37° with five different concentrations of radiolabeled steroids. The concentration of binding sites was determined graphically by plotting the concentration of unbound steroid against the corresponding total steroid concentration (PETERSON et al., 1960). Fig. VIII-3 shows an example of this plot for

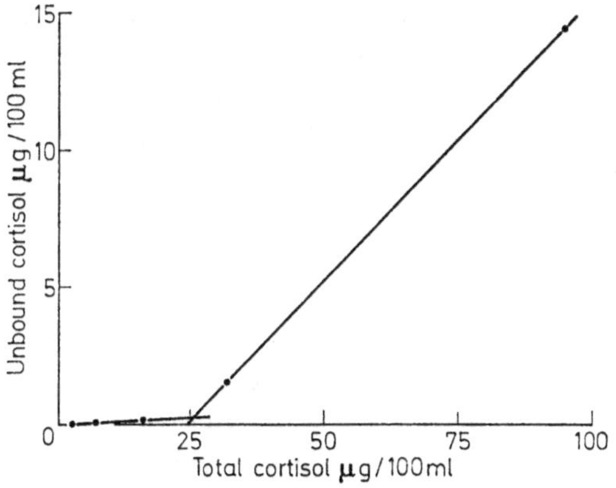

Fig. VIII-3. Determination of concentration of binding sites of CBG in human serum. Cortisol-4-^{14}C, 4°, 0.05 M phosphate, pH 7.4. One μg cortisol per 100 ml equals 2.76×10^{-8} M. From WESTPHAL (1967)

cortisol binding in human serum; the concentration of unbound and total cortisol is expressed in μg per 100 ml which equals 2.76×10^{-8} M. It is apparent that at very low concentrations of the added radioactive cortisol, the unbound cortisol fraction increases very little, due to the high binding affinity of CBG. When the available binding sites of CBG become saturated, the unbound portion of cortisol rises at a much greater rate with increasing concentration of total cortisol. Since the steroid concentrations applied cover several orders of magnitude, the plot describes high-affinity binding to CBG at the low concentrations of total steroid, and low-affinity binding to serum albumin at the high steroid levels (YATES and URQUHART, 1962). The intersection of the two parts of the curve indicates the concentration of high-affinity binding sites, n[CBG], where n is the apparent number of binding sites per mole CBG, and [CBG] is the concentration of CBG. Since n has been found to equal one for the isolated human CBG (see XI. 4), it may be omitted. The value of [CBG] is expressed in moles per liter. This graphic determination of [CBG] is approximate since it disregards the binding to albumin at the low steroid concentrations; however, the procedure appears justified for the present purpose in view of the 10^3 to 10^5 times larger association constants of the CBG-steroid complexes compared with those of the albumin complexes.

The association constants were determined by the method of reciprocal plots (V. 4 c) in a similar way as reported by BOOTH et al. (1961). The following formulas indicate the calculation of k, the apparent association constant for the complex between steroid hormone and the high-affinity protein, CBG.

$$\frac{1}{[S_{bd}]} = \frac{1}{k\,n\,[CBG]} \cdot \frac{1}{[S]} + \frac{1}{n\,[CBG]} \qquad \text{(VIII-2)}$$

$$y = m \cdot x + a \qquad \text{(VIII-3)}$$

where $[S_{bd}]$ and $[S]$ denote the molar concentrations of bound and unbound steroid, respectively. With the values obtained for the lowest steroid concentrations applied, the reciprocal of the concentration of bound steroid, y, was plotted against the reciprocal of the unbound concentration, x. The apparent association constant, k, was then found from the slope m according to

$$m = \frac{1}{kn[CBG]}\,; \quad k = \frac{1}{mn\,[CBG]}\,. \qquad \text{(VIII-4)}$$

The values of [CBG] used in this equation were obtained from the graphic determination described above. As far as comparative data are available, this procedure (WESTPHAL, 1967) yields values which agree with those obtained with pure CBG (Table VIII-12).

LEBEAU et al. (1969) have recently reported an equilibrium dialysis technique for the determination of high-affinity binding parameters for corticosterone, corticosterone sulfate, testosterone and estradiol in plasma. The method is closely related in principle to the one discussed on the preceding pages (WESTPHAL, 1967). The French investigators plot the logarithm of the bound fraction versus the logarithm of the total steroid concentration and calculate the best fitting parameters with the aid of a computer. Examples are given for systems consisting of up to two high-affinity binding proteins and one protein binding with low specificity (affinity).

A number of equilibrium procedures have been applied in various laboratories for determination of CBG activity. To eliminate albumin as a binding protein, equilibrium dialysis has been carried out against an albumin solution of equal concentration (SANDBERG et al., 1957). A similar approach was used by BOOTH et al. (1961) who dialyzed 1+4 diluted plasma against 1% HSA; the binding data were evaluated by a graphic procedure to obtain CBG concentration and affinity constants for human sera. Another technique (DAUGHADAY et al., 1962) to distinguish between binding to CBG and to albumin is based on selective thermal inactivation of CBG at 60°, a temperature which does not significantly affect cortisol binding by HSA. The CBG capacity is measured by double equilibrium dialysis as differential binding to native and heat-inactivated plasma samples. It should be considered, however, that rate and degree of this inactivation are dependent on the concentration of corticosteroids present which exert a protective effect on the heat denaturation of CBG (DOE et al., 1964). GOLDIE et al. (1969) have described an equilibrium dialysis method for a simplified assay of CBG capacity and affinity in human plasma. The gel equilibration procedure of PEARLMAN and CRÉPY (1967), although not designed for the determination of CBG activity, should be remembered in this context since it also measures high-affinity binding of a steroid to a serum protein. This technique has been described in detail in Chapter IV. 15.

In a thorough investigation from YATES' laboratory (KELLER et al., 1966), an elaborate procedure has been worked out to determine the concentration of binding sites and their association constants for CBG and albumin in undiluted plasma or serum of the rat by ultrafiltration (MILLS et al., 1959, 1960) at 37°. Eight different concentrations of corticosteroid hormone, including tritiated tracer, are equilibrated with the plasma pool; the binding data obtained are subjected to computer analysis using multiple mass action theory. The binding data obtained with this sophisticated method on plasma of rats in various physiological and endocrinological states are in general agreement with and thus confirm the results arrived at in our laboratory (see Chapter IX) by use of the standard equilibrium dialysis procedures outlined above. A graphic method based on equations of multiple binding equilibria is used by ROSENTHAL (1967) in SANDBERG and SLAUNWHITE's group (SANDBERG et al., 1966) to obtain the physicochemical parameters of a binding system which consists of one ligand and one or more binding macromolecules. Additional techniques by BAULIEU and RAYNAUD (1969) and by MILGROM and BAULIEU (1970) have been described in Chapter V. 4 g; a new procedure by GUÉRIGUIAN and CRÉPY (1969) is mentioned in Chapter IV. 15.

b) CBG in Competitive Binding Analysis: Quantification of Binding Sites and of Steroids

MURPHY and PATTEE (1963) have developed a procedure for determination of CBG which is based on the amount of cortisol required to saturate the CBG binding sites. This quantity is found in a series of equilibrium dialyses at increasing cortisol concentrations at 9°, and the CBG capacity is obtained in μg cortisol bound per 100 ml plasma. Using the same principle in reverse, MURPHY and her coworkers devised a method for determination of cortisol and corticosterone in human plasma. A standard plasma is calibrated by measuring the decrease of cortisol-4-^{14}C binding when increasing known amounts of cortisol are added to the equilibrium dialysis system (MURPHY et al., 1963); the known quantity of added cortisol is then substituted by the unknown sample and the result is read from a standard calibration curve.

Application of this method of steroid determination by competitive protein binding to clinical studies has been reported (MURPHY et al., 1964). The time required to perform this analysis was shortened from two days to two hours when the equilibrium dialysis was replaced by gel filtration over Sephadex G-25 (MURPHY, 1964; MURPHY and PATTEE, 1964). MURPHY (1967 c) reported later that the sensitivity of the method could be increased 100-fold by use of tritiated steroids and other modifications. Separation of bound and unbound cortisol is now mostly accomplished by adsorption of the unbound steroid to insoluble particles, such as Fuller's earth or Lloyd's reagent, Florisil, dextran-coated charcoal, or polystyrene resin (SLAUNWHITE and SANDBERG, 1970).

This competitive protein-binding radioassay has been extended to urinary cortisol determinations (MURPHY, 1968 a). Measurement of free cortisol in urine by the binding competition method (BEARDWELL et al., 1968) confirmed the results of MURPHY (1967 c); an accuracy of ±10—15% for all except the lowest values was assessed from duplicate agreement and cortisol recovery experiments. A comparison of these CBG-based procedures with other available methods has been published (MURPHY, 1967 a). Similar principles were applied by NUGENT and MAYES (1966) for the determination of plasma corticosteroids by use of CBG and dextran-coated charcoal, and by NAKAJIMA et al. (1968) for an assessment of blood corticoids. According to LECLERCQ et al. (1969), one person can perform 70 cortisol determinations in a day's work, using 0.2 ml of plasma per analysis in a modified MURPHY technique. Additional modifications have been applied by ITURZAETA et al. (1970).

BOWMAN and DE LUNA (1969) made a critical evaluation of the protein-binding method for cortisol determination in monkey plasma. The values obtained were statistically indistinguishable from those given by a double-isotope derivative method (BOWMAN, 1967). Application of the competitive binding analysis for the assay of cortisol in monkey plasma had been previously reported (JONES and MASON, 1966). The technique was applied by BASSETT and HINKS (1969) for measurement of corticosteroid concentration in ovine plasma. Sensitivity, specificity and precision of the competitive binding method for cortisol determination in human plasma and urine has been studied by KOLANOWSKI and PIZARRO (1969). Patients under dexamethasone (16α-methyl-9α-fluoro-11β,17,21-trihydroxy-pregna-1,4-diene-3,20-dione) treatment showed extremely low cortisol values in plasma (0—3 μg/100 ml at 9 AM); dexamethasone itself had no effect on the assay. This agrees with the observation (PEETS et al., 1969) that dexamethasone and betamethasone (16β-methyl-9α-fluoro-11β,17,21-dihydroxy-pregna-1,4-diene-3,20-dione) are not being bound to CBG and do not compete with cortisol for its high-affinity binding sites. Both of these prednisolone derivatives associate with the albumin fraction.

Since progesterone is bound with high affinity to CBG, the competitive binding assay of MURPHY (1967 c) has also been applied to the progestational hormone. NEILL et al. (1967 a) confirmed MURPHY's procedure by independent validation and determined progesterone in plasma during the normal menstrual cycle, in the human and in the rhesus monkey (NEILL et al., 1967 b); the progesterone was purified by thin layer chromatography. This method was modified by JOHANSSON et al. (1968) who used a selected petroleum ether with the specific properties of extracting essentially progesterone. MARTIN et al. (1969) found paper chromatography preferable to thin layer for the purification of progesterone. Further development of this approach led to the elimination of the thin layer chromatography step (JOHANSSON, 1969 a); only 0.25 ml of luteal phase plasma was needed in this simplified procedure. The modified rapid competitive protein binding technique was applied by JOHANSSON (1969 b) for the determination of progesterone in pregnancy plasma. From experience with more than 8,000 plasma progesterone determinations, JOHANSSON (1970) reports a sensitivity limit of 0.1 ng. Additional refinement of the competitive binding technique made it possible to measure plasma progesterone levels in men and ovariectomized women (YOSHIMI and LIPSETT, 1968). An improved method for the assay of plasma progesterone by competitive protein binding was described by REEVES et al. (1970); further assessment of interfering factors and evaluation of the technique was reported by the same group (DE SOUZA et al., 1970).

A similar competitive binding method has been worked out by STROTT and LIPSETT (1968) for the determination of 17-hydroxyprogesterone in human plasma. A sensitivity of 0.06 ng/ml and satisfactory precision and reproducibility has been reported for the procedure (LIPSETT et al., 1970). STROTT et al. (1969) described the measurement of plasma 11-deoxycorticosteroids by competitive binding. A competitive protein binding assay for aldosterone was developed by ROBINSON and FANESTIL (1970) who utilized the aldosterone-binding protein present in the cytosol fraction of renal homogenates from adrenalectomized rats.

Related in principle to the competitive protein binding analysis is a procedure for measurement of steroid-binding capacity in which the steroid solution is equilibrated with a binding protein and a solid adsorbent (HEYNS et al., 1967; DE MOOR and HEYNS, 1968); protein and adsorbent compete for the steroid according to their binding affinities and concentrations. Fuller's Earth, Norit A, Florisil or Sephadex have been used in this competitive adsorption, and a graphic analysis has been developed (HEYNS et al., 1968). The method has some similarity with the earlier estimation of cortisol binding based on the uptake of radiolabeled cortisol by red blood cells (FARESE and PLAGER, 1962); it also is related to the gel equilibration technique (PEARLMAN and CRÉPY, 1967 b). TRAPP and WEST (1969) reported recently a convenient assay for corticosteroid-binding serum proteins by such an adsorption method. Free steroid is separated from the protein-bound fraction by adsorption to dextran-coated magnesium silicate (Florisil). CBG is determined by subtraction of the albumin-bound portion. The authors state that the adsorbent has a very small effect on the binding equilibrium although the results for CBG concentration are somewhat on the low side. The method should be useful for the analysis of large numbers of samples. PEGG and KEANE (1969) have modified the MURPHY procedure to allow the simultaneous determination of plasma cortisol, CBG capacity, and apparent association constant.

In the preceding paragraphs, a short account has been given of the application of competitive binding analysis, either to measure the concentration of binding sites, i. e., CBG capacity, or to determine extremely low quantities of steroids that are bound to CBG with high affinity. Use of similar procedures for the analysis of minimal concentrations of other steroids, i. e., testosterone, estradiol and related steroids which have a high affinity for the sex steroid binding globulin, or for specific receptor proteins of target tissues, will be discussed at the end of Chapter XII. 3. A comprehensive evaluation of the competitive protein binding assay has been given by MURPHY (1969).

New developments in this applied field of steroid-protein interactions have been the topic of the Second Karolinska Symposium on Research Methods in Reproductive Endocrinology, which was held in Geneva in March, 1970 and was concerned with Steroid Assay by Competitive Protein Binding. Expert discussions of the more general aspects were given by MURPHY (1970) and by SLAUNWHITE and SANDBERG (1970). Theoretical aspects of saturation analysis or displacement analysis, the all-inclusive terms for the methods based on these principles (MURPHY, 1970), were presented by EKINS and NEWMAN (1970). REEVES and CALHOUN (1970) outlined criteria of reliability for saturation analysis of steroids by competitive protein binding, including calculations and statistical control. References to additional discussions on competitive protein binding assay of individual steroids will be found in connection with the given compounds in the present chapter or in Chapter XII. 3. The transactions of the symposium have been edited by DICZFALUSY (1970). For radioimmunoassay see Chapter XV.

c) Gel Filtration

Measurement of CBG activity by Sephadex gel filtration (DE MOOR et al., 1962; BOULOUARD and FONTAINE, 1963; MIDDO et al., 1963; QUINCEY and GRAY, 1963; DOE et al., 1964; SMITH and MUEHLBAECHER, 1969) is a simple method which is being widely used, especially in clinical laboratories. GOLIKOV (1969) reports filtration over biogel P-10 with similar results. The principle of the gel filtration technique has been discussed in Chapter IV. 14. Addition of corticosteroid to the test serum in excess of its CBG capacity and equilibration with the endogenous corticoid results in saturation of the high-affinity binding sites with the steroid; subsequent determination of the protein-bound steroid yields the binding capacity directly in μg/100 ml serum. The association constant of the corticosteroid-albumin complex is low enough, and the equilibration rate high enough to allow dissociation so that the eluted albumin is practically free of corticosteroid. The gel filtration technique described by WESTPHAL (1969) follows published procedures (DE MOOR et al., 1962; DOE et al., 1964) with only minor modifications.

The determination of CBG capacity by gel filtration is a convenient method for the routine analysis of clinical serum or plasma samples. It gives reproducible results which are comparable within one species if the same experimental conditions are strictly maintained. The Sephadex column can be regenerated by washing with the elution buffer and may be reused an indefinite number of times.

The procedure has certain limitations, however, which have to be recognized in the interpretation of the results. Possible adsorption of the steroid to the dextran matrix (PEARLMAN and CRÉPY, 1967 b) has to be considered. Most of the restrictions, however, originate from the fact that the method is not based on attainment of equilibrium between unbound and protein-bound steroid but rather on a separation of the two entities in a continuously changing system. The component parts of the CBG-corticosteroid complex, held together by non-covalent bonds, are in a binding equilibrium, subject to the law of mass action. The complex therefore undergoes continuous dissociation as it moves down the Sephadex column. Due to a relatively high association constant and a slow rate of equilibration, actual dissociation is small at 4°; at higher temperatures, the binding affinity decreases and the equilibration rate becomes greater so that at 46° complete dissociation has taken place when the CBG emerges from the column (DOE et al., 1964). This is a temperature at which the equilibrium dialysis technique still indicates definite binding, although at a fraction of the affinity observed at 4°. Another consequence of the dissociation is the impossibility to demonstrate albumin binding of corticosteroid hormones with the gel filtration column, a deficiency turned to advantage in the very determination of CBG capacity.

The following observations by R. G. ALLEN, G. B. HARDING, and N. RUST in our laboratory (WESTPHAL, 1969) provide experimental evidence for the limitations of the gel filtration method. It has been reported (DE MOOR et al., 1962) that stepwise changes of the elution rate between 0.2 and 2.5 ml/min resulted in a slight increase of the measured binding capacity, i. e., from 22.1 to 26.1 μg cortisol bound per 100 ml normal human plasma at room temperature. Greater differences in CBG binding capacity with changing elution rates were observed (QUINCEY and GRAY, 1963) at 37°. The height : diameter ratio of the column appears to be without measurable influence on the results as demonstrated by experiments in which the length of the column was varied in the approximate ratio of 1 : 2 : 4 while the same volume was maintained (Table VIII-2).

The above considerations of the CBG-corticosteroid complex undergoing continuous dissociation during passage over the Sephadex column were tested by experiments recorded in Table VIII-3. The height (volume) of the column was changed in the ratio 1 : 2 : 3 at constant diameter. It is evident that the value for μg cortisol bound per 100 ml serum obtained with the single volume is about 60% higher than that given by the triple volume. Although such drastic differences in experimental design are unlikely to occur in practice, the results show the limitation of the method; the binding capacities obtained even under technically adequate conditions must be considered minimum values and interpreted with caution. An approximation to the correct values can be obtained by gel filtration at three different heights (volumes) such as those shown in Table VIII-3, and extrapolation to height (volume) zero. Proportionality between height (volume) and amount of Sephadex should be verified in this extrapolation procedure.

Based on this principle, an improved method for determination of CBG concentration has been developed recently (HOFFMANN and

Table VIII-2. *Filtration of corticosteroid-loaded sera over Sephadex columns of different lengths but equal volumes*[a]

Species	Endogenous corticosteroid μg/100 ml	Radiocorticoid added	Columns used[b]		
			S	M	L
Human[c]	11.0 F	F	15.9 (15.0; 16.7)	15.6 (15.6; 15.5)	15.5 (15.6; 15.4)
Guinea pig	81.8 F	F	18.5 (19.6; 17.4)	24.5 (27.3; 21.7)	20.3 (18.2; 22.3)
Rat	59.0 B	B	66.4 (63.9; 68.9)	69.5 (70.2; 68.7)	70.7 (72.9; 68.5)
		F	—	48.7[d]	—
Rabbit	11.2 B	B	28.8 (28.3; 29.3)	31.5 (30.8; 32.1)	26.3 (24.7; 27.9)
		F	—	28.3[d]	—

[a] Unpublished experiments with R. G. ALLEN (WESTPHAL, 1969) in which 0.5 ml serum containing 100 μg/100 ml added radioactive cortisol (F) or corticosterone (B) was filtered over Sephadex G-50, coarse, unbeaded. Figures indicate μg/100 ml bound corticosteroid. Average values from two experiments (in brackets). Phosphate buffer, 0.05 M, pH 7.4; 4°.

[b] The home-built jacketed columns with medium fritted glass discs had the following dimensions:

	Inner diameter mm	Height mm	Volume ml
S	20	170	53
M	14	330	50
L	10	670	53

Note the slightly smaller volume of column M, resulting in somewhat higher capacity values measured.

[c] Diluted serum, 3,85% protein. [d] Average value from 4 experiments.

WESTPHAL, 1969). Three jacketed columns (Pharmacia, type K 25/45) were packed with 3.5, 7.0 and 10.5 g, respectively, of Sephadex G-50, coarse, unbeaded. The quality and homogeneity of the gel beds were checked by Blue Dextran 2000 (Pharmacia). Temperature equilibration was obtained by circulating water of $4° \pm 0.02°$ through the column jackets. Serum was equilibrated with radiolabeled cortisol, corresponding to 48—52 μg and 98—102 μg per 100 ml for

Table VIII-3. *Filtration of cortisol-loaded human serum over Sephadex columns of different lengths and volumes* [a]

Column		
Height mm	Volume ml	Cortisol bound μg/100 ml
103	72	14.7 (14.4; 15.0)
205	145	11.5 (11.1; 11.8)
310	219	9.2 (8.5; 9.9)

[a] Unpublished experiments with R. G. ALLEN (WESTPHAL, 1969) in which 0.5 ml diluted (3.85% protein) serum containing 50 μg/100 ml added cortisol 4-^{14}C was filtered over Sephadex G-50, coarse, unbeaded. Phosphate buffer, 0.05 M, pH 7.4; 4°.

normal and pregnancy serum, respectively. The serum had been stripped of endogenous steroids by gel filtration at 45°, and had been reconstituted to the original protein concentration. A 0.05 M phosphate buffer of pH 7.4 was used as solvent and eluting medium. A volume of 0.5 ml of the equilibrated serum was placed in the sample applicators of the three columns and the gel filtration performed at a flow rate of 1 ml/min. Fractions of approximately 1 ml were collected and analyzed for radioactivity and absorbance at 280 nm.

As an example of the results obtained with this procedure, Fig. VIII-4 shows the separation of protein-bound and unbound cortisol-4-^{14}C in a normal human serum from which the endogenous steroids had been removed by gel filtration at 45°. The peaks of free cortisol are not drawn on the graph. However, it is evident that the interval between protein-bound and unbound cortisol can be clearly distinguished; because of continuous dissociation of the cortisol-CBG complex, even at 4°, the cortisol concentration in this interval never reaches zero.

Under the conditions of the experiment, the total quantity of dissociated cortisol will be proportional to the amount of Sephadex used for the filtration. Plotting the bound cortisol as determined in the eluates from the three columns versus the amount of Sephadex

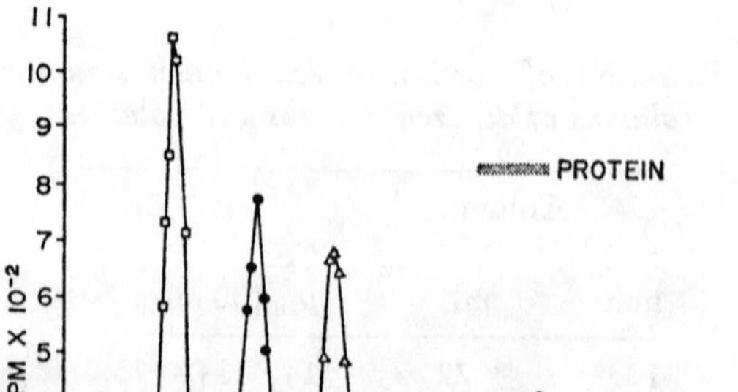

Fig. VIII-4. Separation of protein-bound and unbound 4-^{14}C cortisol after equilibration at 37° for 3 hours with stripped human serum (N_1). Three different amounts of Sephadex G-50 were used. □ 3.5 gm Sephadex G-50 (Column 1); ● 7.0 gm Sephadex G-50 (Column 2); △ 10.5 gm Sephadex G-50 (Column 3). The shaded bars indicate the presence of protein in the three eluates. From HOFFMANN and WESTPHAL (1969)

employed and extrapolating to a zero quantity of Sephadex is therefore expected to give a binding value which is corrected for loss by dissociation and possible reversible adsorption. This is shown in Fig. VIII-5 which summarizes the results obtained for two normal sera (N_1 and N_2) and two pregnancy sera (P II and P III). The binding capacity values obtained for these four sera by extrapolation to zero grams of Sephadex are 29.5, 21.9, 40.8, and 47.2 μg per 100 ml, respectively. The straight lines connecting the three experimental points for the different sera in these and other experiments indicate validity of the assumption that the extent of dissociation and amount of Sephadex are directly proportional.

The results obtained with the sera N_1, N_2, P II and P III (Fig. VIII-5) were compared with those derived from equilibrium dialysis experiments. Fig. VIII-6 shows the graphic evaluation of the

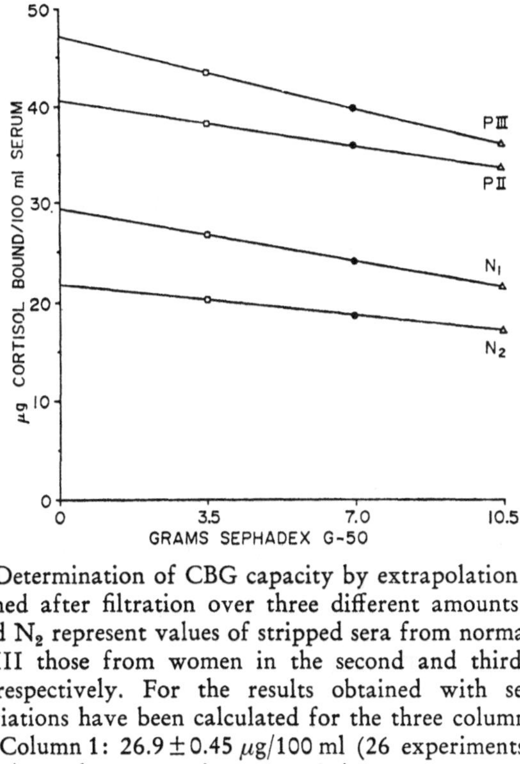

Fig. VIII-5. Determination of CBG capacity by extrapolation of the bound values obtained after filtration over three different amounts of Sephadex G-50. N_1 and N_2 represent values of stripped sera from normal individuals, P II and P III those from women in the second and third trimester of pregnancy, respectively. For the results obtained with serum N_1 the standard deviations have been calculated for the three columns specified in Fig. VIII-4: Column 1: 26.9 ± 0.45 μg/100 ml (26 experiments); Column 2: 24.0 ± 0.61 μg/100 ml (8 experiments); Column 3: 21.3 ± 0.57 μg/100 ml (8 experiments). From HOFFMANN and WESTPHAL (1969)

binding data by the method described in Chapter VIII. 2 a (Fig. VIII-3). It is apparent that at very low concentrations of added cortisol the unbound cortisol fraction increases very little due to the high binding affinity of CBG; this portion of the curves cannot be distinguished from the abscissa and is therefore not indicated in Fig. VIII-6. When all available binding sites of the CBG become saturated the unbound portion of cortisol rises at a much greater rate with increasing concentration of total cortisol, indicating binding to albumin. The intersection of the two curves provides a measure for the concentration of CBG.

The dotted lines in Fig. VIII-6 extend the curves to the points on the abscissa which represent the binding capacity values obtained

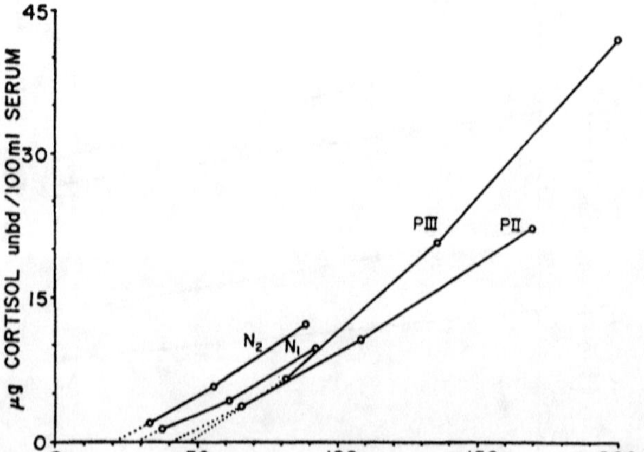

Fig. VIII-6. Determination of CBG capacity in the four stripped human sera described in Fig. VIII-5 by equilibrium dialysis according to WESTPHAL (1967). The unbound (unbd) cortisol values for the high-affinity binding system (CBG) are too low to be distinguished from the abscissa. The three points on each line represent low-affinity binding to albumin. Intersection of the lines with the abscissa provides a measurement of the CBG binding capacity. The dotted lines are extensions to the values obtained by the extrapolation shown in Fig. VIII-5. From HOFFMANN and WESTPHAL (1969)

from Fig. VIII-5. These extensions appear as acceptable extrapolations of the curves N_1, N_2 and P II. In the case of P III, the extrapolation would intersect with the abscissa at about 56 μg/100 ml instead of 47.2 μg/100 ml resulting from Fig. VIII-5. This discrepancy illustrates the uncertainty of binding values obtained by equilibrium dialysis with a limited number of concentrations. Evidently, many more experimental points would be necessary in order to obtain a well-defined intersection between the lines for the CBG and the albumin binding systems.

Table VIII-4. *Sephadex filtration*[a] *of serum containing different amounts of added cortisol*

Radiocortisol added[b] μg/100 ml	Cortisol bound in	
	Human serum[c] μg/100 ml	Serum of pregnant guinea pig[d] μg/100 ml
50	14.7 (14.4; 14.9)	381 (378; 384)
100	15.8 (avg. of 6 exp.)	414 (422; 405)
200	18.7 (17.7; 19.6)	402 (409; 394)
400	28.2 (27.1; 29.3)	470 (488; 456)
600	—	527 (530; 524)
800	—	539 (540; 538)

[a] Unpublished experiments with R. G. Allen and N. Rust (Westphal, 1969). Column 14 mm i.d., 330 mm high; Sephadex G-50, coarse, unbeaded; 0.05 M phosphate of pH 7.4; 4°.

[b] Total of labeled plus unlabeled cortisol.

[c] 0.5 ml diluted serum (3.85 g protein, 11.0 μg cortisol per 100 ml), containing different amounts of added mixtures of cortisol 4-^{14}C and unlabeled cortisol, was applied to column.

[d] 0.6 ml, composed of 0.5 ml serum (4.03 g protein, 432 μg cortisol per 100 ml) and 0.1 ml aqueous solutions of mixtures of cortisol 4-^{14}C and unlabeled cortisol, was applied to column.

The equilibrium dialysis data obtained with the sera N_1, N_2, P II and P III have also been analyzed by the graphic method of Rosenthal (1967). The values resulting from this technique were in agreement with the results shown in Fig. VIII-6.

It has been pointed out before that for a complete, thermodynamically valid determination of binding parameters, i. e., association constants and concentration of binding sites, a procedure is required that measures the bound and unbound portion at equilibrium. Pertinent methods, such as equilibrium dialysis, cannot be easily replaced. However, equilibrium dialysis is time consuming and exposes the solutions for many hours to potentially harmful conditions. In contrast, gel filtration is rapid. Combining the gel filtration technique with an extrapolation procedure eliminates the main disadvantage caused by dissociation and possible reversible adsorption. The preceding method thus provides a reliable and convenient determination of the concentration of binding sites. It should prove

Table VIII-5. *CBG activity of sera determined by equilibrium dialysis and by Sephadex filtration*[a]. From WESTPHAL (1969)

Species	Endogenous corticosteroid[a, b] μg/100 ml	Radioactive corticosteroid added for assay	Equilibrium dialysis C-value[c] at 37°	Equilibrium dialysis Corticosteroid actually bound at 37° μg/100 ml	Gel filtration binding capacity at 4° μg/100 ml
Human[d]	18.6 F	F	0.16	17.0	26.7
Rat	59.0 B	B	0.46	57.2	69.5
Rabbit	11.2 B	B	0.14	10.1	31.5
Guinea pig	81.8 F	F	0.03	50.0	25.7

[a] Average values from duplicate experiments.
[b] F = Cortisol; B = Corticosterone.
[c] Multiple equilibrium dialysis for the two sera having the same endogenous corticosteroid.
[d] Diluted serum; values corrected for normal concentration of 6.5% protein.

useful in all cases where knowledge of the unbound portion of the ligand-macromolecule system is not essential.

Another experimental variable which influences the result is the quantity of corticosteroid added to "overload" the serum. Table VIII-4 shows that even at a ratio of bed volume : sample volume of about 100 : 1, the differences in CBG capacity values obtained are considerable. This is evident for the human serum as well as the guinea pig serum tested. It should be noted that CBG concentration and serum cortisol level in the pregnant guinea pig (see Chapter X) are extremely high.

In Table VIII-5, a comparison is made of the CBG activity values of sera of four species determined by equilibrium dialysis and by gel filtration. A fair proportionality is seen between C-values at 37° and binding capacity values obtained at 4° for human, rat and rabbit serum. For the guinea pig serum, however, the C-value appears low in relationship to the binding capacity. This is presumably caused by the relatively small association constant of guinea pig CBG for cortisol which is only about one third to one tenth that of the

CBG's of the other species (see Table VIII-21). In other experiments, the C-values obtained for two pools of rabbit serum (see Table VIII-19) were compared with CBG capacities. It was found by Sephadex G-50 gel filtration at 4° that Pool I had approximately 3.0 times the corticosterone-binding capacity as the serum from Pool II, whereas it showed a 2.9-fold greater C-value (CHADER and WESTPHAL, 1968 b).

3. Removal of Endogenous Steroid from Serum

Early attempts in our laboratory to extract the endogenous corticosteroids from serum with organic solvents at low temperature without loss of CBG activity were on the whole unsuccessful. This agrees with the experience of other investigators (DE MOOR et al., 1963). Better results were obtained when advantage was taken of the considerable increase of dissociation of the steroid-CBG complex with increasing temperature. This temperature effect is seen in human serum (BUSH, 1957; DAUGHADAY, 1959; SLAUNWHITE and SANDBERG, 1959; MILLS et al., 1960), and has also been demonstrated in other species (WESTPHAL et al., 1962; SEAL and DOE, 1963; WESTPHAL, 1967). Extended dialyses in various systems at elevated temperature provided conditions of slow removal of CBG-bound steroids as has been described elsewhere (WESTPHAL, 1964). As an example of this approach, Fig. VIII-7 shows the dialysis of a diluted rat serum against phosphate buffer at 40°, with frequent changes of the outside solution. As to be expected, the C-values increase with continued removal of corticosteroid from the serum, indicating that previously occupied CBG binding sites become available for the added radioactive cortisol. Fig. VIII-8 (upper curve) illustrates the gradual removal of endogenous corticosteroid, determined as corticosterone, from a rabbit serum by dialysis at 37° against several changes of phosphate buffer. The major portion of the steroid has left the inside solution after 12 hrs; CBG activity is maintained. The rate of dialysis increases when the outside solution contains 1% HSA in the same phosphate buffer. DE MOOR et al. (1963 c) reported removal of 63% of the unconjugated corticoid by dialysis of human plasma against a 1% BSA solution for 12 hrs at 40°, also without alteration of CBG. Apparently, the albumin molecules with their affinity for steroids facilitate the uptake of the steroid molecules from the outer surface of the dialysis membrane and its transport into the outside solution. A similar phenomenon was observed when serum from adrenalectomized rats (diluted 1 : 5) with its high CBG concentration (see Chapter IX. 1) was used as outside solution in an

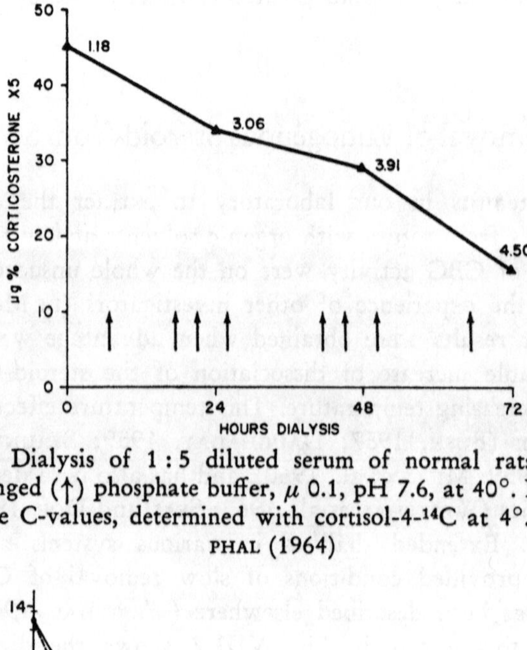

Fig. VIII-7. Dialysis of 1 : 5 diluted serum of normal rats against frequently changed (↑) phosphate buffer, μ 0.1, pH 7.6, at 40°. Numbers over curve denote C-values, determined with cortisol-4-^{14}C at 4°. From WESTPHAL (1964)

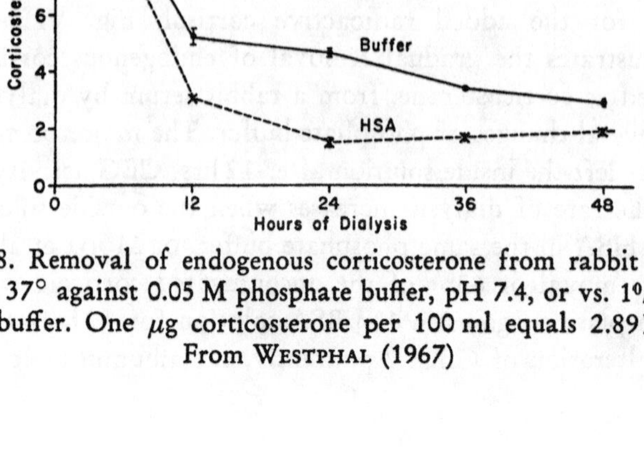

Fig. VIII-8. Removal of endogenous corticosterone from rabbit serum by dialysis at 37° against 0.05 M phosphate buffer, pH 7.4, or vs. 1% HSA in the same buffer. One μg corticosterone per 100 ml equals 2.89×10^{-8} M. From WESTPHAL (1967)

experiment to remove corticosterone from normal rat serum (diluted 1 : 5) by dialysis at 37° (WESTPHAL, 1964). The mechanism of this increase of permeation rate by the binding proteins in the outside phase of the dialysis system should be of interest to the student of membrane transport. We assume that the presence of steroid-binding proteins on one side of a cell membrane or a capillary wall facilitates the transport of the steroid hormone by maintaining a greater and more constant concentration gradient. This may occur, for example, when the newly biosynthesized hormone enters the circulating blood, and may be an important function of the steroid-binding proteins.

Removal of the endogenous steroids from the serum by dialysis is time consuming and exposes the binding protein for a long time to a comparatively high temperature. A more convenient procedure is that of gel filtration at 45°, utilizing increased dissociation and higher rate of equilibration (DOE et al., 1964) of the CBG-steroid complex at the elevated temperature. This "stripping" technique is based on the original gel filtration method of DE MOOR and associates (1962) and similar methods used in other laboratories (QUINCEY and GRAY, 1963; DOE et al., 1964). Sephadex G-25 or G-50 can be used in a chromatographic column; neither type of gel retains serum proteins which appear immediately following the void volume in the effluent of the column. The proteins are eluted in a volume four or more times the original one, depending on amount of serum filtered and the dimensions of the column. For a detailed description of the gel filtration technique at 45° see WESTPHAL (1969). Gel filtration at low temperature, at pH 5.0—5.5, has been suggested as a possibly safer procedure for removal of endogenous steroids (SLAUNWHITE and SANDBERG, 1970).

Figure VIII-9 gives an example of the procedure in which endogenous corticosterone was removed from rat serum by Sephadex G-25 filtration at 45°. Only the protein-free eluate is shown. After the protein has left the column, a volume of several milliliters can be measured before the corticosterone, detached from the binding serum proteins, appears in the effluent. It should be noted that the residual concentration of corticosterone in the stripped serum was determined (SILBER et al., 1958) as 3.5 μg per 100 ml. This is a value which has been found also in completely adrenalectomized rats a few days after operation; it is not considered to indicate the presence of corticosterone, but rather of non-specific fluorogens which interfere with the fluorometric determination of corticosteroids (DALY and SPENCER-PEET, 1964; EECHAUTE, 1966 a; KENDALL et al., 1968). In view of the uncertainties involved, we are not subtracting this "background fluorescence" from the corticosterone values in our laboratory (WESTPHAL et al., 1962).

Results of gel filtration experiments to strip endogenous corticosteroids from sera of several species are given in Table VIII-6. Colorimetric determination (PETERSON et al., 1957) of cortisol in human and guinea pig serum shows practically complete removal;

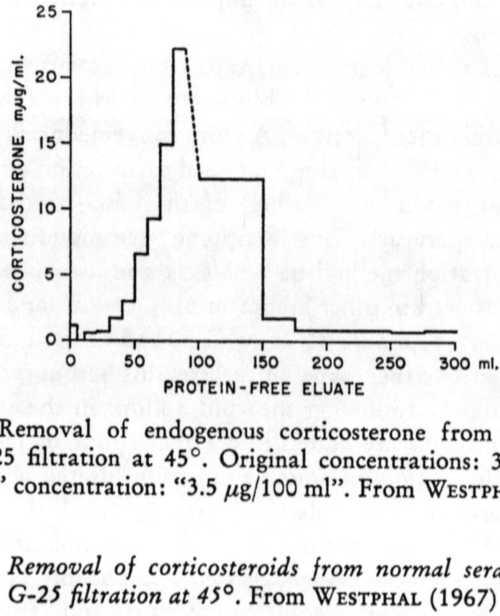

Fig. VIII-9. Removal of endogenous corticosterone from rat serum by Sephadex G-25 filtration at 45°. Original concentrations: 35.4 μg/100 ml; "residual" concentration: "3.5 μg/100 ml". From WESTPHAL (1966)

Table VIII-6. *Removal of corticosteroids from normal sera by Sephadex G-25 filtration at 45°.* From WESTPHAL (1967)

Species	Steroid [a]	Before filtration μg/100 ml	Before filtration 10^{-8} M	After filtration μg/100 ml
Human	Cortisol	14.5	41.4	0.14
Rat	Corticosterone	35.3	102	4.1 [b]
Rabbit	Corticosterone	9.3	26.9	1.3 [b]
Guinea pig	Cortisol	81.8	226	0.13
	Corticosterone	13.1	37.9	2.2 [b]

[a] Cortisol and corticosterone were determined by a colorimetric (PETERSON et al., 1957) and fluorometric (SILBER et al., 1959) method, respectively.

[b] Presumably "residual" fluorescence; see text.

less than 1% of the steroid remained after the filtration at 45°. The corticosterone values indicate the residual or background fluorescence discussed in the preceding paragraph. Overlapping specificities of the methods used for the determination of cortisol and corticosterone make the relative concentrations of these two steroids in the guinea pig serum uncertain.

Because of the ambiguity of the fluorometric corticosterone determinations in the sera after gel filtration, another more stringent test was applied to measure the degree of steroid removal by the Sephadex procedure. As will be discussed in Chapter VIII. 5 a, the association constant of the complex human CBG-progesterone at 37° is about three times that of the corresponding complex with cortisol. Stripped human serum was therefore equilibrated with a limited amount of progesterone-4-^{14}C to give a concentration of 10 μg/100 ml (3.18×10^{-7} M). This level of progesterone is below the saturation of the high-affinity binding sites. The serum was then filtered over Sephadex G-25 at 45° and the progesterone-4-^{14}C content of the protein eluate was determined. Fig. VIII-10 is a composite graph averaged from three experiments, the results of which were in close agreement. It shows that more than 99.5% of the total progesterone-4-^{14}C was removed from the serum protein by the gel filtration; 99.7% of the total available binding sites were cleared of progesterone by this procedure. There is a wide gap which separates the residual protein-bound radioactivity from the unbound progesterone-4-^{14}C, which shows some retention by the Sephadex gel. It is concluded from these studies that the removal by the Sephadex filtration procedure of a steroid from its protein complex is practically complete even if its association constant is as high as that of the CBG-progesterone complex, assuming that the rate of dissociation at elevated temperature is of high order as observed for the cortisol-CBG complex in human serum (Doe et al., 1964).

The CBG in human plasma has been reported to be inactivated by heating at 60° (Daughaday, 1959) for 10—15 minutes (Daughaday et al., 1962). It was therefore necessary to test the stability of CBG in the sera subjected to the gel filtration procedure at 45°. Table VIII-7 shows that a second gel filtration of human and rat serum at 45° does not affect the binding affinity measured after the first filtration; the C-values do not differ significantly from each other. These results confirm observations (Doe et al., 1964) on thermal stability of human CBG subjected to gel filtration at 46°. The combining affinities or C-values at 4° and 37° for human and rat serum (Table VIII-7) increase after the first Sephadex filtration as would be expected as a consequence of the removal of the endogenous corticosteroids, and possibly other steroids bound by CBG. It is concluded from these experiments that CBG in human or rat serum is stable to the gel filtration procedure at 45° under the conditions applied. It should be

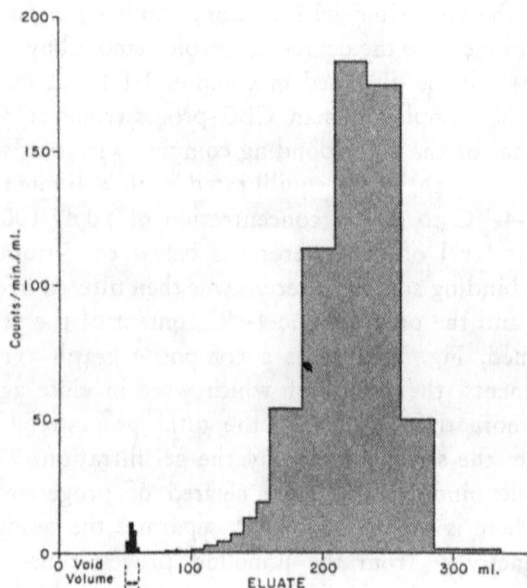

Fig. VIII-10. Removal of progesterone-4-^{14}C from human serum by Sephadex G-25 filtration at 45°. Original concentration of progesterone-4-^{14}C added to "stripped" serum, 10.0 μg/100 ml $= 3.18 \times 10^{-7}$ M; after filtration, <0.05 μg/100 ml. From WESTPHAL (1967)

Table VIII-7. *Stability of CBG to gel filtration at 45°. Binding affinity determined as C-values in 0.05 M phosphate, pH 7.4, at 4° and 37°.* From WESTPHAL (1967)

Serum sample	Human serum		Rat serum	
	4°	37°	4°	37°
Before filtration	1.7	0.33	20.0	0.70
After first filtration	5.9	0.40	24.3	0.83
After second filtration	5.7	0.39	22.9	0.86

noted, however, that this thermal stability is not an intrinsic property of the CBG molecules. When the pure, homogeneous corticosteroid-CBG complexes prepared from serum of man (MULDOON and WESTPHAL, 1967), rabbit (CHADER and WESTPHAL, 1968 b) and rat (CHADER and WESTPHAL, 1968 c) are subjected to similar gel filtration at 45°, the glycoproteins become inactivated. This would indicate a protective effect of other serum proteins, present in comparatively massive amounts, on the steroid-binding properties of the CBG molecules.

The gel filtration at 45° is considered the method of choice when complete removal of endogenous steroids is to be achieved. The technique is not practical, however, in the routine analysis of many samples, as for instance in clinical studies or in animal experimentation, especially when the original protein concentration has to be reconstituted. A simpler, less time-consuming method to remove the major portion of endogenous steroids from plasma or serum is available in the adsorption to powdered charcoal (HEYNS et al., 1967). The plasma samples are gently shaken for 30 min at room temperature with 50 mg of Norit A per ml. The adsorbent is removed by centrifugation and filtration. This treatment reduces the endogenous corticosteroid levels to very low values (2.6 ± 1.0 (S.D.) μg per 100 ml). Other steroids are also removed from plasma. The portion remaining in the plasma depends on concentration and affinity of the steroid-binding proteins which form part of a binding equilibrium; the carbon adsorbent and the binding proteins compete for the steroids. This is different in principle from the gel filtration technique at 45° in which the equilibrium is continuously disturbed by dissociation of the complex during the migration over the length of the Sephadex column.

4. General Properties of CBG in Serum

a) Influence of Temperature on Binding Affinity

Long before CBG preparations of man or other species became available in pure form, various properties of the binding macromolecule were recognized which were confirmed after isolation of the active protein. Early studies showed that the portion of unbound cortisol in undiluted human plasma was extremely low (approximately 1%) at 4° (Daughaday, 1958 b) or even at room temperature (Upton and Bondy, 1958); however, the percentage of unbound steroid at 37° was about 5 to 10% measured either by equilibrium dialysis (Bush, 1957) or by ultrafiltration (Mills and Bartter, 1959). Slaunwhite and Sandberg (1959) clearly demonstrated greater dissociation of the human CBG-cortisol complex at 37° than at 4°;

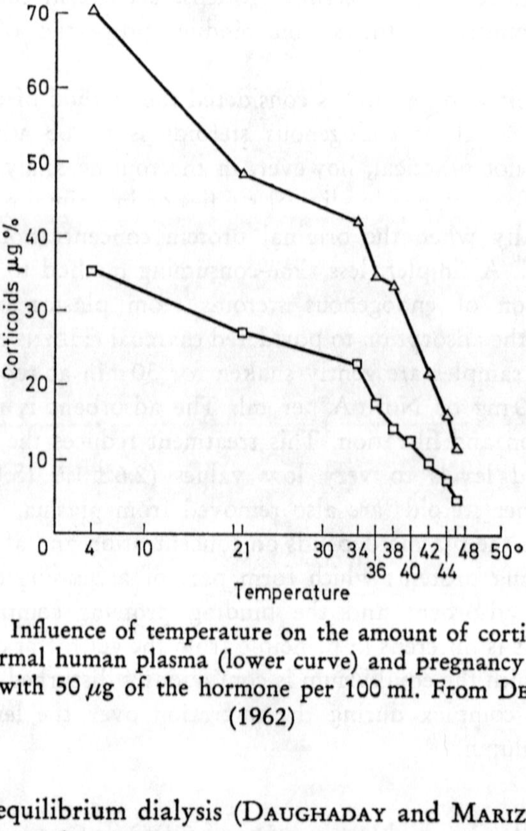

Fig. VIII-11. Influence of temperature on the amount of cortisol bound by 100 ml of normal human plasma (lower curve) and pregnancy plasma after overloading with 50 µg of the hormone per 100 ml. From De Moor et al. (1962)

subsequent equilibrium dialysis (Daughaday and Mariz, 1961) and gel filtration studies (de Moor et al., 1962; Doe et al., 1964) confirmed this result. Fig. VIII-11 shows the temperature effect on CBG binding in normal human plasma and in pregnancy plasma.

Verification was later obtained with the purified CBG preparations of man (Seal and Doe, 1962 b; Muldoon and Westphal, 1967), rabbit (Chader and Westphal, 1968 b), and rat (Chader and Westphal, 1968 c). In a number of other species, the binding activity as measured by gel filtration (Seal and Doe, 1963) was also found to be considerably lower at 37° than at 4° (see Table VIII-18). Fig. VIII-12 shows that a similar relationship between temperature and binding affinity exists also for the complex between corticosterone and human CBG, as well as for the corticosterone interaction with CBG of rat and rabbit serum. In all cases, the C-values at a temperature of 37° or higher were only small fractions of those observed

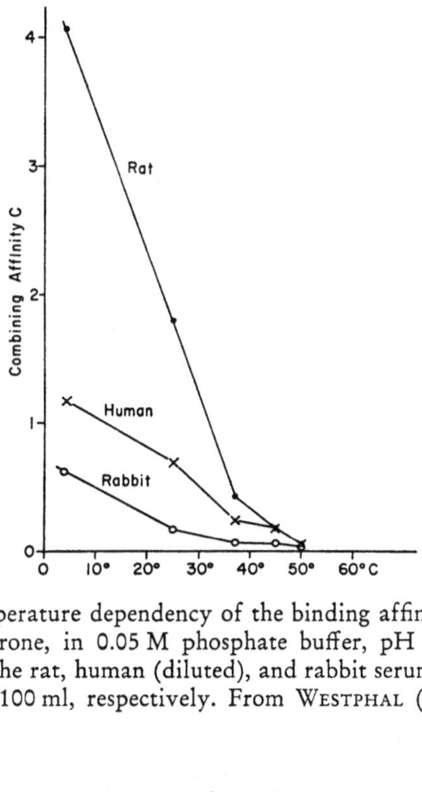

Fig. VIII-12. Temperature dependency of the binding affinity of "stripped" sera for corticosterone, in 0.05 M phosphate buffer, pH 7.4. The protein concentrations of the rat, human (diluted), and rabbit serum are 6.8, 4.7 and 6.5 mg/100 ml, respectively. From Westphal (1967)

at 4° (see Table VIII-21). High-affinity binding of cortisol in salmon (Salmo salar) plasma was found to have a similar temperature dependency when measured by gel filtration between 4° and 40° (Freeman and Idler, 1966); however, the decrease between 4° and 20° was much less than observed in man and other mammalian species.

Mills (1961) has raised the interesting question of whether the rapidly increasing dissociation of the CBG-cortisol complex at temperatures above 37° has physiological implications. Rise of general body temperature during acute illness may result in release of more cortisol to the tissues. In a similar way, local elevation of temperature of the skin and superficial structures as a consequence of inflammation, may contribute to greater availability of cortisol to these tissues. The physiological significance of these possibilities remains to be established.

The substantial decrease of the combining affinity at the high temperatures seen in Fig. VIII-12 is not completely reversible. Table VIII-8 shows that exposure of steroid-stripped human and rat serum

Table VIII-8. *Heat lability of CBG in human, rat and rabbit serum* [a]

Species	4°	50°	4° [b]	Percent of original activity lost by exposure to 50°
Human	2.62	0.03	0.19	93
Rat	7.92	0.04	0.15	98
Rabbit	0.60	0.04	0.12	80

[a] Average C-values from duplicate multiple equilibrium dialysis experiments with corticosterone-4-^{14}C in 0.05 M phosphate, pH 7.4. The sera were cleared of endogenous steroids by gel filtration at 45°. From WESTPHAL (1969).

[b] Equilibrium dialysis at 50° for 48 hours, then for an additional period of 48 hours at 4°.

to 50° for 48 hrs, the usual duration of the equilibrium dialysis, destroys the CBG activity almost completely. Similar treatment of rabbit serum appears to inactivate about 80% of the affinity. The final C-values after heat inactivation are smaller for the three species and the difference in percent loss of activity may be apparent only since the proportional contribution of albumin to the binding of corticosterone is higher in the rabbit serum; the greater thermal stability of serum albumin is well recognized (DAUGHADAY et al., 1962). As mentioned above, CBG in human serum is inactivated by heating at 60° for about 20 minutes (DAUGHADAY et al., 1962; DAUGHADAY, 1967). It is of interest for the mechanism of this inactivation that it is dependent on the concentration of cortisol present. The results on heat inactivation summarized in Table VIII-8 were obtained with sera from which the endogenous steroids had been removed. Fig. VIII-13 shows the protective effect of increasing levels of the corticosteroid on rate and degree of the thermal inactivation (Doe et al., 1964). In comparison to the heat lability of CBG at 50° and 60°, Slaunwhite and Sandberg (1959) observed CBG in human serum to be stable when exposed to 37° for at least 72 hrs. The same result was obtained for the CBG activity in rat serum (Westphal, 1961). Freezing and thawing does not affect the corticosteroid-bind-

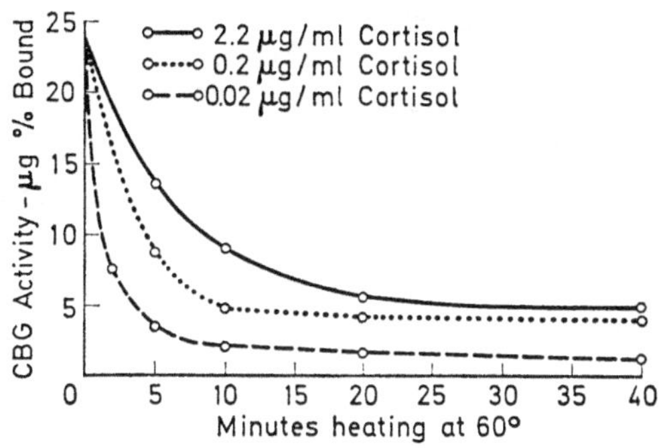

Fig. VIII-13. Protective effect of cortisol upon heat inactivation of human CBG. From Doe et al. (1964)

ing affinity (Slaunwhite and Sandberg, 1959), and no time limit has been noticed in our laboratory to storage of serum or solutions of partially purified CBG at —85° C.

Doe et al. (1964) in establishing the conditions for the gel filtration procedure have examined the equilibration time of CBG in human serum with endogenous bound corticosteroid and added radiolabeled cortisol. They found, in agreement with observations by Murphy (1964), that the exchange equilibrium was essentially instantaneous at 37.5°; complete exchange was observed after one minute. However, at 4°, equilibration had not yet been achieved after one hour. More recent studies of Dixon (1968) confirmed these results, and provided quantitative data on the equilibration times. Human plasma was incubated with a trace quantity of ^{14}C-cortisol for binding to CBG. A larger amount of ^{3}H-cortisol was then added, and the kinetics of the exchange equilibrium between the two isotopic cortisol species was studied by Sephadex gel filtration at several temperatures.

It was found that association is a rapid process at 4°, being complete in minutes, and virtually instantaneous at 37°. Dissociation of the cortisol CBG complex, however, was slow; the half-dissociation times were determined to be 25 min at 4°, 1 min at 22° and about 10 sec at 37°. For quantitative equilibration by dissociation, or for qualitative equilibrium which is necessary whenever radiolabeled corticosteroids are used with sera containing their original corticoid level, several hours are required at 4°, 6 min at 22°, and about 1 min at 37° to be essentially complete. These valuable studies (DIXON, 1968) have been performed with a normal human plasma of unknown endogenous corticosteroid content and of unknown CBG capacity. The reasonable assumption is made that all CBG binding sites are homogeneous, and that all of the endogenous steroid participates in the equilibration. A confirmation of the results with a stripped serum, free of endogenous steroids, may be a desirable supplementation of these important observations.

b) Influence of pH on Binding Affinity

The binding affinity between the corticosteroid hormones and CBG is dependent on the hydrogen ion concentration; a maximal value has been observed (MILLS, 1959; DE MOOR et al., 1963 c) at approximately pH 8. At pH 5 and below, irreversible denaturation appears to take place with loss of the binding affinity (DAUGHADAY and MARIZ, 1961; SEAL and DOE, 1962 b). Table VIII-9 shows the influence of pH on the CBG activity in human and rat serum, measured as corticosterone binding. Fig. VIII-14 illustrates that maximal C-values for CBG in both species are reached at pH 7.9; the apparent affinity is generally higher in the rat serum. The C-values are slightly lower in the ammediol than in the citrate-phosphate buffer; similar observations will be discussed below for the progesterone-AAG interaction. It is evident from Table VIII-9 that some irreversible inactivation takes place at pH 2.8. However, the very low binding affinity at this pH is not entirely the result of this inactivation; the CBG molecule appears to be somewhat protected while in the serum. The decrease of binding affinity from the maximum near pH 8 to the value observed at pH 9.8 is completely reversible.

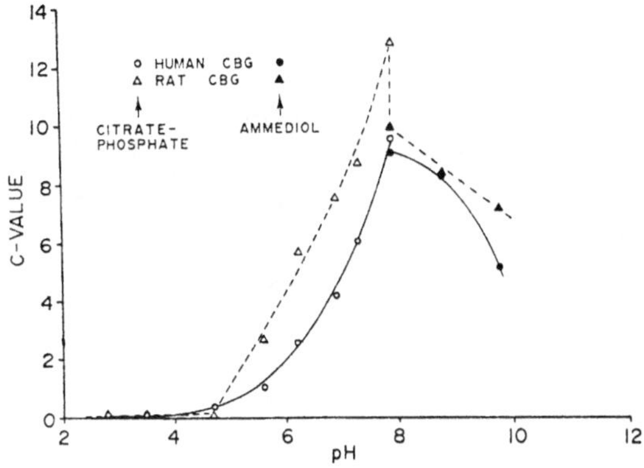

Fig. VIII-14. Corticosterone-binding activity (C-value) in human and rat serum at 4°, at different pH values. The break in the curves at pH 7.9 results from use of different buffers. For further details see Table VIII-9. From WESTPHAL (1970 a)

Table VIII-9. *Corticosterone-binding affinity of human and rat serum at different pH values* [a]

pH	Buffer [b]	C-values for serum of Human	Rat
2.8	CP	0.11	0.09
3.5	CP	0.13	0.13
4.7	CP	0.36	0.08 [c]
5.6	CP	1.05	2.68
6.2	CP	2.55	5.68
6.9	CP	4.14	7.51
7.3	CP	6.05	8.72
7.9	CP	9.66	12.8
7.9	A	9.08	9.92
8.8	A	8.21	8.38
9.8	A	5.10	7.15
2.8 → 7.35 [d]	CP	0.46	1.63
9.8 → 7.35 [d]	A	6.52	9.35

[a] From WESTPHAL (1969). Average C-values from duplicate multiple equilibrium dialysis experiments for 48 hours at 4°. Sera cleared of endogenous steroids by gel filtration at 45°.

[b] CP = 0.05 M citrate-phosphate; A = 0.05 M 2-amino-2-methyl-1,3-propanediol (ammediol).

[c] Reason for low value unknown.

[d] Samples dialyzed for 48 hours at pH 2.8 or 9.8, at 4°, then equilibrated for 24 hours at pH 7.35, 4°; C-values determined at this pH.

c) *Electrophoretic Characteristics*

The first demonstration of CBG being an α-globulin was given by DAUGHADAY (1958 c) when he applied the principle of equilibrium paper electrophoresis to continuous flow electrophoresis on the hanging curtain. At pH 8.8 in Veronal buffer, the major portion of cortisol-4-^{14}C or corticosterone-4-^{14}C was transported with the α-globulin fraction whereas only a minor quantity was associated with albumin (Fig. VIII-15). Similar results were reported by SLAUNWHITE and SANDBERG (1959). Binding of endogenous 17-hydroxycorticosteroids to the α_1-globulin component was observed in human plasma by starch block electrophoresis at 5° in Veronal buffer, pH 8.6 (DOE et al., 1960). OSORIO and SCHATS (1964) have reported that radiolabeled cortisol migrates with the α-globulin component in conventional paper strip electrophoresis. They were unable to demonstrate binding of cortisol to albumin with their technique, in contrast to earlier observations (DAUGHADAY, 1956 c) by equilibrium paper electrophoresis which clearly showed corticosteroid binding to serum albumin. The reasons for the discrepancies of these results with those

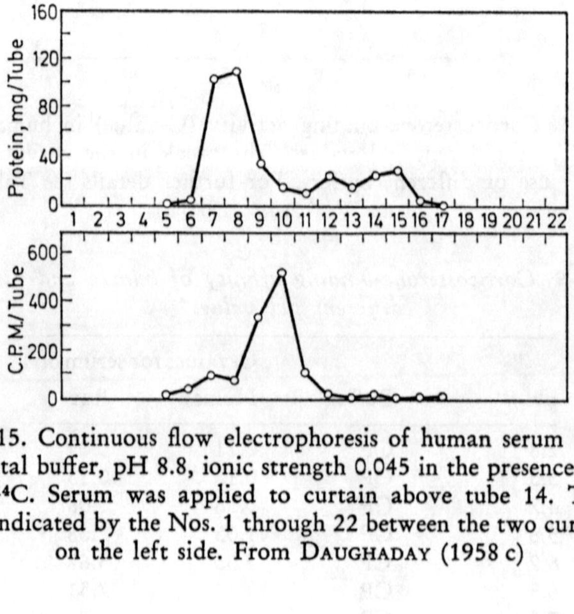

Fig. VIII-15. Continuous flow electrophoresis of human serum carried out with barbital buffer, pH 8.8, ionic strength 0.045 in the presence of corticosterone-4-^{14}C. Serum was applied to curtain above tube 14. The elution tubes are indicated by the Nos. 1 through 22 between the two curves. Anode on the left side. From DAUGHADAY (1958 c)

obtained in other laboratories (WESTPHAL et al., 1955; DAUGHADAY, 1956 c; UPTON and BONDY, 1958) are not obvious; the authors (OSORIO and SCHATS, 1964) used Tris-maleate buffer of pH 7.4 and 8.6 in their experiments. Results with the isolated human CBG confirmed the electrophoretic properties of an α_1-globulin (SEAL and DOE, 1962 b; MULDOON and WESTPHAL, 1967).

The α-globulin nature of CBG was demonstrated in other mammalian species by equilibrium paper-strip electrophoresis (WESTPHAL and DE VENUTO, 1966). Fig. VIII-16 shows the migration of the major portion of added radiolabeled corticosteroids in the sera of rat, rabbit, ox and horse. No difference was found in the paper-electrophoretic properties between CBG from normal and adrenalectomized rats. The highest CBG activity in the sera of rat and rabbit was found in the α_1 rather than the α_2 region, in agreement with the observations on human CBG. Recent results with pure, homogeneous CBG preparations from rabbit (CHADER and WESTPHAL, 1968 b) and rat serum (CHADER and WESTPHAL, 1968 c) also indicated the α_1-globulin nature of these glycoproteins. No distinction between α_1 and α_2-globulins was possible in the bovine and horse sera (Fig. VIII-16).

The paper-electrophoretic procedure used is not precise, and the results are at best "semiquantitative". No significance is attributed, therefore, to differences in specific activity of the peaks of CBG in different species. Similarly, the low activities seen close to the origin on Fig. VIII-16 are not considered indicative of other binding protein components. Spreading of radioactivity from the α_1-globulin area to an adjacent protein component may be caused by the discontinuous nature of the analysis resulting from cutting the paper strip into segments. Within these limitations, however, the equilibrium paper-electrophoresis technique proves to be a reliable method to allocate high-affinity binding proteins to distinct electrophoretic components.

Continuous-flow electrophoresis on the hanging paper curtain has been applied to determine whether the high-affinity binding of cortisol and corticosterone is associated with one and the same electrophoretic component in human serum (WESTPHAL and DE VENUTO, 1966). The serum proteins were separated in Veronal buffer at pH 8.6 into 26 fractions which were dialyzed and lyophilized. Multiple equilibrium dialysis at 37° in the presence of radiolabeled cortisol or corticosterone showed that the same fraction of the α_1-globulin area (Fig. VIII-17, Fraction No. 19) has the strongest bind-

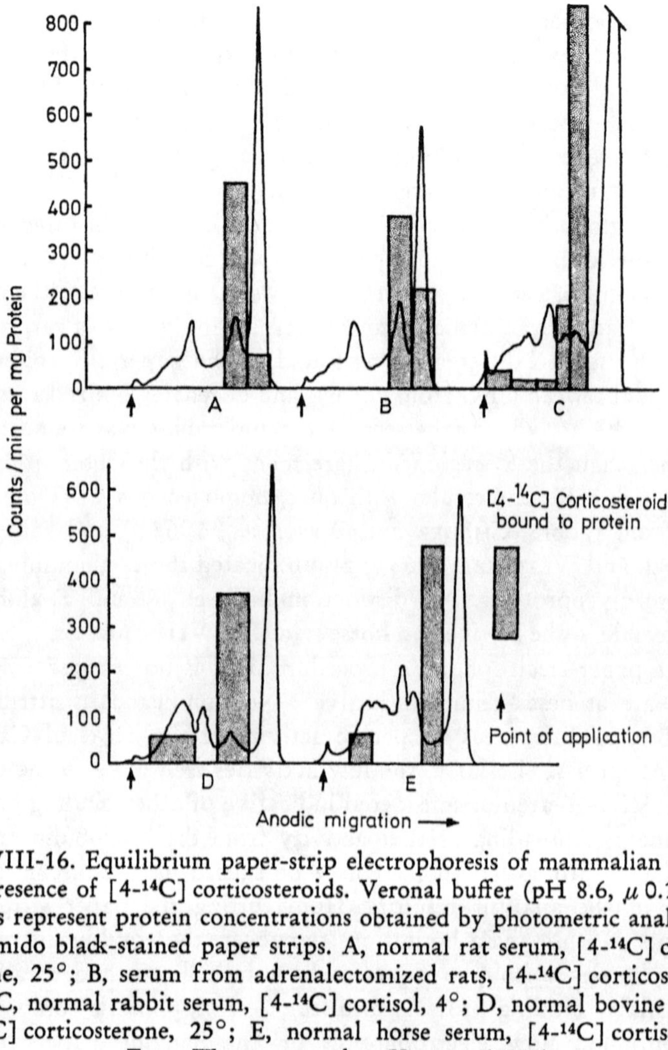

Fig. VIII-16. Equilibrium paper-strip electrophoresis of mammalian sera in the presence of [4-^{14}C] corticosteroids. Veronal buffer (pH 8.6, μ 0.1). The curves represent protein concentrations obtained by photometric analysis of the amido black-stained paper strips. A, normal rat serum, [4-^{14}C] corticosterone, 25°; B, serum from adrenalectomized rats, [4-^{14}C] corticosterone, 25°; C, normal rabbit serum, [4-^{14}C] cortisol, 4°; D, normal bovine serum, [4-^{14}C] corticosterone, 25°; E, normal horse serum, [4-^{14}C] cortisol, 4°. From WESTPHAL and DE VENUTO (1966)

ing affinity for either of these corticosteroids. No distinction can be made, under these conditions of electrophoretic separation, between possibly different CBG entities specific for cortisol and corticosterone, respectively. This result is in agreement with earlier studies on binding competition with the two corticosteroid hormones in human serum (DAUGHADAY, 1958 b; SLAUNWHITE and SANDBERG,

1959). It may be seen in Fig. VIII-17 that the binding affinity (C-value) for corticosterone is slightly higher than that for cortisol. A somewhat stronger binding for corticosterone than for cortisol has also been reported for human plasma, Fraction IV-4 (DAUGHADAY, 1958 b) and for human serum at 37° (WESTPHAL et al., 1962). As will

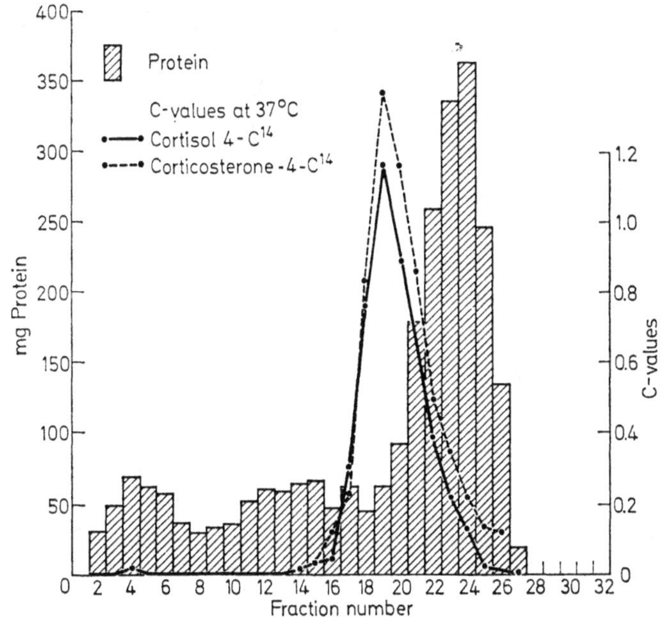

Fig. VIII-17. Continuous paper electrophoresis of human serum and CBG activity of lyophilized fractions determined by multiple equilibrium dialysis in the presence of [4-^{14}C] cortisol and [4-^{14}C] corticosterone. Electrophoresis in Veronal buffer of pH 8.6, μ 0.02; 4°. 100 ml 1:1 diluted dialyzed serum, 640 V, approx. 40 mA, 2.0 ml/h. Anode on the right side. CBG activity determined in 0.05 M phosphate buffer of pH 7.4, 37°. 10 ng [4-^{14}C] cortisol or 10 ng [4-^{14}C] corticosterone per ml outside solution; 2 mg protein per ml inside solution. From WESTPHAL and DE VENUTO (1966)

be discussed in Section 5 a of this chapter, the apparent association constant for corticosterone in human serum has also been found slightly greater than that for cortisol (WESTPHAL, 1967). However, it should be noted that all these experiments were carried out in the presence of serum albumin which has a somewhat higher affinity for corticosterone than for cortisol. An unequivocal comparison of the binding affinities for these closely related steroids can be made only with the isolated CBG.

DAUGHADAY (1958 c) as well as SLAUNWHITE and SANDBERG (1959) have reported that human CBG migrates as a "prealbumin" in continuous-flow electrophoresis at pH 5.2. DAUGHADAY (1958 c) also published a figure which shows clearly that progesterone is bound to the same electrophoretic component, with an affinity considerably higher than that for albumin. A subsequent illustration (Fig. 5 in DAUGHADAY, 1958 c) demonstrates displacement of the major portion of progesterone-4-^{14}C by unlabeled cortisol; the progesterone is shifted to the albumin peak, and little remains at the CBG site. The results clearly suggest identical binding sites for cortisol and progesterone. These important observations, reported in 1958, were ignored for many years by most of us working in this field—showing how easily the state of mind of an investigator can be preoccupied, resisting the adaptation to new results. It is for the psychologist to speculate to what extent the term "corticosteroid-binding globulin" or "transcortin" may have had a quasi-hypnotic influence on the approach to the associated problems.

The principle of competitive displacement has been applied more recently (WESTPHAL and DE VENUTO, 1966) in a study on binding of aldosterone in human serum. The protein components were separated by continuous-flow paper electrophoresis, and the dialyzed fractions analyzed for binding activity toward [7-^{3}H] aldosterone alone and in mixture with [4-^{14}C] cortisol; the molar ratio of cortisol to aldosterone was about 50 : 1. When both of these corticoid hormones were present in the equilibrium dialysis system, the highest binding affinity for either hormone was found in the same electrophoretic fraction (Fig. VIII-18, Fraction No. 21). When cortisol was omitted from the system, approximately 6-times more aldosterone was bound to the CBG fraction (Fig. VIII-18, curve B). The shift of maximal aldosterone binding by one electrophoretic fraction (to No. 22) is not considered significant because of the small difference in the C-values between fractions No. 21 and 22. It is concluded that aldosterone in human blood is associated, besides with albumin, with a protein component indistinguishable from CBG by the electrophoretic method used. These results confirm or corroborate binding studies with various human sera or plasmas, HSA and COHN's fraction IV-4, which had suggested that aldosterone is bound to a serum protein

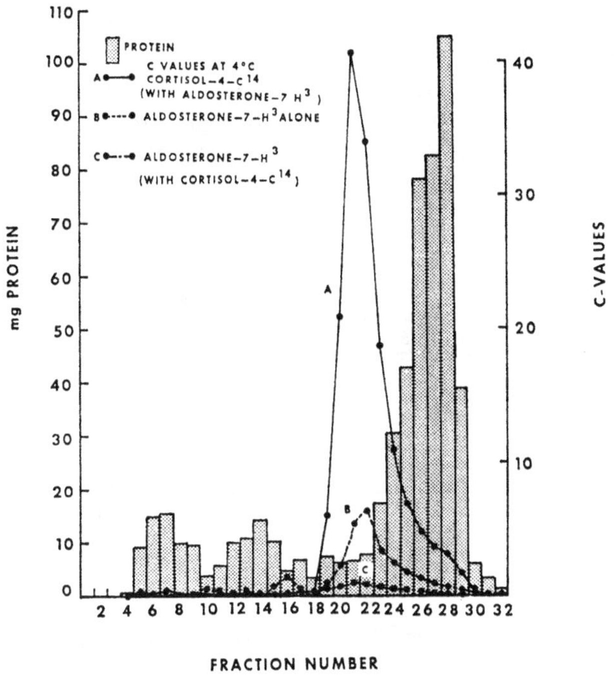

Fig. VIII-18. Continuous paper electrophoresis of human serum and determination of CBG activity in dialyzed fractions by multiple equilibrium dialysis in the presence of [7-^{3}H] aldosterone alone (curve B) and of both [7-^{3}H] aldosterone (curve C) and [4-^{14}C] cortisol (curve A) simultaneously. Electrophoresis in Veronal buffer, pH 8.6, μ 0.02; 4°. 18 ml dialyzed serum, 830 V, approx. 63 mA, 3.5 ml/h. Anode on the right side. CBG activity determined in 0.05 M phosphate, pH 7.4, 4°; 0.054 ng [7-^{3}H] aldosterone and 2.67 ng [4-^{14}C] cortisol per ml outside solution. From WESTPHAL and DE VENUTO (1966)

other than albumin, most likely to CBG (SANDBERG et al., 1960; DAUGHADAY et al., 1961; MEYER et al., 1961; DAVIDSON et al., 1962).

Whereas large quantities of spirolactones displaced aldosterone from its protein binding sites of COHN fraction IV-4 (DAVIDSON et al., 1962), no such competition by spirolactone could be demonstrated on kidney cell membranes (DAVIDSON et al., 1963). The question of aldosterone binding to CBG has been raised again recently (GUÉRIGUIAN et al., 1969) in connection with studies on chromatographic fractionation of human pregnancy serum. No separation was obtained between CBG and the aldosterone-binding activity.

Continuous-flow paper electrophoresis and subsequent analysis of the separated protein fractions for CBG activity was also applied to normal rat serum (WESTPHAL and DE VENUTO, 1966). Fig. VIII-19 shows that cortisol-4-^{14}C and corticosterone-4-^{14}C are again bound maximally to one and the same fraction (No. 27). The CBG activity

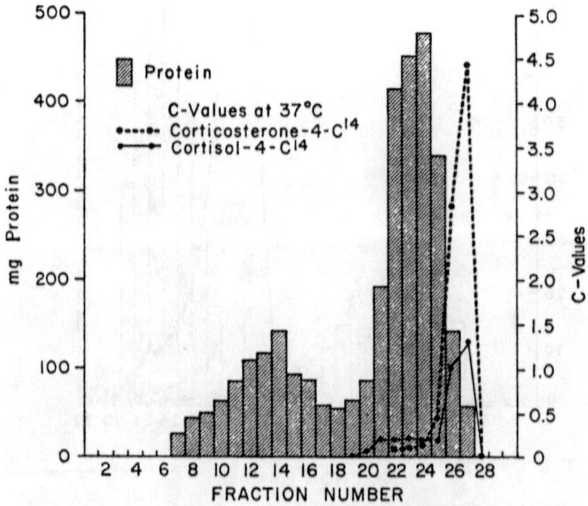

Fig. VIII-19. Continuous paper electrophoresis of normal rat serum and determination of CBG activity of lyophilized fractions by multiple equilibrium dialysis in the presence of [4-^{14}C] corticosterone or [4-^{14}C] cortisol. Electrophoresis in Veronal buffer, pH 8.6, μ 0.02; 4°. 100 ml 1:1 diluted dialyzed serum, 700 V, approx. 49 mA, 1.5 ml/h. Anode on the right side. CBG activity determined in 0.05 M phosphate buffer, pH 7.4, 37°; 10 ng [4-^{14}C] corticosterone or [4-^{14}C] cortisol per ml outside solution; 5 mg (Fractions 7—24) or 2 mg (Fractions 25—27) protein per ml inside solution. From WESTPHAL and DE VENUTO (1966)

is located in the prealbumin area as will be discussed below. Clearly the binding affinity for the less polar corticosterone is considerably higher than that for cortisol. This is not the result of mixed binding of corticosterone to CBG and albumin since fractions No. 26 and 27 (Fig. VIII-19) were free of a demonstrable amount of serum albumin (see below).

In independent experiments, the apparent association constant of rat CBG has been found greater for the complex with corticosterone than for that with cortisol (Chapter VIII. 6, Table VIII-21). The higher binding affinity of rat corticosteroid-binding globulin for corticosterone, as compared to that for cortisol, may be the reason for the relatively low physiological activity of corticosterone in the rat. It has been observed under different test conditions that in the presence of CBG cortisol has a higher biological activity than corticosterone (TALWALKER et al., 1961; DORFMAN, 1962 a), whereas in a system *in vitro*, i. e., in the absence of CBG, the biological potency of the two corticosteroids appears to be about the same (RIVERA, 1964).

Serum of adrenalectomized rats which has been shown to have a higher CBG activity than normal rat serum (WESTPHAL et al., 1962) was subjected to similar electrophoretic fractionation (see Fig. 5 in WESTPHAL and DE VENUTO, 1966). The highest binding activity for [4-^{14}C] corticosterone was found in the fraction which had the same electrophoretic properties as the most active fraction of normal serum (Fig. VIII-19). From the viewpoint of electrophoretic mobility under the conditions applied, no difference could thus be detected between the CBG fractions of normal and adrenalectomized rat sera.

In the experiment illustrated in Fig. VIII-19, and in a number of similar rat serum experiments not described here, the protein fraction with the highest binding activity for the corticosteroids was found in front of the albumin peak. This was in contrast to the paper-strip electrophoretic studies (Fig. VIII-16) which had shown that the corticosteroid-binding protein in rat serum migrated as an α_1-globulin. To clarify this discrepancy, the electrophoretic properties of the individual fractions obtained by continuous-flow electrophoresis of rat sera were determined by paper-strip electrophoresis in Veronal buffer, pH 8.6, μ 0.1 (WESTPHAL and DE VENUTO, 1966). The electrophoretic analysis showed that Fraction 24 (Fig. VIII-19) consisted of albumin and α-globulin in the approximate ratio of 3 : 2; in Fraction 25, the α-globulin component exceeded the amount of albumin present by a factor of at least 5. Fractions 26 and 27 migrated as α-globulins; no albumin could be detected in these fractions by the paper-strip electrophoresis techniques applied. Characterization of Fractions 25 and 26 by electrophoresis on cellulose-acetate strips in Veronal buffer (pH 8.6, μ 0.1), verified these results; the α-globulin band gave intensive staining for glycoprotein.

Since the nature and pH of the buffer used in these paper-strip electrophoreses were the same as those used in the continuous-flow fractionation, additional experiments were performed to clarify the reasons for the "paradoxical" electrophoretic behavior of rat CBG on the hanging curtain.

DAUGHADAY (1958 c) as well as SLAUNWHITE and SANDBERG (1959) have demonstrated that human CBG migrates as a "pre-albumin" in continuous-flow electrophoretic fractionation at pH 5.2.

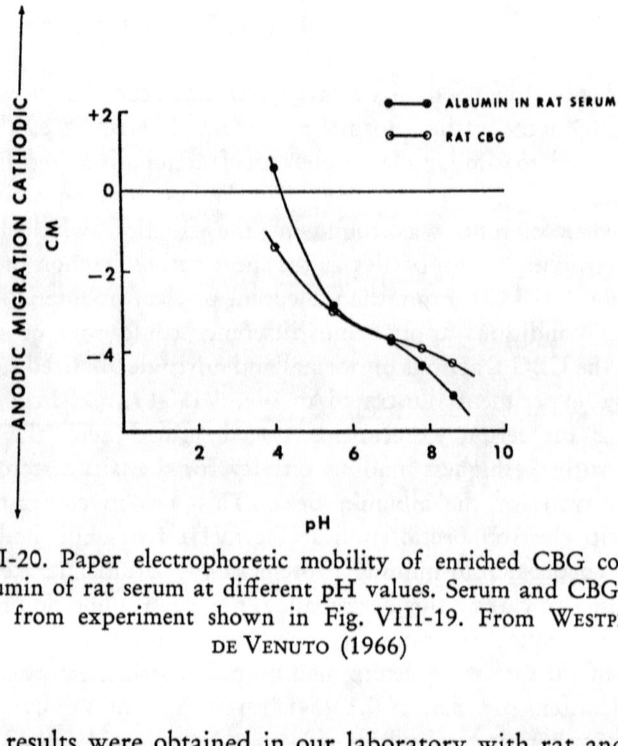

Fig. VIII-20. Paper electrophoretic mobility of enriched CBG component and albumin of rat serum at different pH values. Serum and CBG fraction (No. 26) from experiment shown in Fig. VIII-19. From WESTPHAL and DE VENUTO (1966)

Similar results were obtained in our laboratory with rat and guinea pig serum by paper curtain electrophoresis in acetate buffer at pH 5.5 (WESTPHAL and ALLEN, unpublished results). This reversal of the electrophoretic relationship to albumin is characteristic for certain glycoproteins (PHELPS and PUTNAM, 1960); it has also been demonstrated for α_1-acid glycoprotein (POULIK and SMITHIES, 1958) at pH 8.6 in starch-gel electrophoresis.

Paper-strip electrophoretic analysis of fractions No. 25 and 26 from the experiment illustrated in Fig. VIII-19, at different pH values, demonstrated that the mobility curve of the purified CBG fraction crossed over the mobility curve of rat serum albumin between pH 7 and 5.5 (Fig. VIII-20). At pH 4, the CBG fraction migrated far ahead of albumin, indicating a relatively low isoelectric point of this fraction. It is assumed that the low isoelectric point of the CBG fraction of rat serum, in connection with the low buffer capacity of the Veronal buffer used (μ 0.02) causes the paradoxical migration of the CBG fraction as a prealbumin (Fig. VIII-19).

5. CBG Activity in Human Plasma or Serum

a) Concentration of Binding Sites and Association Constants of Human CBG

The initial observations by DAUGHADAY (1958 b) indicated that the high-affinity binding sites for cortisol in normal human plasma became saturated when the total cortisol concentration exceeded a value of about 15 μg/100 ml. Subsequent studies and calculations (DAUGHADAY, 1959) gave 20 μg of cortisol per 100 ml as the level at which nearly all of the hormone was bound to CBG. A similar capacity of high-affinity binding for cortisol had been found by BUSH (1957) in his equilibrium dialysis experiments at 37°. The value of 20 μg cortisol bound per 100 ml plasma corresponds to a concentration of corticosteroid-binding sites of 5.5×10^{-7} M. A minimum concentration of CBG sites in normal human serum was calculated from chromatographic studies to be 2.5×10^{-7} M (WESTPHAL, 1961). The distribution of cortisol and its metabolites between plasma and red cells has been studied *in vivo* by KORNEL et al. (1970). Investigations on the interaction of progesterone with human erythrocytes have been reported by DE VENUTO (1968 a, 1968 b) and by DE VENUTO et al. (1969).

Determination of CBG concentration in normal human plasma or serum by equilibrium dialysis, gel filtration and other procedures (Table VIII-10) gave binding capacity values equivalent to approximately 25 μg cortisol bound per 100 ml. No significant differences were found between males and females. The values did not vary significantly between the ages of 15 and 60 years. This is seen in Table VIII-11 which gives results of gel filtration studies at room temperature. The errors arising from high dissociation at this tem-

Table VIII-10. *Concentration of cortisol and CBG in normal human plasma or serum*

No. of Subjects and Sex	Total peripheral cortisol [a]		CBG Concentration as cortisol bound		as CBG	Method [b]	References [c]
	μg/100 ml	μM	μg/100 ml	μM	mg/liter [d]		
10 M	13.8	0.38	29 ± 5 (21—38)	0.80	41	ED	BOOTH et al. (1961)
9 F	10.2	0.28	42 ± 13 (24—71)	1.16	60	ED	GRAY et al. (1961)
10 M	12.5	0.34	21	0.58	30	DED	DAUGHADAY et al. (1962)
7 F	9.2	0.25	18	0.51	26	DED	
combined	12.3	0.34	20 ± 5.2 (14—32)	0.55	29 ± 7.3	DED	
83 M	—	—	26	0.71	37	GF [e]	DE MOOR et al. (1962) (see Table VIII-11)
32 F	—	—	26	0.72	38	GF [e]	
combined [f]	12.1	0.33	26 ± 3.8	0.71	37	GF [e]	
4 F	—	—	23	0.63	33	UF	MILLS (1962)
100 M + F [f]	—	—	25 ± 0.2	0.69	36	GF [e]	DE MOOR et al. (1963 c)
19 M	—	—	24	0.67	35 ± 3.6	GF	DOE et al. (1964)
16 F	—	—	23	0.63	33 ± 4.2	GF	DOE and SEAL (1963)
12 M	15.2 ± 5.1	0.42	32 ± 5.5	0.88	46	GF [g]	MURPHY and PATTEE (1963)
12 F	15.4 ± 4.6	0.42	31 ± 4.5	0.85	45	GF [g]	
6 M + F	—	—	15	0.43	22	UF	BEISEL et al. (1964 a)
unknown	—	—	23 ± 3.6	0.63	33 ± 4.2	PE	OSORIO and SCHATS (1964)
3 M	13.7	0.38	30	0.82	43	ED	SANDBERG et al. (1964)

400 M + F	—	—	24 ± 3.3	0.67	35	GF[e]	DE MOOR et al. (1966 b)
50 F	—	—	24 ± 2.3	0.67	35	GF[e]	DE MOOR et al. (1966 b)
10 M	—	—	20 (15—28)	0.55	29	ED	DIXON et al. (1967)
9 F	—	—	20 (18—25)	0.55	29	ED	DIXON et al. (1967)
23 M	14.5	0.40	26 ± 3.4	0.70	37	EA	LEYBOLD et al. (1967)
8 M + F	12 ± 6	0.33	21	0.58	30	UF	MURRAY (1967)
Pool[h]	14.9	0.41	26	0.72	37	ED	WESTPHAL (1967)
9 M	13.5 ± 3.4	0.37	19 ± 5.5	0.53	28	GF[e]	DE MOOR et al. (1968 d)
4 F	15.0 ± 4.0	0.41	21 ± 3.1	0.57	30	GF[e]	

[a] The corticosteroid composition of the "cortisol" measured may vary somewhat according to the method applied; ± in this table indicates S.D., except when marked [f].

[b] Abbreviations used: DED, double equilibrium dialysis; EA, enzymatic assay, ED, equilibrium dialysis; GF, gel filtration; PE, paper electrophoresis; UF, ultrafiltration.

[c] Some values in this table were calculated from data given in the references.

[d] Molecular weight assumed 52,000 (SEAL and DOE, 1966; MULDOON and WESTPHAL, 1967).

[e] The gel filtration was carried out at 22° resulting in low binding capacity; the fluorometric assay of cortisol gave high values. For a critical discussion of these compensating errors see DOE et al. (1964)

[f] ± S.E.M.

[g] at 9°.

[h] Serum prepared from outdated blood from blood bank.

Table VIII-11. *Influence of age and sex on the cortisol-binding capacity of human plasma* [a]

Age (yrs):	Cortisol-binding capacity in μg %					
	Umbilical cord blood	1—14	15—29	30—44	45—59	60
Males	10.9 ± 0.5 [20]	27.2 [4]	25.1 ± 1.3 [17]	25.8 ± 0.9 [30]	26.2 ± 0.8 [36]	24.5 ± 0.9 [12]
				25.8 [83]		
Females		27.1 [6]	23.4 ± 1.0 [9]	27.6 ± 1.2 [12]	26.8 ± 1.3 [11]	
				26.1 [32]		

[a] Figures in [ ] indicate number of subjects in each group. Values are means ± S.E.M. From DE MOOR et al. (1962). Later studies by DE MOOR and HEYNS (1968) showed constancy of the CBG affinity over the same range of age; no sex difference was observed.

perature are compensated by the high cortisol values obtained by the fluorometric technique applied; the CBG levels are therefore comparable to those obtained in other laboratories. Pertinent comments on the compensating errors (DOE et al., 1964) and comparative analyses (DE MOOR, 1965) have been published; any such errors do not affect the relative capacity values for the various age groups shown in Table VIII-11. Umbilical cord blood has a distinctly lower CBG activity as had been found before (DAUGHADAY et al., 1959; SANDBERG and SLAUNWHITE, 1959; DE MOOR et al., 1963c). No diurnal variation was observed by DE MOOR et al. (1962) whereas DOE et al. (1964) reported a significant decrease of the CBG level at 3 AM. The CBG capacity in the normal human blood plasma exceeds the resting cortisol concentration by a factor of approximately 2 (Table VIII-10).

The concentration of CBG in the lymph is only slightly lower than that in the blood (SANDBERG et al., 1960); the participation of lymph in the transport of steroid hormones has been discussed by LINDNER (1966). DAUGHADAY et al. (1959) found rather high CBG concentrations in synovial fluids and in pleural effusions, whereas spinal fluid and amniotic fluid seemed to contain but little activity. No high binding affinity for cortisol was observed in human semen (STEENO et al., 1965).

The binding capacity of 25 μg cortisol per 100 ml corresponds to a concentration of corticosteroid binding sites of about 7×10^{-7} M. This would also be the CBG concentration if one assumes one binding site per CBG molecule. Accepting a value of 52,000 as the molecular weight of CBG (SEAL and DOE, 1966; MULDOON and WESTPHAL, 1967), this concentration would equal about 36 mg CBG per liter of plasma or serum. A factor of 1.44 applied to the cortisol binding capacity value in μg per 100 ml would thus give the concentration of human CBG in mg per liter (DOE et al., 1964; WESTPHAL, 1969). Normal ranges for CBG concentration in human plasma would

Table VIII-12. *Association constants, k, of steroid-complexes with human CBG*

Steroid	Temp.	k, M^{-1}	Method [a]	References
		A. Determinations in plasma or serum		
Cortisol	4°	7×10^8	ED	DAUGHADAY (1960) DAUGHADAY and MARIZ (1961)
Cortisol	37°	5.2×10^7	UF	MILLS (1962)
Cortisol	37°	4.7×10^7	UF	BEISEL et al. (1964 a)
Cortisol	4°	1×10^9	ED	SANDBERG et al. (1964)
Cortisol	25°	7.2×10^7	CA	HEYNS et al. (1967)
Cortisol	37°	4.3×10^7	UF	MURRAY (1967)
Cortisol	4°	6×10^8	ED	WESTPHAL (1967)
	37°	3×10^7	ED	WESTPHAL (1967)
Corticosterone	4°	1×10^9	ED	WESTPHAL (1967)
	37°	3×10^7	ED	WESTPHAL (1967)
Progesterone	4°	7×10^8	ED	WESTPHAL (1967)
	37°	9×10^7	ED	WESTPHAL (1967)
Cortisol	10°	4.4×10^8	UF	GOLDIE et al. (1968)
	10°	2.6×10^8	UF	GOLDIE et al. (1968)
	25°	2.6×10^8	UF	GOLDIE et al. (1968)
	37°	1.1×10^8	UF	GOLDIE et al. (1968)
	37°	0.9×10^8	UF	GOLDIE et al. (1968)
Aldosterone [b]	4°	2.6×10^7	GE	GUÉRIGUIAN et al. (1969)

Table VIII-12 (continued)

Steroid	Temp.	k, M^{-1}	Method[a]	References
B. Determinations with isolated or purified CBG				
Cortisol	4°	6×10^{8}	ED	SEAL and DOE (1962 b)
Corticosterone	37°	6.4×10^{6} [c]	ED	DOE and SEAL (1963)
Progesterone	37°	9.5×10^{7}	ED	DOE and SEAL (1963)
Aldosterone	37°	6.5×10^{6}	ED	DOE and SEAL (1963)
Testosterone	37°	1.4×10^{6}	ED	DOE and SEAL (1963)
Estradiol	37°	2×10^{4}	ED	DOE and SEAL (1963)
Cortisol	4°	5×10^{8}	ED	SEAL and DOE (1966)
	37°	3×10^{7}	ED	SEAL and DOE (1966)
Progesterone	4°	3×10^{8}	ED	SEAL and DOE (1966)
	37°	8×10^{7}	ED	SEAL and DOE (1966)
Cortisol	4°	5.2×10^{7}	ED	MULDOON and WESTPHAL (1967)
	37°	2.4×10^{7}	ED	
Corticosterone	4°	8.3×10^{8}	ED	LEBEAU et al. (1969)
Corticosterone sulfate[d]	4°	4.0×10^{8}	ED	LEBEAU et al. (1969)

[a] CA, competitive adsorption; ED, equilibrium dialysis; GE, gel equilibration; UF, ultrafiltration.
[b] CBG-enriched chromatographic fractions from pregnancy serum were used.
[c] This relatively low value needs confirmation.
[d] Corticosterone sulfate is bound at the same site as corticosterone. Testosterone and testosterone sulfate, but not dehydroepiandrosterone sulfate, also appear to compete with corticosterone at the same binding site (LEBEAU and BAULIEU, 1970).

then be 23 to 45 mg/l (DOE et al., 1964), for both males and females. These values have been substantiated by the isolation of CBG from human blood (SEAL and DOE, 1962 b; SLAUNWHITE et al., 1966; MULDOON and WESTPHAL, 1967).

Association constants of the corticosteroid-CBG complexes were determined before isolation of the active glycoprotein. Table VIII-12 A gives these values and shows the remarkably close agreement with the affinity constants measured with the isolated CBG (Table VIII-12 B). Various procedures have been used (see references in Table VIII-12 A) for the assessment of the association constants of the CBG complexes with cortisol, corticosterone and progesterone in the unfractionated serum; the problem is to separate the binding contribution of other serum proteins, especially albumin. The method applied in our laboratory for determination of the CBG concentrations and association constants of steroid-CBG complexes in the sera of various species, including man, has been described in Chapter VIII. 2 a and Fig. VIII-3. It should be noted that DE MOOR et al. (1963 b) have suggested that the high-affinity binding of progesterone in human plasma is not due to CBG, but possibly to a β-lipoprotein. The existence of a specific progesterone-binding protein different from CBG has been concluded by DIAMOND et al. (1969) for the pregnant guinea pig (see Chapter X).

b) CBG Concentration in Humans Under Conditions Other than Normal

α) Hereditary Familial Decrease of CBG Level

DOE, LOHRENZ and SEAL (1965) have investigated members of a family who showed a hereditary deficiency in CBG concentration. The CBG level was reduced to about half the normal level; this was not caused by any disease state. The study was extended to include a four-generation kindred (LOHRENZ et al., 1967). All members of the second generation tested were found to have low CBG values; of a total of 33 descendants examined, 25 had the deficiency. The mean CBG concentration in the fourth generation was significantly higher than the average value for the affected members of the older generations (Fig. VIII-21). Total plasma 17-OHCS level was significantly below normal, whereas the unbound 17-OHCS was in the normal range. The trait was not sex linked. Other estrogen-responsive serum proteins, including TBG, ceruloplasmin and haptoglobin were present in normal concentrations. The mode of inheritance has not been clarified, and it is not known whether the CBG in the deficient subjects has the same physicochemical properties as that from normal individuals. Low cortisol binding capacity has also been observed by DE MOOR et al. (1965 b; 1965 c) who suggest control by an x-linked gene (DE MOOR et al., 1965 a, 1966 a, 1967). The genetically controlled abnormality is analogous to familial decrease (NICOLOFF et al., 1964; NIKOLAI and SEAL, 1966, 1967) and increase (BEIERWAL-

TES and ROBBINS, 1959; JONES and SEAL, 1967) of thyroxine-binding globulin, to familial absence of serum albumin (analbuminemia) (BENNHOLD et al., 1954) and to bisalbuminemia (SCHEURLEN, 1955; BENNHOLD et al., 1958), to familial decrease in ceruloplasmin concentration (GITLIN and JANEWAY, 1960) and to other hereditary serum protein deviations (BENNHOLD et al., 1958; GITLIN and JANEWAY, 1960; DOE et al., 1965; HELINSKI and YANOFSKY, 1966).

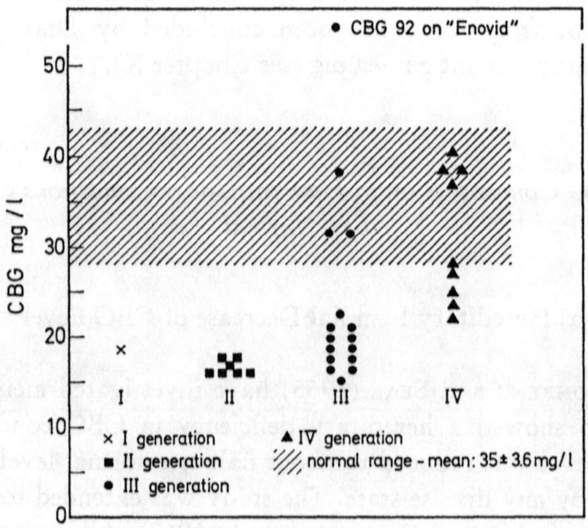

Fig. VIII-21. Serum CBG levels of the low CBG family members grouped according to generation, with the normal range for adults indicated by the shaded area. From LOHRENZ et al. (1967)

Elevated CBG levels were recently observed in the sera of a brother and sister pair (LOHRENZ et al., 1968); the increase was not caused by estrogen administration or other external conditions. A genetic relationship remains to be established.

β) CBG Level in Pregnancy and after Estrogen Administration

In 1959, SLAUNWHITE and SANDBERG (1959) published the important observation that the transcortin capacity in human pregnancy was more than twice that in non-pregnant women. Since no differences in the equilibrium constants were found, the authors assumed that no qualitative changes had occurred in CBG as a result of pregnancy. Soon afterwards, SANDBERG and SLAUNWHITE (1959) gave an explanation for the increase in CBG when they demonstrated a rise of the transcortin level following the injection of estrogens. The progressive increase of the total cortisol level in pregnancy or after estrogen administration was confirmed in many laboratories (TALIAFERRO et al., 1956; MIGEON et al., 1957; WALLACE et al., 1957; COHEN et al., 1958; BONGIOVANNI et al., 1958; CHRISTY et al., 1959; ROBERTSON et al., 1959; YATES and URQUHART, 1962). This finding answered a question that had puzzled earlier researchers: elevations in peripheral corticosteroid concentrations (VENNING, 1946) had been observed in human pregnancy (GEMZELL, 1953), up to about four times the normal level during the last trimester; however, in spite of the high 17-OHCS concentration, there were no classical signs of hypercorticism (GEMZELL, 1953; ROBINSON et al., 1955; YATES and URQUHART, 1962). The hypothesis was advanced (SANDBERG and SLAUNWHITE, 1959) that the transcortin-bound cortisol was biologically inactive. The authors furthermore postulated that high estrogen levels, either endogenous during pregnancy or exogenous, would produce an increase of plasma transcortin, followed by a rise of the 17-OHCS. These interpretations have been amply confirmed in subsequent years (BOOTH et al., 1961; DE MOOR et al., 1962).

The increase of CBG activity as a result of estrogen administration is not a rapid process. A period of several days appears necessary for the elevated transcortin level to develop; for example, three to seven days after intake of large amounts of diethylstilbestrol by male subjects were required (DOE et al., 1960) for an increase of cortisol binding which exceeded that of a pregnancy group. SANDBERG et al. (1964) concluded from plasma clearance studies with ^{125}I-labeled CBG that the rise in transcortin concentration after estrogen treatment results from increased synthesis rather than from decreased turnover. The CBG level returned to pretreatment values within 7 to 10 days after cessation of therapy (SANDBERG et al., 1960). Addition *in vitro* of estrogens to normal human plasma and incubation at 37° for 24 hrs did not significantly increase the CBG activity (DOE et al., 1960). The mechanism of the stimulation of CBG by estrogens will be discussed in Chapter IX. 7.

The changes in CBG concentration during pregnancy and after administration of estrogens have been extensively studied in numerous laboratories and reported in additional publications (MILLS and BARTTER, 1959; PETERSON, 1959; DAUGHADAY and MARIZ, 1960; HERRMANN et al., 1960; MILLS et al., 1960; PETERSON et al., 1960; WALLACE and CARTER, 1960; MARKS et al., 1961 a; MARKS et al., 1961 b; MILLS, 1961; DAUGHADAY et al., 1962; KOPETZ et al., 1962; MILLS, 1962; SEAL and DOE, 1962 a; SEAL and DOE, 1962 b; MIDDO

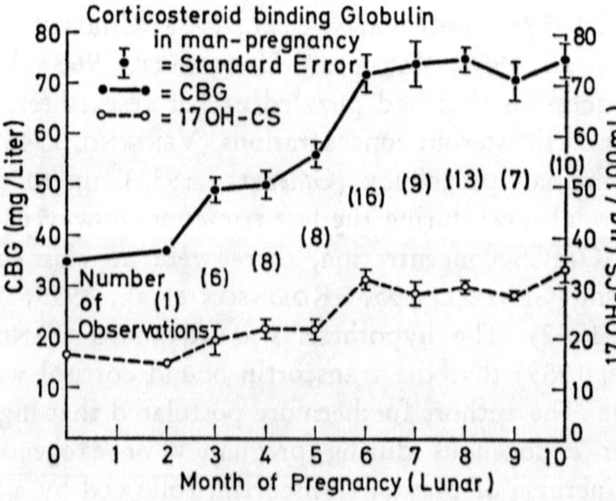

Fig. VIII-22. Concentrations of CBG and cortisol during human pregnancy. From SEAL and DOE (1967)

et al., 1963; MURPHY and PATTEE, 1963; NELSON et al., 1963; QUINCEY and GRAY, 1963; DOE et al., 1964; MILLS and BROOKS, 1964; CUSHMAN et al., 1965; DE MOOR and STEENO, 1965; BRADLEY and WATERHOUSE, 1966; DE MOOR et al., 1966 b; SANDBERG et al., 1966; SLAUNWHITE et al., 1966; DAUGHADAY, 1967; DOE et al., 1967; KATZ and KAPPAS, 1967; LEYBOLD et al., 1967; MURRAY, 1967; MUSA et al., 1967 a; MUSA et al., 1967 b; DE MOOR et al., 1968 a). DRAY et al. (1970) reported increased CBG activity after administration of clomiphene citrate (clomid).

Fig. VIII-22 illustrates the rise of CBG concentration and of Porter-Silber chromogens during human pregnancy. High values of approximately 70 mg CBG per liter of plasma are reached during the last trimester; they decrease within one or two weeks after delivery to non-pregnant levels. Fig. VIII-23 shows the dependency of the CBG increase on the estrogen dose given. Oral intake of enovid, norethynodrel and ethynylestradiol 3-methylether, the constituents of contraceptive pills, resulted also in elevated CBG levels (LAYNE et al., 1962; LAYNE and MEYER, 1965). Similar increases in pregnancy and after estrogen administration have been observed for serum β-glucuronidase (COHEN and HUSEBY, 1951; MUSA et al., 1965 b), ceruloplasmin (RUSS and RAYMUNT, 1956; MUSA et al., 1965 b), TBG

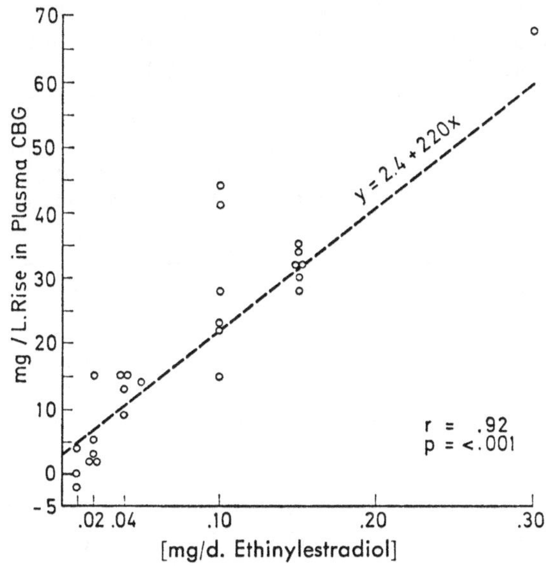

Fig. VIII-23. Dose-response relationship of CBG increase following six days' administration of graded doses of ethinylestradiol. From SEAL and DOE (1967)

(DOWLING et al., 1956; ROBBINS and RALL, 1957), and other serum protein components (DOE et al., 1967; MUSA et al., 1967 a, b). In contrast to estradiol and other highly active estrogens, estriol did not influence the CBG or TBG level of human serum (KATZ and KAPPAS, 1967). The plasma concentration of unbound cortisol has been found elevated during pregnancy and possibly after estrogen administration (PLAGER et al., 1964; SANDBERG et al., 1967; SEAL and DOE, 1967; SANDBERG et al., 1968; SEAL and DOE, 1968); the enhanced biological potency of unbound cortisol is considered responsible for the aminoaciduria which occurs in the third trimester of pregnancy, and which can also be produced by administration of

Table VIII-13. *Components of a pregnancy serum.* According to WESTPHAL (1966)

	Molecular weight	Concentration mg/l	10^{-7} M
HSA	69,000	38,000	5,500
AAG	41,000	750	180
CBG	52,000	74	14
Progesterone	314	0.15	4.8
Cortisol	362	0.30	8.3

Table VIII-14. *Association constants for steroid-protein complexes of human serum; 37° C. These values of k for n=1 were used for calculation of Table VIII-15.* From WESTPHAL (1966)

Protein	Progesterone	Cortisol
CBG	9×10^{7} M^{-1}	3×10^{7} M^{-1}
AAG	1.6×10^{5} M^{-1}	1×10^{3} M^{-1}
HSA	6×10^{4} M^{-1}	3×10^{3} M^{-1}

cortisol (SEAL and DOE, 1960; ZINNEMAN et al., 1963, 1965, 1967). Other authors (BOOTH et al., 1961; DIXON et al., 1967) did not observe a significant change of non-protein-bound cortisol during pregnancy.

Calculation of the distribution of cortisol and progesterone among different binding proteins (see Chapter V. 4 g) in a pregnancy serum on the basis of concentrations (Table VIII-13) and association constants (Table VIII-14) of the components involved gave the values compiled in Table VIII-15. It is evident that progesterone associates to about equal parts with CBG and with HSA; binding to AAG is very low. Cortisol interaction is relatively higher with CBG.

It should be mentioned in this context that we have to be aware of certain limitations in the application of these calculations. They are based on physicochemical parameters obtained with the isolated, highly purified proteins. We know that the association constants of the isolated proteins depend greatly on the precise state of their purity; they can be increased by factors of 2 to 4 (see Chapter IV. 5 b), in case of the progesterone complex with chromato-

Table VIII-15. *Distribution of 15 μg progesterone and 30 μg cortisol per 100 ml among transcortin, albumin and orosomucoid in pregnancy serum.* From WESTPHAL (1966)

Steroid bound to	Progesterone %	Progesterone nM	Cortisol %	Cortisol nM
CBG	48.1	231	59.8	496
HSA	49.6	238	32.6	271
AAG	1.0	5	—	—
unbound	1.3	6	7.6	63
Total	(100)	480	(100)	830

graphically pure orosomucoid even up to 10 (see Chapter XIII) when the proteins are delipitated under mild conditions. This process is reversible. In the circulating blood, the proteins are certainly not present in a delipitated form; their exact affinity constants are therefore not known. Perhaps some of these theoretical discrepancies cancel out because of a general lipid content of all binding proteins involved, so that the effects compensate each other; this would be difficult to analyse. But these limitations have to be recognized, and we have to try to understand the consequences.

Concentrations of CBG and of cortisol in its different forms, and related parameters have been given by SANDBERG et al. (1967) for human pregnancy and estrogen treatment (Table VIII-16 and VIII-17). These authors assume that plasma clearance and hepatic extraction of cortisol reflect the concentration of nonprotein-bound and possibly albumin-bound corticosteroid. ROSENTHAL et al. (1969 a) have extended these studies to an analysis of more than 60 sera from different stages of pregnancy for the binding distribution of cortisol and progesterone. For the 3rd trimester, they obtained interaction values of progesterone with CBG, and with proteins other than CBG, which were similar to those of Table VIII-15. For the binding equilibria of cortisol, association with HSA seemed relatively low in comparison with the data in Table VIII-15, even considering the somewhat smaller cortisol concentrations and higher CBG levels (ROSENTHAL et al., 1969 a). This may be a consequence of reduced steroid affinity of HSA in the serum which contains numerous types of lipids that may inhibit steroid binding to albumin. The concentra-

Table VIII-16. *Mean values for concentrations of cortisol in various plasma compartments and for some metabolic parameters of cortisol.* From SANDBERG et al. (1967)

	Normal	Pregnancy (third trimester)	Estrogen therapy
Plasma cortisol	12—14 μg-%	35 μg-%	40 μg-%
[I]	1.1—1.25 μg-%	2.2 μg-%	3.2 μg-%
[I_A]	1.4—1.6 μg-%	2.9 μg-%	4.2 μg-%
[I_T]	9 μg-%	30 μg-%	33 μg-%
Transcortin (T) [a]	17—25 μg-%	45—55 μg-%	40—60 μg-%
T 1/2	70—90 min.	120—150 min.	> 240 min.
Daily secretion	16—28 mg	11—14.5 mg	8—16 mg
6β-Hydroxycortisol in 24 hr urine	100—200 μg	200—400 μg	200—400 μg
Cortisol in urine	50 μg	200—500 μg	30—50 μg

[a] The transcortin concentration has been expressed in terms of its capacity to bind cortisol and was determined by the method of DE MOOR et al. (1962). [I], unbound cortisol; [I_A], albumin-bound cortisol; [I_T], transcortin-bound cortisol.

Table VIII-17. *Transcortin (T) and cortisol (F) levels in pregnancy and in cord plasma and binding values for cortisol and progesterone (P). Each value represents the average of determinations on 3 plasmas.* From SANDBERG et al. (1967)

	T [a]	F μg-%	% F unbound	[I] [b]	% P unbound
1st trimester	28	22	5.6	1.2	5.0
2nd trimester	43	28	5.7	1.6	5.4
3rd trimester	50	35	6.3	2.2	5.9
Cord blood (newborn)	10	19	24.3	4.6	8.1

[a] Expressed as capacity for cortisol in μg-%.
[b] μg-% unbound cortisol.

tions of unbound progesterone or cortisol determined by ROSENTHAL et al. (1969 a) are very close to those given in Table VIII-15. As an additional result of the studies on pregnancy sera, the ratios of bound: unbound cortisol or progesterone were found to be nearly constant throughout pregnancy. Since the levels of both hormones increase

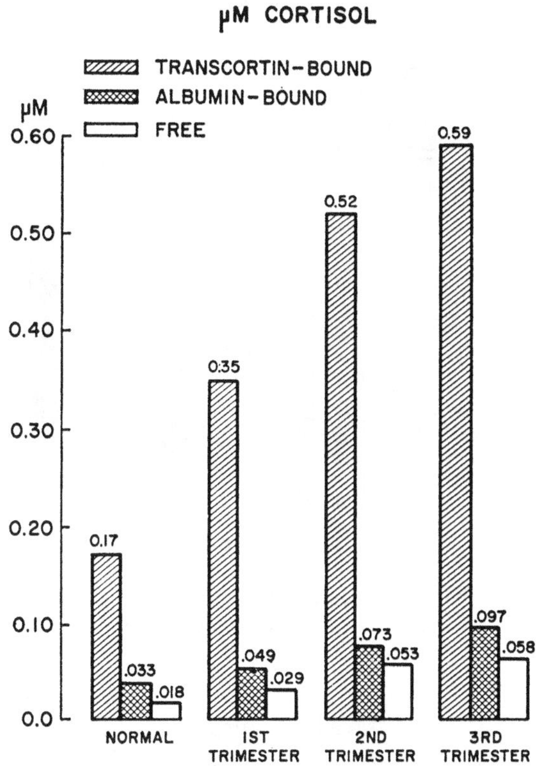

Fig. VIII-24. Distribution of bound and unbound cortisol in pregnancy plasma. From Rosenthal et al. (1969 a)

during gestation, the absolute concentration of the unbound portion goes up for both steroids. This is illustrated in Figs. VIII-24 and VIII-25 for cortisol and progesterone, respectively. In confirmation, Yannone et al. (1969) observed maximal values of unbound, biologically active progesterone at the onset of labor. Ultrafiltration studies by O'Connell and Welsh (1969) indicated a significantly higher concentration of unbound cortisol in pregnant (third trimester) women (2.47 μg/100 ml) than in nonpregnant (1.00 μg/100 ml) and Enovid-treated (2.5 mg 17α-ethynyl-5(10)-estren-17β-ol-3-one and 0.1 mg 17α-ethynylestradiol-3-methyl ether daily) women (1.31 μg/100 ml). Butler and Dixon (1969) measured the binding distribution of cortisol by equilibrium dialysis utilizing formulas of multiple bind-

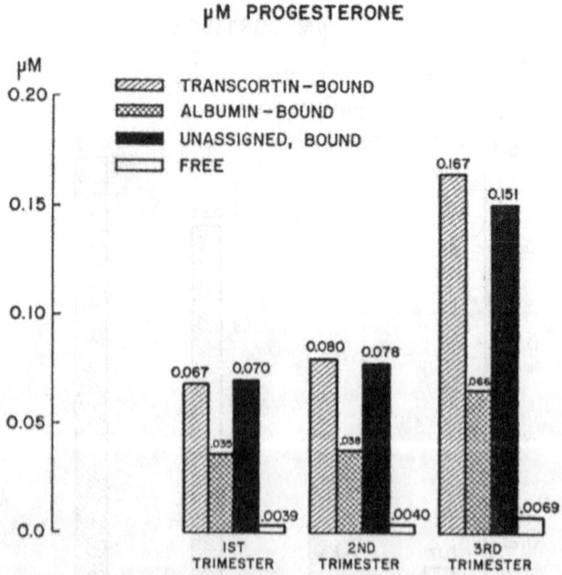

Fig. VIII-25. Distribution of bound and unbound progesterone in pregnancy plasma. From ROSENTHAL et al. (1969 a)

ing equilibria. They observed an increase of unbound cortisol in pregnancy, but did not consider the rise outside the normal non-pregnant range.

γ) CBG Level in Pathological States

Patients with liver cirrhosis, nephrotic syndrome and hypoalbuminemias of other etiologies showed significant decreases in CBG concentration (DAUGHADAY, 1958 a; SANDBERG and SLAUNWHITE, 1959; MURPHY and PATTEE, 1963; DOE et al., 1964; LEONARD and MACWILLIAM, 1964; LEONARD and D'ARBELA, 1966). A low serum albumin level, however, is not consistently associated with low CBG values; DOE et al. (1964) observed in cooperative studies with H. BENNHOLD and P. G. SCHEURLEN normal to elevated CBG activity in a patient with congenital analbuminemia. Increased CBG levels were found in some cases of untreated adrenal insufficiency (DOE et al., 1964; confirmed by MÜLLER-BARDORFF et al., 1967; HOCHHEU-

SER et al., 1968, 1969). Elevated levels were also reported in lung cancer (DOE et al., 1964). Low CBG values occurred in certain forms of Cushing's disease (DOE et al., 1964; DE MOOR et al., 1965 b, 1968 b; SHOLITON et al., 1962; KONO et al., 1966; RAMIKH et al., 1969; HOCHHEUSER et al., 1969). Subnormal levels were also observed in adrenocortical hyperplasia and disorders of male and female puberty (DE MOOR et al., 1968 b). RIEKSTINYA et al. (1969) noted low CBG capacity in some cases of diencephalohypophysial insufficiency; normal values were found in primary myxedema (ROZEN et al., 1967). Other pathologies showed normal values (DOE et al., 1964). Observations in diabetic patients (DE MOOR and HEYNS, 1967) and general clinical features associated with extreme CBG values have been discussed by DE MOOR and his coworkers (DE MOOR and STEENO, 1965; DE MOOR et al., 1968 a). Patients with toxemia of pregnancy had lower CBG levels than normal pregnant women (MEYER, 1965). Abnormally low CBG capacity was observed in gastrointestinal bleeding of stress origin (SHENKIN and FELDMAN, 1968). Metabolic clearance rate, metabolism rate constant, and volume of distribution of aldosterone, cortisol, and corticosterone have been determined in normal subjects and in patients with various diseases by KONO et al. (1966); daily secretion rates, mean plasma concentrations, and plasma protein binding were measured. Influences of steroid-protein interaction on the metabolism rate constants, and on the metabolic clearance rate through variations of the volume of the inner pool were discussed.

6. CBG Activity in Species Other than Human

In accordance with the ubiquitous occurrence of corticosteroid hormones in all vertebrate classes (JONES et al., 1959), corticosteroid-binding serum proteins have been observed in virtually all vertebrate species examined. CBG activity was demonstrated early in rat plasma (DAUGHADAY, 1958 a; SLAUNWHITE and SANDBERG, 1959) and serum (WESTPHAL, 1961), as well as in plasma of rabbit, guinea pig and alligator (SLAUNWHITE and SANDBERG, 1959). The activity in plasma of dog (DAUGHADAY, 1958 a; SLAUNWHITE and SANDBERG, 1959; FLORINI and BUYSKE, 1961; DE MOOR et al., 1962; PLAGER et al., 1963), was found to be low, as it was in sheep (PATERSON and HILLS, 1967), goat and ox (LINDNER, 1964). SEAL and DOE (1963; 1965; 1966)

Table VIII-18. *High-affinity binding of cortisol (F) and corticosterone (B) in vertebrate sera, determined by gel filtration at 4°, as μg per 100 ml. Some values of Porter-Silber chromogens included.* Compiled from SEAL and DOE (1963, 1965, 1966)

Species	F	B	P.S.Chr.[a] μg/100 ml
Fish			
Lamprey	0.5	1.9	—
Bonnet shark	1.5	2.1	—
Stingray	2.1	2.0	—
Garfish	1.0	0	—
Bowfin	4.3	0	—
Carp	1.5	0	19.3
Grey snapper	2.5	7.1	—
Bass	6.8	1.0	—
Eel	4.4	1.2	—
Mudfish	6.4	—	13.3
Porcupine fish	2.4	45.0	—
Amphibians			
Amphiuma	10.7	5.3	—
Leopard frog	4.8	5.5	1.6
Bullfrog	1.0	3.1	—
Marine toad	18.0	26.0	—
African toad	12.6	27.0	—
Reptiles			
Gopher turtle	13.0	2.0	—
Painted turtle	2.7	0	—
Lizard	9.2	8.8	—
Alligator	43.2	10.2	1.8
Crocodile	9.3	9.5	—
Iguana (green)	110.0	0	—
Water snake	6.9	6.5	—
Rattlesnake	60.0	62.0	—
Cobra	4.4	1.7	—
Garter snake	33.0	20.0	—
Anaconda	14.3	7.8	—
Birds			
Penguin	3.5	16.1	—
Chicken	6.0	1.5	2.0
Turkey	6.4	5.5	0.2
Pigeon	19.2	6.7	4.0
Duck	14.6	12.8	3.0
Goose	22.4	—	6.7
Flamingo	10.5	2.0	—

Table VIII-18 (continued)

Species	F	B	P.S.Chr.[a] µg/100 ml
Pelican	0.9	1.0	—
Hawk	19.2	17.0	—
Owl	6.9	5.0	—
Crane	2.1	2.2	—
Ibis	14.2	15.8	—
Egret	11.2	7.0	—
Mammals			
Man	22.0	20.0	16.0
Monkey[b]	36.8	32.4	37.8
Ox	14.0	—	7.9
Goat	10.4	—	3.8
Sheep	8.8	—	3.0
Pig	3.5	—	7.0
Horse	10.0	8.2	2.2
Deer	7.1	7.6	4.7[c]
Llama	5.7	3.8	—
Tapir	7.0	7.8	—
Kangaroo	8.9	10.2	—
Dog	6.5	3.8	6.5
Fox	8.2	3.6	—
Cat	6.4	1.0	13.7
Raccoon	32.0	16.0	—
Skunk	12.0	—	—
Ferret	8.9	6.2	6.9
Rat	45.0	52.0	—
Mouse	18.0	33.0	—
Guinea pig	20.0	11.0	8.2
Porcupine	6.3	11.2	—
Squirrel	92.0	63.0	—
Hedgehog	64.0	38.1	—
Rabbit	20.0	23.0	2.7
Hamster	14.6	1.0	1.8
Opossum	10.0	1.0	—
Bat	12.4	13.6	—
Armadillo	2.8	2.2	—
Sloth	2.1	3.2	—
Anteater	3.7	11.6	—
Dolphin	5.9	6.0	—
Seal	17.0	—	—
Whale	2.9	2.1	—
Elephant (Afr.)	4.0	—	—

Footnotes see p. 228

have determined in extensive gel filtration studies the CBG capacities in 131 species representing all the vertebrate classes. Table VIII-18 shows the cortisol-binding capacity in the sera of numerous vertebrate species, including peripheral Porter-Silber chromogens in some cases. Compared to most other species reported in Table VIII-18, the ratio of cortisol-binding capacity to peripheral corticoid concentration appears to be low in the fishes. Even lower ratios have been observed (FREEMAN and IDLER, 1966) in plasma of the Atlantic salmon *(Salmo salar)* and Atlantic cod *(Gadus morhua L.)*. When relating CBG capacity to endogenous corticosteroid level in all these species investigated, one should remember the uncertainty of whether or not resting values were obtained for the hormone concentrations.

The corticosteroid-binding proteins demonstrated in the sera of various species resembled that of man in their thermolability; heating for 15 min at 60° destroyed about 30 to 90% of the binding activity in the different sera (SEAL and DOE, 1966). The binding activity at 37° was only a fraction of that measured at 4°, also in accordance with observations in human serum (SEAL and DOE, 1966).

In view of the divergent ratios of cortisol to corticosterone in different species, a comparison of the high-affinity binding of these two corticosteroids in various sera is of interest. Table VIII-18 indicates that the relative binding affinities for the two steroids can differ widely. This is seen particularly with the iguana serum in which no binding of corticosterone but high cortisol-binding capacity is demonstrable by gel filtration; a reversed relationship was observed with the porcupine fish. A similar comparison of high-affinity binding for cortisol and corticosterone has been made in our laboratory for 20 vertebrate species by the method of equilibrium dialysis (Table VIII-19). Peripheral cortisol: corticosterone ratios are included. The agreement between the two methods on the relative binding is only fair; however, it should be realized that sera from the same species but not necessarily from the same strain may differ in their CBG make-up. It is evident from Table VIII-19 that the binding affinities of all sera

---

[a] Porter-Silber chromogens, determined with cortisol as the standard.

[b] For other species and subspecies of monkeys, binding capacity values (4°) between 1.4 and 27 μg cortisol per 100 ml have been observed (SEAL and DOE, 1966).

[c] From SEAL and ERICKSON (1969).

Table VIII-19. *Relative binding activities (C-values) for cortisol (F) and corticosterone (B) in sera of various species, determined by equilibrium dialysis at 4° and 37°.* From CHADER and WESTPHAL (1968 b)

Species	F : B in blood (SEAL and DOE, 1963)	C-values			
		4°		37°	
		F	B	F	B
Rabbit					
Pool I	0.05	1.11	0.74	0.49	0.27
Pool II	0.05	0.34	0.26	0.15	0.14
Rat	0.05	1.22	1.86	0.22	0.59
Human	10	0.92	0.77	0.32	0.39
Monkey	20	0.65	1.00	0.14	0.48
Guinea pig	>1	1.20	0.79	0.17	0.15
Dog	1—2	0.30	0.23	0.13	0.13
Cat	4—6	0.16	0.22	0.07	0.10
Sheep	15—20	0.27	0.25	0.17	0.18
Steer	1	0.29	0.26	0.15	0.13
Pig	—	0.16	0.16	0.09	0.08
Horse	—	0.51	0.26	0.26	0.17
Goat	—	—	—	0.10	0.10
Ferret	1.5	—	—	0.19	0.19
Starling	—	1.34	1.24	0.20	0.26
Turkey	<1	—	—	0.19	0.19
Goose	—	—	—	0.13	0.18
Duck	—	—	—	0.20	0.27
Chicken	—	—	—	0.16	0.15
Pigeon	—	—	—	0.27	0.38
Carp	>1	—	—	0.08	0.10

are higher at 4° than at 37°. No correlation was observed between affinity to the two corticosteroids and their concentration ratio in the serum.

If one considers the quantitative distribution of corticosteroid-binding proteins in the sera of the 131 species investigated (SEAL and DOE, 1965), it is noteworthy that approximately two thirds of the species have binding capacities of less than 10 μg per 100 ml of either cortisol or corticosterone, whichever has the higher affinity (Table VIII-20). Only 29 species, i.e., 22%, show binding capacities over 20 μg per 100 ml. Twelve of these were mammals, including man, rodents, raccoon, hedgehog, and other old-world primates. Of 32 reptile species studied, twelve had similarly high binding values (snakes, lizards and alligators). Transcortin-type binding of cortisol is generally low in fish; according to IDLER and FREEMAN (1968), the values in salmonids and Atlantic cod average about 5% of those reported for human plasma at 4°. Considering the mammals listed in Table VIII-18, accumulation of high and low binding values in certain groups is evident, although the studies are not yet extensive enough for a general evaluation of the differences observed with respect to their adaptive significance. It should be noted that in general the concentrations of the corticosteroid-binding serum proteins and the resting values of the circulating corticoid hormones at the peak of the circadian cycle are closely related (SEAL and DOE, 1966).

Table VIII-20. *High-affinity binding capacity for corticosteroids in vertebrate sera, determined by gel filtration at 4°.* From SEAL and DOE (1965)

	Species frequency distribution (as µg cortisol or corticosterone bound per 100 ml serum)					
Group	< 5.0	5—10	11—20	21—40	> 41	Total
Fish	18	2	1	0	2	23
Amphibia	4	1	1	2	0	8
Reptiles	6	11	3	6	6	32
Birds	3	5	9	1	0	18
Mammals	17	15	6	8	4	50
Total	48	34	20	17	12	131

SEAL and DOE (1963) have suggested, on the basis of their results involving a great number of different species, that the basic phenomenon of serum protein binding of corticosteroids appeared early in the history of the vertebrates, and that a high structural specificity of the proteins which offers strong binding forces for the corticosteroid hormones has been retained. The first appearance of vertebrates dates back to the paleozoic era, about 500 to 600 million years ago. It should also be remembered that biogenic steranes have been traced back more than 2 billion years (BURLINGAME et al., 1965).

A major study of the corticosteroid-binding capacity in mammalian and avian species has been carried out by STEENO and DE MOOR (1966). Endogenous corticoid concentrations were determined, and a comparison between cortisol and corticosterone binding was made for many sera. The binding capacity was assayed by gel filtration at 20—22° and reported as μg bound per 100 ml; to what extent these values are lowered by dissociation of the complexes at room temperature is not known. Studies of cortisol-binding affinity of CBG in animals (DE MOOR and HEYNS, 1968) at 25° gave for

Dahomey cows	$10.6 \times 10^7 M^{-1}$
Chickens	2.3
Guanaco, female	18.2
Chimpanzee, female	4.4
Antelope, male	6.2

in comparison with $7.3 \pm 1.2 \times 10^7 M^{-1}$ as an average value of 268 normal human plasma samples.

Concentration of CBG binding sites and association constants of CBG-steroid complexes have been determined (WESTPHAL, 1967) in the sera of five mammalian species (Table VIII-21). The pooled normal sera were first cleared of endogenous steroids (Chapter VIII. 3) and the binding parameters for interaction with cortisol, corticosterone and progesterone determined at 4° and 37° according to methods described in Chapter VIII. 2 a). It is evident from Table VIII-21 that the CBG's from different species show considerable differences in their binding affinity to a given steroid. Furthermore, the relative binding strength to the steroid hormones with 5, 4 and 2 oxygen functions, respectively, differs markedly among the species studied. For example, the most highly polar cortisol, in accordance with the polarity rule, is bound least firmly to the CBG of human, monkey, rat and guinea pig serum, whereas rabbit CBG shows the opposite behavior in that its steroid complexes do not follow the polarity rule; the binding affinity increases from progesterone to corticosterone to cortisol. Gel filtration studies with rabbit serum on Sephadex G-200 also gave a higher binding value for cortisol than for corticosterone (CHADER and WESTPHAL, 1968 b). It is noteworthy that the association constant of guinea pig CBG with progesterone is seemingly approximately 100-times greater than that of the complex with cortisol; however, results have been obtained which indicate that two different proteins are responsible for the binding of the two steroids (see Chapter X). For all other sera reported in Table VIII-21, the association constants for the different steroid-CBG complexes are of the same order of magnitude. The k values given in Table VIII-21 are in agreement with the association constants determined with the

Table VIII-21. *Association constants, k, and concentrations of binding sites, [CBG], of steroid-CBG complexes in mammalian sera.* From WESTPHAL (1967)

Species	k in 10^8 M^{-1} for						Serum Levels		
	Cortisol (F)		Cortico-sterone (B)		Progesterone		Protein g per 100 ml	F µM	B µM
	4°	37°	4°	37°	4°	37°			
Human	6	0.3	10	0.3	7 [a]	0.9 [a]	6.5	0.55	—
Monkey	3	0.3	—	1.4	—	—	7.3	—	—
Rat	3	0.1	5	0.3	3	—	6.8	—	1.02
Rabbit	10	0.4	8	0.2	4	—	6.5	—	0.27
Guinea pig	0.5	0.04	1.1 [b]	0.14 [b]	48 [b]	3.9 [b]	5.1	2.26	0.38
	[CBG], µM							Average [CBG] µM	
Human	0.69	0.69	0.72	0.64	0.76 [a]	0.83 [a]		0.72	
Monkey	0.94	0.86	—	0.98	—	—		0.93	
Rat	1.41	0.69	1.36	0.98	1.21	—		1.13	
Rabbit	0.33	0.36	0.32	0.35	0.35	—		0.34	
Guinea pig	1.05	—	0.64 [b]	0.66 [b]	0.48 [b]	0.48 [b]		0.57 [b]	

[a] Pool II, 6.5% protein, 0.27 µM cortisol.
[b] Pool II, 6.1% protein.

isolated homogeneous CBG preparations in those cases in which the pure proteins have been studied (see Chapter XI. 4). ROSENTHAL et al. (1969 b) have recently reported for guinea pig transcortin an apparent association constant of $k = 7.4 \times 10^7 M^{-1}$ at 5°; normal plasma concentrations were 0.6 and 1.0 μM in male and higher in female guinea pigs, in confirmation of the data of Table VIII-21. The concentration of binding sites, [CBG], for each of the five sera investigated (Table VIII-21) was similar in the interaction with cortisol and corticosterone. In most cases, about the same concentration of high-affinity binding sites was found for progesterone, except for guinea pig serum where this value was approximately 25% lower than [CBG] for the corticoids (WESTPHAL, 1967). Although the concentrations of high-affinity binding sites for the three different steroids have been designated as [CBG] and averaged in Table VIII-21, it should be noted that no definite evidence is available for the identity of these binding sites, or of the binding proteins. This problem will be discussed in Chapter X.

A relatively high association constant ($0.9 \times 10^8 M^{-1}$ at 37°) has been reported for the cortisol-CBG complex of ovine plasma (PATERSON and HILLS, 1967); the CBG capacity was low (0.07 μM), as was the peripheral cortisol concentration (1.8 μg/100 ml). A very low corticosteroid-binding capacity in sheep serum was also observed by REICH (1960). Mouse and rat plasma have been reported to bind deoxycorticosterone with higher affinity than 1α-hydroxycorticosterone (IDLER et al., 1967). HOLLANDER and CHIU (1968) reported on the purification of a corticosteroid-binding protein from P1798 lymphosarcoma of the mouse; its properties resembled those of a binding protein from thymus and lymph node (HOLLANDER and STEVENS, 1970). Protein from spleen and liver of mouse and rabbit behaved more like serum CBG.

In view of the increased CBG activity of human serum in pregnancy or after estrogen administration (see Chapter VIII. 5 b. β), SEAL and DOE (1966) examined the effects of pregnancy or egg-laying activity in various species. Exploratory experiments indicated that hormonal effectiveness of estrogens did not parallel CBG changes in pregnancy, even in species that showed marked increase of transcortin activity during gestation. Whereas no changes in the CBG level were found in gravid serum of a few amphibians and reptiles tested, the gravid chicken had a 50% higher cortisol-binding value (12 μg per

Table VIII-22. *Effect of pregnancy on CBG concentration in mammals.* From SEAL and DOE (1966, 1967)

Species	Placental type	Cortisol binding (µg/100 ml) Normal	Gravid	G/N
Chiroptera				
Pteropus giganteus	Epitheliochorial	105	90	0.9
Marsupialia				
Didelphia marsupialis	Choriovitelline	8	10	1.2
Primates				
Homo sapiens (20)	Hemochorial	22	55	2.5
Saimiri sciureus (4)	Hemochorial	5	80	16.0
Cercopithecus aethiops sabeus (2)	Hemochorial	28	72	2.5
Nycticebus coucang (2)	Epitheliochorial	26	15	0.6
Rodenta				
Rattus norvegicus (6)	Hemoendothelial	100	170	1.7
Mus Musculus (12)	Hemoendothelial	45	190	4.2
Cavia porcellus (12)	Hemoendothelial	20	600	30.0
Lagomorpha				
Oryctolagus cuniculus (3)	Hemoendothelial	20	140	7.0
Carnivora				
Felis domesticus (3)	Endotheliochorial	5	4	0.8
Canis familiaris (6)	Endotheliochorial	6	6	1.0
Procyon lotor (3)	Hemochorial	30	120	4.0
Mephitis mephitis (2)	Endotheliochorial	12	10	0.8
Mustela vison (2)	Endotheliochorial	12	12	1.0
Artiodactyla				
Bos taurus (3)	Syndesmochorial	5	3	0.6
Ovis aries (3)	Syndesmochorial	5	5	1.0
Sus scrofa (3)	Epitheliochorial	4	4	1.0
Perissodactyla				
Equus caballus (3)	Epitheliochorial	10	11	1.1
Edentata				
Dasypus novemcinctus (2)	Hemochorial	2.8	4.2	1.5

( ) Number of animals in parentheses.

100 ml) than the normal hen (8 μg per 100 ml). The rooster showed only half that concentration (4 μg per 100 ml). It would be desirable that these interesting results be corroborated by other methods and extended to additional avian species.

SEAL and DOE (1966) studied the effect of pregnancy on serum CBG in mammals, including 17 species from 8 orders. Elevated levels up to 30-fold were observed (Table VIII-22); all species exhibiting this increase were of the hemochorial type of placentation. The biological significance of the CBG-affected corticosteroid function in

Table VIII-23. *Maternal, fetal, and placental CBG capacity (μg cortisol bound) and 17-OHCS in man and guinea pig.* From SEAL and DOE (1967)

Sample	CBG µg/100 ml	Porter-Silber chromogens µg/100 ml
Human		
Normal serum	25	15
Maternal serum	75	35
Fetal serum	10	14
Placenta (µg/100 g)	14	60
Guinea pig		
Normal serum	20	15
Maternal serum	620	700
Fetal serum	173	100
Placenta (µg/100 g)	56	188
Isolated maternal guinea pig CBG = 140 mg/100 ml.		

hemochorial placentation remains to be elucidated. No rise in CBG activity was observed in the pregnant white-tailed deer (SEAL and ERICKSON, 1969).

Considerable increases of CBG activity were observed in the pregnant mouse, rabbit and guinea pig (GALA and WESTPHAL, 1967); the C-values were 10 to 20 times greater than in the normal animals. These alterations during pregnancy and subsequent lactation period will be discussed in Chapter X. A particularly high increase was seen in the pregnant guinea pig (GALA and WESTPHAL, 1967; SEAL and DOE, 1967; DIAMOND et al., 1969; ROSENTHAL et al., 1969 b). Table VIII-23 shows the CBG capacity in the maternal and fetal sera, together with the 17-OHCS values; human data are included for comparison (SEAL and DOE, 1967). Injection of estrone to normal guinea pigs increased the CBG capacity several fold (YUDAEV et al., 1964). It is noteworthy that the pregnant guinea pig shows, concomitantly with the elevation of CBG, a massive increase in the high-affinity binding of progesterone and testosterone. This will be discussed in Chapter X.

It should be noted at the conclusion of the present chapter that many of the properties of CBG, which have been discussed in some detail, may be considered characteristic of high-affinity binding serum proteins. Similarities can be found with the thyroid hormone-binding TBG (WESTPHAL, 1970 b), and also with the steroid-binding β-globulin which interacts strongly with testosterone and estradiol, as will be discussed in Chapter XII. Chapter IX and X will summarize experimental results on various factors which affect CBG in laboratory animals, and Chapter XI will review our present knowledge of the physicochemical properties of CBG. It is hoped that some of the problems discussed in these pages with reference to CBG may be useful to future investigators in their efforts to elucidate new areas of high-affinity interaction between steroids and proteins or other macromolecules.

IX. Influence of Various Endogenous Factors on CBG Activity in the Rat

1. Initial Observations; Effect of Adrenalectomy

Soon after the discovery of CBG in the human blood, it was observed that the plasma of the rat also contains macromolecules capable of binding cortisol with high affinity; the binding capacity was found to be higher than that of human plasma (DAUGHADAY, 1958 a; SLAUNWHITE and SANDBERG, 1959). Rat CBG activity, which can also be demonstrated in serum, was found to be stable at 37° for at least 72 hrs (WESTPHAL, 1961) or 96 hrs (WESTPHAL et al., 1962), resembling human CBG in that respect (DAUGHADAY et al., 1959). These basic observations on a high-affinity binding system for corticosteroids in the rat, similar to transcortin in the human, made possible the general investigation of the corticosteroid-binding protein by experimental approaches, in a thoroughly studied laboratory animal (WESTPHAL, 1961). The general characteristics of the corticosteroid-binding protein in rat serum have been discussed in Chapter VIII. 4 in connection with the properties of human CBG.

Initial studies in our laboratory on distribution of injected radiolabeled cortisol (FIRSCHEIN et al., 1957) revealed that the intravascular half-time of the steroid administered to adrenalectomized rats was prolonged when compared with values obtained in intact control animals (ULRICH and LONG, 1956; GULYASSY et al., 1958). In a similar way, the passage of cortisol into the gut had been found slower in adrenalectomized mice than in normal controls (COPE and SEWELL, 1956). The difference in the turnover rates could be explained when we observed in equilibrium dialysis studies that the corticosteroid-binding activity in the serum of the adrenalectomized rat was markedly higher than in the normal animal (WESTPHAL, 1961; WESTPHAL et al., 1962). This is shown in Table IX-1.

The high-CBG sera of the adrenalectomized rats (Table IX-1) had been obtained 3 days after the operation. An analysis of the increase of CBG with time after adrenalectomy (Fig. IX-1) showed that the CBG activity remains at a normal level for at least 24 hrs. after the operation. Rising C-values are seen after 48 and 72 hrs, for interaction with corticosterone as well as with cortisol. The increase of the corticosterone-binding activity from the 24-hr to the 48-hr

Table IX-1. *Binding affinity of rat serum and human serum for cortisol and corticosterone, before and after elimination of adrenal secretion.* From Westphal et al. (1962)

Serum	Cortico-steroid µg%	C-values for Cortisol 4°	37°	C-values for Corticosterone 4°	37°
Rat, normal male	31 [a]	1.20	0.16	2.1	0.46
		1.25	0.17	2.0	0.45
Rat, adrenal-ectomized male	8 [a]	6.6	0.26	10.7	0.82
		5.5	0.22	13.7	0.84
Human, normal (Ni)	10 [b]	0.65	0.28	0.53	0.34
		0.62	0.25	0.53	0.33
Human, normal (Cha)	19 [b]	0.48	0.25	0.45	0.33
		0.53	0.28	0.41	0.36
Human, normal (Cha) after 0.5 mg dexa-methasone p.o. $\overline{q}$. 6 h×9	0 [b]	1.26	0.35	0.82	0.41
		1.23	0.35	0.80	0.38

[a] Corticosterone concentrations determined according to Silber et al. (1958). The values include the "background fluorescence" discussed in Chapter VIII. 3.

[b] 17-hydroxy corticosteroid concentrations of sera (Ni) and (Cha) determined by modification (Peterson et al., 1957) of Porter-Silber procedure.

value is statistically highly significant as indicated in Fig. IX-2. After 3—4 days, a maximal C-value is reached which is approximately twice normal and remains essentially at this level for at least 3 weeks. The curve obtained for cortisol binding (Fig. IX-2, curve C) has a similar tendency although the CBG activity is lower. Shamadrenal-ectomized control rats (curve B) had normal C-values.

It is further evident from these data, and from numerous results of other experiments, that corticosterone is bound more firmly to rat CBG than cortisol is, contrary to the initial observations in several

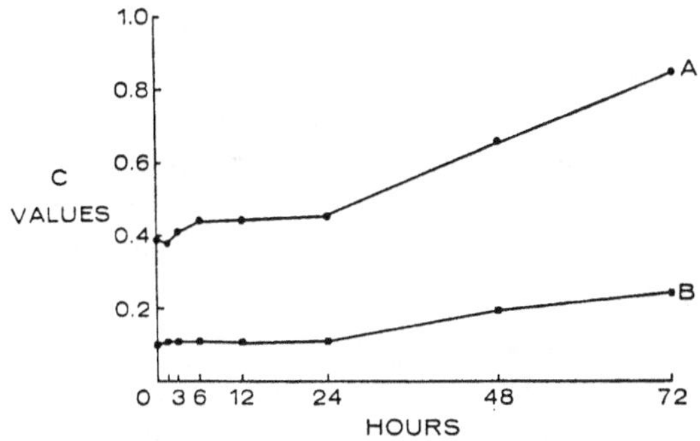

Fig. IX-1. Corticosteroid-binding activity of rat serum 0, 1.5, 3, 6, 12, 24, 48, and 72 hrs after adrenalectomy. The C-values were determined by multiple equilibrium dialysis at 37° with corticosterone-4-^{14}C (curve A) or with cortisol-4-^{14}C (curve B). From WESTPHAL et al. (1963)

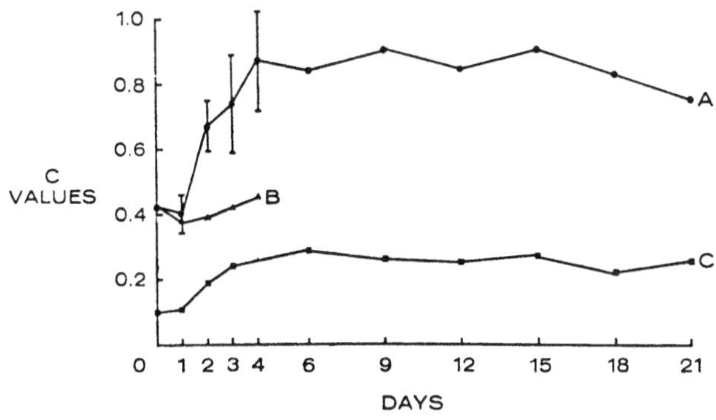

Fig. IX-2. Corticosteroid-binding activity of rat serum 0, 1, 2, 3, 4, 6, 9, 12, 15, 18 and 21 days after adrenalectomy. The C-values were determined by multiple equilibrium dialysis at 37° with corticosterone-4-^{14}C (curve A for adrenalectomized rats, curve B for shamoperated controls) or with cortisol-4-^{14}C (curve C for adrenalectomized rats). The vertical bars on curve A indicate fiducial limits at a probability level of 0.01. From WESTPHAL et al. (1963)

laboratories. Higher association constants for the CBG complexes with corticosterone have been substantiated in more complete studies as has been shown in Table VIII-21. KELLER et al. (1966) have confirmed the higher affinity for corticosterone in ultrafiltration studies on rat plasma. The observation of increased CBG activity in the rat after adrenalectomy has also been confirmed in other laboratories

(SEAL and DOE, 1965; KELLER et al., 1966; ACS et al., 1967; LABRIE, 1967). No similar increases were found in the adrenalectomized guinea pig, rabbit, dog, or cat (SEAL and DOE, 1965).

Elevation of CBG activity in the rat after adrenalectomy on day 12 of pregnancy was almost as high as in the normal non-pregnant control; however, no such increase of CBG activity was found on day 21 of gestation (KOCH et al., 1968 b). Apparently, the normal control of CBG synthesis is disturbed during the final stage of pregnancy in the rat. This is also suggested by the observation that thyroxin administration (see Chapter IX. 7) does not increase the CBG level in the rat pregnant for 21 days, whereas normal CBG stimulation occurs after thyroxin injection on day 12 of pregnancy (KOCH et al., 1968 b).

For comparison, Table IX-1 shows the corticosteroid-binding activity of normal human serum, and of the serum from the same subject after administration of dexamethasone for over 2 days. This treatment depletes the serum of endogenous corticosteroids as the 17-OHCS analysis indicates. Increased CBG activity in the serum of the "chemically adrenalectomized" patient is evident. It is of interest that an increased CBG capacity was also observed (DOE et al., 1964) in some cases of untreated adrenal insufficiency. A different explanation presumably applies to the observation by DE MOOR and HINNEKENS (1961) of a slower removal rate of endogenous corticoids after dexamethasone since the analysis was made only several hours after administration of the synthetic steroid. Longer latent periods appear necessary for measurable changes in CBG concentration to occur. It should be mentioned in this connection that experiments with rats have shown that the route of injection has a definite influence on distribution and excretion of radiolabeled cortisol (GULYASSY et al., 1958).

The increase of CBG activity in the adrenalectomized rat (Figs. IX-1, IX-2) can be suppressed and reversed by administration of corticosterone as seen in Fig. IX-3. Normal and adrenalectomized rats were injected sc once daily for 4 days, starting from the day of the operation, with either saline or 5 mg of a microcrystalline suspension of corticosterone; the serum was analyzed 24 hrs after the last injection. Treatment with this massive dose of glucocorticoid not only suppresses the increase of CBG activity in the adrenalectomized rats, but reduces it to about half the normal level. The C-value thus becomes that of the corticosterone-injected intact rat which also has about half the corticosterone-binding activity as the normal, saline injected control (Fig. IX-3). The decrease of CBG activity in rat plasma by administration of corticosterone has been confirmed (KELLER et al., 1966; LABRIE, 1967; PELLETIER et al., 1968).

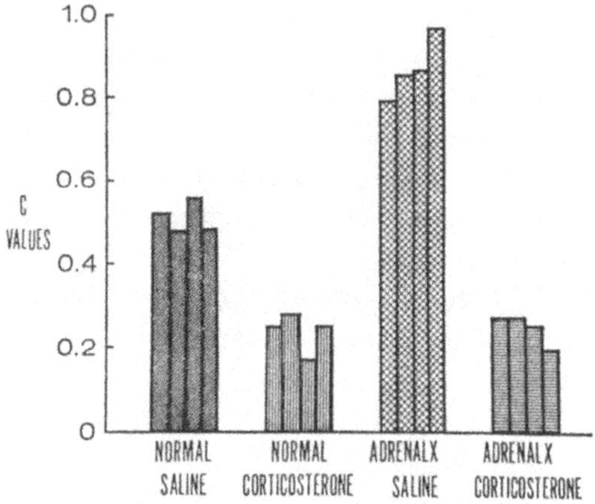

Fig. IX-3. Corticosteroid-binding activity in the serum of normal and adrenalectomized rats after injection of saline or of corticosterone. C-values determined with corticosterone-4-^{14}C by multiple equilibrium dialysis at 37°. Each bar represents average C-values from duplicate determinations with different serum pools obtained from 3—4 rats each. From WESTPHAL et al. (1963)

It is of interest to relate these experimental results, obtained with the rat, to observations on human patients. Several cases of Cushing's syndrome of various etiology (DOE et al., 1964) showed low normal or subnormal CBG concentrations. BROOKS et al. (1963 a) assume a decreased transcortin concentration in patients with adrenocortical hyperfunction. Only half the normal CBG capacity was found by DE MOOR et al (1965 b) in a female patient suffering from Cushing's syndrome; the same author (DE MOOR et al., 1968 a) suggests a general association of adrenocortical hyperplasia (Cushing) with low CBG values. Other examples of normal transcortin concentration in Cushing patients (GRAY et al., 1962) show that the effects of adrenocortical hyperfunction on the plasma binding system cannot be generalized. See also Chapter VIII. 5 b. γ.

2. General Procedures Used in the Endocrine-Related Studies

In our studies on rat CBG to be described in subsequent sections of this chapter, Sprague-Dawley rats from commercial sources were used unless specified otherwise (GALA and WESTPHAL, 1965 a, 1965 b). Prior to initiation of experimental procedures, the animals were maintained for an adjustment period of at least 5—7 days under animal room conditions of uniform temperature (25 ± 1°) and humidity. Lighting was controlled to be turned on from 7 : 30 AM to 8 : 30 PM. Water and Purina Laboratory Chow were available *ad libitum* throughout the studies. All surgical procedures were performed under ether anesthesia.

In a few initial experiments, blood was collected from ether-anesthesized animals by section of the left subclavian artery and bleeding into a skin pouch. However, since it is known that ether anesthesia results in elevated corticosterone levels (BEIGELMAN et al., 1956; KITAY, 1961; ZIMMERMAN and CRITCHLOW, 1967) blood was obtained in all subsequent studies under "resting conditions" by decapitation with a guillotine within 30 sec from the time the cage was handled. The time of sacrifice was between 8 : 30 and 11 : 00 AM when the corticosterone levels are at a minimum (GUILLEMIN et al., 1959 b; MCCARTHY et al., 1960). The blood was allowed to clot overnight at 4°. Serum obtained either from 3—6 adult rats or from 4—12 young animals was pooled into single samples, immediately frozen and stored at —85° C. A total of 2 to 3 serum pools were collected for each experimental group studied.

Pooling of serum was necessary in order to obtain adequate quantities for the various determinations. We have verified in separate experiments (DE VENUTO et al., 1964) that the values for corticosterone-binding activity and corticosterone concentration, determined in individual rat sera and averaged, are the same as those of a pool obtained by combining aliquots of each rat serum. The same agreement between averages calculated from individual serum values and values of the combined pool of the same sera was found for cortisol-binding activity and 17-OHCS in human sera; similar results have also been reported for rabbit serum (CHADER and WESTPHAL, 1968 b). The data obtained with serum pools, therefore, constitute correct average results; in fact, analysis of serum pools is preferable to measurements with limited numbers of individual sera since each pool provides an average value representative of a larger population.

No differences have been observed between CBG activities in serum and in plasma (SLAUNWHITE and SANDBERG, 1959). Similarly, corticosterone levels have been reported the same in serum and plasma (GUILLEMIN et al., 1958); this has been confirmed in our laboratory (GALA and WESTPHAL, 1965 a). Freezing and thawing of the serum does not influence the C-values. Thymus weights were determined in order to make an independent assessment of unbound corticosterone in the various experimental groups. The weight of the thymus gland has been used as a biological indicator of corticosteroid activity and its measurement forms the basis of several bioassays (DORFMAN, 1962 a); a decrease in thymus weight indicates a rising level of physiologically active, i. e., unbound, corticosteroid hormone, and *vice versa*.

The CBG activity of each rat serum pool was determined in duplicate by multiple equilibrium dialysis at 37° and expressed as C-values. The duplicate values for each pool were then averaged, and these averages, obtained from at least two pools for each experimental group, were again averaged to give the final mean C-value listed in the tables for the various experimental groups. Average C-values for the individual pools may be found in the original publications. The contribution of albumin to the C-value was determined by including a pure rat serum albumin sample (3 mg/ml; see WESTPHAL and DE VENUTO, 1966) in the multiple equilibrium dialysis (GALA and WESTPHAL, 1966 b). The result obtained, $C = 0.07$ at 37°, indicated that this contribution is less than 8% for the sera of female rats and less than 15% for those of male animals. Phosphate buffer, 0.05 M, pH 7.4, was used as solvent; the sera were diluted to give a total protein concentration of 5 mg per ml, determined by a biuret technique (GORNALL et al., 1949). Corticosterone concentrations were measured for each serum pool by the fluorometric method of SILBER et al. (1958) and expressed as μg per 100 ml. Limitations of the fluorometric procedure have been discussed in Chapter VIII. 3. The portion of free, i.e., unbound corticosterone was calculated from the C-values and the total peripheral corticosterone concentration under the assumption that, at the corticosterone level present, the binding was due to CBG (see Chapter VIII. 2 a). The ^{14}C-labeled corticosterone was periodically checked for radiochemical purity. Differences in thymus weights were assessed statistically by use of Student's t-test; a 95% level of probability was accepted as the lower level of significance. The standard error of the mean for thymus weights was calculated by the formula

$$SE = \pm \sqrt{\frac{\Sigma x^2 - \frac{(\Sigma x)^2}{n}}{n(n-1)}}.$$

Statistical analysis was not performed for C-values and serum corticosterone levels (GALA and WESTPHAL, 1965 a); individual values for the serum pools together with average figures have been given in the original publications.

3. Influence of Age and Sex on CBG Activity and Corticosterone Level

The CBG activity develops in the growing rat from very low values in the immature animal to the normal adult level (GALA and WESTPHAL, 1965 a, 1965 b). Observations on postnatal rats showed

Table IX-2. *Corticosterone-binding activity (C-values) and peripheral corticosterone levels of suckling female and male rats.* From GALA and WESTPHAL (1965 a)

		C-values at 37°		Corticosterone µg/100 ml [a]	
Age (days)	No. of animals	females	males	females	males
3	70	0.21	0.19	9.8	7.4
12	64	0.11	0.12	8.0	8.2
21	64	0.36	0.33	35.0	34.0

[a] Values from ether-stressed animals.

that the C-value at 3 days of age was higher than that found at day 12 (Table IX-2); the CBG activity then increased from day 12 to day 21. This rise continues as seen in Fig. IX-4 which shows the corticosterone-binding activity in the serum of rats 12 to 57 days old.

Male rats showed an increase in CBG activity from day 12 to day 30; the C-values then remained about the same to at least day 57. The binding level in the serum from female animals continued to rise beyond day 30 to level off at day 48 to 66. A sex difference in binding activity was first apparent at 30 days of age and reached a maximum at approximately 48 days when the female values were 85% above those of the males. More than twice the male binding activity

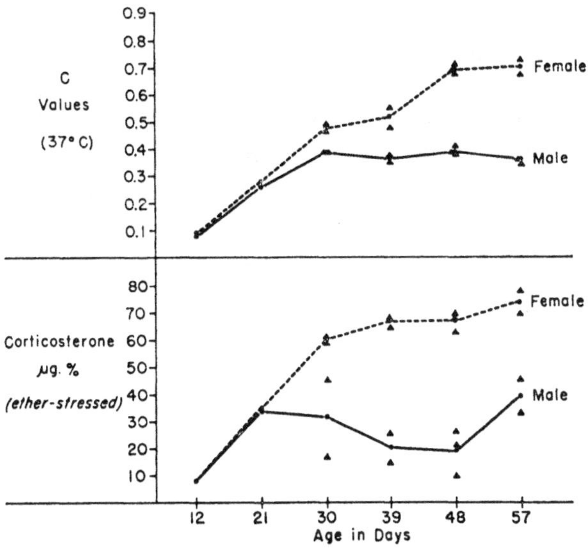

Fig. IX-4. Influence of age and sex on corticosterone-binding activity (C-values) and corticosterone levels of rat serum. Each triangle represents the average of duplicate values from one of several serum pools; dots indicate overall averages. The C-value of female rats at 66 days of age equaled those at 48 and 57 days. Vaginal patency at age 30, 39, 48 and 57 days was 0%, 17%, 82% and 100%, respectively, of all female rats in the group. From GALA and WESTPHAL (1965 b)

was found (GALA and WESTPHAL, 1964) in other experiments with adult females (see also Table IX-8). Vaginal opening occurred between day 30 and 57.

The peripheral corticosterone concentrations, increasing in the growing rat (Fig. IX-4), were similar for both sexes at day 12 and 21, but differed at later periods. The considerable fluctuations in the values of the male rats are probably a consequence of the ether-stressed condition. The general similarity of the parallel changes in C-values and corticosterone level in the female and male rat is evident in Fig. IX-4. Increasing corticoid levels in the growing male rat have been reported by THORNTON et al. (1962). DECKX et al. (1965) have found increases in the corticosterone levels of growing male and female rats very similar to those shown in Fig. IX-4.

Our findings of higher CBG activity in adult female rats as compared to males (GALA and WESTPHAL, 1964), also observed by SEAL and DOE (1965), have been confirmed by KELLER et al. (1966) with Sprague-Dawley rats. These authors also confirmed the lower binding activity in 23-day old rats, and the similarity of binding between male and female animals at the weanling age. Female Wistar rats also had about twice the CBG activity of males (MCDONAGH, 1966; KOCH et al., 1967 a). More than twice the normal male CBG capacity was measured in normal female Sprague-Dawley rats by STEENO and LLOYD (1967). A detailed study in postnatal male rats (KOCH et al., 1967 b) showed a CBG capacity at birth about as high as that of the normal adult male; the binding activity declined rapidly to a minimum at day 6 to 9, then increased to reach normal adult levels at day 20. These results confirm and extend our observations (Table IX-2, Fig. IX-4) with another strain of rats, since KOCH and his coworkers used Wistar rats in their studies.

The lower CBG activity of serum from immature rats compared with adult animals seems to be analogous to the low binding values found in newborn infants (cord blood, see Table VIII-11) when compared with their mothers (DAUGHADAY et al., 1959; SANDBERG and SLAUNWHITE, 1959; DE MOOR et al., 1963 c). However, the CBG activity in human blood increases long before sexual maturation; transcortin levels of children a few months to 14 years of age have not been found significantly different from those of adults (Table VIII-11). The observation of a sex difference in the CBG activities of normal adult rats is in contrast to the results in humans (Table VIII-11).

The finding of typical changes in the peripheral corticosterone concentration of the developing rat (Table IX-2 and Fig. IX-4) has been confirmed and extended in a number of laboratories, in connection with studies on the development of the circadian rhythm (ALLEN and KENDALL, 1967), stress response (HALTMEYER et al., 1966; LEVINE et al., 1967; ZARROW et al., 1967) and other problems (HOLT and OLIVER, 1968, KOCH et al., 1968 a). BOHUS et al. (1963) did not observe sex differences in the *in-vitro* production of corticosterone by the adrenals of rats up to 37 to 39 days of age; glands from 52-days old females gave higher values. KITAY (1961) reported for Sherman rats average resting plasma corticosterone concentrations of 9.0 and 18.3 μg-% for males and females, respectively; CRITCHLOW et al. (1963) also found higher corticoid levels in female than in male rats. Very little sex difference was observed (MCCARTHY et al., 1960) in the peripheral corticosterone in normal rats of the Purdue strain, in contrast to results by D'ANGELO (1968) on Sprague-Dawley rats which showed differences of plasma corticosteroid values between females and males similar to those observed in our laboratory. GLENISTER and YATES (1961) found a shorter half-life for corticosterone in the female rat, and conclude this to be based on a greater capacity for corticosteroid inactivation in the female sex, due to a higher Δ^4-steroid hydrogenase activity (see also TROOP, 1959). The inactivation results in greater ACTH stimulation which then leads to the greater secretion rate in the female adrenal.

4. Influence of Gonadal Hormones

a) *Effects of Gonadectomy*

The observation of a sex difference in the CBG activity of the adult rat led us to investigate the role of the gonadal secretions. In a study of *prepuberal* gonadectomy, groups of male and of female rats were castrated on day 12, and sacrificed when 57 days old. The CBG activity of the serum from the female rats ovarectomized at 12 days of age was not significantly different from that of the controls (Table IX-3). The castrated male rats, however, showed a 70% increase in binding activity when compared with their controls, and approached the female level. The serum corticosterone concentrations of the gonadectomized animals varied greatly, possibly as a result of ether anesthesia; the average values were not found significantly different from those of the intact controls. The CBG activities given in Table IX-3 were determined in the same multiple equilibrium dialysis as those shown in Fig. IX-4; in accordance, the general level of the C-values is closely comparable.

The effect of *postpuberal* gonadectomy was studied in a similar way. Adult male and female rats were gonadectomized at approximately 70 days in experiment A (Table IX-4) and at 76 days in experiment B; they were sacrificed 14 and 24 days later, respectively, under resting conditions. The CBG activity of serum pools from experiments A and B was determined in separate multiple equilibrium dialyses; the results from the two experiments were averaged as shown in Table IX-4.

Rats gonadectomized postpuberally gave a response similar in binding activity to that of the animals castrated prepuberally, i.e., there was no significant change in CBG levels of female animals, whereas the binding activity of gonadectomized male rats increased approximately 50% over their controls (Table IX-4, average values).

Table IX-3. *Effect of prepuberal gonadectomy on corticosterone-binding activity of rat serum. Average values from duplicate pools; individual values given in original publication* (GALA and WESTPHAL, 1965 b)

Age days	Sex	No. of animals	C-values at 37°	Corticosterone µg/100 ml Serum [a]
57-control	F	19	0.70	73.5
	M	15	0.36	39.8
57-gonadx	F	7	0.79	61.0
at 12 days	M	9	0.61	33.0

[a] From ether-stressed animals.

Castration in the male resulted in little change in resting corticosterone concentrations, while the corticosterone levels in the ovarectomized female were significantly decreased and similar to those of the male. The thymus weights when expressed per body weight were not significantly higher for the intact males, whereas after gonadectomy the weights were significantly ($p<0.05$) greater in the females than in the males. In accordance, calculation of the amount of unbound corticosterone revealed a slightly higher value for the intact female animals; after gonadectomy, the values were higher for the males.

The influence of the duration of gonadectomy on the CBG activity was studied in a number of multiple equilibrium dialyses. The data are presented in Fig. IX-5 as the percentage of change in CBG activity from the control; each point represents a single serum pool. There was

Table IX-4. *Effect of postpuberal gonadectomy on corticosterone-binding activity of rat serum. Average values from duplicate pools, obtained in two multiple equilibrium dialyses (A and B). Individual values and body weights given in original publication* (GALA and WESTPHAL, 1965 b)

Exp.	Approx age days	Sex	No. of animals	C-values at 37°	Corticosterone µg/100 ml Serum	Thymus wt in mg Actual	Thymus wt in mg per 100 g body wt
A	84-sham gonadx	F	8	1.33	42.2	277 ± 16	142 ± 8
		M	8	0.58	4.7	449 ± 17	155 ± 6
	84-gonadx for 14 days	F	8	1.33	8.7	562 ± 21	253 ± 11
		M	8	0.79	4.7	608 ± 32	216 ± 11
B	100-sham gonadx	F	8	0.92	16.9	252 ± 9	98 ± 3
		M	8	0.34	13.1	432 ± 16	117 ± 11
	100-gonadx for 24 days	F	8	1.02	7.9	677 ± 26	216 ± 4
		M	7	0.54	16.9	642 ± 36	189 ± 9
Avg values	sham gonadx	F	16	1.12	29.5 (0.38)	265	120
		M	16	0.46	8.9 (0.29)	441	136 [NS]
	gonadx	F	16	1.17	8.3 (0.11)	620	234
		M	15	0.67	10.9 (0.24)	625	204 [a]

± Standard error of the mean.

( ) Average values in parentheses are µg/100 ml unbound corticosterone.

NS Not significantly different from female thymus ($p > 0.05$).

[a] Significantly smaller than female thymus ($p < 0.05$).

little change in the binding activity of male rats four days after gonadectomy; binding activity was not studied in females for this time interval. Male rats gonadectomized for 14, 24 and 45 days showed an increase in binding activity of 41, 68 and 70%, respectively. No significant changes in CBG activity were observed in female rats at the same times after gonadectomy.

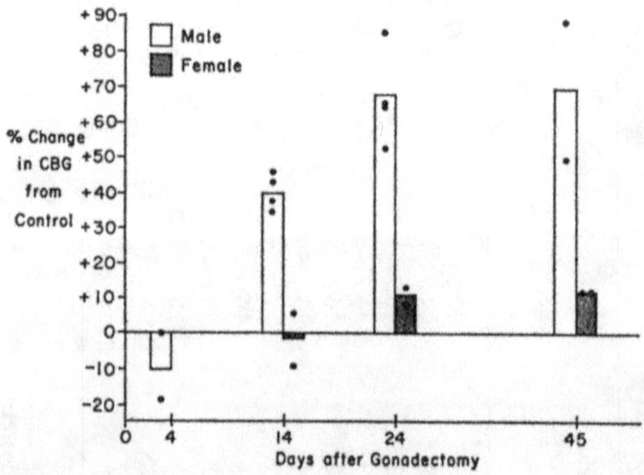

Fig. IX-5. Effect of gonadectomy on corticosterone-binding activity of rat serum. Each dot represents a serum pool from 3—5 animals; the bars indicate average values. From GALA and WESTPHAL (1965 b)

These observations show that gonadectomy either before or after puberty resulted in an increased CBG activity in the male rat but produced no significant CBG change in the female (Fig. IX-5); the increased values in the male, however, did not reach the female level. Rats orchidectomized prepuberally gave C-values rather close to those observed in sera from intact females. Concerning the variation in the C-values of the female control rats, it has been our observation that the CBG activity of adult females housed in groups was lower than that of animals kept in individual cages. Rats maintained in larger groups are known to have higher corticosteroid levels (EECHAUTE et al., 1962); such higher levels, maintained over longer periods of time, may lead to decreased CBG activity in a similar way as corticosterone administration has been shown to lower the C-values in the rat (Fig. IX-3). The CBG activities of the prepuberally and postpuberally gonadectomized males were not significantly different.

A decrease in resting and stressed serum corticosterone levels in the female rat after gonadectomy (Tables IX-3 and IX-4) has been observed by other investigators (CRITCHLOW et al., 1963; KITAY, 1963 b, 1965 a, 1965 b; KITAY et al., 1965). Equalization of the corticosterone levels in male and female Sprague-Dawley rats after gonadectomy has been reported (SKELTON and BERNARDIS, 1966).

Increase of CBG capacity in the male rat after castration has been noted by SEAL and DOE (1965); the same authors also observed unchanged CBG values in the ovarectomized rat. In contrast, ÁCS et al. (1967) found a decrease in the CBG capacity of the female rat after gonadectomy; Wistar rats were used in these studies, and the binding capacity was measured by gel filtration at room temperature. Several reasons, therefore, may account for these discrepancies. STEENO and LLOYD (1967) confirmed with Sprague-Dawley rats the increased CBG capacity after orchidectomy, and observed a statistically significant decline of the CBG values following ovarectomy.

b) Injection of Steroid Sex Hormones

Estradiol was injected into male and female rats at a level of 1 and 10 μg per day for a period of 10 days (GALA and WESTPHAL, 1965 b). Blood was collected from stressed animals on the day after the last injection. Fig. IX-6 shows that the estrogen treatment did not alter the corticosterone-binding activity of the female rats. In the male animals, however, either level of estradiol increased the binding activity.

Serum corticosterone levels of stressed females appeared depressed after estradiol administration, whereas, in the male, slightly elevated peripheral corticosterone concentrations were observed. Thymus weights were reduced in both sexes by the administration of estradiol, to a greater extent in the female rats.

The increased binding activity in the serum of intact male rats after estradiol injection did not reach normal female levels (Fig. IX-6); the activity was similar to that noted for gonadectomized males (Tables IX-3 and IX-4). Estrogen injection has been reported to increase binding activity in male rats (YATES et al., 1963; SEAL and DOE, 1965; KELLER et al., 1966; LABRIE et al., 1968). Analogous observations have been made in humans as discussed in Chapter VIII. 5 b. β.

In an effort to shed light on the mechanism of the somewhat similar increase of CBG activity in the male rat after estrogen and after gonadectomy, a combination of the experimental conditions was studied (GALA and WESTPHAL, 1966 c) in order to test for possible additive effects. The corticosterone-binding activity was compared in the serum of rats after sham gonadectomy, gonadectomy, sham

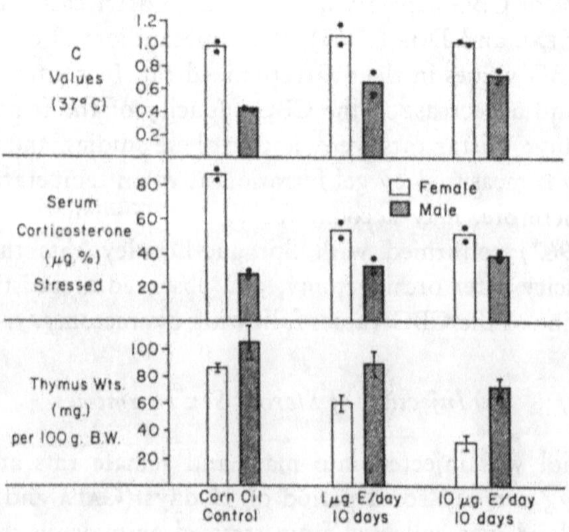

Fig. IX-6. Effect of estradiol (E) injections on corticosterone-binding activity of rat serum. For details see GALA and WESTPHAL (1965 b); graphical explanations in Fig. IX-5

gonadectomy plus estradiol injection, and after gonadectomy plus estradiol injection. The administration of estrogen was initiated on the 14th day after gonadectomy and was continued for the next 10 days. All animals were sacrificed 24 days after operation.

As evident in Table IX-5, and as to be expected from the results given in Table IX-4 and Fig. IX-6, the CBG activity either of gonadectomized male rats, or of sham-operated male rats given estradiol, was significantly higher than that of sham controls. However, the administration of estradiol to the orchidectomized rats did not enhance the binding activity above that noted for the gonadectomy alone. This suggests that the elevated CBG activity in the male rat after castration and after estrogen injection may be mediated

Table IX-5. *Influence of estradiol (E) injection on the corticosterone-binding activity of gonadectomized (gonadx) male rats. Average values from duplicate pools, obtained by multiple equilibrium dialysis. Values for individual pools and body weights given in original publication* (GALA and WESTPHAL 1966 c)

Experimental group	No. of animals	C-values at 37°	Corticosterone $\mu g/100$ ml serum [a]	Thymus wt (mg) Actual	Thymus wt (mg) per 100 g BW
Sham gonadx-24 days	7	0.37	4.7	339 ± 16	93 ± 5
Gonadx-24 days	8	0.59	3.9	630 ± 45	182 ± 10
Sham gonadx + 10 µg E [b]	8	0.61	10.3	299 ± 21	87 ± 6
Gonadx-24 days + 10 µg E [b]	8	0.54	6.1	403 ± 19	120 ± 5

[a] Values from rapidly decapitated animals.
[b] Amount of hormone injected subcutaneously once daily for the last 10 days.
± Standard error of the mean.

by a similar mechanism; an additive effect on CBG increase might otherwise have been expected. Thymus weights increased after gonadectomy; estradiol injection reversed this trend.

In a complementary study (GALA and WESTPHAL, 1965 b), testosterone was administered to male and female intact rats for a period of 10 days; blood was collected on the day after the last injection. The results are shown in Fig. IX-7. The administration of 1 and 5 mg testosterone per day to the female rat lowered the CBG activity by 18 and 30%, respectively; injection of the same amounts to the male did not alter the binding activity. Resting serum corticosterone concentrations were not significantly altered in the male rat, while decreased levels were observed in the female animal. Thymus weights were decreased by testosterone injection; the depression in the female rat was again greater than in the male.

As in the case of estradiol, the effect of testosterone was also examined in male rats which had been previously castrated. The testosterone administration at a level of 5 mg per day was begun 14 days after orchidectomy and continued for the following 10 days.

One group was sacrificed 14 days after castration to indicate the level of CBG at the onset of testosterone injections. All other groups were sacrificed the day after the last injection.

As documented in Table IX-6, the corticosterone-binding activity of serum 14 and 24 days after castration had increased 46 and 78%, respectively, over that of the sham controls. This is in accordance

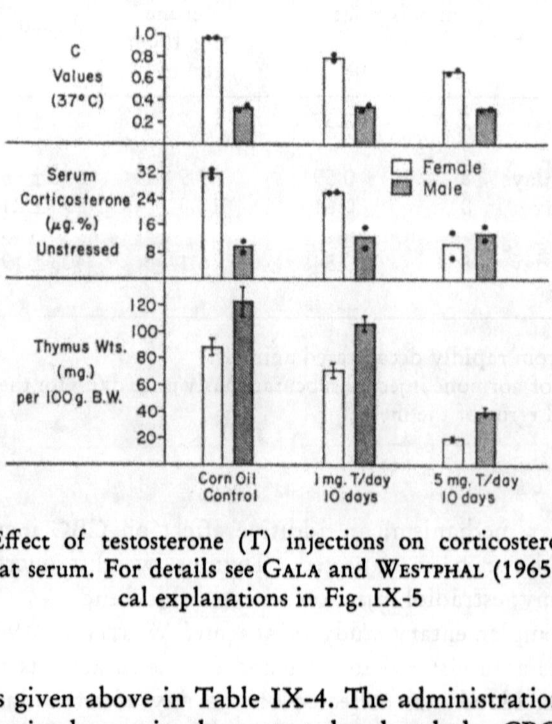

Fig. IX-7. Effect of testosterone (T) injections on corticosterone-binding activity of rat serum. For details see GALA and WESTPHAL (1965 b); graphical explanations in Fig. IX-5

with results given above in Table IX-4. The administration of testosterone to animals previously castrated reduced the CBG levels to control values. The increased thymus weights found after castration were reduced below control values by the injection of testosterone; this result is similar to that seen in Fig. IX-7 and indicates a high level of unbound corticosteroid.

Testosterone administration to the female rat (Fig. IX-7) depresses the CBG activity to values observed in estrogen-treated or gonadectomized males; however, the decrease does not lead to CBG values as low as those found in intact males. It appears that testosterone has a suppressing effect on CBG activity and this may in part explain why the activity in the adult male rat is lower than in the female. Elimination of testosterone by castration did not result in

CBG activities as high as those in intact female rats (Table IX-4). This would seem to indicate that testosterone is not solely responsible for the sex difference in CBG activity, and that additional modes of regulation are operative. The obvious assumption that the estrogenic hormones could play a role as stimulants in a hormonally balanced system is not borne out in our experiments which showed continued

Table IX-6. *Effect of testosterone injections on corticosterone binding activity of gonadectomized (gonadx) male rats. Average values from duplicate pools, obtained by multiple equilibrium dialysis. Individual values and body weights given in original publication* (GALA and WESTPHAL, 1965 b)

Experimental group	No. of animals	C-values at 37°	Cortico-sterone μg/100 ml serum	Thymus wt in mg	
				Actual	per 100 g BW
Control-sham gonadx	8	0.35	13.1	432 ± 16	118 ± 4
Gonadx-14 days	7	0.51	3.3	665 ± 50	—
Gonadx-24 days	7	0.62	16.9	642 ± 29	190 ± 9
Gonadx-24 days, 5 mg testosterone per day last 10 days	8	0.42	20.0	180 ± 12	51 ± 3

± Standard error of the mean.

relatively high CBG activity in the female rat after gonadectomy (Tables IX-3 and IX-4). However, it has been mentioned in a preceding section that decreased CBG levels after ovarectomy were observed in other laboratories. The CBG-depressing effect of testosterone in the female rat has been confirmed (LABRIE et al., 1968).

The suppression of the binding activity by testosterone reduces the elevated CBG values of castrated male rats to essentially those of the controls (Table IX-6), indicating that testosterone may be the physiological factor that maintains the low binding activity in the intact male. Fig. IX-7 shows that normal endogenous secretions can produce maximum suppression of CBG activity since injection of up to 5 mg testosterone per day into the intact male rat does not result in a decrease of the normal male CBG level.

The results obtained by administration of progesterone to intact rats are compiled in Table IX-7. The hormone was injected for 10 days and the animals were sacrificed on the 11th day. The male rat responded to the progesterone treatment at the 2 and 10 mg level with an increase of the CBG activity by 30 and 90%, respectively. The lower dose of progesterone did not alter the CBG activity in the female animal, but injection of the higher amount resulted in a 22%

Table IX-7. *Effect of progesterone injections on corticosterone-binding activity of rat serum. Average values from duplicate pools, obtained by multiple equilibrium dialysis. Individual values and body weights given in original publication* (GALA and WESTPHAL, 1965 b)

Experimental group	Sex	No. of animals	C-values at 37°	Cortico-sterone µg/100 ml Serum	Thymus wt in mg Actual	Thymus wt in mg per 100 g body wt
Control; corn oil for 10 days	F	8	1.08	18.6	294 ± 19	136 ± 8
	M	8	0.41	6.7	359 ± 19	117 ± 5
2 mg progesterone per day for 10 days	F	8	0.97	11.9	307 ± 20	138 ± 9
	M	8	0.53	5.9	397 ± 21	139 ± 7
10 mg progesterone per day for 10 days	F	8	1.31	4.3	340 ± 17	152 ± 9
	M	8	0.78	3.5	428 ± 27	149 ± 10

± Standard error of the mean.

increase of CBG. Progesterone lowered the resting peripheral corticosterone levels in males and females to a point at which no significant difference between the sexes was apparent. In both sexes there appeared to be a trend for the thymus weights to increase with increasing dose of progesterone. The effect of progesterone on CBG activity and peripheral corticosterone concentration in the male rat has been confirmed (LABRIE et al., 1968). It is of interest in this context that VERMEULEN and FERIN (1962) reported increased CBG levels in humans treated with the progestogen 17α-methyl-19-nortestosterone.

Exogenous sex steroids have been reported by a number of investigators to alter the peripheral corticosterone levels. VOGT (1955) found that hexoestrol administration resulted in a reversible inhibition of corticosterone secretion by the left adrenal of adult male Wistar rats. A decrease in corticosterone concentration of female rats and a slight increase in male animals after estrogen administration (Fig. IX-6) has been observed in other laboratories (KITAY, 1963 a, 1964; PINCUS and HIRAI, 1964; KITAY, 1966). Testosterone administration to female rats lowered corticosterone (Fig. IX-7) in accord with other published reports (SAKIZ, 1960; KITAY, 1963 b), but slightly raised the levels in male animals, as also observed previously (KITAY, 1963 b). Progesterone administration to male and female rats considerably decreased the resting corticosterone levels (Table IX-7). This confirms published observations (SINGER et al., 1963; STEINETZ et al., 1965) which showed that progesterone injection depressed corticosterone secretion in rats; the adrenocortical response to ACTH or ether-induced stress was lessened. A possible explanation is the displacement of corticosterone from the binding sites on CBG by the injected progesterone. The initial increase of unbound corticoid would by feedback regulation lower the adrenal secretion, and lead to increased CBG levels. Since the major portion of the corticosteroid-binding sites are occupied by the exogenous progesterone, the equilibrium between bound and unbound corticoid is such that the level of peripheral corticosterone remains low.

The elevated CBG activity after injection of progesterone and the coincidental decrease of the peripheral corticosterone levels in the male and female rat (Table IX-7) deserves special comment. The response of the animal resembles that after adrenalectomy when the corticosterone-binding activity rises following depletion of the circulating corticosteroids (see Section 1 of this chapter). The sequelae of the progesterone administration may represent a second example of a general mechanism by which a lowering of the corticosterone level is accompanied by increased CBG activity. This mode of influence on the binding affinity appears to be fundamentally different from that observed after estradiol injection into the male rat (Fig. IX-6), or after testosterone administration to the female rat (Fig. IX-7). In these cases, the augmentation and reduction of CBG activity, caused by the estrogenic and androgenic steroid hormones, respectively, are accompanied by parallel changes in the corticosterone level; the changes after adrenalectomy and progesterone administration, however, are in opposite directions.

Considerable variability in resting corticosterone levels was noted in some experiments. Care was taken to minimize this lack of constancy by using the precautions described in Chapter IX.2. Some of the variability in female animals could be attributed to different stages of the estrous cycle which are known to cause differences in resting corticosterone levels (CRITCHLOW et al., 1963; GALA and WESTPHAL, 1965 a), as will be discussed below. Occasional low values, comparable to those found in adrenalectomy, were observed in male control rats; the cause of these values is unknown. Any corticosterone determinations that were not in accord with our previous findings or with published observations, or seemed unusual otherwise, were repeated to verify the original determination.

Thymus weights were used to indicate the amount of free biologically active corticosteroid. However, little use could be made of changes in thymus weights when exogenous steroid was administered since it has been reported that certain steroid hormones, i.e., estradiol, testosterone, and progesterone, potentiate the effect of corticosteroids on thymus involution (DORFMAN and DORFMAN, 1962). In the progesterone study, thymus weights were somewhat higher with increased doses of steroid (Table IX-7). The amount of unbound corticosterone for the 10 mg group was calculated to be about one fourth that of the controls; comparable calculations for the experiments with high levels of estradiol and testosterone showed either no change in the unbound corticosterone or a decrease to about one half the control value. It appeared then that in the progesterone study decrease in unbound corticosterone counterbalanced the potentiating effect of progesterone on thymus involution, resulting in a slight increase in thymus weight.

The amount of unbound, biologically active corticosterone is the resultant of total steroid concentration and binding activity, essentially provided by CBG. In an effort to assess the level of unbound corticosterone in female and male rats, we have summarized the data from all experiments in which both intact male and female animals were studied under unstressed conditions. Table IX-8 shows that the amount of unbound corticosterone appeared only slightly greater in intact female rats than in intact males; the difference was not statistically significant. Thymus weight per 100 g body weight for the same groups was somewhat greater in the intact male when compared to the intact female. The lower thymus weight of female rats may be the result of the greater potentiation of corticosteroid-induced thymus involution by estrogens as compared to androgens

(DORFMAN and DORFMAN, 1962). It appears, then, that despite the higher peripheral corticosterone concentration in the female rat, the level of unbound, biologically active corticosterone is only insignificantly higher than that in the male. Although the calculations listed in Table IX-8 cannot claim the accuracy expected of a physico-chemical treatment of a biological problem due to the large range of

Table IX-8. *Summary of sex differences in intact rats.* From GALA and WESTPHAL (1965 b)

Sex	C-values at 37°	Corticosterone μg/100 ml Serum[a] Total	Unbound	Thymus wt per 100 g body wt
Female (32)	1.08	27.4	0.39	116
	±.05	±4.6	±0.06	±5
Male (31)	0.46	8.5	0.33	128
	±.04	±1.2	±0.06	±4

[a] Unstressed.
( ) Number of animals.
± Standard error of the mean.

error of the variables entering the calculations, we believe the correlation of CBG activities and resting corticosterone levels to obtain the approximate concentrations of unbound corticosteroid provides a meaningful insight into the problems involved.

5. CBG Activity during Estrous Cycle, Pregnancy, Lactation and Weaning

When we found the higher CBG activity in the adult female rat, and established the causative influence of the estrogenic hormone in this species, it became of interest to follow this steroid binding parameter through the various reproductive phases (GALA and WESTPHAL, 1965 a). For a determination of the corticosterone-binding activity during the estrous cycle, blood was obtained from rats with known estrous cycle histories on the day of vaginal estrus and on the second day of vaginal diestrus. No difference was found in the binding level of serum obtained in estrus and in diestrus (Table IX-9). The peripheral corticosterone concentrations of the estrous rats were 2.7 times as high as those measured in diestrus. CRITCHLOW et al. (1963) also reported for this species the lowest resting serum corticosterone level in the diestrus phase.

In the preparation of the rats for the study of CBG activity in pregnancy (GALA and WESTPHAL, 1965 a), the day of detection of sperm in two- to three-month-old females after cohabitation was

Table IX-9. *Corticosterone-binding activity and corticosterone levels of female rats during the estrous cycle.* From GALA and WESTPHAL (1965 a)

Phase of cycle	C-values at 37°	Corticosterone μg/100 ml [a]
estrus (8)	0.87 ± 0.02	40.5 ± 9.0
diestrus (8)	0.84 ± 0.02	15.2 ± 1.6

[a] Values from rapidly decapitated animals.
( ) Number of animals in parentheses.
± Estimate of the variation of the mean outlined in Section 2 of this chapter.

considered as day 1 of pregnancy and at this time they were placed into individual cages. Serum was prepared from animals pregnant for 8, 14, and 20 days. A single serum pool from three rats nursing six to eight pups per litter for 12 days was included in this dialysis, as were serum pools from virgin mature animals. Multiple equilibrium dialysis was performed using 0.01 and 0.001 μg of radioactive corticosterone and 0.01 μg of radioactive cortisol per ml outside solution.

The analysis at the higher level of radiocorticosterone revealed an insignificant increase in the binding activity of serum from pregnant rats when compared with virgin animals (Table IX-10). C-values of serum from lactating rats were about half those of pregnant or virgin animals. Peripheral corticosterone levels of rats pregnant for 8 and 14 days were similar and about 60% lower than those found in virgin animals; they were somewhat higher at 20 days of gestation. Serum from lactating rats had corticosterone levels similar to those observed in early pregnancy. Thymus weights at 8 or 14 days of

Table IX-10. *Corticosterone-binding activity, corticosterone levels and thymus weights of pregnant rats.* From GALA and WESTPHAL (1965 a)

Reproductive epochs	C-values at 37°	Corticosterone μg/100 ml [a]	Thymus weight (mg)
virgin (8)	0.95 ± 0.03	33.2 ± 0.4	208 ± 13
8 days pregn (12)	1.03 ± 0.02	13.2 ± 0.8	318 ± 17
14 days pregn (11)	0.99 ± 0.03	14.2 ± 3.2	276 ± 16
20 days pregn (9)	1.05 ± 0.03	19.3 ± 0.4	185 ± 22
12 days lact (3)	0.57 ± 0.02	12.0	191

[a] Resting values.
( ) Number of animals in parentheses.
± Estimate of the variation of the mean outlined in Section 2 of this chapter.

gestation were significantly higher than those of virgin animals; they decreased again in later pregnancy and lactation.

Multiple equilibrium dialysis of these sera using the lower level of ^{14}C-labeled corticosterone, 0.001 μg/ml, seemed to indicate an increase in the binding activity of pregnant rats when compared with that of virgin animals (Fig. IX-8). Under either test condition

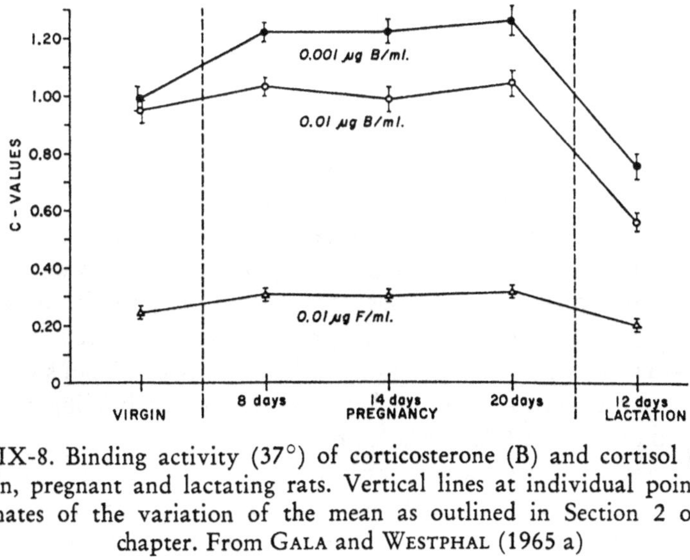

Fig. IX-8. Binding activity (37°) of corticosterone (B) and cortisol (F) in virgin, pregnant and lactating rats. Vertical lines at individual points are estimates of the variation of the mean as outlined in Section 2 of this chapter. From GALA and WESTPHAL (1965 a)

a precipitous decline of CBG activity was observed in lactation. Binding activity for these sera to cortisol was one third to one fourth that observed for corticosterone but followed a similar trend.

The sharp decrease of the CBG activity after parturition prompted an analysis of the C-values, corticosterone concentrations and thymus weights during lactation. Female rats when found by palpation to be pregnant were placed into individual cages. The day of parturition was considered day 1 of lactation and the litters on

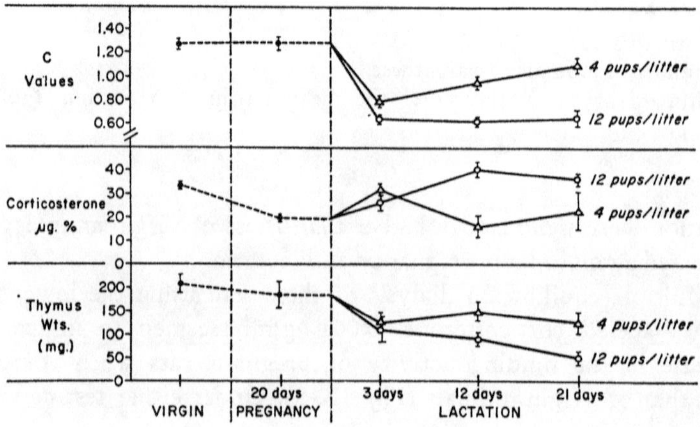

Fig. IX-9. Corticosterone-binding activity, resting corticosterone levels and thymus weights of virgin, pregnant and lactating rats. For other explanations see Fig. IX-8. From GALA and WESTPHAL (1965 a)

this day were adjusted to either four or twelve pups. Blood was collected by decapitation from the mother rats at the end of 3, 12 and 21 days of lactation. A decrease in binding activity was again observed in this experiment (Fig. IX-9); the decline was evident at the third day of lactation. In subsequent lactation periods, the serum binding activity of mothers nursing 4 pups increased until it approached that of the virgin animal; the suckling of 12 pups resulted in the maintenance throughout lactation of the low binding activity observed on the third day of lactation.

Peripheral corticosterone concentrations on the third day of lactation were higher than those of late pregnancy. In the presence of few pups, lower corticosterone levels were found in subsequent lactation periods when compared with the virgin state, whereas high pup numbers resulted in higher levels. Thymus weights on the third day of lactation decreased considerably from those seen during late pregnancy and in virgin animals. In the later lactation periods, the thymus weights of mothers nursing four pups plateaued at the level observed on the third day of lactation, whereas they continued to decrease in the group with 12 pups. It should be mentioned that the decreased CBG level in the lactating rat has been confirmed (KOCH, 1969).

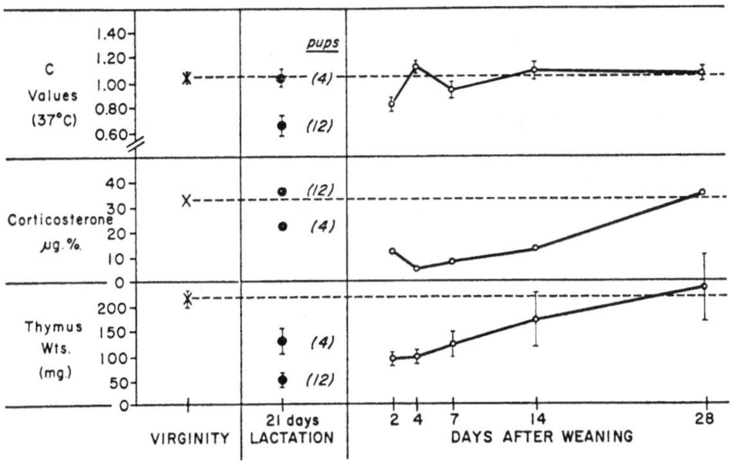

Fig. IX-10. Corticosterone-binding activity, resting corticosterone levels and thymus weights of virgin and lactating rats and mother rats after weaning from 7—9 pups/litter. For other explanations see Fig. IX-8. From GALA and WESTPHAL (1965 a)

The study of the corticosterone-binding activity was continued through the weaning period. On the day of parturition, litters were adjusted to 7—9 pups. After 21 days of lactation, the litters were removed and mothers were sacrificed 2, 4, 7, 14 and 28 days after weaning. Serum pools from virgin rats and from rats lactating for 21 days were again included in the multiple equilibrium dialysis.

Corticosterone-binding activity of the rats with 7—9 pups two days after weaning was, as expected, between the values for animals lactating 21 days and nursing either 4 or 12 pups (Fig. IX-10). On the fourth day after weaning, and in subsequent periods, the binding activity was found to be similar to that observed for the virgin condition. Serum corticosterone levels two days after weaning were considerably lower than those noted for either virginity or late lactation; they decreased further still on the fourth day to a level characteristic of adrenalectomized rats. Subsequently, the resting corticosterone concentration increased continually until the 28th day, when the values became similar to those found in adult virgin females. Thymus weights on the second day after weaning were between those observed late in lactation for animals nursing either 4 or 12 pups; they increased gradually to the level of virgin animals. The normal adult female values for the three parameters measured were thus restored not later than 4 weeks after weaning.

Serum from rats pregnant for 20 days (Fig. IX-9) had a corticosterone-binding activity similar to that of virgin adult animals. The somewhat greater difference between CBG activity of virgin and pregnant rats shown in Fig. IX-8 should be interpreted with caution; such difference could be considered significant only when found consistently in a greater number of assays. This conclusion has been validated by more recent studies of Koch et al. (1967 a) who confirmed our results; the authors found within range of error the same corticosterone-binding capacity for pregnant and non-pregnant female rats. Seal and Doe (1967) have reported a CBG value for the pregnant rat which is higher than that of the normal female (Table VIII-22).

The lack of a marked increase in CBG activity in the pregnant rat, more recently confirmed by Koch (1969), is in contrast to observations in man (see Chapter VIII. 5 b. β) and a number of other species (Tables VIII-22, VIII-23). Whereas the causative role of the estrogens in CBG increase in human pregnancy is well established, the CBG activity of female rats does not rise further after administration of estrogens (Fig. IX-6). Relatively large amounts of estrogens have been found in the urine of pregnant rats (Zarrow, 1961), suggesting high circulating levels; a possible interpretation of this difference between the two species may therefore be that in the rat the estrogen level of the adult female is optimal in relation to production of CBG, whereas in man the normal estrogen concentration is not sufficient for maximal CBG production. Only during pregnancy or after administration of estrogenic hormone are levels of estrogens reached in the woman which are consistent with maximal production of CBG.

The present results (Table IX-10; Fig. IX-9) have led to the assumption that CBG plays a significant role in the initiation of lactation (GALA and WESTPHAL, 1965 a). This suggestion is made in corroboration of the studies of MEITES (TALWALKER et al., 1961; MEITES et al., 1963; VOOGT et al., 1969) which emphasize the importance of corticosteroids in the initiation of lactation. These authors showed that administration of corticosterone or cortisol initiates lactation in pregnant rabbits and rats without disturbing pregnancy, and suggested "that during pregnancy a decreased synthesis and/or increased protein binding of adrenal glucocorticoids" occurs (TALWALKER et al., 1961). In confirmation, FRIESEN (1966) induced lactation in pregnant rabbits by placental lactogen and cortisone. Corticosterone injection has been found to increase milk yield in the lactating rat (HAHN and TURNER, 1966).

In rats a higher dose of corticosterone than cortisol was necessary to obtain a similar response (TALWALKER et al., 1961). This result is in accordance with the previous observation of a higher binding affinity of rat CBG for corticosterone than for cortisol (WESTPHAL et al., 1962) if one assumes in the rat a quantitatively similar physiological effect of these two corticosteroids in their unbound state. This assumption finds support in studies by RIVERA (1964), in which the addition of either corticosterone or cortisol to mouse mammary gland organ cultures resulted in the initiation of secretory activity as evidenced by histological observations. Under these experimental conditions, i. e., in the absence of CBG, corticosterone and cortisol at the same concentrations gave similar secretory responses in the mammary lobules.

Although there was little change in binding activity during pregnancy, the resting serum corticosterone levels were found to be about one third to one half those noted for the nonpregnant rat. This combination resulted in decreased corticosteroid activity, as evidenced by the increase in thymus weights. In late pregnancy (20 days), there was a slight increase in the peripheral corticosterone which was biologically active, as evidenced by the decrease in thymus weights to the nonpregnant level. The cause of the increased corticosteroid activity may be the presence of elevated estrogen, since administration of estrogens has been reported to potentiate the biological effects of corticosteroid (DORFMAN and DORFMAN, 1962; NELSON et al., 1963). This would explain why rats during late pregnancy and non-pregnant animals had similar thymus weights, although their corticosterone levels were different.

The increase in estrogens late in pregnancy may act in conjunction with the elevated corticosteroid levels to bring about the initiation of lactation. The precipitous decrease in CBG activity early in lactation (Fig. IX-9), when considered in connection with the increase in corticosterone levels, would provide the proper conditions for maximum corticosteroid activity. Changes in thymus weight during early lactation reflect increased corticosterone activity, in spite of decreased progesterone concentrations with concomitant availability of CBG binding sites previously occupied by progesterone (DAVIS and LIU, 1969). The low binding activity and high corticosterone levels of mothers nursing high pup numbers may reflect the need for large amounts of free corticosterone, and here again the continuous decrease in thymus weight throughout lactation indicated that biologically active corticosterone was ample.

The finding of high peripheral corticosterone levels in the presence of low CBG activity as observed in lactation is unusual since elevated corticosteroid concentration is mostly connected with high CBG. It should be noted in this connection that McDONAGH (1966) found decreased CBG activity accompanied by high corticosterone levels in the cold-exposed male rat. This again would imply the presence of relatively large amounts of unbound, biologically active corticosterone.

Several days after weaning, the CBG activity had returned to normal values (Fig. IX-10), whereas the peripheral corticosterone concentration and the thymus weights needed several weeks to reach the normal pre-pregnancy level. The regulating factors responsible for the increase of CBG back to the normal activity of the virgin adult rat are not known. A possible interpretation of the parallel and prolonged increase of corticosterone concentration and thymus weights is that the decrease in the portion of unbound, i. e., physiologically active corticosterone triggers increased corticosterone secretion up to normal level, while at the same time normal CBG permits the thymus to grow to normal size.

6. Significance of the Pituitary Gland for CBG Activity

It has been shown in the preceding sections that the steroid hormones of the adrenal cortex and of the gonads influence the CBG activity of the rat in a positive or negative sense. All of these endocrine secretions are functionally related to the pituitary gland. The logical question arose therefore whether the hypophysis was involved in the regulation of the CBG level.

The first problem studied (GALA and WESTPHAL, 1966 a) was that of the role of the pituitary in the increase of CBG activity after adrenalectomy. The initial observations (see Chapter IX. 1) had been obtained with male rats; they were now extended to female animals. Male and female rats were sacrificed 14 days after adrenalectomy; sham-adrenalectomized animals served as controls. These two types of animals in turn were used as controls for a more complete study of the influence of corticoid hormones on the CBG activity. Corticosterone was administered at the levels of 0.4 and 2.0 mg per day for 7 days, to male and female rats which had been adrenalectomized 7 days before the start of the injections.

Absence of the adrenals for 14 days resulted in an increase in binding activity of 53% in female rats and 86% in males (Table IX-11). The injection of 0.4 mg of corticosterone did not alter the binding activity of the female serum, and hardly decreased that of the male serum. The higher quantity of corticosterone returned the CBG levels of male animals to that of sham controls; female rats still had a binding activity 22% higher than sham controls. Thymus weights in both sexes increased with adrenalectomy and decreased after corticosterone injection.

In a parallel study, the effect of cortisol was examined under the same conditions. In this case, the increase in binding activity after adrenalectomy for female rats was 51% and for male rats 118% (Table IX-12). The administration of 0.4 mg of cortisol to male and female animals returned the binding activity to that of the sham controls; the sex difference was still apparent. Injection of 2.0 mg of cortisol decreased the CBG activity markedly below that of the sham controls; the levels attained in females were the lowest ever observed by us for that sex. The higher cortisol dose also abolished the sex difference in CBG activity. Thymus weights increased again

Table IX-11. *Corticosterone-binding activity of serum from adrenalectomized (adx) rats after corticosterone (B) injection. Average values from duplicate pools, obtained by multiple equilibrium dialysis. Individual values and body weights given in original publication* (GALA and WESTPHAL, 1966 a)

Experimental group	Sex	No. of animals	C-values at 37°	Cortico-sterone μg/100 ml serum [a]	Thymus weight	
					actual	per 100 g body wt
control-sham-adx +	F	8	1.04	60.8	236 ± 18	103 ± 7
corn oil	M	8	0.42	8.5	347 ± 17	110 ± 6
adx-14 days; corn oil	F	6	1.59	4.1	360 ± 24	158 ± 13
last 7 days	M	5	0.78	5.4	530 ± 40	185 ± 10
adx-14 days; 0.4 mg	F	8	1.54	4.0	300 ± 20	133 ± 9
B/day last 7 days	M	7	0.63	4.4	499 ± 37	165 ± 12
adx-14 days; 2.0 mg	F	8	1.27	7.5	243 ± 28	105 ± 12
B/day last 7 days	M	7	0.38	8.1	263 ± 37	86 ± 11

[a] Values from unstressed animals [values for female control rats were in the range of those obtained from stressed animals; these rats must have been inadvertently stressed prior to guillotine decapitation].
± Standard error of the mean.

Table IX-12. *Corticosterone-binding activity of serum from adrenalectomized (adx) rats after cortisol (F) injection. Average values from duplicate pools, obtained by multiple equilibrium dialysis. Individual values and body weights given in original publication* (GALA and WESTPHAL, 1966 a)

Experimental group	Sex	No. of animals	C-values at 37°	Cortico-sterone μg/100 ml serum [a]	Thymus weight	
					actual	per 100 g body wt
control-sham-adx	F	7	0.97	36.4	275 ± 24	121 ± 10
+ corn oil	M	8	0.40	5.5	398 ± 37	116 ± 10
adx-14 days; corn oil	F	8	1.46	6.0	393 ± 34	175 ± 14
last 7 days	M	8	0.87	3.8	549 ± 34	179 ± 10
adx-14 days; 0.4 mg	F	8	0.99	4.2	252 ± 17	116 ± 7
F/day last 7 days	M	8	0.47	4.7	414 ± 22	129 ± 7
adx-14 days; 2.0 mg	F	7	0.30	5.6	42 ± 11	21 ± 2
F/day last 7 days	M	8	0.26	5.9	60 ± 6	20 ± 2

[a] Values from animals sacrificed by rapid decapitation.
± Standard error of the mean.

with adrenalectomy and decreased to very low levels after administration of the higher cortisol dose.

The results supplement our previous observations and show that female rats respond to adrenalectomy in a similar manner (Tables IX-11, IX-12 and Tables IX-14, IX-21, IX-23 below). Marked differences were noted in the response of CBG to the two corticoids: cortisol was found to be about five times as potent as corticosterone. The data presented here, combined with the observation that corticosterone administration from the time of adrenalectomy prevented the rise in CBG activity (WESTPHAL et al., 1963), led to the conclusion that CBG responded inversely to the corticosteroid level, i. e., that increased peripheral corticosteroid concentration resulted in decreased CBG activity, and *vice versa*.

It was not evident from our experiments whether the increase in CBG activity after adrenalectomy was caused by the lack of corticosterone *per se* or by some secondary effect of corticosterone deficiency; involvement of the pituitary gland appeared to be a possibility. This led us to examine the effect of hypophysectomy on the response of CBG to adrenalectomy (GALA and WESTPHAL, 1966 a).

Serum samples of four groups of male rats (sham-hypophysectomy plus sham-adrenalectomy, hypophysectomy, adrenalectomy, and both hypophysectomy and adrenalectomy) were analyzed for binding activity. The C-values of adrenalectomized animals were found to be 133% higher than those of sham controls, whereas in the hypophysectomized group the increase was only 40%, and the response was variable (Table IX-13). Hypophysectomy plus adrenalectomy eliminated the increase in CBG activity observed in adrenalectomized animals. The low C-values of the doubly sham-operated controls may be caused by the suppressing effect on CBG activity of increased endogenous corticosteroids immediately after the operation. There was an increase of thymus weight in all operated animals over sham controls, most markedly in the adrenalectomy group.

The data in Table IX-13 indicate that the lack of circulating corticosteroids *per se*, as a consequence of hypophysectomy, did not result in increased CBG activity. The requirement of the pituitary gland itself, or possibly of a pituitary factor, was suggested when it was found that the increase of CBG activity, regularly observed after adrenalectomy, did not occur when the animals were both adrenalectomized and hypophysectomized. Subsequent studies were performed to explore the reasons for the indispensability of the pituitary for the rise in CBG activity.

Since it is well known that adrenalectomy in the rat results in increased ACTH production (BRODISH and LONG, 1956; FORTIER, 1959; ULRICH and SLUSHER, 1964), studies were designed to examine the influence of this hormone on CBG activity. Unilateral adrenalectomy has been reported to increase ACTH secretion while corticosterone levels are either normal or elevated (BRODISH and LONG, 1956; RECK and FORTIER, 1960). In our experiment, male and female rats were unilaterally or bilaterally adrenalectomized and sacrificed at the end of the fourth postoperative day. At autopsy, the weights of the remaining right adrenal from unilaterally operated animals as well as the right adrenal from sham-operated animals were recorded.

Table IX-13. *Corticosterone-binding activity of serum from male rats hypophysectomized (hypox) and hypophysectomized-adrenalectomized (hypox-adx). Average values from duplicate pools, obtained by multiple equilibrium dialysis. Values for individual pools given in original publication* (GALA and WESTPHAL, 1966 a)

Experimental group	No. of animals	C-values at 37°	Corticosterone μg/100 ml serum	Thymus wt actual
control-sham-hypox + sham-adx-4 days	11	0.27	7.2	240 ± 24
hypox-4 days	9	0.38	2.4	279 ± 19
adx-4 days	12	0.63	2.3	426 ± 10
hypox + adx-4 days	6	0.30	2.3	285 ± 21

± Standard error of the mean.

Bilateral adrenalectomy resulted in a 41% increase in the binding activity of female rats and a 71% rise in males (Table IX-14). Unilateral adrenalectomy, on the other hand, did not affect the CBG levels; this indicated that elevated endogenous ACTH did not alter CBG activity during a period of four days—a time sufficiently long to demonstrate an increase in CBG activity (Chapter IX. 1). The significant weight gain of the remaining adrenal gland four days after unilateral adrenalectomy was evidence of an actual augmenta-

Table IX-14. *Corticosterone-binding activity of serum from rats after bilateral and unilateral adrenalectomy (adx). Average values from duplicate pools, obtained by multiple equilibrium dialysis. Values for individual pools given in original publication* (GALA and WESTPHAL, 1966 a)

Experimental group	Sex	No. of animals	C-values at 37°	Corticosterone μg/100 ml serum [a]	Thymus wt. actual	Adrenal weight	
						rt. only	% increase from control
Sham-adx-4 days	F	8	0.91	33.2	213 ± 15	30.3 ± 0.9	—
	M	8	0.41	7.7	459 ± 32	22.7 ± 0.5 [b]	—
bilateral adx for 4 days	F	8	1.28	4.7	267 ± 22	—	—
	M	8	0.70	2.3	633 ± 28	—	—
unilateral adx for 4 days	F	8	0.90	34.1	238 ± 11	37.0 ± 1.0 [c]	22.1
	M	8	0.45	7.8	539 ± 28	25.9 ± 0.7 [d]	14.1

[a] Values from rapidly decapitated animals.
[b] Only 4 values in this group.
[c] Significantly larger than the control gland ($p < 0.01$).
[d] Significantly larger than the control gland ($p < 0.05$).
∓ Standard error of the mean.

tion in ACTH production; it has been reported (GANONG, 1954) that unilateral adrenalectomy of hypophysectomized rats receiving a compensating amount of ACTH does not result in enlargement of the remaining gland. Thus, any increase in adrenal size after unilateral adrenalectomy is due to enhanced ACTH production. Resting serum corticosterone levels were unchanged by the unilateral operation and the sex difference was apparent. Thymus weights were

Table IX-15. *Corticosterone-binding activity of serum from hypophysectomized-adrenalectomized (hypox-adx) male rats after ACTH or estradiol (E) injection for seven days. For details on numbers of determinations see original publication* (GALA and WESTPHAL, 1966 a)

Experimental group	No. of animals	Body weight initial	Body weight final	C-values at 37°	Corticosterone μg/100 ml serum [a]	Thymus weight per 100 g body wt
sham-hypox, sham-adx	12	252	284	0.37	5.7	118 ± 5
hypox-adx control	8	227	216	0.47	3.6	93 ± 4
hypox-adx + ACTH [b]	10	227	207	0.42	4.1	71 ± 7
hypox-adx + 1 μg E [c]	4	226	205	0.45	3.0	79 ± 10
hypox-adx + 10 μg E [c]	5	227	208	0.47	4.8	52 ± 5

[a] Values from unstressed animals.
[b] 3.0 U injected sc twice daily.
[c] Injected sc once daily.
± Standard error of the mean.

significantly higher in bilaterally adrenalectomized rats but only slightly so in unilaterally operated animals.

To examine the possibility that the presence of normal levels of corticosteroids, as observed after unilateral adrenalectomy, might prevent the hypothetical ACTH-induced increase in CBG activity, ACTH was administered to hypophysectomized-adrenalectomized rats. The same animal preparation was used to study the influence of the gland extirpations on the CBG increase caused by estradiol. Table IX-15 shows once more that there was no marked change in the corticosterone-binding activity of the hypophysectomized-adrenalectomized rats when compared with their sham controls, in sharp contrast to the increased CBG activity of the adrenalectomized animal with the pituitary gland intact (Tables IX-11, IX-12, IX-13, IX-14). Exogenous ACTH did not significantly alter the binding activity. These results were duplicated with a second group of male rats; for details see GALA and WESTPHAL (1966 a). We conclude from these findings that ACTH is not the pituitary factor which controls the rise of CBG after adrenalectomy. This is in accordance with observations in humans which showed that ACTH administration did not increase CBG levels (SANDBERG et al., 1960; DEMOOR et al., 1962; MURPHY and PATTEE, 1963).

Once it was established that the pituitary gland is necessary for the CBG increase after adrenalectomy, the question was raised whether the hypophysis was also involved in the estrogen-induced enhancement of CBG activity. Sham-hypophysectomized and hypophysectomized male rats were allocated to four groups: sham-hypophysectomy control, hypophysectomy control, sham-hypophysectomy + 10 μg of estradiol per day, and hypophysectomy + 10 μg of estradiol per day. The duration of the experiment was 12 days; estradiol was administered for the last 10 days. Normal control groups without and with injection of 10 μg of estradiol per day were included.

The results are shown in Table IX-16. The CBG activity of the hypophysectomized male rats did not differ significantly from that of sham-operated or normal controls. Estradiol injection into either normal or sham-control animals resulted in approximately 50% increase in binding activity over the respective controls. Similar estradiol injections into hypophysectomized rats did not alter CBG activity indicating that the pituitary is necessary for the estrogen-induced increase. Resting corticosterone levels of non-hypophysectomized animals after estradiol administration were higher in accordance with the increased CBG activity.

These studies indicate that hypophysectomy prevents the increase in CBG activity noted both after adrenalectomy and after estrogen administration. The necessity of the pituitary gland was also shown when estradiol was given to hypophysectomized-adrenalectomized male rats (Table IX-15); no alteration of the corticosterone-binding activity resulted. Two of various possibilities were suggested to explain this effect of hypophysectomy: 1) It may be due to the loss of a specific pituitary factor that directly controls the biosynthesis of CBG; this factor may or may not be one of the known pituitary hormones exerting a previously unrecognized action. 2) The effect may be due to the loss of a number of pituitary hormones; the lack of a CBG increase would then be a consequence of the generally deficient physical condition of the hypophysectomized animal. The first possibility has been investigated in our laboratory (GALA and WESTPHAL, 1966 b) and will be discussed in Chapter IX. 7.

The cause of the depressing effect of estrogen administration on thymus weight in rats without pituitaries (Tables IX-15 and IX-16)

Table IX-16. *Corticosterone-binding activity of serum from hypophysectomized (hypox) male rats after ten days' injection of estradiol (E). Average values from duplicate pools, obtained by multiple equilibrium dialysis. Values for individual pools, and for an additional experiment, given in original publication* (GALA and WESTPHAL, 1966 a)

Experimental group	No. of animals	Body weight initial	Body weight final	C-values at 37°	Corticosterone μg/100 ml serum [a]	Thymus weight per 100 g body wt
normal control	10	254	294	0.37	4.4	105 ± 8
sham-hypox	10	238	291	0.43	6.8	129 ± 7
hypox-12 days	10	233	222	0.34	3.6	121 ± 11
normal control + 10 μg E/day	10	255	262	0.56	19.8	80 ± 7
sham-hypox + 10 μg E/day	10	237	251	0.63	18.5	122 ± 9
hypox + 10 μg E/day	6	232	211	0.32	3.2	52 ± 7

[a] Values from unstressed animals.
± Standard error of the mean.

is not known. It may be related to observations by DORFMAN and DORFMAN (1962) who reported that estrogen potentiates corticosteroid-induced thymus involution in gonadectomized-adrenalectomized mice but had no effect when administered alone to similarly operated animals.

The preceding results show that increase in CBG activity following adrenalectomy or estrogen treatment requires the participation of the pituitary gland. As discussed below (Table IX-24), the administration of cortisol to hypophysectomized rats decreases CBG activity when compared with untreated hypophysectomized controls. The depressing effect of the corticosteroids on the CBG level, therefore, does not appear to require the presence of the pituitary.

Before the influence of hypophysectomy on the CBG activity was further analyzed, the role of the duration of the glandless state was examined. Serum from male and female rats hypophysectomized for varying lengths of time were studied in a number of different multiple equilibrium dialyses; the data have been presented as the

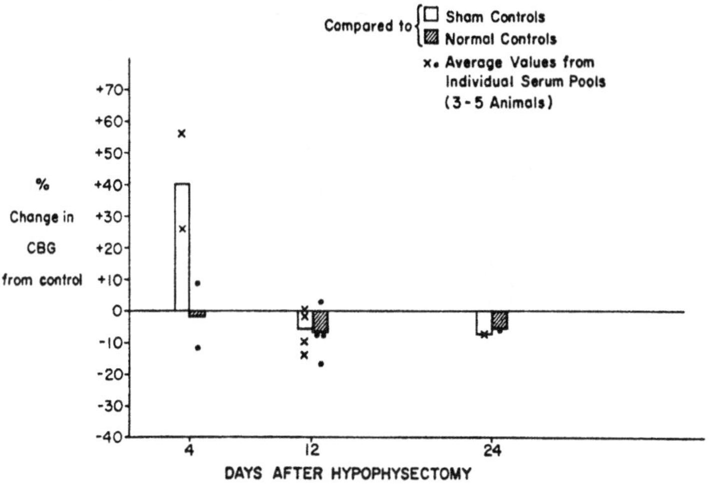

Fig. IX-11. Influence of duration of hypophysectomy on corticosterone-binding activity in male rats. Sera of hypophysectomized rats were compared to those of sham-hypophysectomized controls and normal controls. From GALA and WESTPHAL (1966 b)

percent change in CBG from the controls. Comparisons of CBG activity were made with both sham-hypophysectomized and normal animals; the serum protein concentration of hypophysectomized animals did not differ from control values (sham-hypophysectomized or normal animals). Values for female hypophysectomized rats were taken from various experiments involving hypophysectomized controls. A normal rat serum pool was added for comparison to several multiple equilibrium dialyses since early experiments had indicated that sham-hypophysectomized animals gave lower C-values than normal ones.

Figs. IX-11 and IX-12 summarize the influence of hypophysectomy on CBG activity. In the male rat (Fig. IX-11), hypophysectomy for as long as 24 days does not markedly alter the binding activity. Four days after hypophysectomy there appeared to be a significant increase in CBG activity when compared to sham-operated animals, but this increase was not apparent when compared to normal males. In female rats (Fig. IX-12), hypophysectomy resulted in a marked decrease in CBG activity, reaching maximal deviation from the controls by the ninth day. This low level of binding activity (40% below control values) was maintained through the 24th day; from the 24th to 48th day there appeared to be a gradual lessening of the response, so that by the 48th day the binding activity was only 20% lower than that of the controls. Comparisons with normal controls always gave a greater decrease in CBG activity than comparisons with sham controls. It should be noted in this context that ANTONICHEV and ROZEN (1967) observed a lowering of the CBG activity in female guinea pigs after hypophysectomy; the decrease could not be reversed by estradiol injection.

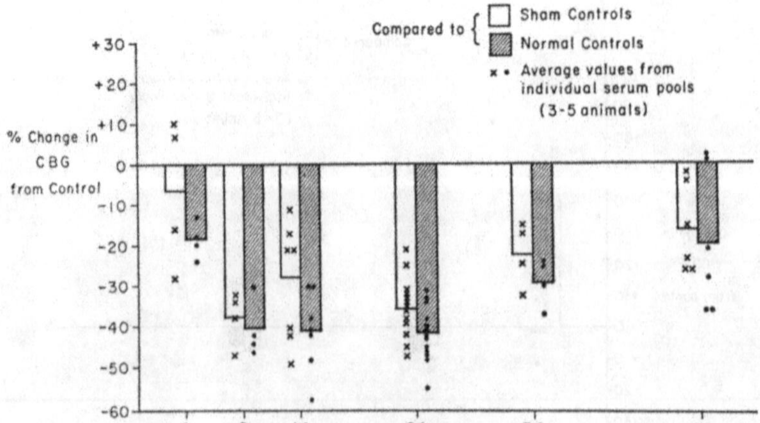

Fig. IX-12. Influence of the duration of hypophysectomy on the corticosterone-binding activity in female rats. The sera of the hypophysectomized rats were compared to those of sham-hypophysectomized controls and of normal controls. From GALA and WESTPHAL (1966 b)

In order to understand the relationship of the difference in CBG activities between hypophysectomized rats and their normal and sham-operated controls, respectively (Figs. IX-11 and IX-12),

Table IX-17. *Influence of sham-hypophysectomy (hypox) on the corticosterone-binding activity of rat serum. Recording of thymus weights incomplete since body and thymus weights were not determined for some pools.* From GALA and WESTPHAL (1966 b)

Sex	Condition	No. serum pools [a]	C-values at 37°	Corticosterone μg/100 ml serum [b] Total	Unbound	No. of animals	Thymus wt (mg) per 100 g BW
F	normal	20	1.01 ± 0.04	31.3 ± 2.8	0.477 ± 0.052	56	102 ± 3
	sham hypox [c]	14	0.85 ± 0.03 [d]	33.5 ± 2.7	0.610 ± 0.064	70	93 ± 2 [e]
M	normal	26	0.39 ± 0.01	7.4 ± 0.7	0.306 ± 0.042 [f]	74	110 ± 3 [g]
	sham hypox [h]	6	0.37 ± 0.03	7.1 ± 0.5	0.336 ± 0.039	20	118 ± 5

[a] Each pool from 3—5 animals.
[b] Values from rapidly decapitated animals.
[c] Values determined 12—24 days after sham hypophysectomy.
[d] Values of sham-hypophysectomized female rats significantly lower ($p < 0.004$) than those of normal females.
[e] Values of sham-hypophysectomized female rats significantly lower ($p < 0.02$) than those of normal females.
[f] Values of normal male rats significantly lower ($p < 0.02$) than those of normal female rats.
[g] Values of normal male rats not significantly different ($p > 0.05$) from those of normal female rats.
[h] Values determined 14—24 days after sham hypophysectomy.
± Standard error of the mean.

Table IX-17 was compiled. It contains a comparison of the C-values of all normal unstressed and sham-hypophysectomized animals obtained in our laboratory over a period of two years. The C-values, corticosterone levels both total and unbound, and thymus weights of sham-hypophysectomized male rats (determined at least 2 weeks after the sham operation) were found not to differ significantly from those of normal male animals. The C-values of sham-hypophysectomized female animals, however, were significantly lower ($p < 0.004$) than those of their normal controls. Since the corticosterone concentration was found not to be different in these two groups, a higher portion of corticoid should be in the unbound, physiologically active state. The thymus weights are concordant with this assumption: a significantly lower ($p < 0.02$) average thymus weight was observed in the sham-hypophysectomized group.

The concentration of unbound corticosterone in the normal female rat was found to be significantly greater ($p < 0.02$) than that in the normal male (Table IX-17). It should be noted, however, that the total corticosterone level found in the female rats was relatively high. Although the handling and rapid decapitation of the female rats was done in the same way as with the males, the corticosterone values of the female animals do not appear to indicate an unstressed state.

7. Recognition of the Thyroid-Stimulating Hormone as the CBG-Controlling Factor

The crucial question of whether the decrease of CBG activity caused by hypophysectomy in the female rat could be reversed by administration of pituitary factors, was examined in the following studies (GALA and WESTPHAL, 1966 b). With the objective first to provide the total of all pituitary factors, anterior pituitaries from 4- to 5-month-old male and female rats were homogenized in saline at 4° to give a suspension corresponding to one gland per 0.2 ml. Hypophysectomized female rats were divided into three experimental groups: hypophysectomy 14 days, hypophysectomy 24 days, and hypophysectomy 24 days plus the injection of the equivalent of one anterior pituitary per rat per day for the last ten days. A group of sham-hypophysectomized animals served as controls.

Hypophysectomy resulted in a marked decrease in CBG activity (Table IX-18). The administration of the anterior pituitary suspension produced a 50% increase in the binding activity when compared with that of hypophysectomized animals. The weights of all endocrine organs examined as well as thymus and body weights increased (see Table 3 in GALA and WESTPHAL, 1966 b).

In the further analysis, the whole anterior pituitary suspension was replaced by individual hormones. The administration of growth

Table IX-18. *Influence of the injection of an anterior pituitary (AP) suspension on the corticosterone-binding activity of hypophysectomized (hypox) female rats. Average values from duplicate pools, obtained by multiple equilibrium dialysis. Values for individual pools given in original publication by* GALA *and* WESTPHAL *(1966 b)*

Experimental group	No. of animals	C-values at 37°	Corticosterone µg/100 ml serum [a]
Sham hypox, 24 days	10	0.97	23.7
hypox, 14 days	12	0.53	3.3
hypox, 24 days	12	0.53	3.8
hypox, 24 days plus AP [b]	12	0.80	3.7

[a] Values from rapidly decapitated animals.
[b] Equivalent of 1 anterior pituitary/day for last 10 days.

hormone and prolactin did not significantly alter the CBG activity from that of hypophysectomized controls. Detailed results of these experiments may be found in Table 4 of the original publication (GALA and WESTPHAL, 1966 b). The injection of luteinizing hormone did not alter the CBG activity from that of the hypophysectomized control (Table IX-19). Ovarian weights were not significantly changed; uterine weights, on the other hand, were increased significantly. These and other details may be found in Table 5 of the original publication (GALA and WESTPHAL, 1966 b).

The administration of a human pituitary gonadotropin preparation containing predominantly follicle-stimulating hormone did not alter the CBG activity of hypophysectomized animals from that of their saline-injected hypophysectomized controls. Details of this experiment are recorded in Table 6 of the original publication (GALA and WESTPHAL, 1966 b).

The injection of thyroid-stimulating hormone significantly increased the CBG activity. In fact, the highest dose enhanced the binding activity slightly above that of sham-hypophysectomized controls (Table IX-19). A somewhat higher corticosterone level was noted for one of the serum pools after injection of the larger dose of

Table IX-19. *Influence of the injection of luteinizing hormone (LH) and thyroid-stimulating hormone (TSH) on the corticosterone-binding activity of hypophysectomized (hypox) female rats. Average values from duplicate pools, obtained by multiple equilibrium dialysis. Values for individual pools given in the original publication by* GALA *and* WESTPHAL *(1966 b)*

Experimental group	No. of animals	C-values at 37°	Corticosterone μg/100 ml serum [a]
Sham hypox-24 days	10	0.83	35.3
Hypox-24 days	9	0.56	1.8
Hypox-24 days + 80 μg LH [b]	9	0.61	1.9
Hypox-24 days + 400 μg LH [b]	10	0.60	1.8
Hypox-24 days + 0.1 U TSH [b]	9	0.70	1.9
Hypox-24 days + 0.5 U TSH [b]	9	0.90	4.5

[a] Values from rapidly decapitated animals.
[b] Total amount per day in 2 sc injections for the last 10 days.

thyroid-stimulating hormone. The significance of this value is not apparent, since all other parameters to assure completeness of hypophysectomy were satisfied as may be seen from the organ weights compiled in Table IX-20. The lack of a rise in thyroid weight upon the injection of the high dose of thyroid-stimulating hormone is in accordance with the observation that changes in cellular histology and iodine uptake and release occur at much lower dosages than those needed to increase thyroid weight (ZARROW et al., 1964).

In answer to our original question, the evidence given in Table IX-19 shows that the thyroid-stimulating hormone alone, of the anterior pituitary hormones examined, is capable of returning the CBG activity of the hypophysectomized rat to the level of the

Table IX-20. *Influence of the injection of luteinizing hormone (LH) and thyroid-stimulating hormone (TSH) on organ weights of hypophysectomized (hypox) female rats.* From GALA and WESTPHAL (1966 b)

Experimental group	No. of animals	Body weight (g)			Ovarian wt (mg) per 100 g BW	Uterine wt (mg) per 100 g BW	Thyroid wt (mg) per 100 g BW	Thymus wt (mg) per 100 g BW
		Initial	14 days	Final				
Sham hypox-24 days	10	218	242	241	36.0 ± 1.7	155 ± 10	7.3 ± 0.4	88 ± 6
Hypox-24 days	9	193	195	196	17.9 ± 1.3	41 ± 1	5.2 ± 0.3	109 ± 6
Hypox-24 days + 80 μg LH [a]	9	194	200	194	22.0 ± 1.8	114 ± 7	5.3 ± 0.3	108 ± 7
Hypox-24 days + 400 μg LH [a]	10	192	196	187	27.0 ± 1.9	93 ± 15	5.8 ± 0.3	106 ± 8
Hypox-24 days + 0.1 U TSH [a]	9	193	193	192	21.5 ± 0.9	50 ± 4	5.6 ± 0.3	114 ± 10
Hypox-24 days + 0.5 U TSH [a]	9	192	192	195	19.7 ± 1.8	51 ± 5	5.8 ± 0.2	140 ± 13

[a] Total amount per day in 2 sc injections for the last 10 days.
± Standard error of the mean.

sham-hypophysectomized controls. The results are summarized in Fig. IX-13.

The effect of the thyroid-stimulating hormone is assumed to be a stimulation of thyroid hormone production, which in turn results in an enhanced CBG activity. Support for such mechanism was presented by LABRIE et al. who found independently of our results that thyroidectomy decreased and thyroxine administration increased

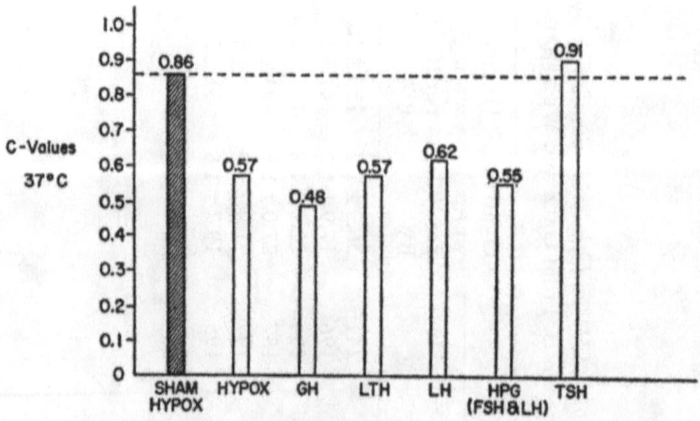

Fig. IX-13. Influence of anterior pituitary hormones on the CBG activity in hypophysectomized female rats. HYPOX, hypophysectomy; GH, growth hormone; LTH, prolactin; LH, luteinizing hormone; HPG (FSH+LH), human pituitary gonadotropin (mixture of follicle-stimulating and luteinizing hormone)

the CBG activity in rats (LABRIE et al., 1965; LABRIE, 1967; LABRIE et al., 1968). These observations are in accordance with our conclusion that the hypophysial factor involved in the control of CBG activity is the thyroid-stimulating hormone.

The effect of the thyroid-stimulating hormone in reversing the reduction of CBG activity by hypophysectomy has been demonstrated in the female rat (Table IX-19). Since no sex difference in thyroid secretory activity has been observed in the rat (HARRIS et al., 1964), one would expect hypophysectomy in the male rat to cause a decrease of the CBG level similar to that seen in the female. This has not been found, however; the reduction of CBG activity in the hypophysectomized male rat (Fig. IX-11) is minimal compared to that of the female animal (Fig. IX-12). This observation has been recently confirmed (LABRIE et al., 1968).

As an explanation it may be suggested that hypophysectomy also removes an inhibitory factor in the male rat: the suppressing action of testosterone on CBG activity has been discussed in detail (Fig. IX-7 and Table IX-6). Loss of both thyroid hormone and testosterone in the hypophysectomized male rat may therefore result in normal C-values, since the elimination of the inhibitory action of the androgen may counterbalance the loss of the stimulatory effect of the thyroid hormone. This is different from the situation in the hypophysectomized female rat; no suppressing agent is removed by hypophysectomy, so that the lack of the stimulatory effect of the thyroid hormone becomes manifest.

The decrease in CBG activity after sham hypophysectomy in the female rat (Table IX-17) may also be due to changes in thyroid activity. Sham hypophysectomy in rabbits has been reported to diminish the response of the thyroid gland to TSH administration, as evidenced by a decrease in the release of ^{131}I from the thyroid (GARCIA et al., 1964).

After hypophysectomy in the female rat the binding activity was still considerably greater than given by albumin binding alone (C-value for hypophysectomized female rats between 0.46 and 0.62, and for albumin 0.07). It may be that the liver, the presumed source of CBG, is capable of synthesizing a basal level of CBG and that hormonal influences may either increase or decrease this level. A residual thyroid secretion has been reported for hypophysectomized rats to be approximately 13% of that of intact rats (RANDALL et al., 1951).

The gradual recovery of CBG activity in female rats hypophysectomized for five to seven weeks (Fig. IX-12) was an unexpected finding. It is not known whether the possibility can be considered that, in rats hypophysectomized for these long durations, an increased sensitivity to the residual thyroid secretion may occur as a compensatory reaction of the animal to normalize its metabolism, or whether other compensatory mechanisms are responsible for the recovery.

Some years ago it was demonstrated that the increase in CBG activity precedes the rise in peripheral corticosteroids (SANDBERG et al., 1960). It may be assumed, therefore, that conditions which enhance CBG activity, such as increased thyroid hormone levels, will result in a subsequent elevation of peripheral corticosterone concentration to maintain homeostasis (LABRIE et al., 1965; LABRIE, 1967). The administration of thyroid hormone to hypophysectomized rats does not result in an increase in adrenal weight (STEINETZ and BEACH, 1963; D'ANGELO and GRODIN, 1964) or in plasma corticosterone (D'ANGELO and GRODIN, 1964) as observed in intact animals (STEINETZ and BEACH, 1963). Considering these findings together, it appears possible that stimuli which increase thyroid secretion will raise CBG levels in the serum; the elevated CBG activity will now result in a reduction of biologically active corticosterone available to the animal, and, in a feedback mechanism to maintain homeostasis, ACTH production will be stimulated and the peripheral corticosterone concentration will rise.

In future research on the interrelationship between the thyroid and adrenal glands, it would be of considerable importance to determine the CBG activity along with the usual adrenal and thyroid indices of secretion. Although the observations discussed suggest CBG as a possible link in the thyroid-adrenal interrelationship, direct evidence would be highly desirable (GALA and WESTPHAL, 1966 b). It should be noted that the above interpretations have been confirmed and extended by LABRIE et al. (1968). Recent studies by FLORSHEIM et al. (1969) show that adrenalectomy results in an increase of thyroid hormone secretion in the starved rat.

8. Further Interrelation of Endocrine Effects on CBG Activity in the Rat

The results reported in the preceding sections indicate that the CBG of the rat is capable of undergoing considerable fluctuation as a result of changes in the endocrine state of the animal. In an effort to further delineate the controlling stimuli, the influence of various combinations of endocrine conditions was investigated (GALA and WESTPHAL, 1966 c).

Table IX-21 shows the simultaneous effect of adrenalectomy and gonadectomy on male and female rats. In agreement with our previous observations (Tables IX-4, IX-12, IX-14) adrenalectomy alone enhanced the CBG activity in either sex, whereas gonadectomy did not alter the C-values in the female animal but significantly increased it in the male. Simultaneous removal of the adrenals and the ovary in the female rat resulted in CBG levels slightly lower than those of the adrenalectomized animals. The double operation in the male significantly raised CBG activity above that observed after adrenalectomy alone; the C-values observed were slightly higher than additive. These changes would suggest that the influences of

Table IX-21. *Simultaneous influence of adrenalectomy (adx) and gonadectomy (gonadx) on the corticosterone-binding activity of rat serum. Average values from duplicate pools, obtained by multiple equilibrium dialysis; 8 animals used in each of the 8 groups listed. Individual values and body weights given in original publication by* GALA *and* WESTPHAL *(1966 c)*

Experimental group	Sex	C-values at 37°	Corticosterone μg/100 ml serum [a]	Thymus wt (mg) per 100 g BW
Sham adx-sham gonadx-14 days	F	1.33	42.2	130 ± 8
	M	0.58	4.7	149 ± 6
adx-14 days	F	2.08	4.5	216 ± 18
	M	1.10	4.2	202 ± 9
gonadx-14 days	F	1.32	8.7	250 ± 11
	M	0.79	4.7	220 ± 11
adx-gonadx-14 days	F	1.92	5.5	277 ± 24
	M	1.39	2.8	245 ± 4

[a] Values from rapidly decapitated animals.
± Standard error of the mean.

gonadectomy and of adrenalectomy are independent of each other. This conclusion is in accordance with the changes in thymus weights which increased in both sexes after adrenalectomy or after gonadectomy; the combined operation, however, resulted in larger thymus glands than observed after the single operations. For a discussion of the unusually low corticosterone levels in the male rats (Tables IX-5, IX-21 and IX-22) see the original publication (GALA and WESTPHAL, 1966 c). The relatively high C-values in Table IX-21 may be due to the different source of rats used in this group.

The results of a combination of adrenalectomy and estradiol injection in male rats are given in Table IX-22. Adrenalectomy markedly elevated the CBG activity when compared to that of sham controls; estradiol injection also resulted in higher C-values. The CBG level of the estrogen-treated adrenalectomized animals was three times that of the sham controls. This increase was greater than expected for the additive effects of adrenalectomy and estradiol administration alone; the analogy with the results in Table IX-21 is obvious.

Table IX-22. *Influence of estradiol (E) injection on the corticosterone-binding activity of intact and adrenalectomized (adx) male rats. Average values from duplicate pools, obtained by multiple equilibrium dialysis. Individual values and body weights given in original publication by* GALA *and* WESTPHAL *(1966 c)*

Experimental group	No. of animals	C-values at 37°	Cortico-sterone μg/100 ml serum [a]	Thymus wt (mg) per 100 g BW
Sham adx-24 days	9	0.31	4.5	101 ± 7
Adx-24 days	7	0.69	2.1	155 ± 10
Sham adx + 10 μg E [b]	10	0.47	8.5	92 ± 5
Adx-24 days + 10 μg E [b]	6	1.05	2.1	150 ± 8

[a] Values from rapidly decapitated animals.
[b] Amount of hormone injected sc once daily for the last 10 days.
± Standard error of the mean.

It has been shown in Table IX-5 that the increased CBG activity in the castrated male rat, which is about the same as that of the intact male estrogen-treated animal, cannot be raised further by combining estradiol injection and castration. In contrast, the effect of adrenalectomy plus estrogen treatment is more than additive (Table IX-22). It is concluded from these findings that the elevation of CBG activity resulting from estrogen administration and from gonadectomy, both sex hormone-mediated stimuli, is brought about by a mechanism different from that following adrenalectomy.

As shown in preceding sections, the administration of corticosteroids decreases the CBG activity in intact rats (Table IX-1, Figs. IX-1 and IX-2), prevents its rise after adrenalectomy (Fig. IX-3), and reduces the high CBG level established after adrenalectomy (Tables IX-11 and IX-12). We now investigated whether this corticoid effect in the adrenalectomized rat was reversible. We also examined the role of the pituitary gland in the CBG-reducing influence of cortisol and its reversibility.

Table IX-23 documents, in confirmation of our previous observations, that adrenalectomy increases and cortisol administration reduces the CBG levels in female rats. When the adrenalectomized animals were analyzed ten days after the cessation of cortisol injections, there was a 4-fold rise in CBG activity above that obtained for animals sacrificed immediately after the termination of cortisol injections. The binding activity, however, was still below that of the 34-day adrenalectomy control.

Hypophysectomy markedly decreased the CBG activity below that of the female control rats (Table IX-23). The administration of cortisol further lowered the binding activity to a value that was less than one eighth that of control animals. The termination of cortisol injections resulted again in an increase of CBG activity; at the end of ten days, however, the binding value was still below that of hypophysectomized controls. On a percentage basis, the fall in CBG activity after exogenous cortisol and the recovery ten days after cessation of cortisol injections were similar in both adrenalectomized and hypophysectomized animals. For a discussion of the corticosterone levels and thymus weights see original publication (Gala and Westphal, 1966 c).

The data presented in Table IX-23 indicate that the depressing effect of cortisol on the high CBG levels resulting from adrenalectomy is reversible. Similar reversibility was demonstrated for the suppressing effect of cortisol on the CBG activity in hypophysectomized female rats. This latter observation is considered additional evidence that the influence of adrenocortical hormones on the binding activity is mediated in a different way than that of the gonadal secretions; in the latter instance the response is dependent on the presence of the pituitary (Table IX-16). The effects of cortisol on the CBG activity in adrenalectomized and hypophysectomized animals (Table IX-23) are similar, the main difference being the actual CBG levels at the start of the cortisol injections. The rat appears able to synthesize CBG in the absence of the pituitary.

No definitive evidence is available for the site of CBG production. In analogy to other serum proteins, including glycoproteins

Table IX-23. *Influence of cortisol (F) injection on the corticosterone-binding activity of previously adrenalectomized (adx) or hypophysectomized (hypox) female rats. Average values from duplicate pools, obtained by multiple equilibrium dialysis. Values for individual pools given in original publication by* GALA *and* WESTPHAL (1966 c)

Condition	Experimental group	No. of animals	Body weight (g)				C-values at 37°	Corticosterone μg/100 ml serum [a]	Thymus wt (mg) per 100 g BW
			Initial	14 days	24 days	Final			
Intact	Sham adx-34 days	10	192	216	220	224	1.16	33.2	85 ± 4
Adx	14 days	6	208	230	—	—	1.81	4.5	138 ± 11
	24 days + 2 mg F/day last 10 days	9	207	216	214	—	0.42	10.2	18 ± 2
	34 days + 2 mg F/day days 14—23	10	207	220	212	223	1.64	3.7	111 ± 9
	34 days	9	206	229	235	244	2.03	4.5	158 ± 7
Hypox	14 days	10	194	200	—	—	0.67	5.3	132 ± 7
	24 days + 2 mg F/day last 10 days	9	194	199	165	—	0.14	14.5	19 ± 3
	34 days + 2 mg F/day days 14—23	9	196	199	165	172	0.54	3.7	42 ± 4
	34 days	10	196	198	193	192	0.72	2.9	132 ± 6

[a] Values from rapidly decapitated animals.
± Standard error of the mean.

(ATHINEOS et al., 1964; MOLNAR et al., 1964; ROBINSON et al., 1964; MOLNAR et al., 1965), the liver is a likely organ for its biosynthesis. Preliminary observations in our laboratory, obtained with single rats and summarized in Table IX-24, support this assumption since subtotal hepatectomy in the rat (60—80% of the liver removed) results in a marked decrease in CBG activity. Since the serum protein concentrations of the partially hepatectomized rats are the same as

Table IX-24. *Preliminary results on effect of subtotal hepatectomy (hepatX) on CBG activity.* From GALA and WESTPHAL (1966 c)

Animal Preparation	Sex	Days after operation	C-values at 37°	Serum protein concn mg/ml
HepatX	M	3	0.15	53.4
Sham hepatX	M	3	0.21	67.9
HepatX	F	7	0.28	61.6
Sham hepatX	F	3	0.60	60.4
Adx[a]-hepatX	M	7	0.37	52.3
Adx-sham hepatX	M	7	0.69	61.5

[a] Adx = adrenalectomy.

or smaller than those of the sham-hepatectomized controls, the lower C-values cannot be explained by differences in the serum protein concentrations.

ROZEN and ANTONICHEV (1966) observed a decrease of CBG activity in guinea pigs subjected to liver injury by carbon tetrachloride. Corticosteroid-binding activity has recently been observed (GUIDOLLET and LOUISOT, 1969 c) in cell sap fractions from rat liver cells; about 90% of the activity was associated with the non-particulate cytoplasmic phase (LOUISOT and GUIDOLLET, 1968). The question whether this binding material constitutes CBG or receptor-type proteins remains to be elucidated. The same authors concluded from studies on incorporation of radioactive glycoprotein precursors that the increased CBG activity after estrogen and thyroxin administration does not result from increased glycoprotein biosynthesis but rather from an activation of binding sites present in masked form (GUIDOLLET and LOUISOT, 1969 a, 1969 b).

9. Summary and Conclusions

The results of our studies on the endocrine control of CBG activity in the male and female rat are summarized in Figs. IX-14 and IX-15, respectively. The C-values in these figures have been corrected to make their comparison valid; details of this adjustment have been given in the original publication (GALA and WESTPHAL, 1966c).

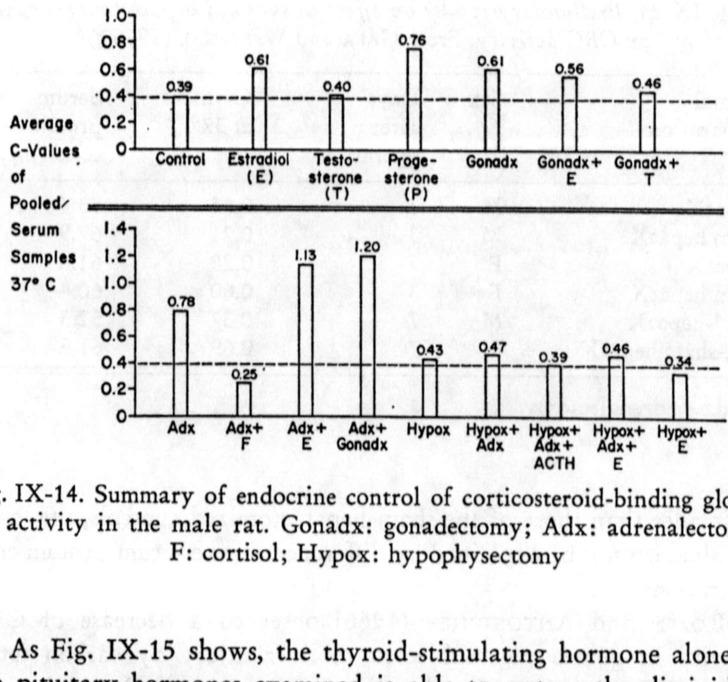

Fig. IX-14. Summary of endocrine control of corticosteroid-binding globulin activity in the male rat. Gonadx: gonadectomy; Adx: adrenalectomy; F: cortisol; Hypox: hypophysectomy

As Fig. IX-15 shows, the thyroid-stimulating hormone alone of the pituitary hormones examined is able to restore the diminished CBG activity of the hypophysectomized female rat to the normal level. This result is in accordance with the observation, made independently of our studies, that thyroidectomy reduces and administration of thyroid hormone enhances the CBG activity in the rat (LABRIE, 1967). It is concluded, therefore, that TSH or thyroid hormone has a controlling influence on CBG activity. This control is not absolute: the data of Fig. IX-14 indicate that the elimination of TSH by hypophysectomy does not affect the CBG activity in the male rat. A certain level of CBG appears to be autonomous of TSH-thyroid control.

Estradiol administration to male rats results in an enhancement of CBG activity (Fig. IX-14). This effect of estrogen is not observed in hypophysectomized rats, suggesting that a pituitary factor is involved in the estrogen effect; we have demonstrated that the only anterior pituitary hormone that influences CBG activity is TSH (Fig. IX-15). Stimulation of thyroid hormone secretion has been reported after estrogen administration (MONEY et al., 1951; FELDMAN and DANOWSKI, 1956), directly and by augmenting TSH release

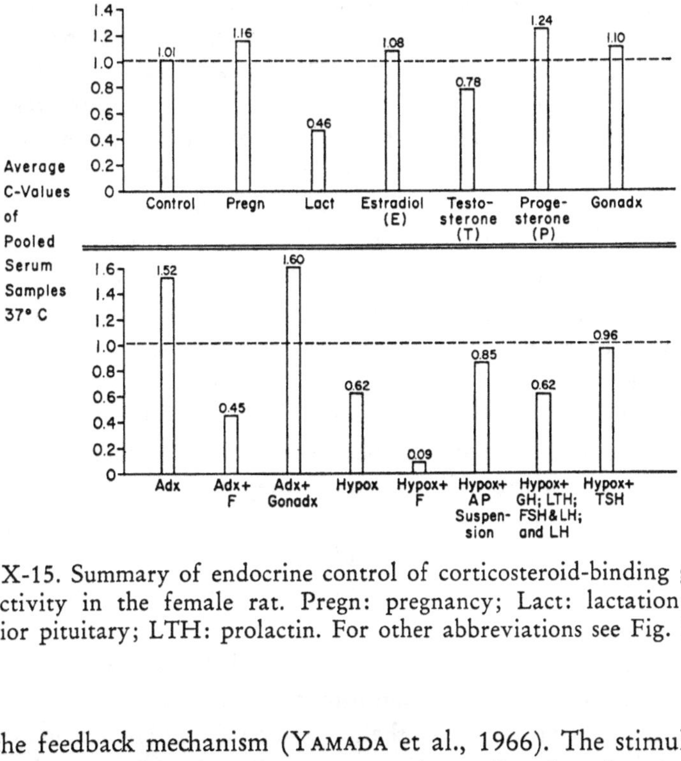

Fig. IX-15. Summary of endocrine control of corticosteroid-binding globulin activity in the female rat. Pregn: pregnancy; Lact: lactation; AP: anterior pituitary; LTH: prolactin. For other abbreviations see Fig. IX-14

via the feedback mechanism (YAMADA et al., 1966). The stimulation of TSH activity by estrogens appears to be mediated at the pituitary rather than at the hypothalamic level since intrahypophysial implants of estradiol elicited the effect in rats with lesions in the median eminence (D'ANGELO and FISHER, 1969). Increased thyroid activity has been observed in the rat at the time of vaginal estrus, possibly due to higher TSH production (BROWN-GRANT, 1962). LABRIE et al. (1967, 1968) have shown that the stimulation by estradiol of CBG activity can be eliminated by thyroidectomy. The same authors observed an increase of TSH within 6 hrs after injection of estradiol benzoate, thus verifying our conclusions (GALA and WESTPHAL, 1966 c) that the estrogen effect on CBG activity is mediated by the thyroid-stimulating hormone of the anterior pituitary.

Testosterone injections into either intact female (Fig. IX-15) or gonadectomized male rats (Fig. IX-14) result in reduced CBG activity. The mechanism of this depressing influence of testosterone on the CBG is not apparent. A direct relationship to TSH or thyroid activity is not evident; reports on the influence of testosterone administration on thyroid activity (MONEY et al., 1951; KOCHAKIAN and EVANS, 1956; FEDERMAN et al., 1958; VAN REES et al., 1965) do not indicate a decreasing effect. It is of interest, however, that administration of androgens to human patients has been found to reduce the concentration of thyroxine-binding proteins (FEDERMAN et al., 1958). There may be a common if unknown basis for the depressing effect of androgens on TBG and on CBG; both are α-glycoproteins of similar chemical nature (see Chapter XIII. 1).

Corticosteroid injection results in a marked lowering of CBG activity in both the male (Fig. IX-14) and female rat (Fig. IX-15); this decrease has been observed in adrenalectomized, hypophysectomized and intact animals. A possible interrelationship between the thyroid and the adrenals has been considered for many years. An increase in corticosteroids appears to depress the secretions of the thyroid gland (MONEY, 1955). There is also evidence that adrenal insufficiency in man and rat results in augmented thyroid activity (MONEY, 1955; GABRILOVE and SOFFER, 1950). It has been concluded (INGBAR and FREINKEL, 1956) that corticosteroids depress thyroid function by inhibiting TSH production. These various observations are in agreement with the assumption that the effect of corticosteroids on CBG activity is related to their action on the TSH-thyroid system.

One last point which is in accordance with the proposed role of TSH-thyroid control of CBG is the significant loss of CBG activity in lactation (Fig. IX-15). It has been reported (IINO and GREER, 1961) that thyroid activity markedly decreases in the rat on the last day of pregnancy, and that this reduced activity continues throughout lactation. This finding parallels our observation on the rapid decline of the CBG level with onset of lactation (Fig. IX-9). The involvement of the thyroid and CBG in the initiation of lactation will be further discussed (Chapter X) in connection with studies on CBG activity in other species (Gala and Westphal, 1967).

In summary, our results are in agreement with the assumption that TSH-stimulated thyroid secretion controls CBG synthesis. We further assume that the liver is the site of this synthesis. It has been postulated (Gala and Westphal, 1966 c) that the increase in CBG after estrogen administration is mediated through the anterior pituitary stimulating TSH production, which in turn increases the secretion of thyroid hormones; recent findings (Labrie et al., 1968) verify this mode of control, and confirm various phases of our studies on endocrine control of CBG in the rat. Testosterone depresses CBG activity by a mechanism yet to be elucidated. There are indications that the inhibitory effect of the corticosteroid hormones on CBG activity is mediated by the thyroid hormone; however, a direct action at the site of CBG biosynthesis or a combination of these possibilities cannot be excluded at the present time. It is hoped that the remaining questions in the endocrine control of CBG activity in the rat will be answered, and that the knowledge obtained in this species will be basis and stimulus to study the analogous relationships in man.

X. CBG Activity in Mouse, Rabbit and Guinea Pig during Pregnancy and Lactation; PBG in the Guinea Pig

1. Observations on the Steroid Binding Activities

The observation of a marked decline of CBG activity in the rat early in lactation (Fig. IX-9) suggested that CBG may play a role in the initiation of lactation (GALA and WESTPHAL, 1965 a). If such a mechanism were of general significance, similar changes in the CBG level should occur in other species. Corticosteroid-binding activity was therefore determined in mouse, rabbit and guinea pig during pregnancy and lactation (GALA and WESTPHAL, 1967).

The corticosteroid-binding protein in the serum of the rat, rabbit and other species has the electrophoretic properties of an α-globulin (WESTPHAL and DE VENUTO, 1966); unpublished experiments have shown that the corticosteroid-binding protein in guinea pig serum also is an α-globulin. No such evidence is available at present for the mouse; however, in view of the α-globulin nature of the corticosteroid-binding serum-proteins in all mammalian species examined so far, we are referring generally to the corticosteroid-binding globulin (CBG) as the serum component responsible for this binding activity.

In the study of CBG activity in the mouse, Swiss Webster Albino mice, 3—4 months old, were housed in individual cages for at least 5—7 days before sacrifice; this procedure was also followed for the rabbits and guinea pigs. Blood was collected between 8 and 11 AM from all species. The mice were rapidly decapitated with scissors within 30 sec from the time the cage was handled. Other procedures have been described in Chapter IX. 2.

Day 1 of pregnancy in female mice cohabiting with males was the day that vaginal sperm and/or a copulatory plug was detected. Animals were placed in individual cages on this day and sacrificed on day 12 and 18 of a 20-day pregnancy period. Another group of female mice was removed from the breeding cages when they showed signs of pregnancy. These animals were sacrificed on day 1, 4 and 14 of lactation; day 1 was the day of parturition. Litters of the lactating mice were adjusted to 8—10 pups on day 1 of lactation. Except for one group of mice lactating 14 days that had 1—5 (avg. 4.0) pups per litter, the serum pools were obtained from animals that

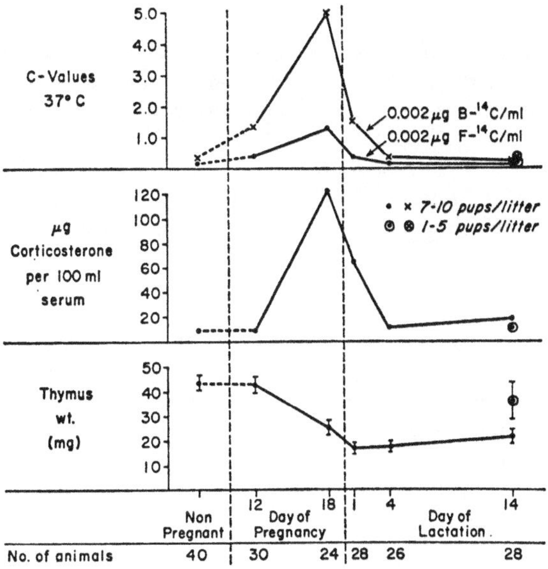

Fig. X-1. Influence of pregnancy and lactation on the corticosteroid-binding activity of mouse serum. Each point on the curve represents the average of two serum pools of 12—20 animals each. For thymus weights, vertical bars indicate the standard error of the mean. B, corticosterone; F, cortisol. From GALA and WESTPHAL (1967)

had 7—10 (avg. 8.8) pups per litter. Serum pools from virgin female mice and from male mice were included in the multiple equilibrium dialyses.

Pregnancy in the mouse resulted in a thirteen-fold increase in corticosterone-binding activity when compared with nonpregnant controls (Fig. X-1). The binding activity for cortisol was approximately one-third that for corticosterone in nonpregnant animals; this ratio was maintained during pregnancy and lactation. On the day of parturition, there was a marked decrease in CBG activity which continued to day 4 of lactation; the level then remained essentially constant to the fourteenth day. Whereas mother rats nursing few pups showed higher C-values than rats suckling a larger number of young (Fig. IX-9), an analogous difference was not marked in the mouse (Fig. X-1).

Serum corticosterone levels on day 12 of pregnancy were not different from those of virgin animals; a marked increase occurred,

Table X-1. *Influence of sex on the corticosteroid-binding activity of mouse serum. Average values from duplicate pools, obtained by multiple equilibrium dialysis. Values for individual pools given in original publication* (GALA and WESTPHAL, 1967)

Sex	No. of animals	Body wt. (g)	C-values at 37°		Corticosterone μg/100 ml serum [a]		Thymus wt. (mg)
			corti-sol	cortico-sterone	total	unbound	per 100 g BW
female	40	28	0.14	0.38	9.7	0.39	156 ± 6
male	28	37	0.05	0.17	8.7	0.71	68 ± 4

[a] Values from unstressed animals.
± Standard error of the mean.

however, by the eighteenth day of pregnancy (Fig. X-1). At parturition, there was a decrease in peripheral corticosterone which continued to the fourth day of lactation. On day 14 of lactation, mice nursing a high number of pups appeared to show a trend toward higher corticosterone levels as compared to mothers having fewer pups (18.2 vs. 10.4 μg/100 ml); however, this difference was not as marked as in the analogous case in the rat (Fig. IX-9). Thymus weights decreased from day 12 of pregnancy to parturition and remained at this low level in the high pup number group (Fig. X-1), indicating continuous high concentration of unbound corticosterone. Mothers nursing a few pups for 14 days had thymus weights similar to those of virgin control animals.

The CBG activity of female mice was two to three times that of the males for both corticosterone and cortisol (Table X-1). The peripheral corticosterone levels, however, were nearly the same. Calculation of the concentration of unbound corticosterone in the male indicated almost twice the female value. This higher level of unbound corticosterone in the male mouse was reflected in lower thymus weights.

The pregnant New Zealand White rabbits were sacrificed in pentobarbital anesthesia on day 21 or 27—28 of a normal 30-day gestation period, and on day 1 and 2, or on days 6, 9 and 12 of lacta-

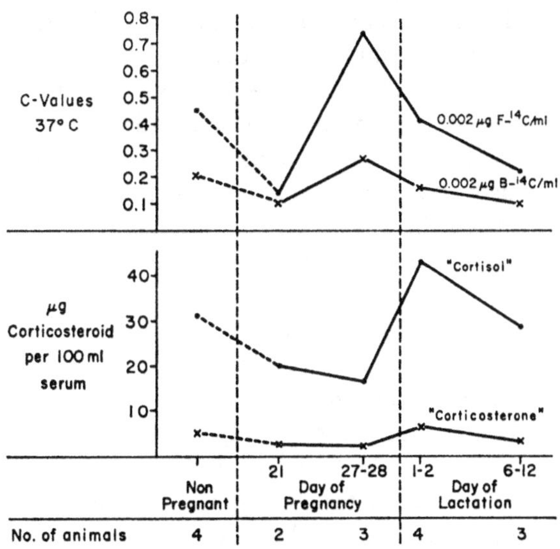

Fig. X-2. Influence of pregnancy and lactation on the corticosteroid-binding activity of rabbit serum. Each point on the curves is the average value for individual serum samples for the number of animals indicated. From GALA and WESTPHAL (1967)

tion, counting from the day of parturition. Since the CBG activities within these two periods were similar, the data are presented in two groups (Fig. X-2). The mother rabbits suckled from 2 to 8 young, and pup numbers were not adjusted. Sera from virgin animals were included in the multiple equilibrium dialyses as controls.

Fig. X-2 indicates that similar trends in CBG activity were obtained when either radioactive corticosterone (B) or cortisol (F) were used in the binding assay. On the 21st day of pregnancy, CBG activity was lower than in the virgin controls; a considerable increase was observed on the 27—28th day. Early in lactation (1—2 days), the binding activity decreased to the nonpregnant level; approximately half that value was observed in later lactation. The corticosteroid concentration decreased during pregnancy; repeated assays indicated higher values for cortisol than for corticosterone. Early in lactation the peripheral corticosteroids increased 2- to 3-fold over the pregnancy values, to return subsequently to virgin levels.

Pregnant guinea pigs of the English Short Hair breed were placed into individual cages and sacrificed by guillotine 7 to 10 days later. On the basis of fetal weight, the following pregnancy groups were established: 10—20 days, 20—30 days, 30—40 days, 40—50 days, and 50—60 days; the duration of pregnancy in the guinea pig is 65—70 days. The lactating animals were sacrificed on day 2 or 3 of lactation as well as on day 12. Litter numbers were not adjusted and mothers nursed the 1—5 animals they delivered. Serum samples from control virgin guinea pigs were included in the multiple equilibrium dialysis.

In the nonpregnant guinea pig, the CBG activity was low, and the difference between binding of cortisol (F) and corticosterone (B) was small (Fig. X-3). During pregnancy the binding activity increased considerably, and a relatively higher affinity for corticosterone became evident. The C-values declined during late pregnancy and the early days of lactation; by the twelfth day the binding activity was further reduced and approached that of the virgin controls. Corticosteroid levels rose greatly during pregnancy and declined to the control level in later lactation. The dramatic increase of the guinea pig CBG during pregnancy has been recently confirmed by Rosenthal et al. (1969 b) who reported concentrations of up to 15.8×10^{-6} M at 50—60 days of pregnancy. The authors were able to increase the CBG level by administration of high doses of estrogens over prolonged periods of time.

The observation in mouse serum of a higher binding affinity for corticosterone than for cortisol at 37° (Fig. X-1) is in agreement with a similar result (Seal and Doe, 1965) at 4° (Table VIII-18). The reverse relationship, i.e., greater affinity for cortisol, was found in rabbit serum, in confirmation of earlier results obtained at 4° by Slaunwhite and Sandberg (1959) and by Westphal (1964); determination of the association constants of rabbit CBG for the two corticosteroids at 4° and 37° (Table VIII-21) has verified the findings (see also Chader and Westphal, 1968 b). The association constants for CBG in guinea pig serum (Table VIII-21) confirm the higher binding affinity for corticosterone compared to cortisol. SEAL and DOE (1963) reported increased cortisol-binding capacity during gestation in the three species studied here, and MUSA et al. (1965) observed particularly high values in the pregnant guinea pig (see also Table VIII-22). Measurement of the progesterone-binding activity in the pregnant guinea pig (Fig. X-4) showed an increase several

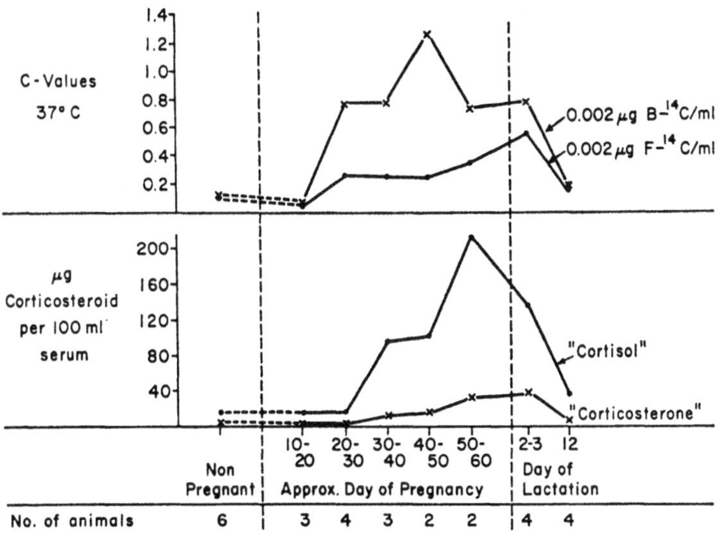

Fig. X-3. Influence of pregnancy and lactation on the corticosteroid-binding activity of guinea pig serum. Each point on the curves is the average value for individual serum samples for the number of animals indicated. From GALA and WESTPHAL (1967)

times greater than that observed for corticosterone or cortisol (Fig. X-3).

Analysis of the steroid-binding activity in the female guinea pig has been extended recently to different phases of the normal cycle, pregnancy and puerperium (DIAMOND et al., 1969). These studies include determination of binding activities for cortisol, progesterone and testosterone, and of the influence of testosterone propionate treatment on the binding parameters. Fig. X-5 and Table X-2 show the C-values and the cortisol levels during the cycle. In both normal and androgen-treated animals, the peripheral cortisol levels appeared to vary systematically with the 16-day estrous cycle, being highest on day 1 and lowest on day 11; the cortisol-binding activity was inversely related to these cortisol levels. Cyclicity of corticosterone levels in the normal rat has been previously reported (CRITCHLOW et al., 1963; GALA and WESTPHAL, 1965 a); the highest corticosteroid concentrations were seen during estrus in the rat (Table IX-9) and guinea pig (Table X-2).

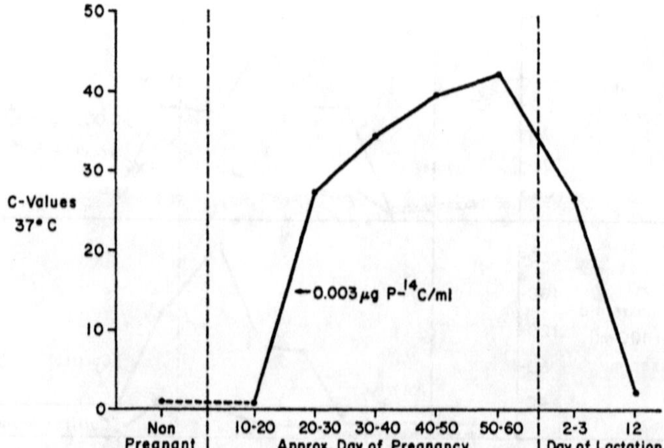

Fig. X-4. Progesterone (P)-binding activity (C-values) in the serum of pregnant and lactating guinea pigs. Unpublished studies with R. R. GALA

Previous observations (GALA and WESTPHAL, 1965 b) have indicated that administration of progesterone to male and female rats induced a significant increase in CBG activity and decrease in resting corticosterone levels (Table IX-7). This inverse relationship of rising corticosteroid binding and lowered corticosteroid levels in response to progesterone was also observed in the present study in the cycling guinea pig. The CBG activity fluctuated in parallel with the peripheral progesterone concentrations which FEDER et al. (1968) reported to be maximal at days 9 and 11 of the estrous cycle; the cortisol values changed inversely to the CBG and progesterone levels (Table X-2). The binding of progesterone was apparently unaltered during the estrous cycle, except in the androgen-treated animals. The fluctuations in testosterone binding with the highest value on day

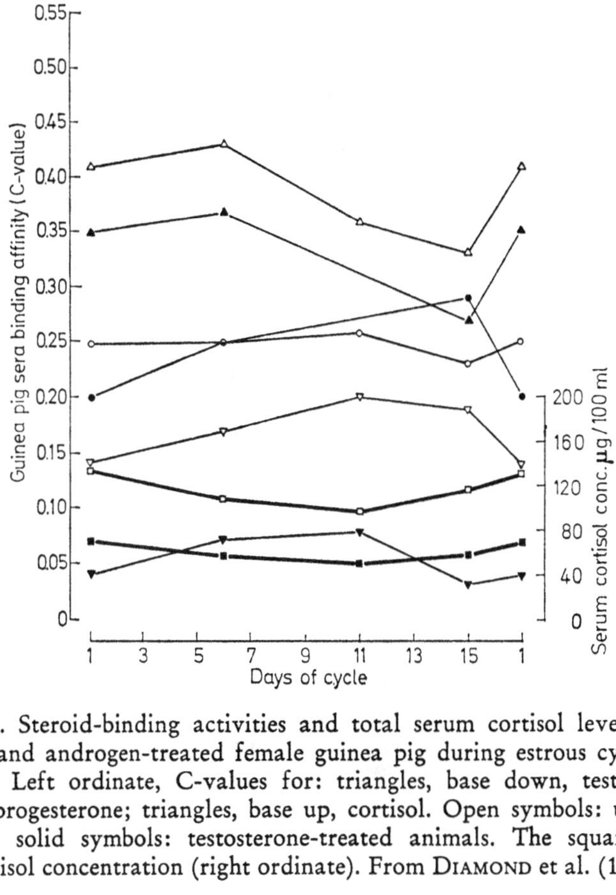

Fig. X-5. Steroid-binding activities and total serum cortisol levels in the normal and androgen-treated female guinea pig during estrous cycle. 37°; pH 7.4. Left ordinate, C-values for: triangles, base down, testosterone; circles, progesterone; triangles, base up, cortisol. Open symbols: untreated animals; solid symbols: testosterone-treated animals. The squares show cortisol concentration (right ordinate). From DIAMOND et al. (1969)

6 were consistent and parallel in both normal animals and those treated with testosterone propionate, but a relation to the estrous cycle was not evident (Fig. X-5).

The cortisol level as well as the cortisol-binding and testosterone-binding activities in the androgen-treated cycling animals paralleled the normal values but were significantly lower in every case. Long term administration of testosterone propionate did not seem significantly more effective in lowering the binding activities than treatment for the days of the cycle (Table X-2).

During pregnancy, the serum cortisol concentrations and the binding activities for cortisol, testosterone and progesterone were

Table X-2. *Steroid-binding activities (C-values at 37°) and cortisol concentrations in guinea pig sera under various reproductive conditions. F, Cortisol; T, Testosterone; P, Progesterone.* From DIAMOND et al. (1969)

Sample	Cortisol μg/100 ml	C-value F	C-value T	C-value P
Normal females				
Estrous cycle				
day 1	133	0.14	0.41	0.25
day 6	111	0.17	0.43	0.25
day 11	99	0.20	0.36	0.26
day 15	118	0.19	0.33	0.23
Pregnancy				
day 22	163	0.45	1.63	9.2
day 44	224	0.69	6.5	22.0
day 55	364	0.76	5.2	20.2
day 66	502	0.84	4.8	20.3
Postpartum				
day 1	323	0.50	3.4	27.7
day 3	271	0.42	1.62	16.8
day 20—22	78	0.13	0.46	0.46
Androgen-treated[a]				
Estrous cycle				
day 1	71	0.04	0.35	0.20
day 6	60	0.07	0.37	0.25
day 11	53	0.08	—	—
day 15	60	0.03	0.27	0.29
Pregnancy				
day 22	118	0.13	2.0	16.7
day 44	188	0.47	5.6	26.6
day 55	292	0.76	6.6	26.2
day 66	432	1.12	8.1	32.1
Postpartum				
day 1	232	0.79	3.8	31.8
Others				
♀ Guinea pigs, castrated over 90 days	70	0.24	0.35	0.32
♀ Guinea pigs, T.P. for 100 days	64	0.07	0.36	0.41
♂ Guinea pigs, adult control	48	0.07	0.37	0.32
♂ Guinea pigs, castrated over 90 days	95	0.10	0.28	0.19

[a] 5 mg testosterone propionate (T.P.) intramuscularly on first day of treatment (day 1 of cycle, or day 18 of pregnancy), 1 mg daily thereafter until day of sacrifice.

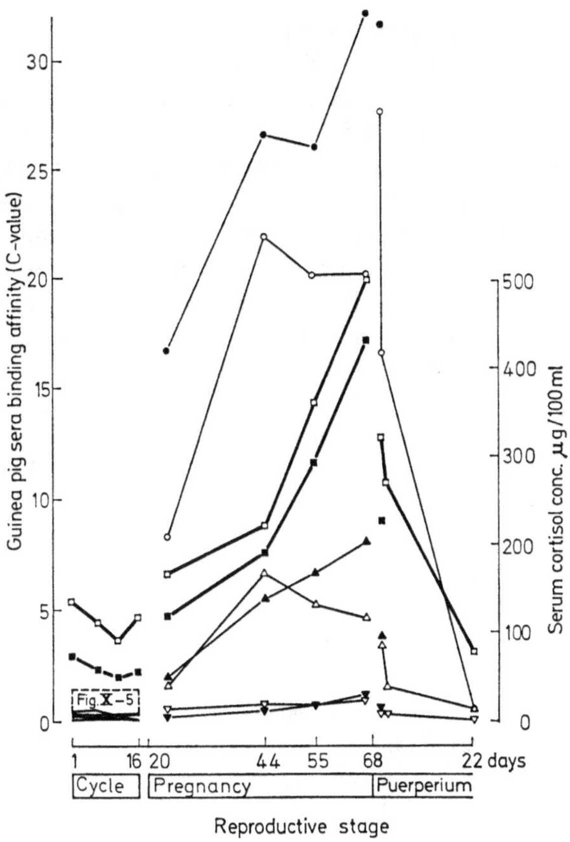

Fig. X-6. Steroid-binding activities and total serum cortisol levels in the normal and androgen-treated female guinea pig during pregnancy and puerperium. 37°; pH 7.4. For explanation of symbols see Fig. X-5. From DIAMOND et al. (1969)

progressively and markedly elevated over normal estrous cycle levels (Fig. X-6, Table X-2). The extremely high concentrations of cortisol and its chemical nature have been verified by isotope dilution techniques (DIAMOND et al., 1969). The almost 100-fold increase in the binding activity (C-value) for progesterone was particularly noteworthy as was the 15-fold increase in testosterone-binding activity. In contrast, the cortisol-binding activity of the pregnant guinea pig was only 3 to 4 times higher than during the cycle. These differences between binding increases for cortisol and for progesterone, and the general lack of parallelism between the curves for CBG and progesterone binding values (Figs. X-5 and X-6), obtained with the same serum samples, cannot be reconciled with the assumption of identical binding sites for cortisol and progesterone in guinea pig serum. The results depicted in Fig. X-5 and especially Fig. X-6 indicate that the high-affinity binding activities for progesterone, testosterone and cortisol in guinea pig serum are given by three unique molecular entities which are independent of each other. Recent studies in our laboratory (BURTON et al., 1970) have proven that the progesterone-binding protein and CBG in guinea pig serum are not identical (see Section 2 of this chapter).

During pregnancy both CBG activities and cortisol levels increase considerably as does the binding capacity for progesterone and testosterone (Table X-2). The extremely high binding activities for progesterone as well as for testosterone found in the pregnant pig are the highest reported for any species. Even more noteworthy is the further increase of testosterone and progesterone binding in the pregnant animals treated with androgen. Testosterone propionate administration tended to lower the binding values during the estrous cycle but increased most values during pregnancy. Increased testosterone binding has also been observed in pregnant women as will be discussed in Chapter XII.

The activities for progesterone and testosterone binding rise to a high about day 44 and decline thereafter. The plasma progesterone levels given by HEAP and DEANESLY (1966) also increase to their highest values about this stage and decline subsequently. A relationship between plasma progesterone concentrations, metabolism, and binding has been suggested by these investigators (HEAP and DEANESLY, 1967).

Following parturition an immediate return toward cycling levels of cortisol and cortisol-binding capacity is seen, but pregestational values are not attained within three days. Progesterone-binding activity in the normal animal is highest immediately after parturition and does not return to normal within 21 days (Table X-2).

Normal female guinea pigs showed at least twice the cortisol concentrations and cortisol-binding values seen in males (Table X-2). The binding activity for testosterone was about the same in both sexes; this is in accordance with observations in humans (Table XII-1). Androgen administration to the normal female guinea pig reduced the cortisol-binding values toward that of the male; ovariectomy did not have a similar effect, at least not at 90 days after operation (Table X-2). Orchidectomy resulted in an increase of peripheral cortisol to approach female levels while the CBG activity showed only slight elevation; the C-values for both testosterone and progesterone were decreased by castration. This shows that some but not all of the effects on CBG activity and peripheral corticosteroid level are similar to those observed in the rat (Chapter IX. 4), and may indicate a species difference; the rat has a 4 or 5 day estrous cycle with a relatively short luteal period compared with the 16-day cycle of the guinea pig characterized by an extensive luteal phase. ROSENTHAL et al. (1969 b) have recently confirmed our findings on the higher CBG activity in female as compared to male guinea pigs.

Concerning the biological significance of the increased steroid-binding activities in the pregnant guinea pig, it might be considered that the hormone-protein complex formation would protect the organism from the effects of excess steroid concentrations. Protection may thus be provided to genital tissues and the behavior-mediating nervous system, and the steroid-binding macromolecules in conjunction with high progesterone levels (DIAMOND and YOUNG, 1963; DIAMOND, 1967) may play a role in guarding the female from androgenic influences. High androgen levels may be present in the maternal organism from aberrant metabolism or from the sexually differentiating male fetus. RIVAROLA et al. (1968) have reported that plasma of pregnant women contains significantly greater levels of androgens than plasma from nonpregnant females. They report that most of the testosterone is protein-bound; unbound androgens would be detrimental (DIAMOND and YOUNG, 1963). Further physiological implications have been discussed by DIAMOND et al. (1969).

To our knowledge, for the species discussed in this chapter, little information has been published from other laboratories on the serum corticosteroid levels during pregnancy and lactation. The corticosteroid found in the serum of the mouse is almost exclusively corticosterone (SOUTHCOTT et al., 1956; WILSON et al., 1958; HALBERG et al., 1959 a, 1959 b).

SOLEM (1966, 1967) using a sensitive method (SOLEM and BRINCK-JOHNSON, 1965) found higher resting corticosterone levels in female than in male mice; the peak values in the circadian rhythm were observed at 6 PM (SOLEM, 1966) or at 4 PM (HALBERG et al., 1959 b; UNGAR and HALBERG, 1963). A marked rise of the resting plasma corticosterone level was noted (SOLEM, 1966) in mice in late pregnancy; the values observed near parturition averaged 105 μg/100 ml as compared to about 10 and 20 μg per 100 ml in control virgin females at 8 AM and 6 PM, respectively. In the serum of the guinea pig, cortisol has been identified as the major corticoid; however, smaller amounts of cortisone as well as four other steroids more polar than cortisol have been observed (FAJER and VOGT, 1963).

In the rabbit, an unusual situation exists in the type of corticosteroid found in the adrenal venous blood (HECHTER and PINCUS, 1954; BUSH, 1962). The ratio of corticosterone to cortisol in the acutely stressed animal was found to be between 20 to 1 and 10 to 1 (BUSH, 1953 a, 1953 b; KASS et al., 1954; KRUM and GLENN, 1965). If, however, ACTH was administered for prolonged periods of time cortisol was secreted in greater (KASS et al., 1954) or about equal amounts as corticosterone (KRUM and GLENN, 1965). In all these studies male rabbits were used and adrenal venous blood was examined. Our results (Fig. X-2) indicate that in the female rabbit cortisol is present in considerably higher concentration than corticosterone. Whether this is due to the lack of specificity of the corticosteroid assays used or to a possible sex difference in the predominant corticosteroid is not known. In an extensive analysis of rabbit organs and tissues (FAZEKAS and FAZEKAS, 1966), corticosterone, cortisol, cortisone, 11-dehydrocorticosterone and other steroids were identified.

2. Identification of PBG in the Guinea Pig

As discussed in the preceding section, studies on high-affinity binding of different steroids in the guinea pig revealed divergencies in the relationship of cortisol binding to progesterone binding that were incompatible with the assumption of identical macromolecules interacting with the two steroids (DIAMOND et al., 1969). This was seen in the normal guinea pig; it was more drastic in the pregnant animal. For example, on day 44 of pregnancy, the progesterone-binding activity was 85 times that of a cycle serum, whereas cortisol association increased only 3.5 fold. In another serum, the corresponding elevations were 120 fold and 2.6 fold, respectively.

Our conclusion that the high affinities for progesterone and cortisol in guinea pig serum are given by two independent molecular entities was confirmed by recent studies on heat and low-pH inactivation (BURTON et al., 1970). Fig. X-7 shows the binding

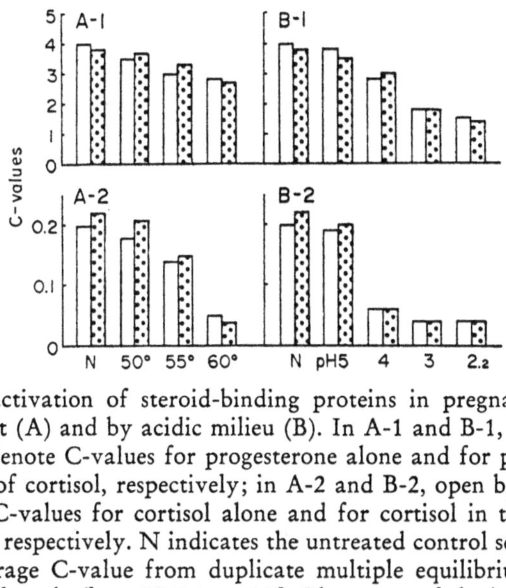

Fig. X-7. Inactivation of steroid-binding proteins in pregnant guinea pig serum by heat (A) and by acidic milieu (B). In A-1 and B-1, open bars and dotted bars denote C-values for progesterone alone and for progesterone in the presence of cortisol, respectively; in A-2 and B-2, open bars and dotted bars denote C-values for cortisol alone and for cortisol in the presence of progesterone, respectively. N indicates the untreated control serum. Each bar gives an average C-value from duplicate multiple equilibrium dialyses in 0.05 M phosphate buffer, pH 7.4, at 37°. Placement of the bars for cortisol binding (A-2, B-2) precisely under the bars for progesterone binding (A-1, B-1) is to indicate that aliquots of the same treated sera have been used for measurement of the binding activities for the two steroids. Other conditions described in the text

activities (C-values) for progesterone (A-1, B-1) and for cortisol (A-2, B-2) of pregnant guinea pig serum that had been exposed to elevated temperatures for one hour and to acidic buffers for 16 hrs at 22°. As evident from A-1 and A-2, open bars, heat treatment decreases the binding activity for progesterone and cortisol to a different degree. For example, increasing the temperature from 55° to 60° destroys only 7% of the progesterone binding activity while over 64% of the activity for cortisol is lost. Fig. X-7 also documents a lower resistance of the cortisol-binding activity to acidic milieu: a sharp decline occurs between pH 5 and 4, conditions which result in a much smaller loss of binding activity for progesterone. The contribution of albumin to these C-values is minimal.

For the conditions used in these experiments, it can be readily calculated that partial inactivation of the binding protein results in a proportional reduction of the C-value, so that the ratio of the C-values for control and experimental samples is equal to the ratio

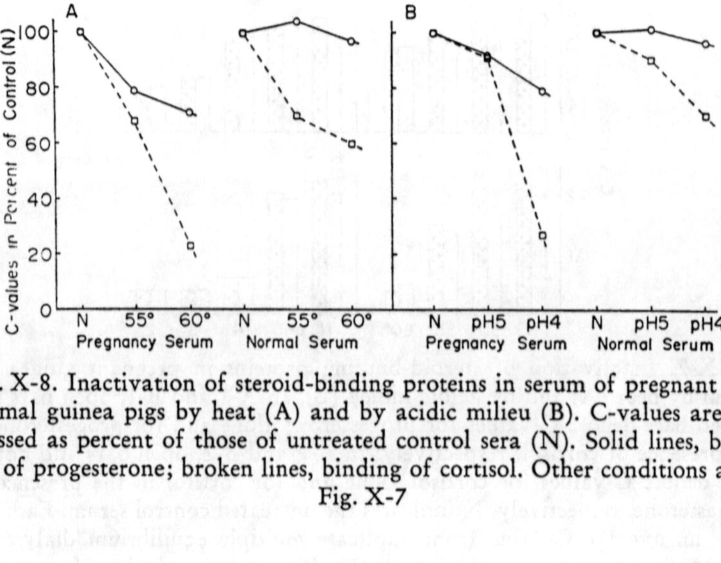

Fig. X-8. Inactivation of steroid-binding proteins in serum of pregnant and normal guinea pigs by heat (A) and by acidic milieu (B). C-values are expressed as percent of those of untreated control sera (N). Solid lines, binding of progesterone; broken lines, binding of cortisol. Other conditions as in Fig. X-7

of concentrations of active protein molecules in the same samples. The decrease of the C-values for progesterone and for cortisol may then be expressed as percent of the normal control values. Fig. X-8 shows the much greater lability of the cortisol-binding in relation to the progesterone-binding system not only for the pregnancy serum but also for normal guinea pig serum. Comparative stability of the progesterone interaction in the latter case is presumably due to the larger albumin concentration relative to that of the progesterone-binding protein.

Exposure of the pregnant guinea pig serum to 55° over extended periods of time (Fig. X-9) resulted in continued decrease of both the progesterone-binding and the cortisol-binding activities. Whereas a minimal value appears to be reached for the cortisol-binding system after 8 hrs, the progesterone-binding activity is still declining after

16 hrs. These findings can best be interpreted as indicating that the inactivation process denatures a fraction of the binding protein molecules, and that this fraction becomes larger with time.

The question of identity or duality of binding sites for progesterone and cortisol was further tested by competition experiments. The dotted bars in Fig. X-7, A-1 and B-1, denote the C-values for interaction of progesterone in the presence of cortisol; their average

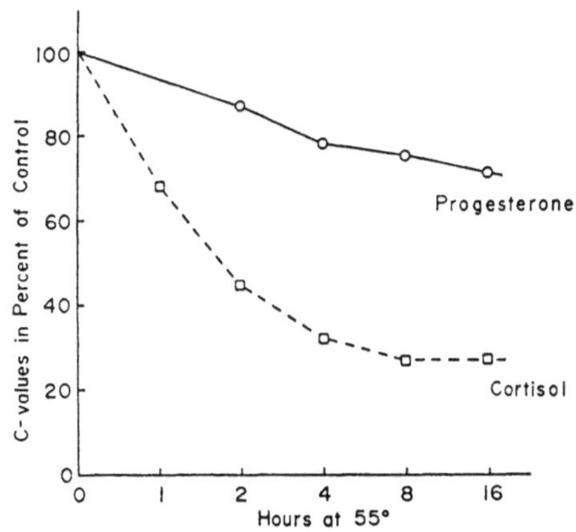

Fig. X-9.Time course of heat inactivation of steroid-binding proteins in serum of pregnant guinea pigs. Other conditions as in Fig. X-7

value is identical with that of the corresponding open bars. No significant difference was found either when the cortisol-binding activity was determined in the absence and presence of progesterone (A-2 and B-2); the amount of added progesterone corresponded to 120 μg per 100 ml serum. Competition could not be demonstrated in additional experiments in which the progesterone-binding activity was measured without and with added cortisol up to 800 μg per 100 ml serum. In all these cases, no competition occurs for the binding sites of one steroid by the other steroid. These results provide further evidence that the progesterone-binding protein is different from the cortisol-binding system.

Unequivocal proof for nonidentity of the two binding proteins was obtained by their chromatographic separation. Hydroxylapatite adsorbed the bulk of the serum proteins, including most of the albumin. Almost all of the radiolabeled progesterone and cortisol, added to the pregnant guinea pig serum, was eluted as bound steroid close to the breakthrough volume, together with a small portion of protein (Fig. X-10). A slight separation of the two steroid-binding systems was consistently found; the progesterone peak emerged ahead of the cortisol peak.

The high-affinity binding proteins contained in the initial fractions (Fig. X-10) were subjected to Sephadex G-200 filtration

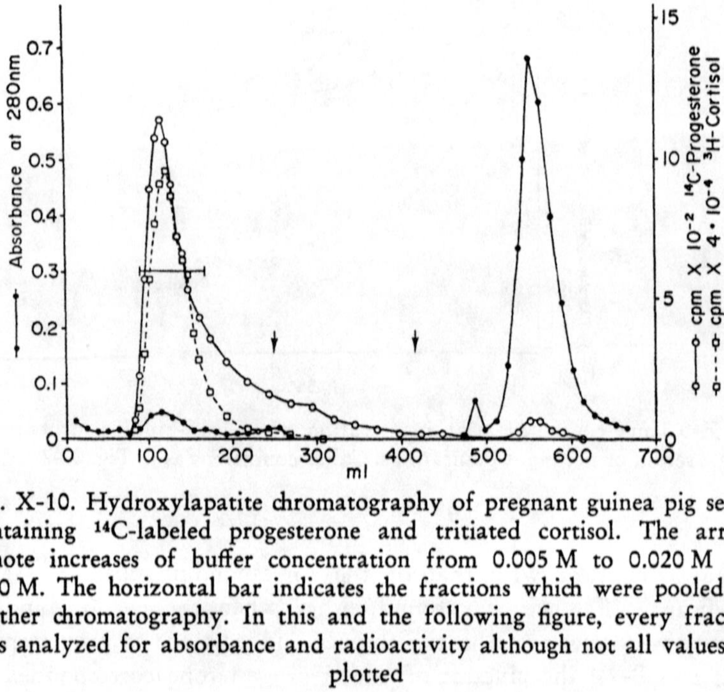

Fig. X-10. Hydroxylapatite chromatography of pregnant guinea pig serum containing ^{14}C-labeled progesterone and tritiated cortisol. The arrows denote increases of buffer concentration from 0.005 M to 0.020 M and 0.20 M. The horizontal bar indicates the fractions which were pooled for further chromatography. In this and the following figure, every fraction was analyzed for absorbance and radioactivity although not all values are plotted

(Fig. X-11). This resulted in clear separation of the progesterone-binding globulin* from the cortisol-binding globulin. Equilibrium dialysis at 4° of the most active PBG and CBG fractions (Fig. X-11, No. 16 and No. 24) showed the expected high binding affinities; the calculation, made under certain assumptions, gave k values of

* Unpublished electrophoretic studies with N. Rust show the globulin nature of the progesterone-binding protein. This is the basis for our use of PBG in abbreviating progesterone-binding globulin.

$2 \times 10^{10}\,M^{-1}$ and $5 \times 10^{7}\,M^{-1}$ for the progesterone and cortisol complex, respectively. It must be emphasized that these association constants are approximations and are not considered definitive values of the pure proteins although they are in general agreement with the affinity constants previously found for guinea pig serum (Table VIII-21). The values are given solely to indicate the presence in the separated fractions of distinct proteins that exhibit high binding affinity for progesterone and cortisol, respectively. In support of this conclusion, very little cortisol was bound to fraction 16. A small but significant binding of progesterone in fraction 24 resulted presumably from contamination by the more abundant PBG which preceded this fraction (see Fig. X-11).

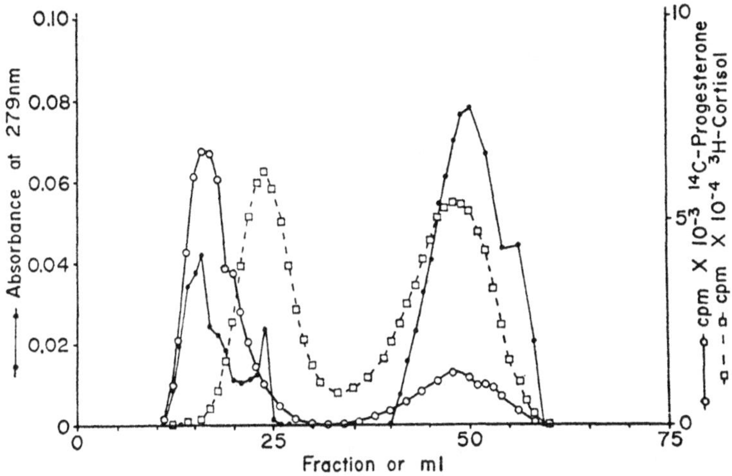

Fig. X-11. Sephadex G-200 filtration of cortisol- and progesterone-binding proteins partially purified by hydroxylapatite chromatography (Fig. X-10). The ultraviolet absorbing material following the free radiosteroids (fractions No. 41—58) does not consist of protein

An assessment of the molecular size (Stokes radius) of the guinea pig PBG and CBG was made by filtration over a Sephadex G-200 column that had been calibrated with pure proteins of different molecular weights. An eluate from hydroxylapatite chromatography, after additional DEAE-Sephadex chromatography, gave separation similar to that seen in Fig. X-11. The elution volumes corresponded to an approximate molecular weight of 100,000 for the PBG and 50,000 for the CBG. These values indicate no more than the order of magnitude.

The discovery in the guinea pig of a progesterone-binding globulin of very high binding affinity, which is independent of CBG, and which does not show overlapping steroid interaction with the cortisol-binding system, adds a new variation to the species differences of steroid-binding proteins. Since the PBG during pregnancy shows the unusually high increase of over 100-fold, its relationship to other progesterone-binding proteins involved in the cellular mechanism of this hormone becomes an important question. The high binding affinity and specificity of PBG resembles the properties of the steroid receptor proteins more than those of the known serum binders.

3. Interpretation of the Possible Role of CBG

In an effort to assess whether the participation of CBG in the initiation of lactation is of general significance, the concentrations of unbound corticosteroids for various reproductive epochs have been compiled in Table X-3 for four species including the rat. A similar trend was observed for the unbound corticosterone in rat and mouse; decrease in midpregnancy was followed by an increase through late

Table X-3. *Concentration of unbound serum corticosteroid (μg/100 ml serum) in several species during pregnancy and lactation.* From GALA and WESTPHAL (1967)

Reproductive epoch	Rat [a, b]	Mouse [b]	Rabbit [b]	Guinea pig [b]
non pregn virgin	0.50	0.39	1.14	2.48
mid pregnancy	0.18 (14)	0.12 (12)	2.66 (21)	6.41 (30—40)
late pregnancy	0.26 (20)	0.48 (18)	0.51 (27—28)	14.92 (50—60)
early lactation	0.60 (3)	0.74 (1)	2.06 (1—2)	4.86 (2—3)
late lactation	1.02 (12)	1.16 (14)	2.59 (6—12)	4.18 (12)

[a] Data taken from GALA and WESTPHAL (1965 a) for calculation of unbound corticosteroid.

[b] Unbound concentrations for rat and mouse calculated for corticosterone; for rabbit and guinea pig calculated for cortisol.

( ) The day of pregnancy or lactation in parentheses.

pregnancy and lactation. In contrast, in the rabbit no reduction of unbound "cortisol" was noted in midpregnancy; the minimum value occurred in late pregnancy prior to a continued increase throughout lactation. It should be noted that the number of observations in the rabbit experiment was limited. In the guinea pig, the calculations seemed to indicate a continued increase in unbound cortisol during pregnancy and a marked decrease during lactation (Table X-3).

In the following paragraphs, an attempt is made to interpret the role of CBG in the initiation of lactation (GALA and WESTPHAL, 1967). Since most of our experimental data have been obtained for the rat, the considerations will mainly apply to this species.

About thirty years ago, FOLLEY and YOUNG (1941) suggested that a lactogenic complex of anterior pituitary hormones was necessary to bring about the initiation of lactation. The two most important hormones in this concept were prolactin and ACTH. The effects of ACTH on the initiation of lactation were subsequently attributed to the glucocorticoid secretions (COWIE and FOLLEY, 1961). It has been discussed in Chapter IX. 5 that the important role of the glucocorticosteroid hormones in the initiation of lactation is well established.

The functional activity of the corticosteroid hormones is dependent on the level of CBG. Since thyroid hormones control the CBG activity in the rat (Chapter IX. 7), it seems pertinent to briefly discuss endocrine changes during pregnancy and lactation in relation to thyroidal secretion. Toward the end of pregnancy and at the time of parturition, the organism is under the influence of estrogens as demonstrated for several species (ATKINSON and LEATHEM, 1946; VENNING, 1957; YOUNG, 1961 a). As pointed out in Chapter IX. 7, estrogens have a stimulatory effect on thyroid hormone secretion. In contrast, administration of corticosteroids results in reduced thyroid activity through decreased TSH production (INGBAR and FREINKEL, 1956).

The thyroid activity of pregnant rats was found not to differ markedly from that of nonpregnant animals (FELDMAN, 1958; IINO and GREER, 1961); a 50 per cent decrease in thyroid activity, however, was observed at the end of pregnancy and throughout lactation (IINO and GREER, 1961). A similar decline in thyroid function shortly after parturition also occurred in the rabbit (BROWN-GRANT, 1956).

The initiation of lactation appears to be the result of a complex interplay between a number of endocrine organs and the mammary gland. The following description of reactions and interactions is offered in an attempt to interpret the role of CBG in the lactating rat, on the basis of the experimental evidence available at the present.

During midpregnancy, the relatively large amounts of progesterone stimulate CBG activity and act synergistically with estrogen, prolactin and other hormones to produce lobular-alveolar development of the mammary gland. The total amount of corticosterone as well as the unbound portion are at a comparatively low level.

Shortly before parturition there is a decrease in the production of progesterone while estrogen secretion is maintained. The uninhibited estrogens increase prolactin production and probably ACTH since it has been demonstrated that estrogen has a stimulatory effect on ACTH formation (GEMZELL, 1952). In accordance, release of pituitary ACTH is increased on the last day of pregnancy, resulting in a sharp rise in adrenal cortical activity (POULTON and REECE, 1957). There is no change in CBG activity but the increase in peripheral corticosterone leads to an enhanced concentration of unbound corticoid.

Thyroid activity is reduced at this time, and is further impaired by the competition of the mammary gland for iodine. After parturition, ACTH is further increased by the suckling stimulus (GREGOIRE, 1947; FOLLEY, 1956) and the corticosteroid level will rise. The elevated peripheral corticosterone in conjunction with low CBG brings about a marked increase in the concentration of unbound corticoid. The assumption is made that this unbound, physiologically active corticosterone acts together with prolactin to initiate full lactation. High prolactin levels are maintained by the nursing stimulus. The same stimulus leads to continued ACTH release so that the peripheral corticosteroids stay at a high level.

Since the CBG concentration remains low during nursing of a large number of pups the level of unbound corticosterone is relatively high. The interpretation of corticosteroid function on the basis of low CBG during lactation appears to have general validity, at least for the species studied in our laboratory.

XI. Isolation and Properties of CBG from Mammalian Sera

1. Isolation of Human CBG

It is apparent from the preceding chapters that the very low blood concentration and the limited stability of CBG make it difficult to isolate. The first investigators to succeed were SEAL and DOE (1961) who reported at the 1961 Federation Meeting the preparation of highly purified CBG from serum of estrogen-treated males. In this pioneering work, an overall recovery of 80% after approximately 1000-fold purification was obtained; physicochemical characterization revealed the properties of a glycoprotein with the characteristics of an α_1-globulin. One of the more important steps for the separation from other serum proteins was the chromatography on hydroxylapatite which did not adsorb CBG but retained most other proteins. A detailed account of the results was given subsequently (SEAL and DOE, 1962 a, 1962 b). Partial purification of human CBG has been briefly reported by ENSINCK and VOLWILER (1962).

A number of years passed before the isolation of human CBG was accomplished in other laboratories (SLAUNWHITE et al., 1966; MULDOON and WESTPHAL, 1967). The starting material used by SEAL and DOE was serum or heparinized plasma from patients with cancer of the prostate who had been receiving 5 mg diethylstilbestrol daily for at least one month; this treatment raises the CBG level approximately 3-fold. The same authors described also the purification of CBG from normal blood bank plasma. SLAUNWHITE et al. (1966) isolated CBG from plasma of diethylstilbestrol-treated patients. Plasma or serum from outdated human acid-citrate-dextrose blood served as starting material in our laboratory. The α-globulin-rich COHN fractions IV-4 and IV-5,6 bound cortisol with much lower affinity than whole human serum did. These serum fractions, which are not obtained under mild conditions, evidently did not contain appreciable amounts of CBG activity (MULDOON and WESTPHAL, 1967).

Prior to fractionation, the serum or plasma was equilibrated with radiolabeled cortisol, preferably in excess of the CBG binding sites. This assured easy recognition and quantification of cortisol-containing protein fractions, and provided protection for the sensitive CBG molecule. Chromatography on DEAE-cellulose, used in the three laboratories as the first step in the purification procedure, had been reported in 1960 to provide efficient separation of human CBG from

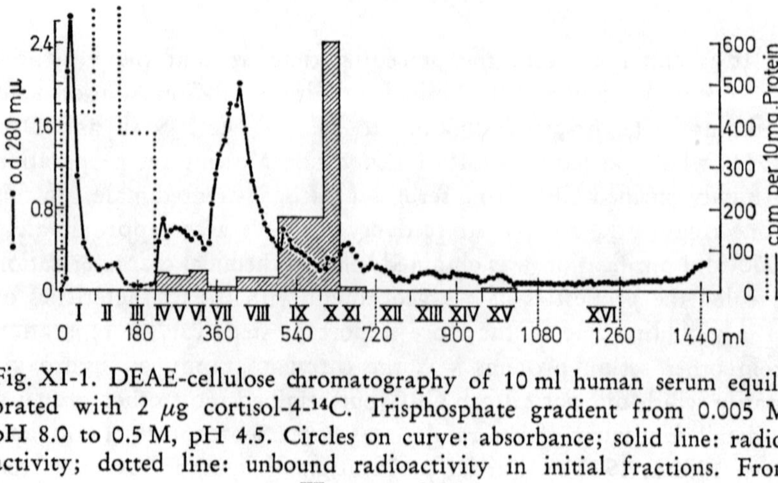

Fig. XI-1. DEAE-cellulose chromatography of 10 ml human serum equilibrated with 2 μg cortisol-4-^{14}C. Trisphosphate gradient from 0.005 M, pH 8.0 to 0.5 M, pH 4.5. Circles on curve: absorbance; solid line: radioactivity; dotted line: unbound radioactivity in initial fractions. From WESTPHAL (1961)

the major portion of the serum proteins (WESTPHAL, 1961). A "cone-sphere" type gradient (SOBER and PETERSON, 1958) was applied. Fig. XI-1 shows this original purification, with the highest radiocortisol concentration in fraction X, in the region of α_1-globulin (FAHEY et al., 1958); the preceding albumin peak (fractions VII and VIII) contains little radioactivity. The same result was obtained when the radiocortisol-free serum was first chromatographed and the separated protein fractions were then subjected to multiple equilibrium dialysis against radiolabeled cortisol. Fig. XI-2 indicates the highest binding affinity in fractions IX and X, and lower binding activity in the albumin fractions. This figure also shows the pH gradient of the elution system. The radioactivity eluted immediately after the void volume (dotted line in Fig. XI-1) represents unbound cortisol which had been added in excess of the CBG binding sites; this was verified by multiple equilibrium dialysis of fractions I—VI (Fig. XI-2) which showed no binding.

SEAL and DOE (1962 b) have applied step-wise elution of the DEAE-cellulose column, using buffers of increasing phosphate concentration and decreasing pH. The purification obtained was 2-fold. The active fraction was subjected to batch adsorption by hydroxylapatite gel which removed 95—96% of the protein and none of the protein-

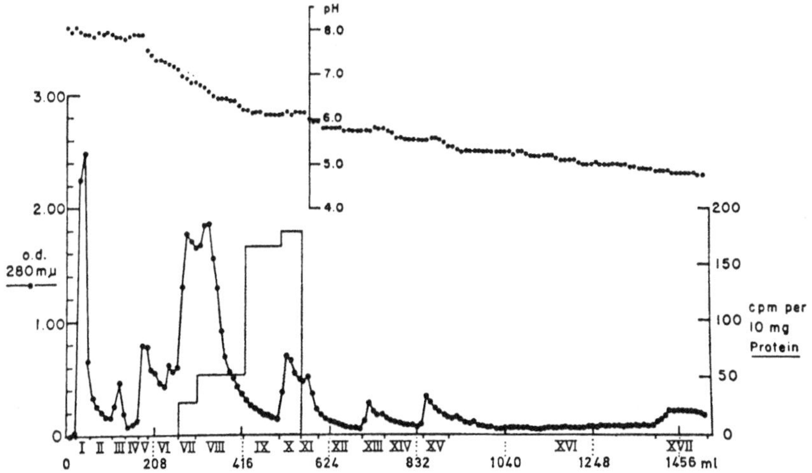

Fig. XI-2. DEAE-cellulose chromatography of 10 ml human serum. Buffer gradient and symbols as in Fig. XI-1; solid circles: pH. Straight-lined area in fractions represents bound radiocortisol (right ordinate) determined by multiple equilibrium dialysis. From WESTPHAL (1961)

bound radioactivity. Several chromatography runs on hydroxylapatite columns completed the purification.

A sodium chloride-sodium phosphate gradient of constant pH 5.0 served SLAUNWHITE et al. (1966) as eluant in the initial DEAE-cellulose chromatography. This step was followed by gel filtration over Sephadex G-25 and repeated hydroxylapatite column chromatography. A Sephadex G-75 gel filtration provided the final CBG preparation.

In the isolation procedure developed in our laboratory (MULDOON and WESTPHAL, 1967), the usefulness of ammonium sulfate fractionation was explored. It was found that precipitation between 40% and 63% saturated ammonium sulfate at pH 6.4 provided a fraction

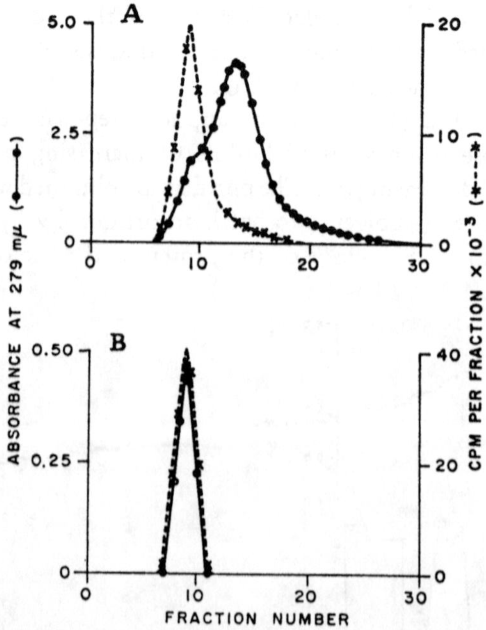

Fig. XI-3. Hydroxylapatite chromatography of active fractions after separation on DEAE-cellulose. Protein was applied at a concentration of 50 mg per ml and eluted with mM phosphate buffer, pH 6.8, at a flow rate of 0.2 ml per min. Fractions of 3-ml volume were collected. A, first chromatography, 140 mg of protein on a column (25×120 mm); B, final chromatography, 10.1 mg of protein on a column (10×120 mm). From MULDOON and WESTPHAL (1967)

containing 85% of the total CBG activity and only 17% of the total protein. However, this salt precipitation was found advantageous only for larger quantities of serum, i. e., more than a liter; smaller volumes were fractionated directly on the chromatographic columns.

It was observed in the initial DEAE-cellulose chromatography that 85—90% of the total protein, free of radioactivity, could be removed with water. Elution of the CBG fraction was achieved with a gradient system composed of 0.005 M Tris-phosphate, pH 8.0, as starting buffer, and 0.5 M Tris-phosphate, pH 5.5, as limit buffer (MULDOON, 1967). This gradient elution was found to give consistently higher purification factors than the stepwise elution used in other laboratories.

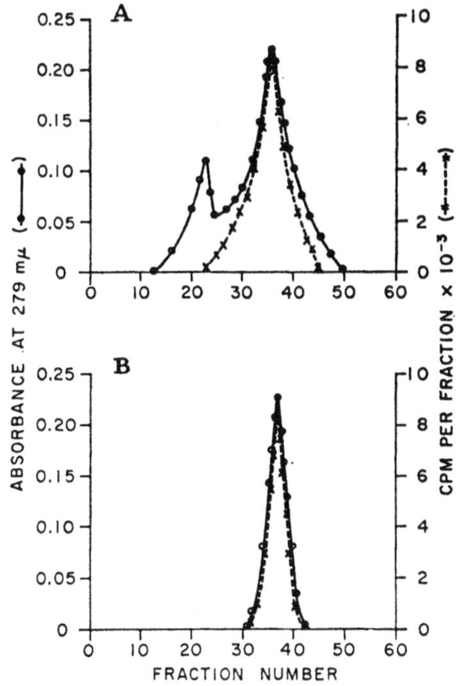

Fig. XI-4. Sephadex G-200 filtration of a partially purified CBG preparation. Protein which appeared homogeneous on hydroxylapatite was applied to a Sephadex G-200 column (25×300 mm) which had been equilibrated with 0.05 M phosphate buffer, pH 7.4. Protein was eluted with the same buffer at a flow rate of 1 ml per min. Each fraction had a volume of 3 ml. A, first gel filtration; B, final gel filtration. From MULDOON and WESTPHAL (1967)

After a single hydroxylapatite gel batch separation according to SEAL and DOE (1962 b), which gave 87% recovery of radioactivity, the active protein was subjected to hydroxylapatite column chromatography (Fig. XI-3). This step was repeated until a product of constant specific activity was obtained; this required up to five runs, depending on the nature of the material. To obtain chromatographic homogeneity and coincidence of protein concentration and radioactivity (Fig. XI-3 B), close adherence to the experimental conditions is necessary. The purification at this step was about 2900-fold, which is similar to that reported by SEAL and DOE (1962 b) for their final product. Immunoelectrophoresis against antihuman serum showed the presence of a second component when the protein was analyzed at a concentration of 8 g/100 ml. SLAUNWHITE et al. (1966) have observed a second band in disc electrophoresis after attainment of transcortin homogeneity on hydroxylapatite. The impurity in our preparation migrated as an α_2-globulin and stained poorly with Amido schwarz.

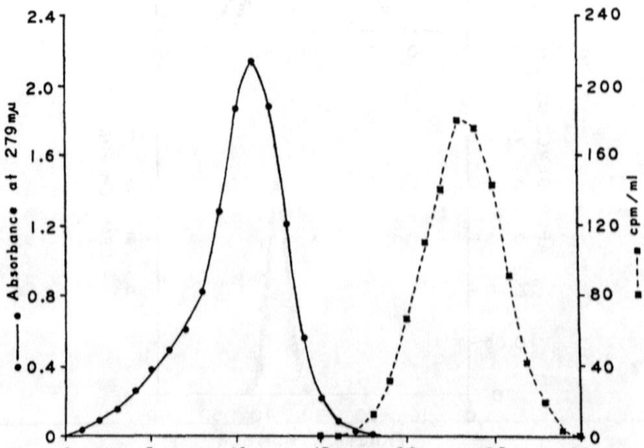

Fig. XI-5. Sephadex G-200 filtration at 4° of 50 mg HSA in 1 ml of 0.05 M phosphate, pH 7.4, after equilibration with 0.2 μg of cortisol-4-^{14}C. From CHADER and WESTPHAL (1968 b)

The contamination seen in immunoelectrophoresis could be removed by gel filtration over Sephadex G-200 (Fig. XI-4). This dextran preparation provides graded exclusion for serum proteins including CBG-corticosteroid complexes, as will be shown in detail in Chapter XI. 2; cortisol remains almost completely bound to CBG during the gel filtration at 4°. In contrast, the lower affinity constant of the cortisol-HSA complex causes dissociation so that the protein is eluted prior to the steroid which is delayed on the Sephadex column (Fig. XI-5). To the solution of the active protein (Fig. XI-3 B) in phosphate buffer, pH 7.4, ascorbate was added to give a concentration of 0.5 mM; ascorbic acid stabilizes the sensitive CBG (SEAL and DOE, 1963). Fig. XI-4 A shows that the contaminating material constituted about 30% of the total protein applied to the column.

The gel filtration was repeated until a single symmetrical peak for both protein and radioactivity was obtained (Fig. XI-4 B). The final CBG preparation represented a purification of approximately 4000-fold, in relation to the original serum protein mixture. The over-all yield was about 50% from plasma samples of 1 liter or less, and about 30% from larger volumes, in which preliminary ammonium sulfate fractionation was employed. Recovery values and purification factors of the principal steps of the isolation procedure are summarized in Table XI-1.

The final preparation of human CBG is considered pure; it was found to be homogeneous by free boundary, agar and paper strip

Table XI-1. *Summary of CBG isolation procedure from 600 ml of normal human plasma.* From MULDOON and WESTPHAL (1967)

Purification step	Protein mg	Specific activity cpm/mg protein	Purification [a]	Recovery Total %	Utilized [b] %
DEAE-cellulose column eluate	3600	24	11.3	92	73
Hydroxylapatite batch fractions	530	141	6.0	87	87
Hydroxylapatite column eluate	10.7	5460	42.2	87	85
Sephadex G-200 column eluate	7.2	7530	1.38	98	91
Final CBG preparation	7.2	7530	3950	68	49

[a] Purification is calculated on the basis of increase in specific activity at each step in the isolation procedure, after correction for losses of radioactivity upon dialysis.

[b] The percentage actually introduced into the next fractionation step. For instance, 100,000 cpm in starting plasma would give 92,000 cpm total recovery in DEAE-cellulose eluate; 73,000 cpm, constituting the fractions of highest specific activity, were subjected to hydroxylapatite batch fractionation. The remaining active fractions (corresponding to 19,000 cpm) were combined from several experiments and rechromatographed. This improves the over-all yield to a value greater than 49% and not higher than 68%. However, this additional yield is not included in the 7.2 mg final preparation listed in this table.

electrophoresis. The mobility was that of a single homogeneous α_1-globulin in Veronal buffer at pH 8.6, in Tris buffer at pH 8.9, or in cacodylate buffer at pH 6.2. The immunoelectrophoretic behaviour against antisera to human serum (Fig. XI-6) and serum fractions indicated purity; no reaction was observed with antisera to HSA, human AAG, and human serum Fractions III-1 and VI. Other immunoelectrophoretic analyses showed lack of reaction with antisera produced against sera from several species other than human. Ultracentrifugation showed a symmetrical peak (Fig. XI-7), and a plot of

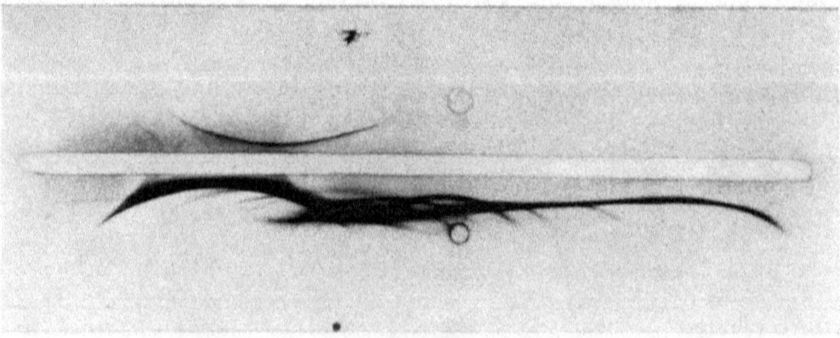

Fig. XI-6. Immunoelectrophoresis of pure human CBG-cortisol complex. Top well contains 8% CBG, bottom well undiluted human serum. The trough contains 70 μl of antihuman horse serum. Electrophoresis proceeded for 60 min at 300 volts in high resolution Tris buffer (LKB 3276-GB), pH 8.9. Migration toward the left is anodic. From MULDOON and WESTPHAL (1967)

the reciprocal sedimentation coefficient versus CBG concentration gave a straight line as another indication of homogeneity (Fig. XI-8). The diffusion pattern confirmed the assumption of homogeneity of the macromolecule (see Fig. 7 in original publication by MULDOON and WESTPHAL, 1967).

From the over-all yield of 49% for the pure CBG preparation described in Table XI-1, it can be calculated that the original acid-citrate-dextrose plasma contained 24.5 mg of CBG per liter. This corresponds to 27.8 mg of CBG per liter of undiluted plasma since the acid-citrate-dextrose plasma contains about 12% diluent. This value of approximately 28 mg of CBG per liter seems somewhat low

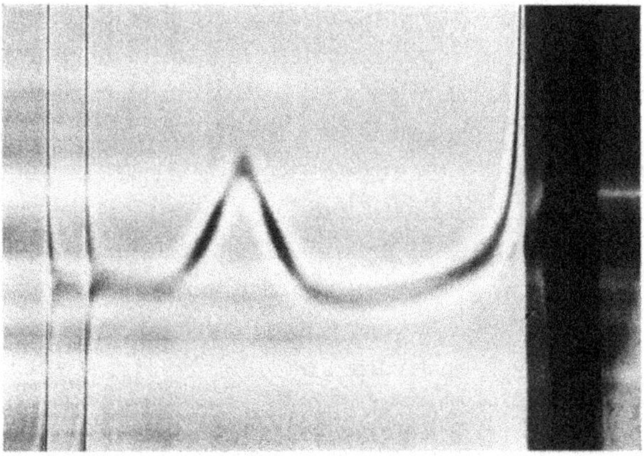

Fig. XI-7. Sedimentation pattern of human CBG, at 4°; 59,780 rpm. Concentration was 9.0 mg per ml in 0.05 M phosphate buffer, pH 7.4. The exposure shown was made 121 min after attainment of maximum speed. Bar angle was 60°. Sedimentation is from left to right. From MULDOON and WESTPHAL (1967)

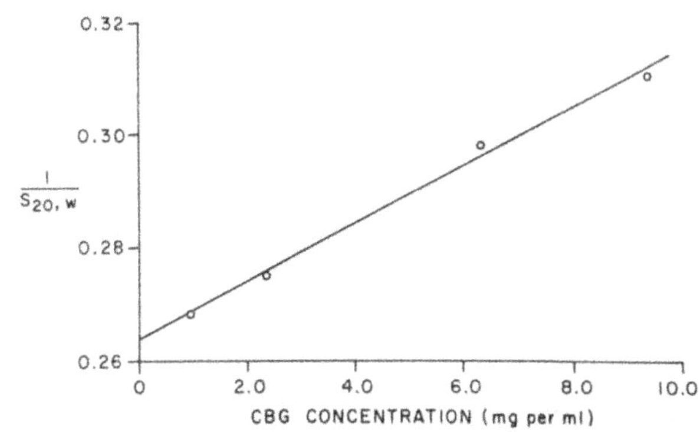

Fig. XI-8. Relationship between reciprocal sedimentation coefficient and CBG concentration. The straight line was calculated by the method of least squares. From MULDOON and WESTPHAL (1967)

considering the level of about 7×10^{-7} M (36 mg per liter) reported for normal human plasma (SANDBERG et al., 1966; SEAL and DOE, 1966). However, analyses of a large number of individual samples show a wide range of 22 to 46 mg per liter as summarized in Table VIII-10. In the earlier work of SEAL and DOE (1962 a,

1962 b), a level of about 20 mg of CBG per liter of normal human plasma has been stated (approximately one-third of 60 mg per liter present in diethylstilbestrol-treated patients). The plasma used as starting material by SEAL and DOE (1962 a), i. e., blood bank plasma, is similar to that used in the present work, whereas the higher values (SANDBERG et al., 1966; SEAL and DOE, 1966) in general refer to fresh plasma. On the basis of SEAL and DOE's and our own experience, plasma samples obtained from outdated blood bank blood may have a somewhat lower CBG concentration than fresh material. MURPHY et al. (1963) described inactivation of CBG by citrate when stored in the frozen state; we have therefore never frozen CBG-containing solutions while citrate was present. However, the possibility remains that extended storage under blood bank conditions results in some destruction of CBG. We do not know whether this assumption can be generalized, especially since the plasma samples obtained for our work varied considerably in storage time. It should be noted that we determined in a large pool of similar acid-citrate-dextrose plasma a CBG concentration of about 7×10^{-7} moles per liter of undiluted plasma (Table VIII-21).

SEAL and DOE (1962 a) have dialyzed blood bank plasma or heparinized plasma from cancer patients against distilled water for a week at 2°, presumably without loss of CBG activity. The same authors reported (SEAL and DOE, 1962 a, 1962 b) that preparations of 5% or greater purity are irreversibly inactivated when dialyzed against distilled water, or when kept at pH 5 or lower. Our experiments confirmed that dialysis of active fractions at different stages of purity against water tended to produce irreversible dissociation of the cortisol-CBG complex; the denaturation increased sharply as the preparations approached homogeneity. Inactivation of pure CBG occurred upon subjection to gel filtration at 45°. Since it had been shown (WESTPHAL, 1967) that removal of steroids from whole serum at 45° does not significantly alter the binding affinity of CBG, it would appear that other plasma components exert a protective influence. Equilibrium dialysis of CBG in the presence of added cortisol at 37° for 48 hours does not result in loss of activity, suggesting that the steroid ligand protects the purified protein in a similar way as observed for CBG in serum (DOE et al., 1964).

2. Isolation of Rabbit and Rat CBG

Our interest in the relationship of the chemical structure of the binding site area to the affinity towards individual steroids prompted the isolation of CBG's from species other than human. In Table VIII-21 the considerable differences existing among mammalian species in their steroid-CBG associations have been documented. Whereas progesterone is bound more firmly to human CBG than is cortisol, the affinity constants of the rat CBG complexes with the two steroids of different polarities are about the same. The polarity rule relationship is reversed for CBG in rabbit serum which forms a cortisol complex of about $2^1/_2$ times higher affinity at 4° than the corresponding progesterone complex. Structural analysis of the pure CBG entities may be expected to shed light on the reasons for the differences.

For the isolation of rabbit CBG (CHADER and WESTPHAL, 1968 b), serum from female animals was used, which was found to have about 20% higher corticosterone-binding capacity than serum pools mixed from males and females (for details see CHADER, 1966). The starting material for the rat CBG was a serum pool collected over several years from male and female Sprague-Dawley rats and stored at —85° C; 1100 ml was subjected to the fractionation procedure (CHADER and WESTPHAL, 1968 c). Although the purification scheme of the CBG's from the two species has a general similarity with that applied to human serum, several differences deserve mention.

In the fractionation of both rabbit and rat serum, equilibration was carried out with radiolabeled cortisol or corticosterone, respectively, in excess of the high-affinity binding sites. DEAE-Sephadex chromatography was used as the first step (Fig. XI-9). This proved advantageous, especially for a more rapid purification of liter quantities of rabbit serum, because of the high anion exchange capacity of the adsorbant. A combination of stepwise and gradient elution was useful in improving purification and yield. Small amounts of radioactive material eluted with the breakthrough volume and with the major protein peak were not strongly bound to protein, as determined by gel filtration and equilibrium dialysis studies; presumably they represented cortisol-4-^{14}C that exceeded the CBG binding capacity. Steroid was bound with high affinity to the protein contained in tubes 250 to 300, as revealed by Sephadex G-200 gel filtra-

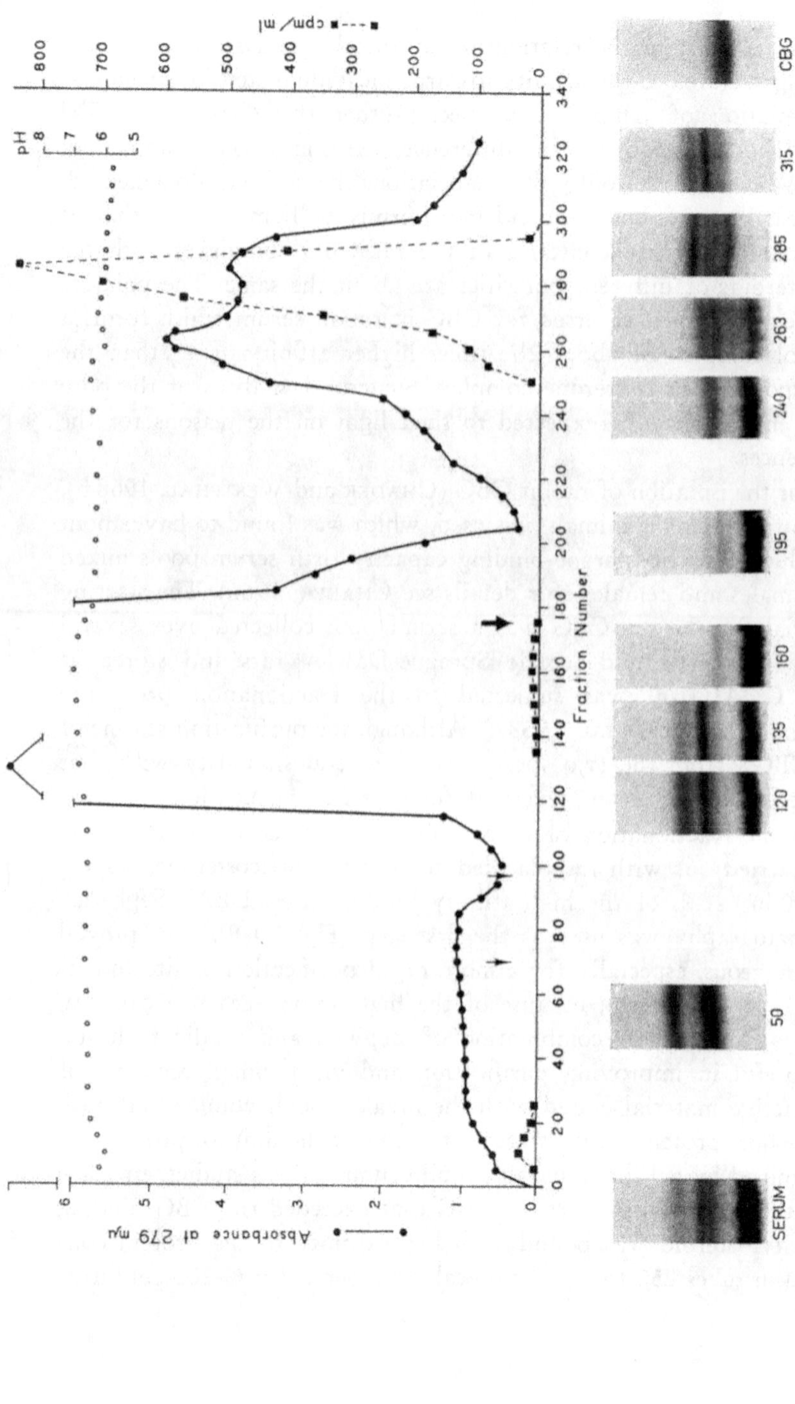

Fig. XI-9. DEAE-Sephadex chromatography at 4° of 1 liter of rabbit serum containing 2.5 μg of cortisol-4-^{14}C. Except for maxima and minima, absorbance and radioactivity were measured for every fifth fraction. Light arrows indicate buffer changes; bold arrow shows start of gradient. Results of paper strip electrophoretic analyses of individual fractions, and of starting serum and pure CBG, are also given. From Chader and Westphal (1968 b)

tion experiments (CHADER, 1966). A 15- to 20-fold concentration was usually achieved in this step.

A rapid preliminary purification of rabbit CBG in serum volumes up to 1 liter was accomplished by ammonium sulfate fractionation at 4°. The fractions were analyzed for protein-bound radioactivity

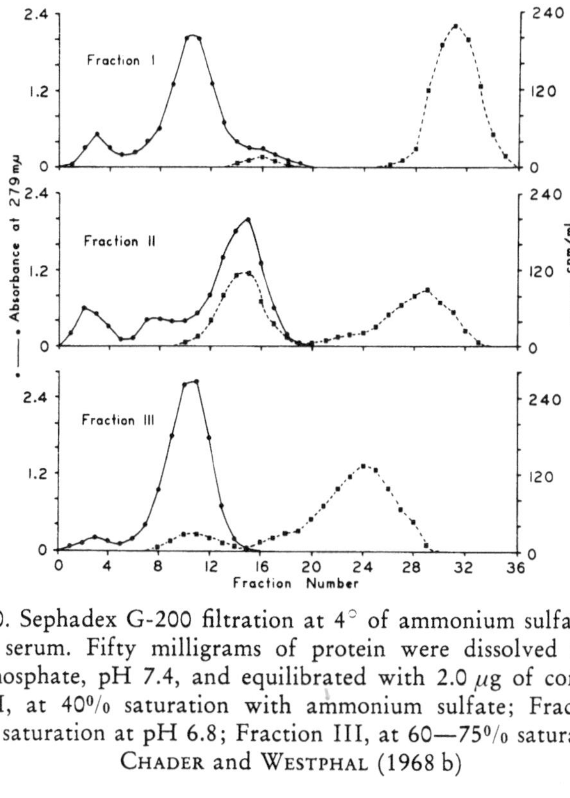

Fig. XI-10. Sephadex G-200 filtration at 4° of ammonium sulfate fractions of rabbit serum. Fifty milligrams of protein were dissolved in 1 ml of 0.05 M phosphate, pH 7.4, and equilibrated with 2.0 μg of cortisol-4-^{14}C. Fraction I, at 40% saturation with ammonium sulfate; Fraction II, at 40—60% saturation at pH 6.8; Fraction III, at 60—75% saturation. From CHADER and WESTPHAL (1968 b)

by gel filtration over Sephadex G-200 at 4° as shown in Fig. XI-10. Fraction II was used for further purification, starting with DEAE-Sephadex chromatography as described for serum.

Filtration over Sephadex G-200 is a useful method also for characterization of the binding proteins in unfractionated serum. Fig. XI-11 A shows cortisol binding to a CBG-containing fraction of rabbit serum in the third peak, the excess of cortisol appears as a later peak of unbound steroid. Fig. XI-11 B shows an analogous

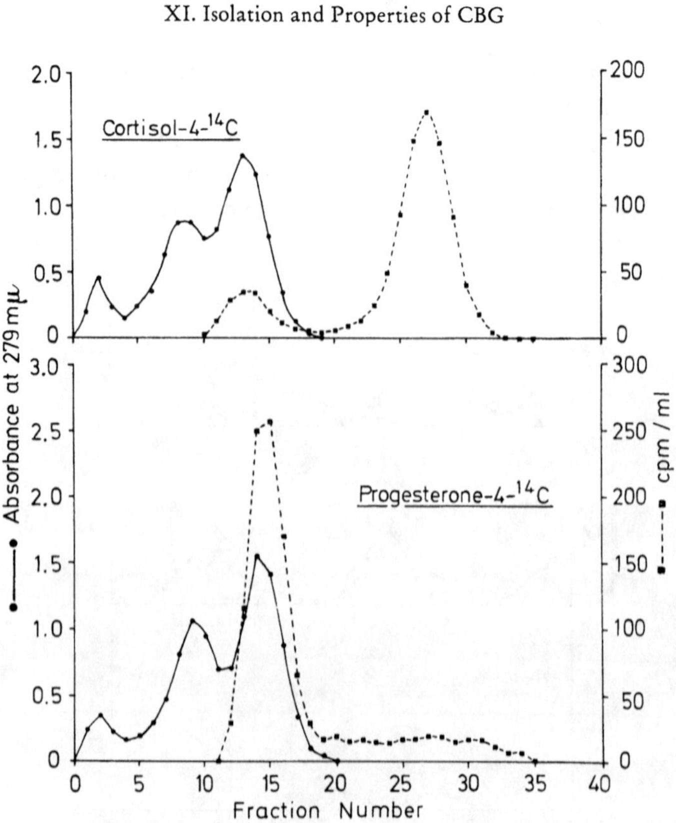

Fig. XI-11. Sephadex G-200 filtration of rabbit serum containing ^{14}C-labeled steroids. One milliliter of serum diluted with 0.05 M phosphate, pH 7.4, to contain 50 mg of total protein was equilibrated with 0.2 μg of cortisol-4-^{14}C (A), or progesterone-4-^{14}C (B), and filtered over a 2.5×28 cm column of Sephadex G-200 at 4°. From CHADER and WESTPHAL (1968 b)

experiment with radiolabeled progesterone; practically all steroid is bound to the same fraction which contains albumin in addition to CBG. Evidence for progesterone binding to rabbit serum albumin under these conditions of Sephadex G-200 filtration is given in Fig. XI-12 B; the association seems particularly strong with the apparently heavier component preceding the main albumin peak. The upper part (A) of the figure shows that under the same conditions cortisol dissociates completely from albumin and is eluted as unbound species by a much larger volume of buffer.

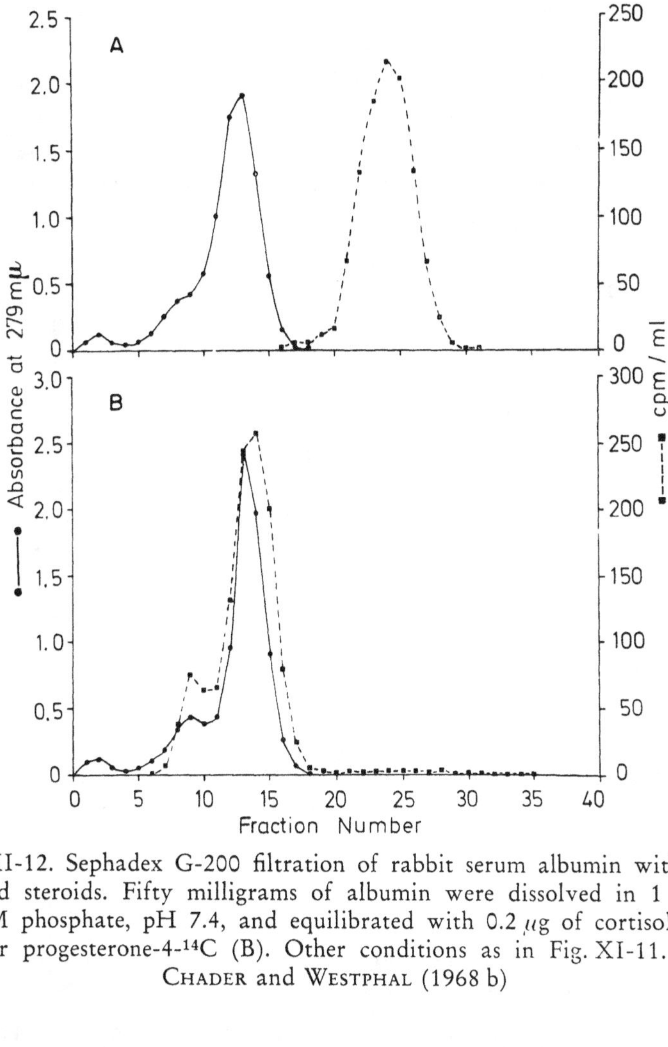

Fig. XI-12. Sephadex G-200 filtration of rabbit serum albumin with ^{14}C-labeled steroids. Fifty milligrams of albumin were dissolved in 1 ml of 0.05 M phosphate, pH 7.4, and equilibrated with 0.2 μg of cortisol-4-^{14}C (A) or progesterone-4-^{14}C (B). Other conditions as in Fig. XI-11. From CHADER and WESTPHAL (1968 b)

The CBG-containing fraction of rabbit serum after DEAE-Sephadex chromatography was further purified by gel filtration over Sephadex G-200. Fig. XI-13 shows that this technique eliminated the major portion of inactive protein; a 2- to 4-fold increase in specific activity was usually obtained. All radioactivity remained protein-bound during the filtration at 4°.

In the case of rat CBG, the initial DEAE-Sephadex chromatography did not approach the 15—20-fold purification usually achieved with rabbit serum; the partially purified material was therefore applied to a DEAE-cellulose column as shown in Fig. XI-14. A small amount of unbound corticosterone-4-^{14}C was observed in the breakthrough volume while over 96% of the radioactivity was eluted with the protein toward the end of the gradient. The active fraction was then filtered over Sephadex G-200 as described for the purification of rabbit CBG.

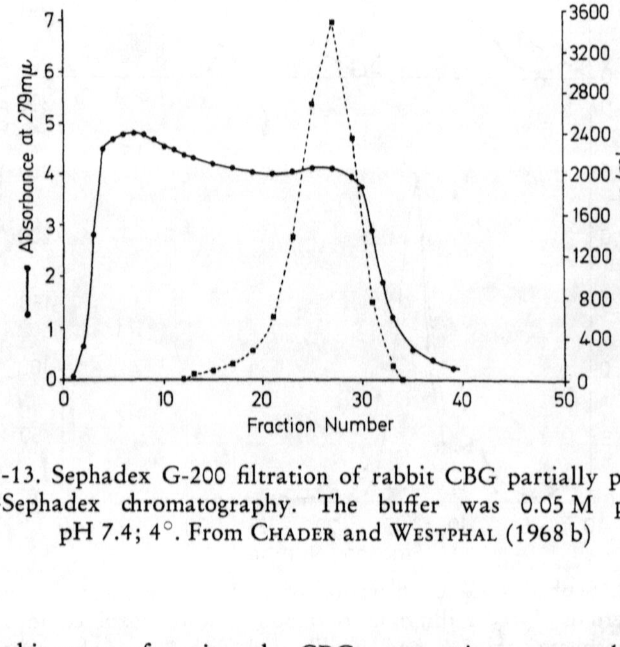

Fig. XI-13. Sephadex G-200 filtration of rabbit CBG partially purified by DEAE-Sephadex chromatography. The buffer was 0.05 M phosphate, pH 7.4; 4°. From CHADER and WESTPHAL (1968 b)

At this stage of purity, the CBG preparations were subjected to hydroxylapatite chromatography. Stepwise elution with phosphate buffers of pH 6.8 and increasing ionic strength was used as illustrated for rabbit serum (see Fig. 7 in original publication of CHADER and WESTPHAL, 1968 b) and for rat CBG in Fig. XI-15. The figure shows that, as previously found for human CBG, the protein-bound radioactivity is eluted in the initial fractions, before the bulk of the protein appears. This rat CBG fraction showed one α_1-globulin band in paper strip electrophoresis but contained three components in the ultracentrifuge. In contrast, less purification was achieved in the first hydroxylapatite chromatography of the corresponding rabbit CBG preparation as shown by Sephadex G-200 filtration (see Fig. 8 in

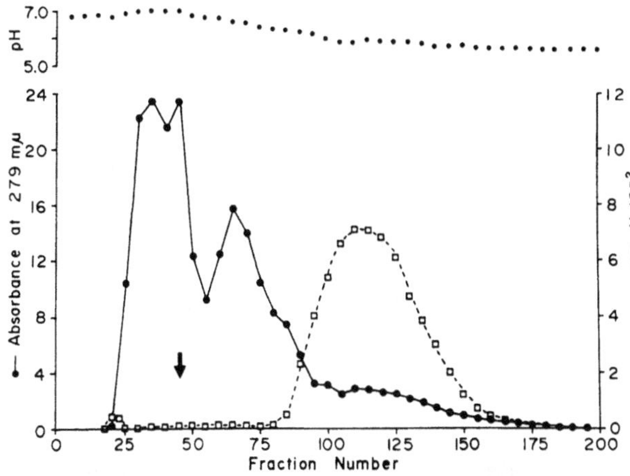

Fig. XI-14. DEAE-cellulose chromatography of the partially purified rat CBG-corticosterone-4-^{14}C complex after DEAE-Sephadex separation. Except for maxima and minima, absorbance and radioactivity were measured for every fifth fraction. Arrow indicates start of gradient. From CHADER and WESTPHAL (1968 c)

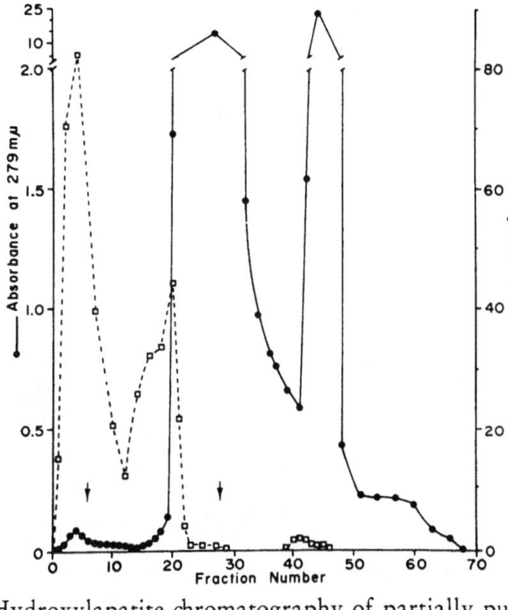

Fig. XI-15. Hydroxylapatite chromatography of partially purified rat CBG after Sephadex G-200 filtration. Arrows indicate changes from 0.005 to 0.05 M and to 0.2 M phosphate buffer of pH 6.8; 4°. From CHADER and WESTPHAL (1968 c)

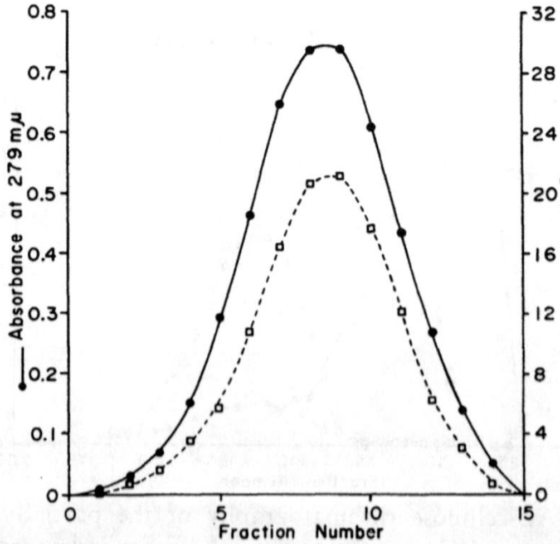

Fig. XI-16. Final Sephadex G-200 filtration of rat CBG-corticosterone complex. Phosphate buffer, 0.05 M, pH 7.4; 4°. From CHADER and WESTPHAL (1968 c)

Table XI-2. *Purification procedure for CBG from 2 liters of rabbit serum*[a]. From CHADER and WESTPHAL (1968 b)

Purification step	Protein applied g	Activity applied cpm	Purified fraction g	Activity recovered cpm	Cumulative recovery %	Specific activity[b] cpm/mg protein
DEAE-Sephadex	156	593,560	8.73	544,980	92	68
Sephadex G-200						
First	8.73	544,980	3.18	524,370	88	169
Second	3.18	524,370	2.17	473,200	80	218
Hydroxylapatite						
First	2.17	473,200	0.890	367,460	62	412
Second	0.89	367,060	0.471	356,880	60	758
Third	0.471	357,280	0.102	340,100	57	3,330
Fourth	0.102	340,100	0.0135	317,720	53	23,500

[a] The procedure outlined starts with 2 liters of rabbit serum; this is divided in half and applied to two DEAE-Sephadex columns. The total resulting CBG fraction (8.73 g) is then applied to a Sephadex G-200 column. The CBG fractions are combined and rechromatographed on Sephadex G-200. The CBG fraction (2.17 g) is again divided in half and applied to two small hydroxylapatite columns; the eluted CBG fractions are combined and subjected to three further hydroxylapatite chromatographies.

[b] Specific activity of the starting material was 3.8 cpm per mg of protein.

original publication by CHADER and WESTPHAL, 1968 b); the hydroxylapatite chromatography was therefore repeated several times.

Final purification of rabbit and rat CBG has been accomplished by gel filtration over Sephadex G-200, although no differences in the properties were found for the pure rabbit CBG when the gel filtrations were performed before the successive hydroxylapatite chromatographies. The last Sephadex G-200 filtration is shown for rat CBG in Fig. XI-16. A single protein peak with a coinciding symmetrical radiocorticosterone pattern indicates homogeneity of the product. A total of 31.5 mg rat CBG-corticosterone complex was obtained from the 1100 ml of pooled serum used as starting material. Assuming a CBG concentration in normal Sprague-Dawley rat serum of 11.3×10^{-7} M (Table VIII-21) and a molecular weight of 53,000 (see below Table XI-3), the isolation procedure gave about 50% yield at a purification factor of 1250 related to total serum protein.

Table XI-2 shows the results of the purification procedure described, starting with 2 liters of rabbit serum. A yield of 13.5 mg of the final CBG preparation was obtained with an over-all 6200-fold purification and a 53% recovery of cortisol-4-^{14}C. When larger amounts of serum were used initially, the recovery was 60—70%. It can be estimated from the data in Table XI-2 that the original female rabbit serum contained about 13.5 mg of CBG per liter, or a concentration of 3.3×10^{-7} M. In good agreement with this value, the concentration of CBG binding sites in the same sample of rabbit serum has been found to be 3.4×10^{-7} M by an equilibrium dialysis technique with whole serum after removal of endogenous steroids (Table VIII-21). The concentration thus is approximately half that found in normal human serum and less than one third the level in normal rat serum. A value of 3.5×10^{-7} M was found for the total endogenous corticosteroids in the starting rabbit serum sample. Assuming one corticosteroid-binding site per CBG molecule, this would indicate that the serum CBG binding sites were saturated before and during the isolation procedure.

The isolated CBG preparations from rabbit and rat serum were homogeneous by several electrophoretic criteria; their mobility was that of an α_1-globulin. Immunoelectrophoretic analysis of two preparations of pure rabbit CBG with antiserum to whole rabbit serum indicated a homogeneous α-globulin (see Fig. 9 in original publication by CHADER and WESTPHAL, 1968 b). No precipitation was obtained

Table XI-3. *Physicochemical properties of the corticosteroid complexes of human, rabbit and rat CBG. Values corrected to dry weight*

Parameter	CBG-Hu (a)	CBG-Hu (b)	CBG-Hu (c)	CBG-Rb (d)	CBG-Rt (e)
$s^0_{20,w}$ (S)	4.0 [f]	4.1 [f]	3.79	3.55	3.56
$D_{20,w}$, ($cm^2\ sec^{-1}$)$\times 10^7$	—	—	6.15	7.02	—
$\bar{v}$ (ml/gm)	0.718	—	0.708	0.695	0.711
Molecular weight	52,000 [g] ± 1500	58,500	51,700	40,700	52,600 ± 3,000 [h] 61,000 ± 1,100 [i]
f/fo	—	—	1.42	1.37	—
$E^{1\%}_{1cm}$ at 279 nm	7.40	—	6.45	8.4	6.2
A_{280} : A_{260}, corticosteroid complex	1.36 [k]	—	1.13	1.38	1.58
A_{280} : A_{260}, complex stripped	1.72	—	1.57	—	1.71
u at pH 8.6 ($cm^2\ volt^{-1}\ sec^{-1}$)$\times 10^5$	—	—	—4.9	—5.1	—
Carbohydrate (%)	14.1 [k, l]	—	26.1	29.2	27.8
Nitrogen (%)	—	—	12.7 [m]	12.1 [n]	—

[a] Seal and Doe (1966). [b] Slaunwhite et al. (1966). [c] Muldoon and Westphal (1967).
[d] Chader and Westphal (1968 b). [e] Chader and Westphal (1968 c).
[f] $s_{20,w}$ value not extrapolated to zero concentration.
[g] Earlier reported values are 43,000 (Seal and Doe, 1962 a) and 45,000 (Seal and Doe, 1962 b).
[h] From corticosterone content of complex. [i] From approach to sedimentation equilibrium. [k] Seal and Doe (1962 b).
[l] 14.1% found in CBG from estrogen-treated patients; 12.7% found in CBG from normal plasma (Seal and Doe, 1962 a).
[m] Nitrogen calculated on the basis of amino acid and carbohydrate composition: 13.0%. Biuret analysis showed that 75% of human CBG was polypeptide in nature.
[n] Calculated value: 12.0%.

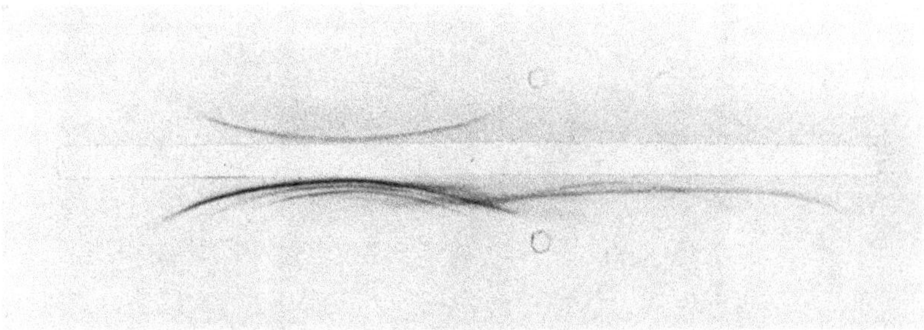

Fig. XI-17. Immunoelectrophoresis of pure rat CBG-corticosterone complex. Top well contains 8% CBG, bottom well normal rat serum. Rabbit antiserum to whole rat serum was added in center trough. High resolution Tris buffer, pH 8.9, 23°. From CHADER and WESTPHAL (1968 c)

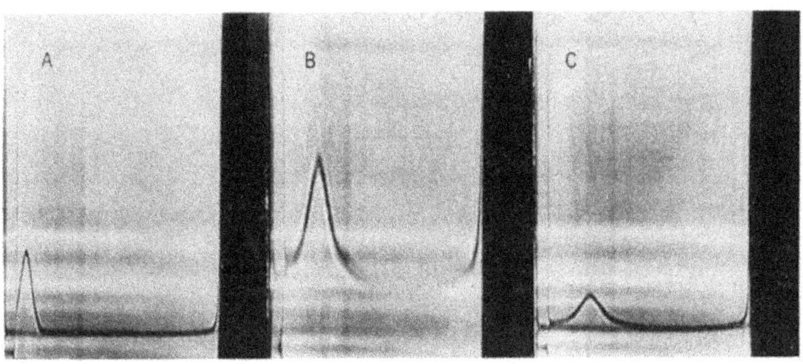

Fig. XI-18. Ultracentrifugal pattern of a 1% solution of rabbit CBG after 29, 88, and 104 min. The buffer was 0.05 M phosphate, pH 7.4; 4°; 59,780 rpm. Bar angle for A, 75°; B, 50°; C, 75°. From CHADER and WESTPHAL (1968 b)

with antisera to human, rat or guinea pig serum. The negative cross reactions of rabbit CBG with anti-human serum, and human CBG with anti-rabbit serum suggest that human CBG and rabbit CBG are immunochemically different. It should be noted, however, that these immunoelectrophoretic studies did not utilize the full potential of the immunological method, mainly because the available quantity of pure CBG was insufficient to permit its application as an antigen for preparation of CBG antiserum. Nevertheless, the results indicate distinct antigenic differences between human and rabbit CBG, as well

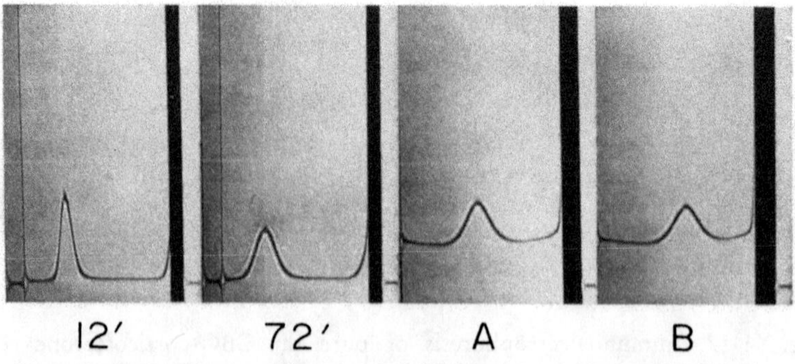

Fig. XI-19. Representative ultracentrifugal patterns of 0.9% solution of rat CBG-corticosterone complex in 0.1 M NaCl during synthetic boundary run after 12 and 72 min; 20°. A and B, sedimentation patterns 18 min (A) and 38 min (B) after speed was increased to 59,780 rpm. From CHADER and WESTPHAL (1968 c)

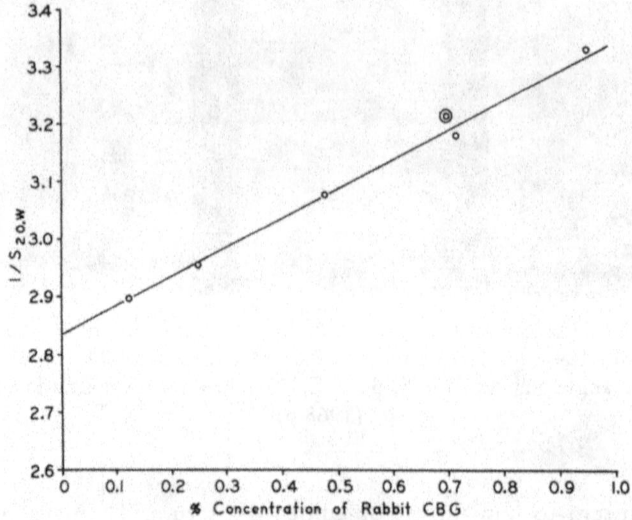

Fig. XI-20. Reciprocal $s_{20,w}$ values plotted against concentration for rabbit CBG-corticosteroid complex. Conditions for 1% concentration are given in Fig. XI-18, all other solutions were in 0.1 M NaCl; 20°. ○, Preparation I; ◎, Preparation II. From CHADER and WESTPHAL (1968 b)

as a probable lack of antigenic similarity of the two CBG proteins to any component of rat and guinea pig serum. Immunoelectrophoretic homogeneity was also found for the pure rat CBG (Fig. XI-17).

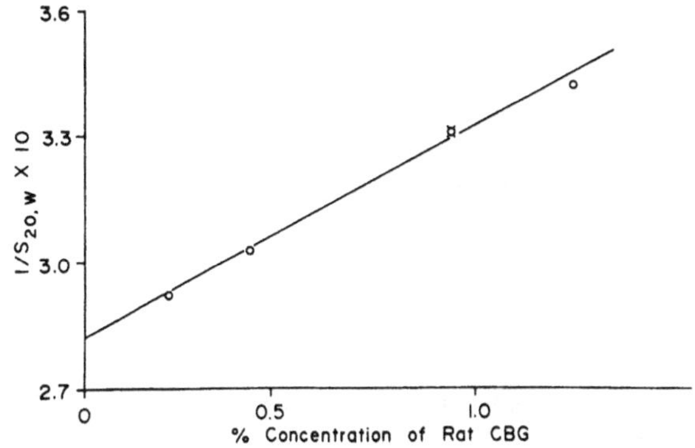

Fig. XI-21. Reciprocal $s_{20,w}$ values plotted against concentration for rat CBG-corticosterone complex in 0.1 M NaCl; 20°. Two runs at 0.9% concentration gave identical $s_{20,w}$ values. From CHADER and WESTPHAL (1968 c)

Ultracentrifugal analysis of the pure rabbit and rat CBG preparations showed symmetrical peaks of homogeneous macromolecules (Figs. XI-18, XI-19). Homogeneity was also indicated when the plots of reciprocal sedimentation coefficients against protein concentration gave straight lines (Figs. XI-20, XI-21). Diffusion studies with rabbit CBG also indicated homogeneity (see Fig. 11 in dissertation by CHADER, 1966).

3. Physicochemical Characterization

The physicochemical properties of human, rabbit and rat CBG are compiled in Table XI-3. Unless noted otherwise, all parameters were determined with the CBG-corticosteroid complexes as isolated. Ultracentrifugal analysis gave single homogeneous peaks in all cases; extrapolation of the sedimentation coefficients, $s_{20,w}$, to zero concentration (Fig. XI-8, XI-20, XI-21) resulted in practically identical values for rabbit and rat CBG, whereas that for the human glycoprotein was somewhat higher. The molecular weight of human CBG was determined in our laboratory from the sedimentation coefficient, diffusion coefficient, and partial specific volume to be 51,700. This is in good agreement with the value of 52,000 reported by SEAL and DOE (1966). An "apparent" molecular weight of 58,500 was obtained by SLAUNWHITE et al. (1966) who used microsedimentation equilibrium data. The value for the frictional ratio, f/fo, was calculated to be 1.42 for human CBG. This value is comparatively high (TANFORD, 1961) and may reflect a high degree of hydration owing to the high carbohydrate content of CBG; however, this conclusion is uncertain in view of lack of information on the shape of the glycoprotein. An even higher f/fo value of 1.78 has been reported for AAG which contains more than 40% carbohydrate (SCHULTZE and HEREMANS, 1966).

The molecular weight of rabbit CBG, on the basis of sedimentation and diffusion coefficients and partial specific volume, was found to be 40,700. This value is in good agreement with the amount of corticosteroid bound in the complex assuming one binding site per CBG molecule and 100% saturation. Steroid determinations on the isolated CBG indicated a content of 8.4 μg of corticosterone and 0.26 μg of cortisol per mg of protein. The accuracy of these values is limited because of some overlap in specificity of the methods applied; nevertheless, the total corticosteroid content is close to the theoretical quantity of 8.5 μg of corticosterone per mg CBG of 40,700 molecular weight. Complete corticosteroid saturation of the isolated rabbit CBG complex was verified by equilibrium dialysis experiments in which small increments of radiocorticoid were added in excess of the approximately 8.5 μg corticosteroid present per mg CBG. The consequence was a sharp increase of the concentration of unbound steroid indicating oversaturation of the binding sites. This result gives an independent confirmation of the molecular weight of 40,700 on the basis of the stoichiometry of the CBG-steroid complex (CHADER and WESTPHAL, 1968 b).

The molecular weight of the pure rat CBG-corticosterone complex was determined by the approach to sedimentation equilibrium method (SCHACHMAN, 1957); a value of 61,000 ± 1100 was obtained. Calculations at the meniscus and bottom of the cell (see Fig. 7 in original publication by CHADER and WESTPHAL, 1968 c) were in good agreement indicating apparent homogeneity of the isolated protein complex. Increasing the speed at the end of the synthetic boundary run resulted in the appearance of a single symmetrical descending boundary pattern and thus demonstrated the maintenance of homogeneity throughout the determination.

In view of the differences obtained in different laboratories for the molecular weight of presumably the same human CBG (Table XI-3), it was desirable to assess the molecular weight of rat CBG by a second, independent method based on steroid content and association constant calculations. The isolated rat CBG complex contained 6.6 μg of corticosterone per mg of protein. If one assumes a single binding site, n, and saturation, a molecular weight of approximately 53,000 would correspond to the observed corticosterone content in rat CBG. The approximate error arising from the steroid analysis would presumably be ±3000.

Evidence for validity of the assumption of one binding site and saturation is given by the following results. In each of the equilibrium dialysis experiments with the pure rat CBG-corticosterone complex, a total of 0.6 mg of protein was employed in association with 4.0 μg of corticosterone bound in the complex. Without further addition of steroid, or after addition of not more than 0.10 μg of radiolabeled corticosterone to the dialysis system, high and consistent association constants were obtained at 4° and 37°. Evidently, the number of available CBG binding sites was not exceeded. When greater quantities of corticosterone (0.20 and 0.40 μg) were added to 0.6 mg of the rat CBG complex in other dialysis experiments, the binding sites were exceeded and the fraction of unbound steroid rose markedly in a manner similar to that described above for rabbit CBG. This rapid increase of unbound steroid when the total corticosterone content of the complex was raised above 4.1 μg/0.6 mg indicated that the high-affinity binding sites were saturated at a value of $n = 1$ per mole of rat CBG of an approximate molecular weight of 53,000.

For a molecular weight of 61,000, the calculated amount of corticosterone bound at equimolar ratio would be 5.7 μg/mg of protein, whereas 6.6 μg was actually found. The following explanations are to be considered. (1) The result of the corticosterone determination is too high. This is not likely on the basis of our experience with the analytical method (Silber et al., 1958) over the years. (2) The number of high-affinity binding sites is greater than one. This assumption has been disproved in the preceding paragraph. (3) Rat CBG possesses, in addition to the one high-affinity binding site, a set of secondary binding sites of lower affinity. This property has been found (Kerkay and Westphal, 1968; 1969) for AAG in its interaction with several C_{21} steroid hormones (see Chapter XIII). This explanation is a possibility for rat CBG, although difficult to assess since the lability of the protein does not permit standard binding analysis by reciprocal or Scatchard plots. (4) The molecular weight of 61,000 determined by approach to sedimentation equilibrium is somewhat different from the size of the molecular species responsible for steroid binding; this latter entity corresponds to a molecular weight of approximately 53,000, which is compatible with the observed sedimentation coefficient of 3.56 S. Possible validation for this fourth explanation may be given by the polymeric nature of rat CBG to be discussed in Section XI. 5.

The explanation of the apparent differences in the molecular weights of rat CBG obtained by the two methods may also apply to the discrepancies in the molecular weights reported for human CBG (Table XI-3). It should be noted that the low molecular weight obtained for human CBG, i. e., 45,000 (Seal and Doe, 1962 b), was calculated from the cortisol content in a similar way as the lower value of 53,000 for rat CBG. The highest molecular weight observed for human CBG, i. e., 58,500 (Slaunwhite et al., 1966), was determined by an approach to sedimentation equilibrium technique similar to that which gave the higher value of 61,000 for rat CBG.

The electrophoretic mobility, determined by free boundary technique in barbital buffer at pH 8.6, was found to be —4.9 and -5.1×10^{-5} cm^2 $volt^{-1}$ sec^{-1} for human and rabbit CBG, respectively. This is well within the range of α_1-serum globulins (Schultze and Heremans, 1966). The absorbance values, $E^{1\%}_{1cm}$, are also similar to those observed for other serum proteins. Removal of corticosteroid from the CBG complexes results in increased absorbance ratios at 280 and 260 nm, $A_{280} : A_{260}$, because the Δ^4-3-ketosteroids at 260 nm have approximately 70% of their maximal absorption which for cortisol in aqueous solution occurs at 248 nm (Westphal, 1957 b).

Carbohydrate analysis of the three CBG preparations in our laboratory gave a total content of about 26—29%; 26.1% was found in human CBG. Since this amount is considerably greater than the 12.7 and 14.1% reported by Seal and Doe for human CBG from normal and estrogen-treated subjects, respectively (see Table XI-3), efforts were made to corroborate the results by an independent analysis of the polypeptide moiety. Kjeldahl determination gave 12.7% nitrogen, in good agreement with the value of 13.0% calculated on the basis of amino acid and carbohydrate composition given below in Table XI-5. Biuret analysis showed that human CBG contained 75% polypeptide, in conformity with the carbohydrate value of 26.1%. A similar indirect confirmation of the carbohydrate content by nitrogen analysis was obtained for rabbit CBG (column d in Table XI-3). We assume, therefore, that 26.1, 29.2 and 27.8% total carbohydrate determined for human, rabbit and rat CBG are of the correct order. The reasons for the approximately 50% lower carbohydrate values are not known; it should be noted, however, that the molecular weight and spectral data given in column a of Table XI-3 may be interpreted under the assumption that the CBG molecule has lost about half of its carbohydrate content. This is readily seen if one calculates the pertinent data for a CBG with 14.1% carbohydrate on the basis of the CBG with 26.1% carbohydrate (column c in Table XI-3) and compares these calculated figures with those actually found (column a in Table XI-3). The calculation is made under the assumption that the carbohydrate components do not contribute to light absorption at 279—280 nm.

	Human CBG	Human CBG with 14.1% carbohydrate	
	(Column c in Table XI-3)	Calculated	Found (Column a in Table XI-3)
Molecular weight	51,700	44,500	45,000
$E_{1cm}^{1\%}$ at 279 nm	6.45	7.49	7.40
$A_{280} : A_{260}$, complex	1.13	1.31	1.36
$A_{280} : A_{260}$, stripped	1.57	1.82	1.72

The amino acid compositions of human, rabbit and rat CBG are given in Table XI-4. The values have been calculated (columns d—f) for the carbohydrate-free polypeptide moiety for ready comparison with the data obtained for the human CBG preparations of low carbohydrate content (column b and c). Amino acid analyses performed in our laboratory on HCl and Pronase (protease from *streptomyces griseus)* hydrolysates of purified CBG gave results which were generally in good agreement. Values for methionine, valine, leucine, and half-cystine (column d) were calculated from analysis of a Pronase digest; all other values are from an HCl digest. It has been reported (SMYTH and ELLIOTT, 1964) that valine, leucine, and isoleucine are released slowly from proteins, especially when they appear in sequence, and that acid hydrolysis for 22 hours is sometimes not sufficient to liberate these residues completely. After Pronase digestion of human CBG, the percentages of valine and leucine obtained were, respectively, 4.91 and 8.84; the corresponding values after acid hydrolysis were 4.20 and 7.73. The higher values are considered more representative of the true content and are used in the calculation of partial specific volume. Values obtained for isoleucine by the two procedures were in good agreement (MULDOON and WESTPHAL, 1967).

Observed percentages of acidic amino acids and ammonia were appreciably higher after HCl hydrolysis than after Pronase digestion. It is assumed that carbohydrate is covalently bound in glycoproteins through N-glycosidic linkages with either asparagine or glutamine (WINZLER, 1960); evidence for a glycosylamine type linkage between aspartic acid and acetylglucosamine has been obtained recently for AAG (WAGH et al., 1969). The lower acidic amino acid and ammonia values observed after Pronase digestion were considered the result of incomplete hydrolysis in accordance with the known specificity of the streptomyces enzyme (NOMOTO et al., 1960); the percentages reported, therefore, are those obtained from acid hydrolysis. Losses of serine and threonine upon hydrolysis were compensated by the use of correction factors, as recommended by HIRS et al. (1954). After these corrections, the values from the two types of hydrolysis were in conformity.

There is fairly good agreement in the amino acid composition of our human CBG (column d in Table XI-4) and that prepared by SLAUNWHITE and coworkers (column c) when calculated on a polypeptide basis. The discrepancies with the values of SEAL and DOE (column b) are somewhat greater, although there is a definite compositional similarity in a general sense. Such general similarity is also evident in comparison with the amino acid composition of rabbit and rat CBG. Some of the percentage figures are almost identical in the three CBG preparations, and the general trend of higher and lower values for the different amino acids is clearly apparent in Table XI-4. However, there are distinct differences in the content of individual amino acids.

Table XI-4. *Amino acid composition of human, rabbit and rat CBG. Calculated for polypeptide moiety*

Amino acid	gm/100 gm CBG Polypeptide				
	CBG-Hu (b)	CBG-Hu[a] (c)	CBG-Hu (d)	CBG-Rb (e)	CBG-Rt (f)
Lysine	3.91	4.05	5.03	4.60	6.64
Histidine	3.75	3.14	3.29	2.89	2.57
Arginine	7.92	5.05	3.99	5.22	4.01
Aspartic acid	8.91	10.81	9.72	9.84	11.89
Threonine	5.41	6.05	5.73	6.70	6.13
Serine	5.62	6.42	6.15	6.21	6.26
Glutamic acid	13.38	12.36	12.77	13.33	14.42
Proline	7.40	5.12	5.48	5.31	4.29
Glycine	4.20	3.61	2.86	3.67	3.16
Alanine	5.49	4.94	4.34	5.12	4.26
Cystine/2	0.32	—	0.54	2.10	2.32
Valine	5.00	7.15	6.66	6.10	5.15
Methionine	1.20	2.41	3.52	1.51	2.70
Isoleucine	3.27	4.57	4.76	3.74	3.90
Leucine	12.97	12.94	11.98	10.05	10.02
Tyrosine	2.91	3.87	4.22	4.41	4.30
Phenylalanine	7.91	6.86	7.29	5.80	6.16
Tryptophan	2.18[g]	—	1.49[h]	3.39[i]	1.82[h]

[a] From estrogen-treated patients.
[b] SEAL and DOE (1962 a).
[c] SLAUNWHITE et al. (1966).
[d] MULDOON and WESTPHAL (1967).
[e] CHADER and WESTPHAL (1968 b).
[f] CHADER and WESTPHAL (1968 c).
[g] Determined by spectrophotometric method.
[h] Determined by method of OPIENSKA-BLAUTH et al. (1963).
[i] Average of values obtained by method[h] and method of BENCZE and SCHMID (1957).

Content variation is particularly marked for half-cystine which is about 4 times as frequent in rabbit and rat CBG compared to the human glycoprotein. Recalculation on a molar basis gives approximately 2, 6, and 9 moles of half-cystine per mole of human, rabbit, and rat CBG, respectively; the molar relationships are shown in Table XI-5 which gives the number of amino acid and carbohydrate

Table XI-5. *Amino acid and carbohydrate composition of human, rabbit, and rat CBG. Calculated as residues per glycoprotein from data in Table XI-4, columns d—f, and Table XI-6*

Residue	Residues per 100,000 g glycoprotein			Residues per mole CBG		
	CBG-Hu	CBG-Rb	CBG-Rt	CBG-Hu 51,700 [a]	CBG-Rb 40,700 [a]	CBG-Rt 52,600 [a]
Lysine	29.0	25.4	37.4	15	10	20
Histidine	17.7	14.9	13.5	9	6	7
Arginine	18.9	23.7	18.5	10	10	10
Aspartic acid	62.4	60.6	74.6	32	25	39
Threonine	41.9	46.9	43.8	22	19	23
Serine	52.1	50.5	51.9	22	21	27
Glutamic acid	73.1	73.1	80.6	38	30	42
Proline	41.7	38.7	31.9	22	16	17
Glycine	37.0	45.5	40.0	19	19	21
Alanine	45.2	50.9	43.3	23	21	23
Half-cystine	3.9	14.6	16.4	2	6	9
Valine	49.6	43.6	37.5	26	18	20
Methionine	19.8	8.2	14.9	10	3	8
Isoleucine	31.1	23.4	24.9	16	10	13
Leucine	78.2	62.9	63.9	40	26	34
Tyrosine	19.1	19.1	19.0	10	8	10
Phenylalanine	36.6	27.9	30.2	19	11	16
Tryptophan	5.9	12.9	7.1	3	5	4
Hexose	70.9	64.1	60.4	37	26	32
Hexosamine	55.8	59.0	59.0	29	24	31
Sialic acid	14.1	29.2	22.0	7	12	12
Fucose	10.3	5.5	14.4	5	2	8

[a] Molecular weight used for calculation

residues per 100,000 g and per mole of CBG. In agreement with results reported by SEAL and DOE (1966), a single titratable sulfhydryl group was observed (MULDOON and WESTPHAL, 1967) by the spectrophotometric method of BOYER (1954) in human CBG. This would suggest the presence of two cysteine residues, one of which is not readily accessible to mercaptide formation. No free sulfhydryl groups were found by the same technique in rabbit CBG (CHADER and WESTPHAL, 1968 b); the question remains open whether the rabbit glycoprotein does not contain thiol structures, or whether such groups are "buried" or otherwise inaccessible. It should be noted, however, that addition of HMB to an equilibrium dialysis system containing rabbit CBG and cortisol-4-^{14}C resulted in a 27-fold decrease of the association constant.

A major dissimilarity between the three CBG's was observed in the ratios of the biosynthetically related, sulfur-containing amino acids methionine and half-cystine. An inverse relationship is seen in

Table XI-6. *Carbohydrate composition of human, rabbit, and rat CBG, in %*

Component	CBG-Hu[a] (MULDOON and WESTPHAL, 1967)	CBG-Rb[b] (CHADER and WESTPHAL, 1968 b)	CBG-Rt[c] (CHADER and WESTPHAL, 1968 c)
Hexose	11.5	10.4	9.8
Hexosamine	9.0	9.5	9.5
Sialic acid	4.1	8.5	6.4
Fucose	1.5	0.8	2.1
Total	26.1	29.2	27.8

[a] The values reported by SEAL and DOE (1962 a) for hexose (equally divided between galactose and mannose), glucosamine, sialic acid, and fucose in purified CBG-Hu are 5.0%, 2.7%, 4.2%, and 0.8%, respectively. The reasons for the differences from the above values are not known.

[b] SEAL and DOE (1963) reported carbohydrate composition of rabbit CBG prepared by "small-scale purification" on hydroxylapatite. Recalculation by use of the $E^{1\%}_{1cm}$ value of Table XI-3 gave 11.3% hexose; 9.2% hexosamine; 7.1% sialic acid, and 0.4% fucose.

[c] In complete analogy to footnote [b], recalculation for rat CBG reported by SEAL and DOE (1963) gave 5.3% hexose, 15.3% hexosamine, 4.4% sialic acid and 0.3% fucose suggesting a lower degree of purity for this preparation.

the percentage of these two residues in human and rabbit CBG, whereas the content of these amino acids in rat CBG is almost equal (Tables XI-4 and XI-5). The increase of half-cystine from human to rabbit to rat CBG may be connected with the polymerization phenomenon to be described below (Chapter XI. 5).

Table XI-6 summarizes the carbohydrate composition of human, rabbit and rat CBG, in g per 100 g glycoprotein. It is apparent that the hexose and hexosamine content is very similar in the three preparations whereas sialic acid and fucose show marked differences among the three species. In confirmation of a postulate by Dische (1963) concerning mammalian glycoproteins in general, there seems to be an inverse relationship between sialic acid and fucose content so that their combined percentage shows less variation than the single constituents. The overall result is a total carbohydrate content for the three CBG's that differs less than ±6% (on basis of 100% carbohydrate) from an average value of 27.7%. Hexose and hexosamine in human CBG have been identified by Seal and Doe (1962 a) as mannose and galactose in a 1 : 1 molar ratio, and glucosamine, respectively. This compositional pattern agrees with that of many other extracellular mammalian glycoproteins, including AAG (Eylar, 1965).

The data in Table XI-6 are in accordance with the general experience that the two more variable carbohydrate residues, sialic acid and fucose, indicate species specificity, while the less variant hexose and hexosamine form the "backbone constituents" of the carbohydrate units in mammalian glycoproteins (Dische, 1963). Complete enzymatic removal of sialic acid from human CBG under mild conditions at 4° did not affect the cortisol binding affinity; the association constants of the cortisol complexes of the native and sialic acid-free glycoproteins (see Chapter XI. 4) were indistinguishable at 4° and 37° (Muldoon and Westphal, 1967). This finding suggests that CBG lability at pH levels below 5 is probably not the result of cleavage of the ketosidic carbohydrate linkage, as had been suggested (Seal and Doe, 1962 a). Thyroxine-binding globulin (Blumberg and Warren, 1961) and human AAG (Ganguly and Westphal, 1968 a) also retain complete binding activity for thyroxin and progesterone, respectively, after removal of sialic acid, indicating that this moiety may not be of general significance for hormone binding of proteins. This interpretation of the significance of sialic acid and possibly the whole carbohydrate moiety of CBG is analogous to the conclusions of Eylar (1965) concerning the relationship of the carbohydrate portion of other glycoproteins to their biological activity. The question whether the sialic acid in CBG has a special function in connection with the transport and transfer of the bound steroids to target cells, possibly nuclear receptor structures, has not been explored. Similarly, it is not known whether the carbohydrate constituents of the hormone-binding glycoproteins are involved in the general function of permeation through cellular membranes, acting as "chemical passports" (EYLAR, 1965). These and other possibilities await elucidation.

4. Steroid Binding Parameters

The binding parameters of the corticosteroid complexes of human, rabbit and rat CBG are compiled in Table XI-7. One primary binding site per molecule has been found in all cases. Some of the evidence for n = 1 has been discussed in the preceding section in connection with molecular weight determinations. Fig. XI-22 shows the analysis

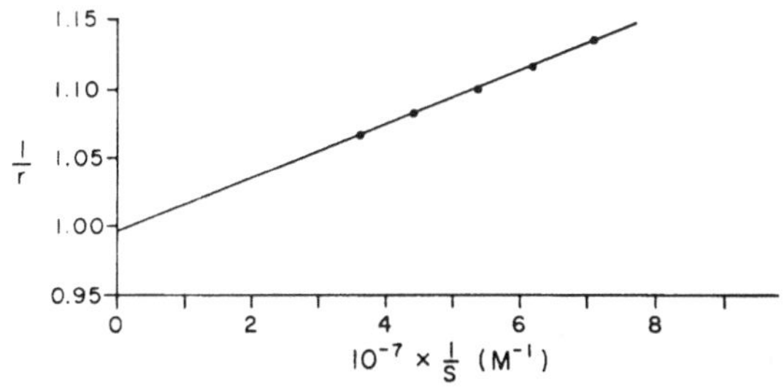

Fig. XI-22. Reciprocal plot analysis of cortisol-human CBG interaction at 4°. S, molar concentration of unbound steroid; r, moles of steroid bound per total moles of protein. Values were obtained from equilibrium dialysis studies with a constant CBG concentration of 0.19 mg per ml. The straight line was calculated by the method of least squares. From MULDOON and WESTPHAL (1967)

of the cortisol-CBG-Hu complex by the method of reciprocal plots; the n-value is slightly higher than one. The question of the presence of secondary binding sites of lower affinity, in addition to the one primary site, has not been explored yet. The association constants, k, of the steroid complexes of the pure CBG's confirm the strong temperature dependency observed with the binding systems in serum (Chapter VIII. 4 a); the affinity at 4° is approximately 20 times that at 37° for CBG of the three species studied (Table XI-7). The numerical values of k are in good agreement with those determined in the unfractionated sera and thus corroborate these earlier values (Table VIII-21). It appears, therefore, that under the conditions of the equilibrium dialysis studies with serum, the corticosteroid-binding affinity of CBG is not significantly influenced by the presence of other serum components. After removal of sialic acid from human CBG to non-detectable levels, the affinity constants for cortisol at

Table XI-7. *Binding parameters of corticosteroid-CBG complexes. References (in parentheses) below table*

Parameter	CBG-Hu [a] (c)	(d, e)	CBG-Rb [a] (e)	CBG-Rt [b] (f)
Number of binding sites, n	1	1	1	1
Association constant, k, 4° (M^{-1})	5×10^8	5.2×10^8	9.0×10^8	5.1×10^8
Association constant, k, 37° (M^{-1})	3×10^7	2.4×10^7	4.7×10^7	2.8×10^7
Free energy, ΔF°, 4° (kcal/mole)	—	—11.0	—11.3	—11.0
Free energy, ΔF°, 37° (kcal/mole)	—	—10.5	—10.9	—10.6
Enthalpy, ΔH°, (kcal/mole)	—	—15.7	—15.0	—14.8
Entropy, ΔS° (cal $mole^{-1}$ deg^{-1})	—	—17	—13	—14

[a] Cortisol complex.
[b] Corticosterone complex.
[c] Seal and Doe (1966). The association constants of the human CBG-progesterone complex determined by these authors at 4° and 37° are 3×10^8 and 8×10^7, respectively.
[d] Muldoon and Westphal (1967).
[e] Chader and Westphal (1968 b).
[f] Chader and Westphal (1968 c).

4° and 37° were found to be 4.9×10^8 M^{-1} and 3.0×10^7 M^{-1}, respectively. This result would indicate that sialic acid is not essential for the steroid-binding properties of CBG.

Observations with unfractionated serum have shown that rabbit CBG differs from human CBG in that its binding affinity increases with the number of polar groups in the steroid molecule (Table VIII-21). This is contrary to the polarity rule. Equilibrium dialysis studies with pure rabbit CBG in the presence of small amounts of radiolabeled steroids at 4° and 37° confirmed the relative binding affinities, which decreased from cortisol > corticosterone > progesterone ≫ aldosterone (Chader and Westphal, 1968 b). Aldosterone may not be directly comparable with the other C_{21}-steroids because of its predominant hemiacetal form.

The binding parameters of the pure CBG preparations (Table XI-7) were determined with the isolated corticosteroid complexes. Removal of the bound steroid by gel filtration at 45° resulted in irreversible inactivation of human CBG as shown by equilibrium dialysis at 4° and 37°; the protein had lost 97% of the associated cortisol. In the case of rabbit CBG, only 48% of the steroid became unbound under the same conditions of filtration over Sephadex G-50; subsequent equilibrium dialysis studies indicated loss of about half of the binding activity by the stripping procedure. Attempts at removing the steroid from the isolated rat CBG-corticosterone complex for a more complete determination of the association constant were unsuccessful. Dissociation of the complex by Sephadex G-50 gel filtration at 45° resulted in complete loss of steroid-binding activity even though such a procedure applied to whole serum removed the steroid without reduction of CBG activity (WESTPHAL, 1967). Filtration of the rat CBG-corticosterone complex over Sephadex G-75 at 23° caused removal of 87% of the steroid and an apparent loss of about 60% of the binding affinity. An association constant of 1.20×10^7 M^{-1} was obtained at 37° with this protein. However, after reconstitution of the steroid-protein complex by readdition of corticosterone (CHADER and WESTPHAL, 1968 c) an association constant of 2.03×10^7 M^{-1} was calculated from dialysis data obtained at 37°. This result shows the reversible nature of the inactivation. A molecular weight of 53,000 was used in these determinations involving rat CBG. The polymerization effect observed after removal of corticosterone from the rat CBG complex will be discussed in Section XI. 5.

Thermodynamic data for the interaction of human, rabbit and rat CBG with corticosteroid hormones are included in Table XI-7. It is evident that the very high values of about 11 kcal per mole for the Gibbs free energy of the complexes is composed of a very high negative enthalpy change, in association with a negative change of entropy. These thermodynamic parameters are interpreted as indicative of a very tight fit of the interacting components, so that the enthalpy is drastically reduced; the total order is much increased, so that a negative entropy change results. As seen in Table VI-11 this is in contrast to the positive entropy changes observed with the progesterone and testosterone complexes of serum albumin. The positive ΔS° values observed for the albumin interactions may be explained by the assumption that the highly structured hydration water is displaced and randomized as solvent water in the complex formation, as discussed in Chapter VI. 5 e. This effect is overcome in the CBG complexes by a far greater stabilization as indicated by the high negative value of the enthalpy change.

5. Observations on Polymerization of Rat CBG

As has been pointed out in the preceding section, the rat CBG-corticosterone complex loses 87% of the steroid, and about 60% of its binding affinity, when it is filtered over Sephadex G-75 at 23°. This inactivation is a reversible reaction. The boundary pattern of the intact CBG-corticosterone complex in each of five sedimentation velocity runs in the ultracentrifuge was found to be symmetrical for concentrations ranging from 1.4 to 0.2%. Fig. XI-23 shows representative exposures taken during the development of the patterns of a 0.9 and a 0.2% solution of the CBG-corticosterone complex in the double-sector cell.

An entirely different type of pattern was obtained when rat CBG was subjected to ultracentrifugation after removal of the bound corticosterone. Fig. XI-24 shows the sedimentation record of the rat CBG preparation described above from which 87% of the steroid had been removed by gel filtration at room temperature. After 77 minutes, four peaks, one major and three minor ones, can be distinguished. The sedimentation coefficients were calculated to be 3.4, 5.4, 6.8, and 8.1, respectively. After prolonged centrifugation, a fifth species appears as a shoulder on the trailing edge of the 3.4 S peak. When this CBG preparation, after recovery from the ultracentrifuge cell, was recombined with approximately 1 mole of corticosterone/53,000 g of protein, the sedimentation pattern changed to that of the original CBG-corticosterone complex (Fig. XI-25). One homogeneous peak was observed with a sedimentation coefficient of 3.4 S, a value in agreement with that found for the original CBG-corticosterone complex at a similar concentration. This phenomenon of reversible polymerization was reproducible; the same polymeric and monomeric sedimentation patterns, respectively, were obtained in a second

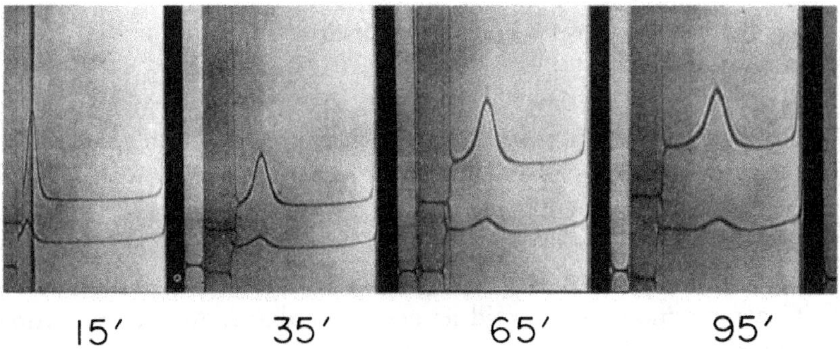

Fig. XI-23. Ultracentrifugation patterns of 0.9 and 0.2% solutions of rat CBG-corticosterone complex in 0.1 M NaCl; 20°; 59,780 rpm. Bar angles, 70, 70, 60, and 55°. From CHADER and WESTPHAL (1968 c)

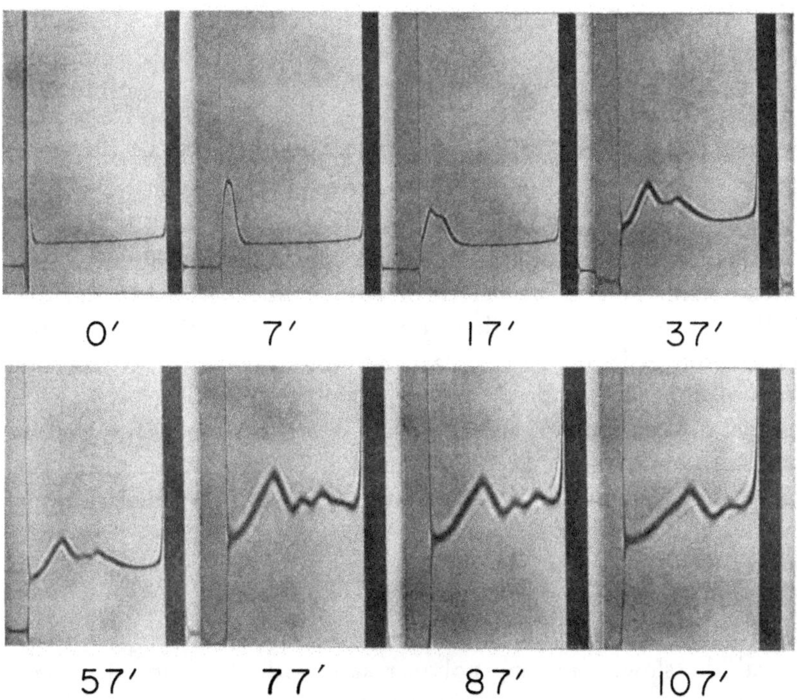

Fig. XI-24. Ultracentrifugation patterns of rat CBG after removal of 87% of the associated corticosterone at various times after speed of 59,780 rpm was reached. 0.1 M NaCl; 20°. Bar angles up to 17 min, 70°; at 37 and 57 min, 50°; remaining exposures, 30°. From CHADER and WESTPHAL (1968 c)

experiment in which rat CBG was stripped and later recombined with corticosterone.

The polymerism observed when steroid-depleted rat CBG was analyzed in the ultracentrifuge, was confirmed by disc gel electrophoresis studies. Fig. XI-26 shows the polymeric pattern developed when samples of rat CBG were subjected to electrophoresis in polyacrylamide gel at room temperature. Tube 4 was a sample of the isolated CBG-corticosterone complex, whereas tube 9 contained rat CBG after removal of steroid as described above. Anodic migration is toward the bottom of the tubes. Microdensitometer tracings are given beside the pictures. The stripped glycoprotein (tube 9) demonstrated at least four distinct bands in a regular stepwise pattern with a somewhat diffuse band (shoulder on microdensitometer tracing) constituting the fastest moving component. By analogy, these bands may correspond to the 8.1 S, 6.8 S, 5.4 S, and large 3.4 S peaks seen in the ultracentrifuge (Fig. XI-24) with the slowly sedimenting shoulder corresponding to the small, fast-moving electrophoretic species.

The ratios of the migration rates of the electrophoretic bands to those of their clearly separated faster migrating neighboring bands for tube 9 were 1.19, 1.21, 1.28, and 1.32; they were 1.26 and 1.32 for tube 4. These ratios are within $\pm 5\%$ in agreement with the theoretical value of relative velocities postulated by Marinis and Ott (1965) on the assumption that each electrophoretic species is followed by a dimer of its own molecular size. According to this concept, the ratio of dimer to monomer velocity in polyacrylamide gel is $1 : \sqrt[3]{2}$ so that each band should migrate with a rate approximately 1.26 times greater than that of the following protein species (Schultze and Heremans, 1966). The average value for the individual ratios of migration rates for each two neighboring, clearly separated bands in tube 9 (Fig. XI-26) is 1.25, in close agreement with the ratio of 1.26 calculated for dimerization.

The isolated complex (tube 4) before removal of steroid does not reveal the slowest moving polymer seen with the stripped protein. The two discernible slow-moving bands appear less concentrated than their counterparts in tube 9; instead, the main band and the fastest moving component contain practically all of the protein. The ratios of migration rates of the three clearly separated bands are 1.26 and 1.32, again in close agreement with the theoretical value of 1.26 for

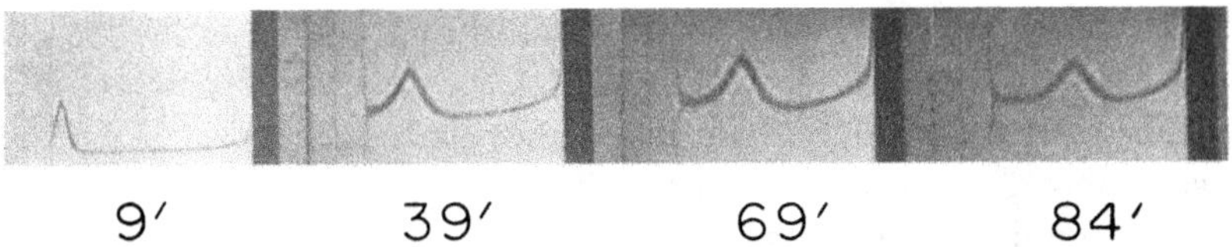

Fig. XI-25. Ultracentrifugation pattern of rat CBG used for study shown in Fig. XI-24 after recombination with corticosterone, at various times after speed of 59,780 rpm was reached. 0.1 M NaCl; 20°. Bar angles were 70, 50, 40, and 40°. From CHADER and WESTPHAL (1968 c)

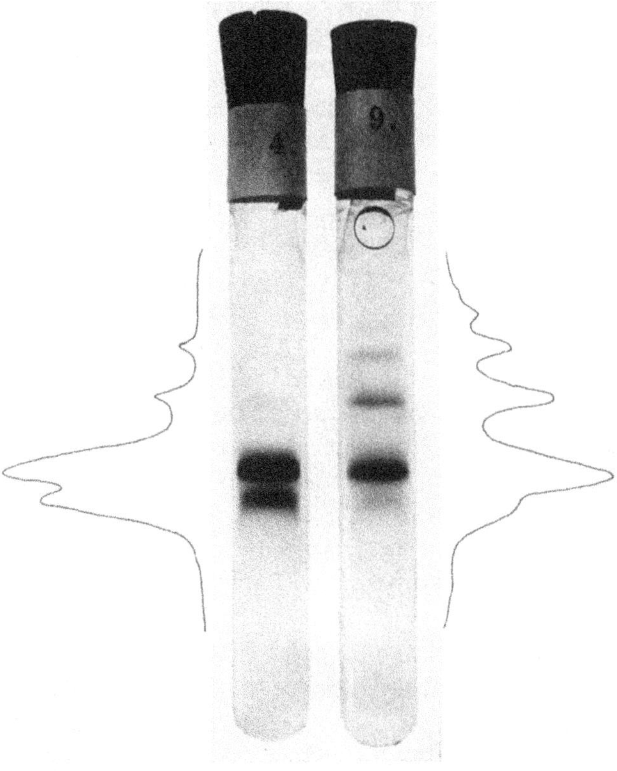

Fig. XI-26. Disc electrophoresis patterns with microdensitometer curves of rat CBG samples. Tube 4, isolated CBG-corticosterone complex; tube 9, the same complex after removal of 87% of the steroid. The peaks of the absorption curves, drawn from projections of the microdensitometer tracings, coincide accurately with the protein discoids in the original tracings. From CHADER and WESTPHAL (1968 c)

dimer formation. The presence of steroid thus appeared to inhibit the formation of the slower moving polymeric forms and enhance the concentration of the smaller, faster moving components. Since the electrophoresis was conducted at room temperature under heat-producing conditions resulting in temperatures of about 35—40° (Davis, 1964), it is assumed that the polymeric pattern observed with the original isolated complex can essentially be ascribed to loss of steroid from the protein due to dissociation during electrophoresis; the particularly high-temperature dependency of the rat CBG-corticosterone complex is well known and has been described in Chapter VIII. 4 a. Polyacrylamide electrophoresis also affords separation of interacting species by gel filtration, thus removing steroid from protein; this would further enhance the dissociation of the steroid-protein complex.

The readiness of rat CBG to form polymeric structures after removal of corticosterone may well explain the apparent discrepancy between the molecular weights found by the approach to sedimentation equilibrium, and that calculated from the corticosterone content. The ultracentrifugation was performed at 20°. Since rate and degree of dissociation of the rat CBG-corticosterone complex increases rapidly with temperature above 4° (Westphal, 1966), it is possible that dissociation of corticosterone and subsequent partial polymerization occurred during the ultracentrifugation. This could result in the calculation of a somewhat enhanced molecular weight if one assumes a rapid dynamic equilibrium between monomer and polymer in the single peak observed (Figs. XI-19 and XI-23). Some of the differences in the reported molecular weights of human CBG which were detailed above, may have a similar basis.

It has been pointed out in Section 3 of this chapter that of the three CBG preparations isolated in pure form, the rat glycoprotein contains the highest amount of half-cystine, and human CBG the lowest (Table XI-5). These differences in content of the amino acid which is intimately involved in the maintenance of conformational structure of proteins may be related to the lower or higher polymeric tendency observed in these glycoproteins. Human CBG apparently does not readily polymerize (Seal and Doe, 1962 b; Muldoon and Westphal, 1967). Indications of polymerism have been obtained for rabbit and guinea pig CBG which will have to be further investigated.

The polymerization of the steroid-free rat CBG can be reversed by recombination with one mole corticosterone. This phenomenon constitutes a first example of the control of the conformational structure of a specific carrier protein by the hormone with which it forms a complex of high affinity. Possible physiological implications for the function of the binding protein in hormone transport through membrane structures are evident and await investigation.

It is apparent from the discussion given in the present chapter that our knowledge of the chemical properties of CBG is scant—although more results have been obtained for this glycoprotein than for any other of the high-affinity steroid binders in the mammalian organism. The main reason for the slow progress is the technical difficulty of preparing the pure protein in sufficient quantity for extensive investigation. This effort is hampered by its small concentration in the blood where it occurs in mixture with the large amount of other serum proteins, and by its lability, which becomes greater with increasing purity. Novel approaches to the fractionation methods may be required, incorporating higher specificity of separation; the principle of affinity chromatography (JENSEN et al., 1967; BURSTEIN, 1969; LINDNER et al., 1970) appears promising.

XII. High-Affinity Binding of Estradiol and Testosterone to Serum Proteins

1. Binding of Estrogens

The problem of estrogen binding to serum proteins has been studied intensively by ROBERTS and SZEGO (1953, 1955), SZEGO and ROBERTS (1953), and SZEGO (1957). The authors reported initially (SZEGO and ROBERTS, 1946; ROBERTS and SZEGO, 1946) that two-thirds of the total estrogens in human blood was associated with protein, especially with COHN fraction III-0 containing lipid-poor β_1-globulins of 7 S and 20 S, in addition to β_1-lipoproteins (COHN, 1947). These early findings, including the estrogens' capability of dialyzing quantitatively, seemed to suggest a reversible binding system similar to that of other steroid-protein interactions. Later studies in the rat (SZEGO, 1953, 1955; SZEGO and ROBERTS, 1956; SZEGO, 1960) led to the assumption that the liver was involved in the binding of estrogens to proteins, and an enzymatically catalyzed biosynthesis of "estroprotein" was postulated. RIEGEL and MUELLER (1954) characterized an enzyme system from rat liver which catalyzed the formation of a protein-bound metabolite of estradiol-16-^{14}C. However, other findings (BISCHOFF and STAUFFER, 1957; SANDBERG et al., 1957; BISCHOFF et al., 1958) did not support the participation of the liver in the origin of the estrogen-protein complexes and assumed albumin to be mainly responsible for estrogen binding in serum. One of the complicating factors in the assumption of liver involvement in the binding of estrogens to protein is the formation of conjugates, especially glucosiduronates (SZEGO, 1957). These and other metabolites make it difficult to interpret the association of the radiolabeled products with the electrophoretically separated serum protein components (SZEGO and ROBERTS, 1956; SZEGO, 1960).

The problem has been clarified considerably by more recent results of HORWITZ et al. (1962) and BROOKS et al. (1963 b). These investigators incubated radiolabeled estrone with rat liver tissue or fractions thereof, and identified the products after electrophoretic separation into three radioactive peaks (origin; albumin peak; second peak migrating faster than albumin). The material at the origin was found to be unconjugated estrone, 2-hydroxyestrone and an unknown metabolite. The radioactivity associated with the albumin peak was shown to consist of the glucosiduronates of the same three compounds, in addition to the glucuronic acid derivative of 2-methoxyestrone. Other conjugates included probably the sulfates of 2-hydroxyestrone and of other more polar steroids; the presence of a phosphate conjugate of unknown structure in the second peak was suggested.

The problems connected with estrogen binding to serum proteins have been actively pursued during recent years. As has been discussed in Chapter VI. 5, serum albumin interacts strongly with estradiol and other estrogenic hormones, and also with the conjugated compounds. This was demonstrated *in vitro* by use of the "overtaking electrophoresis" (Chapter IV. 4) for estrone, estradiol and estriol under conditions suggesting spontaneous association without participation of liver enzymes. The three estrogens, especially estriol, were reported bound also to pure bovine γ-globulin (STRUCK, 1963).

Evidence has been accumulating to show that albumin is not the only serum protein capable of binding estrogens (TAVERNETTI et al., 1964, 1967). ROSENBAUM et al. (1966 a) equilibrated 0.39 ng tritiated estradiol (14.8 c/mmole) with one ml of human plasma and separated the mixture by paper electrophoresis. Fig. XII-1 shows the result for the plasma of a normal woman; the major portion of the radioactivity is associated with the β-globulin fraction, the remainder with albumin. Addition of 1.08 μg unlabeled estradiol to a similar system resulted in a distribution favoring albumin and α_2-globulin in addition to β-globulin, indicating a limited capacity of the strongly binding β-globulin. Other competition experiments demonstrated a lesser displacement effect of estrone; estriol in an equally high excess was without influence on the binding of estradiol to the β-globulin component.

The results indicate the presence in plasma or serum of a β-globulin capable of interacting with estradiol with high affinity but low capacity. The association with estrone was somewhat weaker. Repeated freezing and thawing had no effect on the binding activity; the binding protein was also stable to dialysis and to storage for at least one year in the frozen state at —20°. Normal female subjects had a higher plasma activity than normal males; even higher relative binding to β-globulin was found in the plasma of males suffering from hepatic cirrhosis, not explained solely by lack of binding competition of albumin (ROSENBAUM et al., 1966 a; TAVERNETTI et al., 1967). The observations concerning a β-globulin as estrogen-binding protein may throw new light on the early findings of estrogen association with

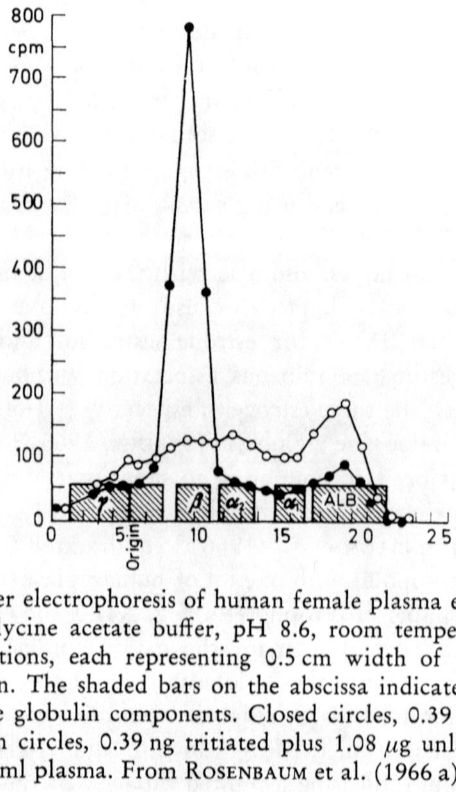

Fig. XII-1. Paper electrophoresis of human female plasma equilibrated with ^{3}H-estradiol. Glycine acetate buffer, pH 8.6, room temperature. Abscissa, number of fractions, each representing 0.5 cm width of paper; ordinate, cpm per fraction. The shaded bars on the abscissa indicate the location of albumin and the globulin components. Closed circles, 0.39 ng ^{3}H-estradiol/ml plasma; open circles, 0.39 ng tritiated plus 1.08 μg unlabeled estradiol/ml plasma. From ROSENBAUM et al. (1966 a)

fraction III-0 (ROBERTS and SZEGO, 1946; MIGEON et al., 1960), fraction III (SANDBERG et al., 1957), and fraction IV-1 (ANTONIADES et al., 1957 a; SANDBERG et al., 1957; MIGEON et al., 1960) of COHN's ethanol fractionation scheme.

In a recent publication, ROSNER et al. (1969) have reported the partial purification of estrogen-binding globulin from human plasma. A globulin fraction, obtained by ammonium sulfate precipitation, was chromatographed repeatedly on DEAE-cellulose. Subsequent filtration over Sephadex G-200 yielded two components, called A and C which on gel filtration showed molecular weights of 420,000 and 135,000 respectively. The binding protein had low stability; a variable loss of activity was observed when the protein was in solution.

Both of these binding proteins were β-globulins. Preliminary determination of the binding parameters by equilibrium dialysis at 4° gave two types of estradiol-binding sites for component A, with $k_1 = 1.7 \times 10^9 M^{-1}$ and $k_2 = 0.12 \times 10^9 M^{-1}$; component C had $k_1 = 4.0 \times 10^9 M^{-1}$; $k_2 = 0.69 \times 10^9 M^{-1}$, and a third type with $k_3 = 0.11 \times 10^9 M^{-1}$. It should be stressed that these values cannot be considered definitive before the purification of the globulins is completed. Nevertheless, it is noteworthy that the binding affinity appears to be of a similar magnitude as that of CBG for corticosteroid hormones.

When whole human serum was filtered over the calibrated Sephadex column, a MW of 115,000 was estimated for the binding protein. ROSNER et al. (1969) assume this value to correspond to the 135,000 weight found for the purified material. It is of particular interest that the 420,000 MW component was detected only after removal from the complex of most of the estradiol by DEAE-cellulose chromatography. The authors assume aggregation of the monomeric species, possibly to a trimer, and suggest that estradiol might be the allosteric determinant for its own binding. These observations and their interpretation are clearly analogous to our findings of the polymerization of rat CBG after removal of cortisol as discussed in Chapter XI. 5.

It must be mentioned at this point that numerous studies in recent years have led to the conclusion that the estradiol-binding globulin is identical with the serum protein which binds testosterone with high affinity. It seems appropriate, therefore, to discuss presently the investigations on testosterone binding, and then return to the steroid-binding β-globulin which appears to have high affinity for both male and female sex steroid hormones.

2. Binding of Testosterone

The work of DAUGHADAY, published in 1958, had clearly shown an increasing steroid-binding affinity of serum proteins with decreasing concentrations of several steroid hormones. The Fig. VIII-1, summarizing these results, revealed a high-affinity and low-capacity binding system for testosterone. As in the case of progesterone (Chapter VIII. 1), this observation was largely ignored; in retrospect, it is fair to say that recognition and investigation of the testosterone-binding protein would have come years earlier if DAUGHADAY's findings would have been fully evaluated.

The relatively non-polar testosterone forms a fairly strong complex with serum albumin (Chapter VI. 5 a, Table VI-9), and the study of this association predominated for many years. However, DAUGHADAY (1958 b) found by equilibrium dialysis that COHN's human plasma fraction IV-4, consisting mainly of α- and β-globulins, bound testosterone more firmly than HSA did. In subsequent years, isolated observations were published which also indicated the presence of (a) serum protein(s) other than albumin with a high binding affinity for testosterone. This was demonstrated by equilibrium dialysis (SLAUNWHITE and SANDBERG, 1959), ultrafiltration (CHEN et al., 1961) and gel filtration (DE MOOR et al., 1962). The latter authors reported displacement of cortisol from CBG by competition with an equal amount of testosterone; this would suggest the ability of CBG to bind testosterone.

It should be noted that the published data of CHEN et al. (1961) show a relatively high capacity for testosterone-binding protein(s) other than albumin; more than 190 μg testosterone were added per 100 ml plasma without saturation of the binding sites, i.e., without decrease of the binding affinity to the level of HSA. This result has been confirmed by LAYNE (1962). Such binding capacity clearly is higher than that of the testosterone-binding globulin (see below this section), and does not correspond to DAUGHADAY's early observations of a high affinity-low capacity binder. It seems likely that the binding protein observed by CHEN et al. (1961), different from albumin and testosterone-binding globulin, is AAG (see Chapter XIII). This glycoprotein binds testosterone with an affinity about 20 and 5 times that of HSA, at 4° and 37°, respectively (see below Table XII-3). Its concentration in normal human serum is approximately 750 mg/l or 18×10^{-6} M. Since AAG has one binding site for testosterone (KERKAY and WESTPHAL, 1968), 518 μg testosterone per 100 ml would be required to saturate the AAG binding sites.

It was not until 1965 that MERCIER et al. (1966) reported results of a systematic investigation of a specific testosterone-binding serum protein and launched the appearance of numerous publications on the subject from various laboratories. BAULIEU and his coworkers separated the components of human plasma by DEAE-cellulose chromatography and obtained a β-globulin fraction with a binding affinity for testosterone higher than that of HSA. PEARLMAN and CRÉPY (1966, 1967 a) postulated the presence of a testosterone-binding protein in human serum with a specific affinity for testosterone considerably higher than that of HSA. The authors used a novel technique of gel equilibration with Sephadex G-25 in a batchwise fashion on a semimicro scale; this method (PEARLMAN and CRÉPY, 1967 b) has been described in Chapter IV. 15. Normal menstruating women were found to have approximately twice the testosterone-binding activity as male subjects. Considerably higher binding levels were observed in late pregnancy.

Exposure of the sera to 60° for one hour destroyed the high-affinity binding for testosterone, whereas the interaction with HSA was little affected. Heat inactivation of the testosterone-binding protein started with one hour exposure to 40° and above, increased at higher temperatures and was practically complete at 60—65°. The thermolability of the binding protein was confirmed in other laboratories: heating of plasma to 57° (VERMEULEN and VERDONCK, 1968 a) or to 60° (STEENO et al., 1968; HORTON and KATO, 1968) destroyed the binding ability; activity was retained up to 45° (VERMEULEN and VERDONCK, 1968 b). Heat inactivation studies for plasma protein binding of testosterone, dehydroepiandrosterone and androstenedione over a range from 37° to 100° have been reported by FOREST et al. (1968); at 60° a maximal effect was obtained after 30 min. The binding values determined at 4° were higher than those found at 37° although the binding properties of the plasma proteins did not change during 15 hrs incubation at 37°. Binding of testosterone was fairly constant at pH values of 6 to 8; lower binding occurred at pH 5, and the interaction was sharply reduced at pH 2.1. Loss of testosterone-binding activity at pH <5 was also observed by HORTON and KATO (1968).

A more detailed study of the testosterone-binding activity in the serum of women (PEARLMAN et al., 1967) showed low and relatively constant values during the normal menstrual cycle. A marked increase occurred in the first trimester of pregnancy, continued gradually, and reached at term a maximal value of 2—3 times the normal level; the binding levels returned rapidly to normal shortly before or after delivery. In parallel, the peripheral testosterone concentration has been found to rise in late pregnancy, and to return to normal after delivery (DEMISCH et al., 1968), or shortly before parturition (RIVAROLA et al., 1968). The increase of the testosterone-binding activity in human pregnancy resembles that observed in the pregnant guinea pig as discussed in Chapter X.

It is not known whether the elevation of the testosterone binding level in pregnancy is the result of increased estrogens or progesterone. A role of estrogenic hormones is suggested by the results of administration of estrogens to patients with prostatic cancer which was followed by a rise of the testosterone-binding activity (PEARLMAN et al., 1967). Women on oral contraceptives also showed elevated binding although not as high as that of pregnant women at term (AUGUST et al., 1969). These changes under various physiological and experimental conditions closely resemble the alterations in the CBG levels observed in comparable situations. However, the proteins responsible for high-affinity binding of testosterone differ from CBG not only in their electrophoretic properties, but they can be separated from CBG by DEAE-cellulose or Sephadex G-200 (GUÉRIGUIAN and PEARLMAN, 1967). Further fractionation studies established the β-globulin nature of the active protein (GUÉRIGUIAN and PEARLMAN, 1968); assessment of the molecular sizes by gel filtration gave values for the Stokes molecular radius of about 47 Å and 36 Å for the testosterone-binding globulin and CBG, respectively.

MERCIER and BAULIEU (1968) observed high-affinity binding for testosterone in COHN's fraction IV, in confirmation of the early observations of DAUGHADAY (1958 b). The binding protein was partially purified by chromatography and separated from albumin; contaminating proteins present, i.e., transferrin, hemopexin, β_2-glycoprotein and an immunoglobulin, did not show testosterone-binding activity. PIZARRO (1969) utilized precipitation by ammonium sulfate, Rivanol (6,9-diamino-2-ethoxyacridine lactate), and polyethylene glycol, in addition to chromatographic procedures, for a partial purification of the testosterone-binding globulin.

It should be mentioned that a first attempt was made (BURSTEIN, 1969) to separate the testosterone-binding globulin from other plasma proteins by affinity chromatography. Covalent bonding of 3β-amino-androstan-17β-ol to the polymeric saccharide agarose (Sepharose) by cyanogen bromide reaction produced an insoluble steroid complex which had the 17β-hydroxy group intact; this substituent in ring D appears to be the most important part of the steroid structure for interaction with the testosterone-binding globulin (see below, Table XII-2). Approximately 75% of the binding activity seemed to be removed from the plasma when filtered over the steroid-agarose complex; however, attempts to dissociate and elute the protein resulted in inactivation so that only 0.4% of the active protein was recovered.

The observation of increased binding capacity of the testosterone-binding β-globulin in the pregnant woman has been confirmed in several laboratories (MERCIER et al., 1967; MURPHY, 1968 b; STEENO et al., 1968; DRAY, 1969). ROSNER and DEAKINS (1968) found increased activity also in men receiving diethylstilbestrol. FOREST et al. (1968) observed in addition that 6α-methyl-17α-acetoxyprogesterone (Provera®) administered to male patients lowered the binding values considerably. Plasma of male patients with liver cirrhosis had higher binding values than normal male plasma (MURPHY, 1968 b); such patients showed also augmented estradiol binding (ROSENBAUM et al., 1966 a, 1966 b), presumably involving the same binding globulin. It would be interesting to explore whether this increase is related to the elevated estrogen levels observed in cirrhotic patients.—High-affinity binding of testosterone could not be demonstrated in human seminal plasma (STEENO and DE MOOR, 1969).

DEAKINS and ROSNER (1968) found by paper electrophoresis in glycine-acetate buffer at pH 8.6 that in plasma of normal women 81% of ^{3}H-testosterone, added as trace amount, migrated with the β-globulin, 7% with the inter-alpha globulin, and 4% with albumin. The corresponding percentages for plasma of normal men were 69%, 14% and 6%, respectively; for pregnant women, about 96%, 2% and 1%; for diethylstilbestrol-treated men, 92%, 3% and 2% (ROSNER and DEAKINS, 1968). The possibility that "inter-alpha globulin" binding is given by AAG has not been investigated. The percentage binding of testosterone in human plasma, determined at 37° by equilibrium dialysis or ultrafiltration, gave about 92%, 95% and 99% for normal males, normal females, and pregnant subjects, respectively (FOREST et al., 1968); careful adjustment of total testosterone concentration in the test systems showed that the difference in binding activity between normal men and women was statistically significant. VERMEULEN and VERDONCK (1968 a) have reported some approximate data which may serve to assess the order of magnitude of plasma concentration and testosterone-binding affinity in normal, estrogen-treated and pregnant humans (Table XII-1).

Early specificity studies by MERCIER (1966), in which plasma was incubated with radioactive steroids and separated by DEAE-cellulose chromatography, showed that 5β-androstan-17β-ol-3-one was also bound to the β-globulin fraction. Androst-4-ene-3,17-dione and cortisol did not show binding to the testosterone-binding β-globulin under

Table XII-1. *Approximate concentration of binding sites and association constant of the testosterone-binding β-globulin.* Data condensed from VERMEULEN and VERDONCK (1968 a)

Subjects	Concentration $M \times 10^8$	Association constant at 25° $M^{-1} \times 10^{-8}$	Association constant at 37° $M^{-1} \times 10^{-8}$
Normal [a]	4— 6	4—16	—
Estrogen treated [a]	19—31	1—12	—
Pregnant, I[st] Trimester	7—14	7—15	—
Pregnant, IInd Trimester	14—36	6—10	—
Pregnant, IIIrd Trimester	19—30	4—10	—
Plasma Pool I	6.3	15	7.8
Plasma Pool II	5.9	13	4.5
Plasma Pool III	7.1	16	7.4

[a] No significant difference was observed between males and females. DRAY (1969) reported 5.5×10^{-8} and 8.0×10^{-8} M for normal male and female serum, respectively, whereas PEARLMAN (1970) found approximately 1×10^{-7} M and 8×10^{-7} M in adult male and in late pregnancy serum, respectively.

these conditions. Ability to displace testosterone from the high-affinity binding sites in human plasma was demonstrated by HORTON et al. (1967) for androstane-3α,17β-diol and for estradiol-17β; negative results were obtained with cortisol, corticosterone, progesterone, androstenedione, dehydroepiandrosterone, epitestosterone, androsterone, etiocholanolone, estrone, and the 17-acetate or the 3-thiosemicarbazone of testosterone. In contrast, ROSNER and DEAKINS (1968) observed displacement by dehydroepiandrosterone, and also by 17α-hydroxyprogesterone. Using partially purified plasma fractions, MERCIER and BAULIEU (1968) confirmed some of these results and found that 17α-methyltestosterone was bound with equal or even higher

Table XII-2. *Relative ability of various steroids to displace ^{3}H-testosterone from testosterone-binding globulin*

Displacing steroid	VERMEULEN and VERDONCK (1968 a)	KATO and HORTON (1968)
5α-Androstan-17β-ol-3-one	300	240
6α-Methyl-5α-androstan-17β-ol-3-one	—	189
5α-Androstane-3β,17β-diol	160	141
5α-Androstane-3α-17β-diol	160	98
Testosterone	100	100
7α-Methylandrost-4-en-17β-ol-3-one	—	100
Androst-4-ene-3β,17β-diol	93	—
Androst-5-ene-3α,17β-diol	—	83
1α-Methyl-19-norandrost-4-en-17β-ol-3-one	80	—
17β-Estradiol	62	28
Androsta-1,4-dien-17β-ol-3-one	—	58
Androst-5-ene-3β,17β-diol	50	—
6β-Methyl-5α-androstan-17β-ol-3-one	—	48
17α-Methyltestosterone	38	65
6α-Methylandrost-4-en-17β-ol-3-one	—	46
17α-Ethylandrost-4-en-17β-ol-3-one	—	46
1α-Methyl-5α-androst-1-en-17β-ol-3-one	37	—
17α-Methyl-5α-androstane-3β,17β-diol	33	—
5β-Androstane-3α-17β-diol	24	—
19-Nortestosterone	22	—

affinity than testosterone itself. HORTON and KATO (1968) reported higher binding affinity for androstan-17β-ol-3-one and androstanediol than for testosterone.

As might be expected of a pregnane derivative the antiandrogenic steroid cyproterone (1,2α-methylene-6-chloropregna-4,6-dien-17α-ol-3,20-dione) has only 0.5 to 1% of the affinity of testosterone for the binding globulin (GUILLEMANT et al., 1969). No increase of testosterone-binding globulin capacity was observed after administration of 200 mg cyproterone although the time of 20 hrs may be too short to detect such an effect.

Requirement of a 17β-hydroxy group for interaction with the testosterone-binding globulin has been emphasized by VERMEULEN and VERDONCK (1968 a) as well as by KATO and HORTON (1968).

These authors determined the relative binding affinities of some 30 steroids by comparing their testosterone-displacing ability with that of testosterone itself. Table XII-2 shows a selection of these results; all compounds possess the important 17β-hydroxy group in ring D. More recent specificity studies by HEYNS et al. (1969 a), involving some 30 steroids of various structures, are in general qualitative agreement with the data of Table XII-2. Displacement of plasma-bound testosterone by C_{19}- and C_{21}-steroids has also been reported by FRICK and KINCL (1969), for C_{18}- and C_{19}-steroids by MAEDA et al. (1969). BLAQUIER and SEGRE DE CAMEO (1970) measured binding capacity and affinity for androst-5-ene-3β,17β-diol in human pregnancy plasma and in a globulin fraction thereof.

3. Steroid-Binding β-Globulin

It is evident from specificity studies described in the preceding section that the testosterone-binding β-globulin has a high affinity for estradiol, although not as high as that for testosterone. DE MOOR et al. (1968 b) have called this protein the steroid-binding β-globulin (SBβG), indicating the identity of the estradiol-binding and the testosterone-binding globulin. Other terms used are sex steroid-binding protein (SBP), sex hormone-binding globulin (SHBG), and 17β-hydroxysteroid-binding protein. A recent review has been given by CRÉPY and GUÉRIGUIAN (1970).

The conclusion of identity of the binding proteins has been corroborated by DE MOOR et al. (1969) who determined correlation coefficients for testosterone and estradiol binding in 59 plasmas of blood donors. STEENO et al. (1968) found that estrone also displaces testosterone from the binding sites, whereas estriol, added in equal amounts as testosterone to steroid-free plasma, did not have such influence. Similar results on the relative affinity of estradiol, estrone and estriol to the testosterone-binding sites of low capacity have been reported by MURPHY (1968 b) who also concludes that testosterone and estradiol are bound to common binding sites. This was also reported by PEARLMAN and FONG (1968). The affinity of the β-globulin for both testosterone and estradiol clearly distinguishes it from the estradiol-binding receptor proteins of the uterus (ROCHEFORT and BAULIEU, 1969; see also E. V. JENSEN's monograph in this series).

Table XII-3. *Affinity of human serum proteins for testosterone and estradiol*

Protein	Temp.	Association constant for Testosterone M^{-1}	Association constant for Estradiol M^{-1}	Reference
Steroid-binding β-globulin	4°	1.7×10^9	0.6×10^9	MERCIER-BODARD and BAULIEU (1968)
	4°	1.2×10^9	0.46×10^9	LEBEAU et al. (1969)
	4°	—	4.0×10^9 [a]	ROSNER et al. (1969)
	20°	6×10^8	—	VERMEULEN and VERDONCK (1968 a)
	25°	4.5×10^8	1.3×10^8	PEARLMAN et al. (1969)
	37°	5×10^8	—	DRAY (1969)
AAG	4°	6.0×10^5	1.0×10^5	KERKAY and WESTPHAL (1968)
	37°	2.4×10^5	0.6×10^5	KERKAY and WESTPHAL (1968)
HSA	5°	3.0×10^4	1.6×10^5	SANDBERG et al. (1957)
	25°	4.1×10^4	—	WESTPHAL (1961)

[a] Two additional association constants, $k_2 = 0.69\times10^9\ M^{-1}$, $k_3 = 0.11\times10^9\ M^{-1}$ have been reported by ROSNER et al. (1969).

MURPHY (1968 b) studied a number of C_{18}-, C_{19}- and C_{21}-steroids for their testosterone- and estradiol-displacing qualities, and obtained results in general agreement with those described above. She observed a weak effect of cholesterol which she considers of possible importance in view of the relatively high blood concentration of this sterol.

An analysis of the binding data published by MURPHY (1968 b) suggests the participation of AAG in the plasma protein binding of testosterone, but not in that of estradiol. With the highest amounts of testosterone added to pooled human plasma, 28.5% was bound, whereas an equivalent or slightly greater than equivalent concentration of HSA bound 22.4%. In case of estradiol, 61% was bound in the same pooled plasma, and 64.7% was bound in the same HSA solution which was used in the testosterone experiment. The different binding behavior of testosterone and estradiol follows from the association constants (Table XII-3): the affinity of the testosterone-

AAG complex is markedly higher than that of the testosterone-HSA complex, whereas the reverse is true for the estradiol complexes of these two binding proteins.

Equilibrium dialysis studies using a partially purified steroid-binding β-globulin (MERCIER-BODARD and BAULIEU, 1968) gave the same number of binding sites for testosterone and estradiol, corroborating the assumption of identical binding proteins. Dissociation constants of $K = 0.6 \times 10^{-9}$ M and 1.7×10^{-9} M were found for testosterone and estradiol, respectively. Table XII-3 gives the association constants calculated from these data, in addition to those from other publications. Partial purification by gel filtration and ion-exchange chromatography has been reported by BARDIN et al. (1968).

Partial purification of the steroid-binding β-globulin has also been reported from DE MOOR's laboratory. VAN BAELEN et al. (1968) fractionated late pregnancy serum by Sephadex G-200 filtration, ammonium sulfate precipitation, and DEAE-cellulose chromatography. The binding protein was precipitated at ammonium sulfate concentrations below 2 M. An approximate molecular weight of 100,000 was suggested by the elution volume in gel filtration. About half this molecular size, i.e., 52,000, has been reported by MERCIER-BODARD et al. (1970). Of methodological interest is the observation of VAN BAELEN et al. (1968) that estradiol interacted with DEAE-cellulose so that *in-vitro* addition of the radiosteroid to the serum resulted in incorrect localization of the steroid-binding globulin; the binding activity had to be measured after chromatographic separation of unlabeled serum.

MERCIER-BODARD et al. (1970) purified the sex steroid-binding protein (SBP) from COHN fraction IV of human plasma by chromatography on DEAE-cellulose, DEAE-Sephadex, and gel filtration over Sephadex G-75. The final preparation was found homogeneous by polyacrylamide gel electrophoresis, ultracentrifugation and binding behavior. A molecular weight of 52,000 was obtained by sedimentation equilibrium; Sephadex G-200 filtration gave a larger molecular size, close to the values of 100,000 or 110,000 reported by VAN BAELEN et al. (1968) and GUÉRIGUIAN and PEARLMAN (1968), respectively.

Preliminary studies on the chemical properties of the steroid-binding β-globulin (PEARLMAN et al., 1969) suggest the importance of sulfhydryl and disulfide groups for the binding activity: reaction with N-ethylmaleimide, p-mercuriphenylsulfonate, dithiothreitol, and sodium sulfite resulted in a marked reduction of testosterone-binding activity. Isoelectric focusing experiments (VAN BAELEN et al., 1969) appeared to indicate microheterogeneity which was ascribed to differences in sialic acid content; however, the separated fractions have not yet been fully characterized. Enzymatic removal of sialic acid did not affect the estradiol-binding activity. This result is analogous to observations with CBG, AAG, and TBG, proteins in which the sialic acid content is also not necessary for full ability to interact with ligand hormones.

Table XII-4. *Influence of sex and age on the estradiol binding index (EBI) ± standard deviation. Number of subjects in parentheses.* From DE MOOR et al. (1969)

Age	0—11 Years	12—19 Years	20—59 Years	> 60 Years
Women	162 ± 99 (19)	110 ± 50 (14)	158 ± 82 (27)	168 ± 128 (10)
Men	232 ± 93 (41)	158 ± 85 (24)	107 ± 81 (16)	101 ± 49 (20)

For convenient assessment of the steroid-binding β-globulin activity in plasma, DE MOOR et al. (1969) applied the index of net binding (or the binding index) as defined by TAIT and BURSTEIN (1964). This index is the dimensionless product of the concentration of binding sites (or binding capacity) and their association constants (or binding affinity). It may be determined with radiolabeled 17β-estradiol as estradiol binding index (EBI) or with testosterone as testosterone binding index (TBI). Correction is made for albumin binding, resulting in a selective (partial) binding index as a measure of the β-globulin activity. Comparative determination of EBI and TBI for a large number of plasma samples from normal and hospitalized subjects showed a very good correlation as another indication of identity of estradiol- and testosterone-binding sites.

Table XII-4 gives EBI values for men and women of different age groups. There seems to be a trend of decreasing binding activity with progressing age in men although the individual biological variation is very high. Fig. XII-2 shows the increase of the EBI during pregnancy. Since the within-person variation has been found much smaller than that between persons, the values connected by dotted lines, representing three women, are of particular interest. Similarly increased EBI values were obtained during estrogen treatment. DRAY et al. (1970) reported elevated binding activity for testosterone and estradiol after administration of clomiphene citrate (clomid).

It has been discussed in Chapter IX. 7 that the CBG activity is under the regulatory control of the thyroid hormone. As another

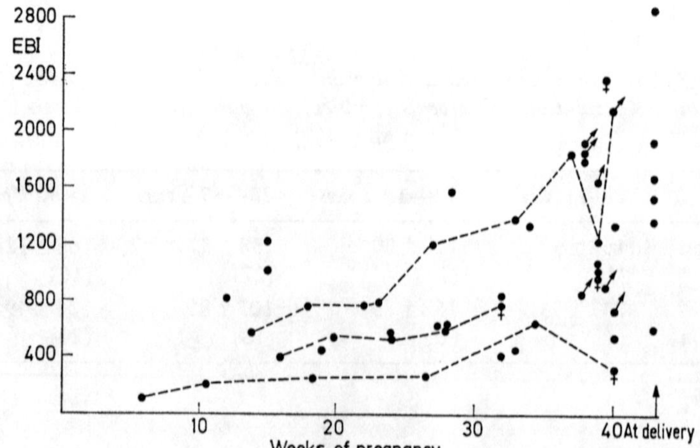

Fig. XII-2. Maternal estradiol binding index (EBI) values at various stages of pregnancy. Each point refers to a different subject except when connected by a dotted line. The sex of the baby is indicated. Values for non-pregnant women lay around 100. From DE MOOR et al. (1969)

analogy between CBG and the steroid-binding β-globulin, the testosterone-binding activity has been reported to be significantly increased in the serum of thyrotoxic male and female patients (CRÉPY et al., 1967; DRAY et al., 1967). This finding may explain the high plasma level of testosterone and its low metabolic clearance in thyrotoxicosis; extended studies (DRAY et al., 1969) with 19 male patients showed testosterone concentrations of 1.66 $\mu g \pm 0.27$ $\mu g/100$ ml as compared to the normal 0.80 ± 0.12 $\mu g/100$ ml. The 1/P values at 50% binding, corresponding to the C-values (Chapters IV. 15 and VIII. 2 a), measured for the testosterone-binding activity was 2.05 ± 0.40 in the thyrotoxic males versus 0.90 ± 0.30 in normal subjects. The metabolic clearance rate was found about half normal. The 1/P values in female patients were 2.25 ± 0.29, also elevated over the normal female binding activity of 1.40 ± 0.30. The testosterone-binding capacity in female thyrotoxicosis was assessed by equilibrium dialysis against HSA solutions to average 36×10^{-8} M (DRAY, 1969), as compared to 6 to 8×10^{-8} M and 40 to 44×10^{-8} M in normal and pregnant women, respectively. No significant increase was observed in the female testosterone levels which are less than 1/10 of the male values. The testosterone-binding activity decreased or became normal after radical treatment of the hyperthyroid condition.

The observations in the thyrotoxic patients were corroborated by results obtained after administration of thyroid hormones (DRAY et al., 1969). Increase of testosterone-binding activity was found in male and female subjects with normal or low binding values after treatment with L- or D-thyroxin (T_4) or triiodothyronine (T_3); the binding values were generally measured from week to week. These studies included cases of idiopathic hirsutism which had decreased testosterone-binding activity; their testosterone levels were significantly elevated (DRAY et al., 1968). It was determined that the effect of the thyroid hormones on the testosterone-binding activity was independent of their metabolic actions.

Increase of steroid-binding β-globulin activity under the influence of thyroid hormone was also observed by DE MOOR et al. (1969). In 12 hyperthyroid patients EBI values varied from 339 to 1035, averaging 669 ± 251 (S.D.), greatly elevated from normal values as seen in Table XII-4. The EBI decreased after treatment of the thyroid condition.

It has been discussed in Chapter IX. 4 that testosterone lowers the CBG activity in the rat (see Fig. IX-7 in Chapter IX. 4 b). In an analogous observation, androgenic hormones have been found to reduce the TBG activity in humans (BRAVERMAN and INGBAR, 1967). DE MOOR et al. (1969) conclude from EBI differences between men and women, from hypogonadal male patients and their treatment that testosterone has a decreasing effect on the steroid-binding β-globulin activity. These observations corroborate the finding in hirsutic women mentioned above. VERMEULEN et al. (1969) made similar observations on the various endocrine effects. Reduced testosterone-binding activity was found in the polycystic ovary syndrome (SOUTHREN et al., 1969).

The high binding values obtained after estrogen treatment and in pregnancy have been reported from several laboratories (BARLOW et al., 1969). DESSYPRIS et al. (1969) observed higher testosterone-binding activity in twin pregnancies than in normal pregnancies. DE MOOR et al. (1969) found in umbilical cord plasma about half the EBI values (83 ± 39) seen in prepubertal subjects; no difference was noted between boys and girls. No change from normal values was seen during the menstrual cycle or after menopause.

Results published by MURPHY (1968 b) show high-affinity binding of estradiol in human, bovine and frog plasma. No activity was found in plasma of rat, rabbit, dog and duck. These latter four species showed high-affinity, low-capacity binding for testosterone, presumably due to CBG. DE MOOR et al. (1969) reported that male chimpanzees had EBI values (121 ± 36) similar to those of humans, while two female chimpanzees showed EBI figures of 613 and 858. In another female primate *(cercopitheus mona)* an even higher value of 1600 was found. In confirmation of MURPHY's observation, no steroid-binding β-globulin activity was detected in the rat. Plasma samples of seven ruminant species were also negative. The testosterone-binding activity of the sheep was weaker than that in normal men; no difference was found among ram, normal, pregnant, and lactating ewe (DRAY, 1969). It has been reported in Chapter X. 1 (Fig. X-6) that the high-affinity binding of testosterone in the guinea pig increases during pregnancy many times over normal values (DIAMOND et al., 1969). ROYCE (1970) observed in the serum and urine of male rats, and at lower concentration in females, a low molecular weight α-globulin that binds testosterone weakly. No relationship to other serum proteins is known.

High-affinity binding of testosterone has been found in thorny skate and Atlantic salmon (IDLER and FREEMAN, 1968). The strong testosterone-binding activity in these fishes may explain their high testosterone concentrations which are up to 100 times greater than in humans. There was no apparent difference in binding activity between the sexes in thorny skate (IDLER and FREEMAN, 1969). In contrast to steroid interaction with human glycoproteins (CBG, AAG) and with human sex steroid-binding globulin, the binding affinity decreased but little as the temperature was elevated from 3° to 40°. Only slight loss of testosterone-binding activity was observed on heating of diluted skate serum at 60° for 45 min. The active protein was isolated from the serum of the mature male thorny skate by DEAE-cellulose chromatography and Sephadex G-200 filtration (FREEMAN and IDLER, 1969); it was found homogeneous by polyacrylamide gel electrophoresis. The molecular weight was estimated by Sephadex G-200 techniques to be about 80,000. During gel filtration over Sephadex G-200, polymerization was observed with loss of steroid-binding activity. This behavior resembles similar polymer formation of rat CBG on removal of corticosterone (Chapter XI. 5). The association of the binding protein from skate could be prevented and reversed to some extent by 2-mercaptoethanol or dithiothreitol. The relative binding affinities of the purified protein for testosterone, estradiol, progesterone, 17α-hydroxyprogesterone, and 1α-hydroxycorticosterone were found to be 100, 125, 118, 108, and 2, respectively (FREEMAN and IDLER, 1969).

The concept of MURPHY (1964) to utilize the steroid-binding properties of plasma for determination of small quantities of steroid hormones by competitive displacement (MURPHY et al., 1963) has been applied in a number of laboratories to the quantification of testosterone and estradiol in plasma and other biological fluids. MURPHY (1967 b) reported on measurement of subnanogram amounts of testosterone and estrogens following their separation by thin-layer chromatography. FRITZ and KNOBIL (1967) determined 3—11 ng/ml testosterone in normal human male peripheral plasma, using 0.5 ml. Details on a similar method including studies on specificity, precision (coefficient of variation 5.8%) and accuracy were given by HORTON et al. (1967). Late pregnancy plasma was used in these and most other laboratories as source of testosterone-binding protein, separation of the unbound steroid was accomplished by Florisil or dextran-coated charcoal. HEYNS et al. (1968) have studied the steroid-adsorbing qualities of several solids.

An amount of only 0.2 ml of male plasma was used by MAYES and NUGENT (1968) to determine testosterone in their technique based on competitive binding. A critical evaluation of this and other procedures for testosterone assay by competitive protein binding has been given by NUGENT and MAYES (1970). An analogous method for testosterone assay in biological fluids has been described by HALLBERG et al. (1968). More recent reports on procedures for measurement of plasma testosterone by competitive protein binding have been published by HEYNS et al. (1969 b), by ROSENFIELD et al. (1969), by

Pizarro and Arredondo (1969), by Frick and Kincl (1969), by August et al. (1969), Maeda et al. (1969), and by Vermeulen and Verdonck (1970). Crépy et al. (1969) in a critical study have evaluated the various aspects of the competitive binding technique for the determination of testosterone in plasma. Rudd et al. (1969) developed a method for assay of urinary testosterone glucosiduronate after enzymatic hydrolysis of the urine followed by extraction and purification of the free testosterone. The values obtained by competitive binding correlated well with those determined by a double isotope technique. Rosenfield (1969) and Moshang et al. (1970) determined androst-4-ene-3,17-dione by the competitive binding assay for testosterone; this 17β-hydroxysteroid was generated either from the extracted and purified dione by enzymatic reduction with 17β-hydroxysteroid dehydrogenase, or by borohydride reduction. The competitive binding method was applied by Nieschlag and Ismail (1970 a, 1970 b) to study diurnal variations of plasma testosterone in men. Lowest values were found at 8 PM, highest at 4 PM or 12 noon.

The preceding paragraphs indicate that utilization of the steroid-binding β-globulin for competitive binding analysis has been essentially limited to the determination of testosterone. Murphy (1968 b) pointed out that the same plasma protein may be applied for quantification of estradiol. Such a method was developed by Mayes and Nugent (1970) for an estradiol assay in plasma of men and women. However, in most laboratories specific binding proteins from other origin were preferred. Baulieu et al. (1967 b) used the estradiol receptor proteins from the endometrium of the immature sow (Baulieu et al., 1967, Baulieu, 1967, 1968). A similar assay procedure for estradiol using a protein fraction from the high speed supernatant of pregnant rabbit uterus has been reported by Korenman (1968, 1969). Estriol and estrone had lesser, diethylstilbestrol greater displacing activity for the tritiated 17β-estradiol. Further improvements of the method and additional results were published by Korenman et al. (1969, 1970). The rabbit uterine cytosol preparation was also utilized by Corker et al. (1970) for estradiol assay in plasma, and by Moxham and Nabarro (1970) for determination of estradiol in urine; the authors stabilize the binding protein for at least 4 months by storage at —15°. Shutt (1969) used a protein fraction from the uteri of ovariectomized ewes which specifically binds 17β-estradiol.

XIII. Interaction of Progesterone and Other Steroids with α_1-Acid Glycoprotein

1. General Properties of the Progesterone-AAG Complex; Relation to CBG Binding

Some time before CBG was isolated and the glycoprotein nature of high-affinity hormone binders in blood serum established, it was found that the α_1-acid glycoprotein (AAG or orosomucoid) formed a dissociable complex with progesterone whose binding affinity was higher than that between progesterone and HSA (WESTPHAL et al., 1961). Other Δ^4-3-ketosteroid hormones were also bound, and the strength of association followed the polarity rule (WESTPHAL, 1964).

The isolation of AAG (SCHMID, 1950; WEIMER et al., 1950) had resulted in a glycoprotein which could be crystallized as lead salt (SCHMID, 1953) and was found to be homogeneous by the physicochemical criteria applied at the time (SMITH et al., 1950). Analogous orosomucoids have been prepared from serum of beef, horse, rabbit, guinea pig, and rat (WEIMER and WINZLER, 1955); also from pig (GRANT et al., 1967), canine (ATHINEOS et al., 1964), avian (GRANT et al., 1967), sheep (CAMPBELL et al., 1961; DAS, 1962) and chimpanzee (LI and LI, 1970) plasma. Particularly the bovine glycoprotein has been thoroughly investigated (BEZKOROVAINY and DOHERTY, 1962; BEZKOROVAINY, 1965 a, 1965 b; GRANT et al., 1967), and the AAG from rat serum has also been further characterized (SARCIONE, 1963; RICHMOND, 1963; KAWASAKI et al., 1966). AAG is synthetized in the liver (ATHINEOS et al., 1964); perfusion studies with isolated rat liver have shown a dependency upon the presence of cortisol (JOHN and MILLER, 1969).

AAG is a very soluble protein — more soluble than any other plasma glycoprotein. It is unusually stable to heat, for example in boiling water. These qualities, and its occurrence in the relatively high concentration of approximately 750 mg/liter human serum have facilitated its chemical investigation so that the AAG is one of the best characterized serum glycoproteins. Comprehensive reviews of its physical, chemical and biological properties are available (WINZLER, 1960; GOTTSCHALK and GRAHAM, 1966; JEANLOZ, 1966; SCHULTZE and HEREMANS, 1966; SCHMID, 1968); the present discussion will be limited to aspects relevant to the steroid-binding ability of AAG. Work on the elucidation of its chemical structure is being actively pursued in several laboratories (MARSHALL and PORATH, 1965; MARSHALL, 1966; KROTOSKI and WEIMER, 1967; SCHMID et al., 1968 b; YAMAGAMI et al., 1968; BEZKOROVAINY et al., 1969; ISHIGURO et al., 1969; WAGH et al., 1969; YAMAUCHI and YAMASHINA, 1969).

As our knowledge of the interaction between corticosteroid hormones and CBG increased it became apparent that there is considerable similarity between the steroid complexes with CBG and those with AAG. In view of the inherent difficulties in obtaining sufficient quantities of pure CBG for chemical studies (see Chapter XI), the investigation of the progesterone-AAG complex was expected to provide answers to many questions concerning the relationship of the chemical structure of a protein to its steroid-binding ability. Table XIII-1 gives some physicochemical parameters of human AAG. A comparison with the data listed in Table XI-3 for human, rabbit, and rat CBG shows the close correspondence. Molecular weights and related properties are of the same order of magnitude; the carbohydrate content is almost as high for CBG as for AAG. The electrophoretic mobility is that of an α-globulin in both cases; at pH 8.6 it is greater for AAG, in accordance with a lower isoelectric point resulting from a high sialic acid content. It will be of interest to include AAG's from other species in future studies on steroid binding in relation to protein structure. Bovine AAG has been found to interact with progesterone with about one-fourth the affinity of human AAG (WESTPHAL and CARNIGHAN, 1964).

The amino acid composition of AAG (Table XIII-2) also has a general resemblance with that of CBG, in spite of distinct differences in certain residues. The serine, proline and glycine content is lower, as is that of valine and leucine. Three disulfide groups and no free sulfhydryl have been found in AAG. The number of methionine residues is smaller than that in human CBG, whereas AAG has more half-cystine. However, a comparison with rabbit CBG (Table XI-5) shows a much greater similarity in the content of the sulfur-containing amino acids. Except for sialic acid, there is considerable likeness in the carbohydrate composition; this fact may be of interest in future exploration of the biosynthetic origin of the two glycoproteins in the human liver. Table XIII-2 includes for comparison the composition of TBG and HSA; the general resemblance of the hormone-binding glycoproteins applies also to TBG.

Table XIII-1. *Physicochemical properties of human AAG* [a]

Parameter		Reference
Sedimentation coefficient, $s^0_{20,w}$ (S)	3.08	GANGULY et al. (1967)
Diffusion coefficient, $D^0_{20,w}$ ($cm^2\ sec^{-1} \times 10^7$)	6.3	KALOUS (1965)
Partial specific volume, $\bar{v}$ (ml/gm)	0.675	SMITH et al. (1950)
Molecular weight	41,000	WINZLER (1960)
Frictional ratio, f/fo	1.42	BEZKOROVAINY (1965 a)
Electrophoretic mobility at pH 8.6 ($cm^2\ volt^{-1}\ sec^{-1}) \times 10^5$	—6.2	GANGULY et al. (1967)
Isoelectric point (phosphate buffer, μ 0.1)	2.7 [d]	SCHMID (1953)
$E^{1\%}_{1cm}$ at 278 nm	9.33	KERKAY and WESTPHAL (1969)
Cotton effect [b] Minimum (nm)	230	SARKAR and DOTY (1966)
Cotton effect [b] Maximum (nm)	204	SARKAR and DOTY (1966)
Carbohydrate %	41.9 [c]	GANGULY et al. (1967)
Nitrogen %	9.8	GANGULY et al. (1967)

[a] For additional data from other laboratories see JEANLOZ (1966).
[b] The same wavelengths for peak and trough were observed by GANGULY et al. (1967).
[c] Hexosamine calculated as acetylhexosamine.
[d] Values of 3.0, 3.2, 3.4, 3.5, and 3.6 for AAG variants have been found recently by RYAN (1970).

One of the similarities between the steroid complexes of AAG and CBG is the temperature dependency of the binding affinity (WESTPHAL and CARNIGHAN, 1964). Fig. XIII-1 shows the apparent association constant, $\bar{\nu}/[S] = n\,k$, of the progesterone-AAG complex at different temperatures. The resemblance with the corresponding temperature effect on the stability of the corticosteroid-CBG complex (Fig. VIII-12) is obvious; the binding affinities at 4° are several times greater than at 37° or 50°. In contrast, the association constant of the progesterone-HSA complex (Fig. XIII-1) is only slightly re-

Table XIII-2. *Amino acid and carbohydrate composition of human AAG, CBG, TBG, and HSA. Calculated as residues per mole protein*

Constituent	AAG 41,000 [b] GANGULY and WESTPHAL (1968 b) [c]	CBG [a] 51,700 [b] MULDOON and WESTPHAL (1967)	TBG 58,000 [b] GIORGIO and TABACHNICK (1968)	HSA 69,000 [b] SCHULTZE and HEREMANS (1966)
Lysine	16	15	25	56
Histidine	4	9	10	15
Arginine	10	10	16	23
Aspartic acid	26	32	38	52
Threonine	19	22	25	28
Serine	8	22	34	24
Glutamic acid	35	38	52	81
Proline	9	22	30	25
Glycine	8	19	25	12
Alanine	11	23	31	62
Half-cystine	5	2	8	33
Valine	10 [d]	26	23	41
Methionine	2	10	5	5
Isoleucine	11	16	9	8
Leucine	17	40	44	60
Tyrosine	13	10	10	16
Phenylalanine	11	19	18	30
Tryptophan	4 [e]	3	4	1
Hexose	37 [d]	37	24	0
Hexosamine	30 [d]	29	15 [f]	0
Sialic acid	16 [d]	7	5	0
Fucose	4 [d]	5	<1	0

[a] Values for rabbit and rat CBG are given in Table XI-5.

[b] Molecular weight used for calculation.

[c] The amino acid composition is in excellent agreement with the values of MARSHALL and PORATH (1965) when calculated for the same MW of AAG. Agreement is also satisfactory with the analyses of WALBORG and WARD (1963) and of HEIMBURGER et al. (1964), whereas deviations occur for several amino acids in comparison with earlier determinations (reviewed by JEANLOZ, 1966).

[d] Value from GANGULY et al. (1967).

[e] Value from MARSHALL and PORATH (1965).

[f] Glucosamine, 12; galactosamine, 3.

duced at elevated temperature; this has been discussed in Chapter VI. 5. e (Table VI-11). It has been verified in the experiments recorded in Fig. XIII-1 that the low binding affinities at the highest temperatures are not the result of (irreversible) thermal denaturation. Solutions of AAG were exposed to 45° for 48 hrs and to 60° for up to 24 hrs without loss of binding activity (GANGULY and WESTPHAL, 1968 a); after 75 hrs at 60°, slight inactivation was observed (GANGULY et al., 1967). Table XIII-3, Exp. 1 documents some of these results. The heat stability of the HSA preparation was checked by incubating duplicate samples at the same high temperature and for the same length of time as the equilibrium dialysis samples used for determination of n k ($\bar{\nu}/[S]$ in Fig. XIII-1), cooling them to 4° and measuring their $\bar{\nu}/[S]$ values at this temperature. No heat inactivation was observed up to 45°, the highest temperature tested.

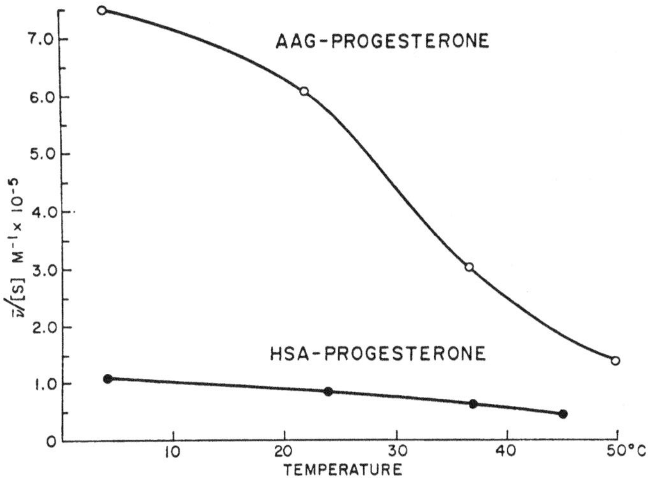

Fig. XIII-1. Effect of temperature on the apparent progesterone-binding affinity ($\bar{\nu}/[S] = n\,k$) of AAG and of HSA, in 0.05 M phosphate, pH 7.4. From GANGULY et al. (1967)

Another property common to the steroid complexes of CBG and of AAG is the change of binding affinity with pH. This is illustrated in Fig. XIII-2 for the progesterone-AAG interaction which has a pH optimum at approximately 8. A comparison with Fig. VIII-14 shows the close correspondence to the pH dependency of the CBG

Table XIII-3. *Apparent binding affinity of progesterone-AAG complex under different environmental conditions. AAG concentration was 0.5 mg per ml; all equilibrium dialyses were carried out at 4°.* From GANGULY and WESTPHAL (1968 a)

Experiment No. and conditions	n k $M^{-1} \times 10^{-5}$
1-1. 0.05 M phosphate, pH 7.4	10.5; 11.0 [a]
1-2. Exposure to 37° [b]	8.4
1-3. Exposure to 45° [b]	8.6 9.8 [a]
1-4. Exposure to 60° [c]	9.2 [a]
2-1. 0.05 M phosphate, pH 5.5 [d]	7.3
2-2. 0.05 M phosphate, pH 10.0 [d]	11.1
2-3. 0.05 M phosphate, pH 11.0 [d]	5.2
3-1. 0.05 M ammediol, pH 10.0 [e]	6.8
3-2. 0.05 M ammediol, pH 11.0 [e]	0.6
3-3. Solvent as in experiment 3-2 [f]	3.2
3-4. 0.05 M ammediol-4 M NaCl, pH 10.0 [e]	34.3
4. Veronal buffer, $\mu = 0.1$, pH 8.5 [g]	8.0
5-1. 0.05 M sodium acetate, pH 4.1	4.0
5-2. 0.05 M sodium acetate, pH 4.1 [f]	4.6
5-3. As in experiment 5-2, followed by precipitation with alcohol-acetone	4.6
5-4. 0.05 M sodium acetate-4 M NaCl, pH 4.1	75.6
6-1. 0.15 M NaCl, adjusted to pH 2.0 with HCl	0.2
6-2. Solvent as in experiment 6-1 [f]	4.3

[a] Second AAG preparation.
[b] In 0.05 M phosphate buffer, pH 7.4, for 48 hours, followed by equilibrium dialysis at pH 7.4.
[c] For 24 hours; other conditions as given in footnote [b].
[d] pH adjusted with H_3PO_4 and NaOH.
[e] pH adjusted with HCl and NaOH.
[f] Exposed for 48 hours at 4°, then adjusted to 0.05 M phosphate, pH 7.4, by dialysis.
[g] Exposed for 24 hours at 4°, then dialyzed against water.

complexes. Further studies on the influence of pH on the progesterone-AAG-complex (Table XIII-3, Exps. 2 to 6) revealed maximal binding affinity up to pH 10 when phosphate buffer was used instead of ammediol; the stability of the complexes decreased at pH 11 in either buffer system. It was generally observed that the n k ($\bar{\nu}/[S]$)

values at a given pH were higher in phosphate than in the ammediol buffer in which chloride was the anion (Table XIII-3, Exps. 2 and 3). This may be a specific, affinity-increasing effect of the phosphate ion as will be discussed in Chapter XIII. 5. However, the possibility has not been excluded that the ammediol used contained traces of heavy metal ions which inhibit the progesterone-AAG interaction (see Chapter XIII. 3).

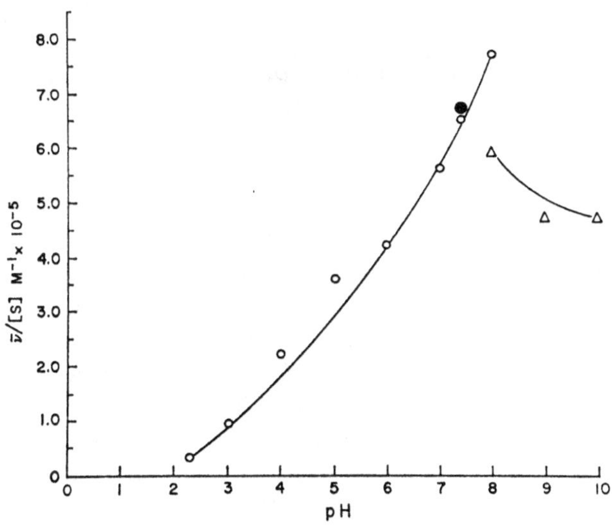

Fig. XIII-2. Effect of pH on the apparent progesterone-binding affinity of AAG, at 4°. Open circles: citrate-phosphate, 0.05 M; closed circle: phosphate, 0.05 M; triangles: 2-amino-2-methyl-1,3-propanediol (ammediol), 0.05 M. From GANGULY et al. (1967)

Table XIII-3 gives the results of exposure of AAG at 4° to solutions of different pH values, mostly for a period of 48 hours, followed by readjustment to neutral pH. This treatment at a pH close to neutral did not markedly affect the progesterone-binding ability. However, in the experiments conducted at pH 2.0, 4.1, and 11.0, the final n k ($\bar{\nu}/[S]$) was decreased; the 2-day exposure apparently resulted in some irreversible denaturation since only part of the original binding affinity was restored. This recovery resulted in n k values about one third to one half original. In contrast, exposure of human and rat CBG in serum to pH 2.8 for 48 hrs led to inactivation of 85—95% of the binding power (Table VIII-9); similar treatment at pH 9.8 did not affect the CBG activity. These results illustrate the difference in the stability of the two glycoproteins, under conditions favorable for CBG; it should be noted that the irreversible denaturation of CBG at pH 2.8 was greater than that of AAG at pH 2.0, although CBG was protected by the presence of all other proteins of the blood serum.

2. Reversible Inhibition by Lipid of the Binding Affinity between AAG and Progesterone

Our studies on interaction of progesterone with AAG were hampered for a long time by inconsistencies in the association constants measured by equilibrium dialysis as $\bar{\nu}/[S] = n\,k$ (Eq. V-8). In spite of rigorous adherence to identical conditions, various highly purified AAG preparations gave n k values as low as $1 \times 10^5\ M^{-1}$ and as high as $8 \times 10^5\ M^{-1}$. Such discrepancies were experienced even with the same AAG samples in different experiments without any apparent changes in the conditions. Two problems were finally recognized as causing the difficulties (WESTPHAL et al., 1966). First, traces of heavy metal ions were found to reduce the binding affinity; this will be discussed in Chapter XIII. 3. Secondly, it was observed that lipid-extractable contaminations present in the purified AAG preparations caused drastic lowering of the apparent association constants. This problem and its solution will be described in the present section.

The AAG used for binding studies was purified by a procedure developed in our laboratory (GANGULY et al., 1967). Double-distilled water, deionized over exchanger columns, was used throughout this work, and efforts were made to completely eliminate contaminating ions, especially those of heavy metals, from all components applied. Plasma was prepared from outdated bloodbank blood and subjected to the phenol precipitation procedure of MICHON (1962): an equal volume of 5% aqueous phenol was added at 4° to the plasma made 20% in sodium chloride. The protein in the supernatant fraction was chromatographed over CM-cellulose, filtered over Sephadex G-200, and chromatographed over DEAE-cellulose. The AAG obtained was homogeneous by chromatographic criteria (AAG Type C) and had the physicochemical properties reported in the literature. However, it did not give reproducible binding parameters.

For removal of lipid-soluble contaminations, this AAG Type C was dissolved in 0.1 M acetate buffer of pH 4.1 and precipitated by addition of 4 volumes of an ethanol-acetone mixture (9 : 1, v/v) at 4°. The glycoprotein was centrifuged, washed with cold ethanol and cold ether, and dried. The resulting delipidated AAG preparation (Type P) was identical in the physicochemical parameters measured (see Table I and Fig. I in Ganguly et al., 1967). The only difference observed was an increase in the progesterone-binding affinity. This is shown in Table XIII-4 which gives representative examples from a large number of similar experiments. Simple washing at 2° with 10% acetone in ethanol increases the n k value of AAG, Type C, from 1.2×10^5 to 2.6×10^5 M^{-1}. Precipitation at pH 4.1 is much more effective; the n k value rises to 10.1×10^5 M^{-1}. Repetition of this alcohol-acetone precipitation at pH 4.1 does not result in further significant increase of the apparent affinity constant. Recombination of a precipitated AAG preparation (Type P) with the material recovered from the supernatant of the precipitation reaction again decreased n k. The value obtained in this case, i. e., 0.3×10^5 M^{-1}, is

Table XIII-4. *Influence of delipidation of chromatographically pure AAG on progesterone-binding affinity*[a]

AAG Type	Treatment	n k $M^{-1} \times 10^{-5}$
C	None	1.2
C	Washed with ethanol-acetone (9 : 1)	2.6
C	Precipitated with ethanol-acetone (9 : 1) at pH 4.1.	10.1
C	Precipitated twice with ethanol-acetone (9 : 1) at pH 4.1.	10.5
P	None	9.9
P	Recombined with excess residue from ethanol-acetone (9 : 1) supernatant	0.3

[a] Apparent association constants determined by equilibrium dialysis as $\bar{\nu}/[S] = n\,k$ (Eq. V-8) in 0.05 M phosphate, pH 7.4, 4°. Abbreviated from Table II in Ganguly et al. (1967).

lower than that of AAG, Type C, because the extracted lipid originated from over 3 times more AAG than used for the recombination.

The n k values given in Table XIII-4 have been obtained as $\bar{\nu}/[S]$ at single concentrations. A complete binding analysis by the method of reciprocal plots (Fig. XIII-3) confirms the results of the simplified binding measurements, and provides some information on the mechanism of the lipid effect. In agreement with an earlier

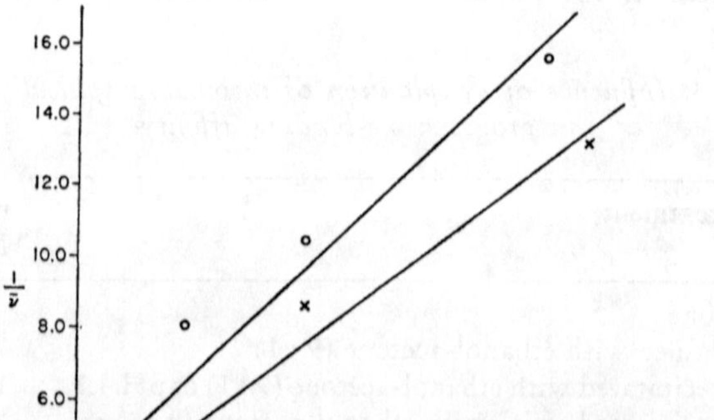

Fig. XIII-3. Interaction of progesterone with different AAG preparations, in 0.05 M phosphate, pH 7.4, 4°. Triangles: AAG, Type P; crosses: AAG, Type C; circles: AAG, Type P recombined with extracted lipid. From Ganguly et al. (1967)

analysis of a progesterone complex with an ethanol-precipitated AAG preparation (see Fig. 6 in Westphal, 1964), the number of binding sites for AAG (Type P) was found $n \cong 1$. Evaluation of the data for number of binding sites according to Scatchard's procedure confirmed this result. It also showed that the chromatographically pure AAG (Type C) had a fractional value of $n = 0.20$, close to that measured for a type P preparation which had been recombined with the extracted lipid (circles in Fig. XIII-3). These n values are considered more reliable than those assessed from the reciprocal plots; it is evident that the straight extrapolation of the binding data for the lipid-containing AAG preparations (two upper curves in Fig. XIII-3) does not account for the nonlinear trend of the values at high concentrations of [S].

These results are interpreted by the assumption that a lipophilic contamination, originally present in AAG, Type C, firmly occupies the binding site for progesterone in the major percentage of the molecules. It is not removed during the various procedures that result in the chromatographically pure preparation Type C. The blockage of the binding site apparently is so strong that the increasing amounts

Table XIII-5. *Numbers of binding sites, n, and corresponding apparent association constants, k, for progesterone-AAG complexes at pH 7.4, 4°*

AAG preparation [a]	n	k $M^{-1} \times 10^{-5}$
I-117, ethanol precipitated	1.0	9.2
Type P	0.90	10.8
Type C	0.20	9.4
Type P, relipidated [b]	0.17	8.5

[a] Data from GANGULY et al. (1967), except those for Prep. I-117 which were taken from WESTPHAL (1964).
[b] Type P recombined with lipid extracted from Type C.

of progesterone used in the binding analysis do not completely displace the interfering substance; the result is a fractional value of n (Table XIII-5, Fig. XIII-3). On the average, (1—n), i. e., approximately 80% of the AAG molecules of the Type C preparation appear inactive or much less active in their binding affinity for progesterone.

If the assumption is correct that the fractional value of n for AAG (Type C) results from inactivation of the major portion of the AAG molecules, then the association constant of the "active" portion of the AAG molecules should be that of Type P. Table XIII-5 shows that this is essentially the case. Removal of the lipid by organic solvent precipitation of AAG, Type C, transforms the 80% inactive (or much less active) molecules into the Type P species. This process is reversible; recombination of the alcohol-acetone precipitated AAG with the lipid material extracted from AAG, Type C, by the precipi-

Table XIII-6. *Progesterone-binding affinity* [a] *of different AAG preparations*

Prep.	Method of preparation	Ref. [b]	n k $M^{-1} \times 10^{-5}$
1	From fraction VI of COHN's alcohol fractionation	(c)	6.6
1 ppt	Prep. 1 precipitated with ethanol-acetone (9 : 1) at pH 4.6	(d)	10.6
2	Chromatography, pH 4.1	(e)	2.6
3	Chromatography, pH 3.7, starting directly from serum	(f)	1.2
4	Chromatography, pH 3.7, starting from fraction VI of COHN's alcohol fractionation	(f)	4.2
AAG, C	Chromatography	(d)	3.0
AAG, Cstr	AAG, Type C, filtered over Sephadex G-25 at 45°, pH 7.4	(d)	2.5

[a] Apparent association constants determined by equilibrium dialysis as $\bar{\nu}/[S] = n\,k$ (Eq. V-8) in 0.05 M phosphate, pH 7.4, 4°. The preparations were obtained from the investigators named in the references. From GANGULY et al. (1967).
[b] References given as footnotes.
[c] SCHMID (1953).
[d] GANGULY et al. (1967).
[e] BEZKOROVAINY and WINZLER (1961).
[f] KALOUS and PONCOVÁ (1965).

tation re-forms an AAG preparation with the binding properties characteristic for Type C (Table XIII-5, Fig. XIII-3).

It should be noted that these observations on non-competitive inhibition of progesterone binding to AAG by contaminating lipid are analogous to similar findings with HSA, described in Chapter VI. 5. b. In that case, increase of number of binding sites and of binding affinity is also obtained by mild delipidation; the effect can also be reversed by addition of lipid (Figs. V-6, V-8, VI-4).

If the low progesterone-binding affinity of purified AAG preparations results from small quantities of contaminating lipids, differences in steroid binding qualities observed in various samples should reflect the mode of their purification. Table XIII-6 bears this out. The smallest n k values were found with AAG preparations 2 and 3 which were obtained by chromatographic fractionation starting directly from serum. Preparations 1 and 4 have a higher binding affinity; they had been subjected to alcohol fractionation which presumably removed some of the contaminating lipid. It is relevant to mention in this context that WEIMER et al. (1950) extracted 3.6% lipid material from their original AAG preparation by hot alcohol. Alcohol-acetone precipitation of preparation 1 resulted in the highest apparent association constant (n k) observed under the conditions of our test procedure, i. e., that of AAG Type P (Table XIII-4). Filtration of the lipid-containing AAG, Type C, over Sephadex G-25 at 45°, pH 7.4, did not increase the binding affinity. Gel filtration under these conditions dissociates and separates progesterone from its relatively strong complex with CBG (Chapter VIII. 3, Fig. VIII-10). Failure to increase n k of AAG, Type C, by the stripping technique at 45° indicates that the low affinity is not caused by other endogenous steroid molecules occupying the binding sites of AAG. Evidently, the lipidic material adheres very tightly to AAG. Another explanation of the difficulties with the presence of lipid in AAG is assumed by VAN BAELEN and DE MOOR (1969) to be transfer of lipid contaminations from the Sephadex gel to the glycoprotein during the chromatographic fractionations.

VAN BAELEN and DE MOOR (1970) recently reported on progesterone binding to AAG, using the gel equilibration technique with a preparation obtained by MICHON precipitation and isoelectric separation. They confirmed our results on reversible inhibition of the interaction by lipid soluble contaminants which they assume originated from the Sephadex applied. The apparent association constant, at about 0°, of the progesterone complex with ether-extracted AAG was found to be $9.8 \times 10^5\ M^{-1}$. In further agreement with our observations, removal of sialic acid did not affect the binding affinity.

Elimination of lipid and of heavy metal ions (Chapter XIII. 3) provided AAG preparations that generally had progesterone-binding affinities of the same order as AAG, Type P. In spite of all precautionary measures, however, n k values were occasionally observed (see for example Table XIII-8 below) that fell short of the usual value of $10 \times 10^5\ M^{-1}$ (Table XIII-4). The reason for such discrepancy is not known in each case; we suspect the influence of trace contaminations, possibly introduced secondarily into the system. This difficulty emphasizes the importance of adequate controls in all binding studies with AAG.

3. Reversible Inhibition by Heavy Metal Ions of the Binding Affinity between AAG and Progesterone

Metal ion interference was first suspected when the n k value (4°) of the progesterone complex with a purified AAG preparation was raised from 3.2×10^5 to $8.9 \times 10^5 M^{-1}$ by use of glass-redistilled deionized water instead of ordinary distilled water (KERKAY and WESTPHAL, 1969). The AAG sample was therefore dialyzed against 60 vol. of 0.1% aqueous EDTA solution for 32 hrs at 4° to eliminate any heavy metal ions. Subsequent dialysis for 60 hrs at 4° against frequent changes of deionized double distilled water and lyophilization yielded an AAG preparation that interacted with progesterone at high affinity; the n k value measured at 4° was $10.5 \times 10^5 M^{-1}$. These results prompted an investigation of metal ion effects on the progesterone-binding affinity of AAG.

In equilibrium dialysis experiments at 4°, at an AAG concentration of 0.5 mg/ml, the influence of the metal ions was studied first at a level of 10 ions per molecule AAG, calculated under the assumption of complete equilibration of metal ions through the dialysis system. The minimal quantities of anions (sulfates or chlorides, nitrate in case of silver) introduced by the metal salts into the systems do not measurably influence the steroid-binding affinity of AAG (see Chapter XIII.5). The dialysis experiments showed that the n k values were practically not affected by the presence of Mg^{++}, Ca^{++}, Mn^{++}, Fe^{+++}, Co^{++}, Ni^{++}, Cu^{++}, Zn^{++}, Sn^{++}, Hg^{+}, and Pb^{++}. Definite inhibition of the apparent binding affinity was brought about by Hg^{++}, Ag^{+}, Cu^{+}, and Fe^{++}. The eleven ions that did not show an effect at the above concentration were not further investigated.

The relative efficiency of the four effective metal ions was determined at metal salt concentrations varying from 0 to 100 moles/mole AAG, assuming equal distribution of the ions throughout the dialysis system. To the test systems containing cuprous and ferrous ions ascorbic acid was added to give mM concentration. Table XIII-7 shows the results. In spite of some inconsistencies in the concentration series, it may be concluded that the approximate order of inhibiting power at the level of 10 ions per molecule AAG is $Hg^{++} > Ag^{+} > Cu^{+} > Fe^{++}$. It is also apparent that increase of the metal ion concentration above a few moles per mole AAG has very little additional effect. This behavior is contrary to what would be expected if the metal ions acted as competitive inhibitors of progesterone at the binding site.

A more complete analysis by Scatchard's procedure (Chapter V. 4 d) of the influence of 4 Hg^{++}, 4 Ag^{+}, 10 Cu^{+}, and 10 Fe^{++} per molecule AAG is presented in Figs. XIII-4, 5, 6 and 7. Straight lines were obtained by the method of least squares in all cases. A quantitative assessment of the inhibitory effects of the metal ions on the basis of $\bar{\nu}/[S] = n\,k$, taken from the intersections of the lines with the ordinates in Figs. XIII-4 to 7, gave 81% and 31% reduction of n k, respectively, for Hg^{++} and Ag^{+} at a ratio of 4 ions per molecule AAG; the inhibition by the addition of 10 Cu^{+} and 10 Fe^{++} was 86% and 25%, respectively. Within the range of error given by the relatively small number of progesterone concentrations used, the lines of the metal-free systems converge approximately at the $\bar{\nu}$ values for the one primary binding site established for the progesterone-AAG com-

Table XIII-7. *Apparent binding affinity of AAG for progesterone in the presence of inhibiting metal ions. In 0.05 M phosphate, pH 7.4, 4°.* From KERKAY and WESTPHAL (1969)

Metal ions per molecule AAG [a]	n k $M^{-1} \times 10^{-5}$			
	Hg^{++}	Ag^{+}	Cu^{+}	Fe^{++}
0	12.2 [b]	9.1	9.8	10.2
2	—	6.2	8.0	9.1
3	5.5	6.8	7.2	8.6
4	3.1	6.6	7.0	9.0
5	4.2	—	10.2 [b]	9.5
10	5.1	6.7	7.3	8.6
15	5.4	—	9.5	9.4
20	4.4	6.6	7.0	—
30	—	—	8.3	9.2
40	4.6	—	5.9	—
50	—	6.6	7.6	8.4
60	4.2	—	4.8	—
80	3.6	—	4.3	—
100	3.9	6.3	6.1	8.8

[a] Assuming even distribution throughout dialysis system.
[b] Reason for high value unknown.

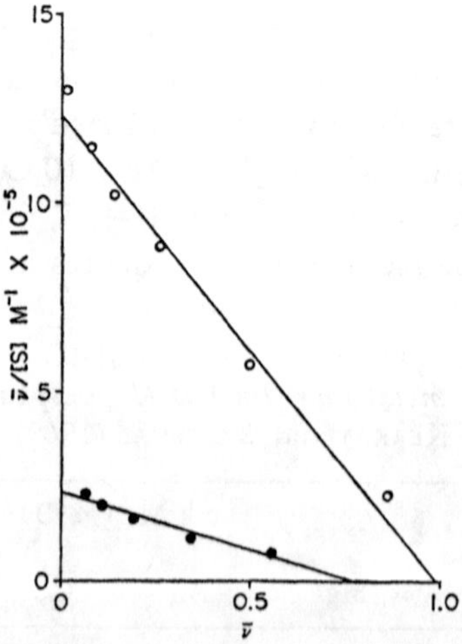

Fig. XIII-4. Influence of Hg^{++} on the progesterone-AAG interaction in 0.05 M phosphate buffer, pH 7.4, 4°. Protein concentration, 2.0 mg/ml (4.88×10^{-5} M); progesterone concentration, 6×10^{-8} to 3.18×10^{-5} M; metal ion concentration, 4 Hg^{++} per molecule AAG, assuming equal distribution of metal ions throughout the equilibrium dialysis system. Open circles, absence of metal ions; closed circles, presence of metal ions. From KERKAY and WESTPHAL (1969)

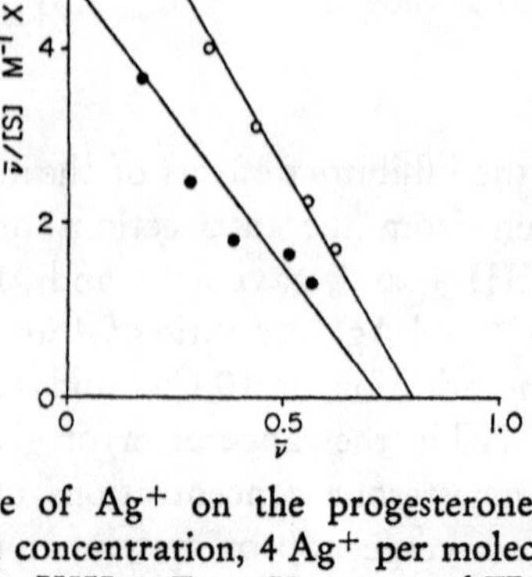

Fig. XIII-5. Influence of Ag^{+} on the progesterone-AAG interaction at pH 7.4, 4°. Metal ion concentration, 4 Ag^{+} per molecule AAG. Other conditions as in Fig. XIII-4. From KERKAY and WESTPHAL (1969)

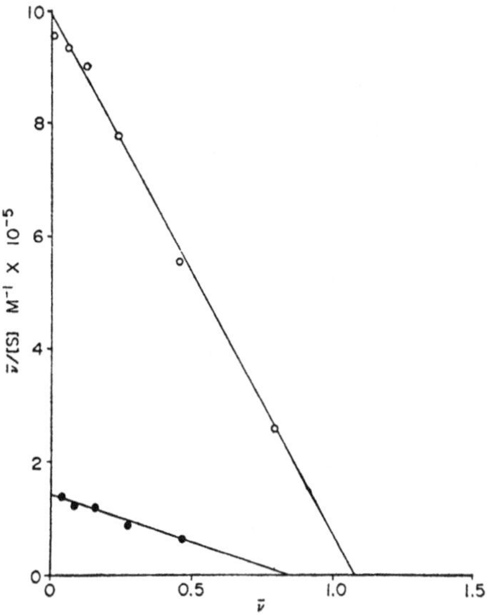

Fig. XIII-6. Influence of Cu^+ on the progesterone-AAG interaction at pH 7.4, 4°. Metal ion concentration, 10 Cu^+ per molecule AAG. Other conditions as in Fig. XIII-4. From KERKAY and WESTPHAL (1969)

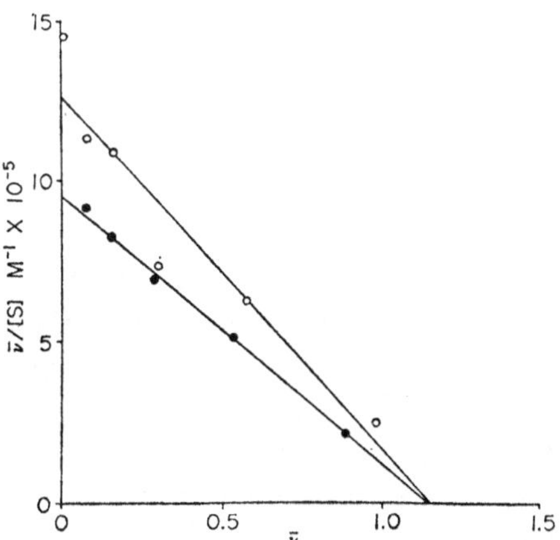

Fig. XIII-7. Influence of Fe^{++} on the progesterone-AAG interaction at pH 7.4, 4°. Metal ion concentration, 10 Fe^{++} per molecule AAG. Other conditions as in Fig. XIII-4. From KERKAY and WESTPHAL (1969)

plex. Somewhat lower n values are indicated in the presence of mercuric, silver and cuprous ions (Figs. XIII-4 to 6) whereas the binding curves in the absence and presence of ferrous ions (Fig. XIII-7) intersect with the abscissa at the same point corresponding to $n = 1.1$.

The effect of EDTA on the metal inhibition of the progesterone-AAG interaction was examined by either including mM EDTA

Table XIII-8. *Effect of EDTA on metal ion inhibition of progesterone-AAG interaction at 4°, in 0.05 M phosphate buffer, pH 7.4.* From KERKAY and WESTPHAL (1969)

Metal ions per molecule AAG [a]	n k $M^{-1} \times 10^{-5}$	
None (control)	8.6	8.1
4 Hg^{++}	1.7	—
4 Hg^{++} + mM EDTA	6.4	—
4 Hg^{++}; then mM EDTA [b]	—	6.0
4 Ag^{+}	4.3	—
4 Ag^{+} + mM EDTA	6.7	—
4 Ag^{+}; then mM EDTA [b]	—	6.9
10 Cu^{+}	5.0	—
10 Cu^{+} + mM EDTA	8.3	—
10 Cu^{+}; then mM EDTA [b]	—	7.9
10 Fe^{++}	7.4	—
10 Fe^{++} + mM EDTA	9.0	—
10 Fe^{++}; then mM EDTA [b]	—	8.3

[a] Assuming even distribution throughout dialysis system.

[b] AAG and metal ions incubated for 48 hr at 4°, followed by usual equilibrium dialysis in the presence of mM EDTA.

directly with AAG, progesterone and the ions in the equilibrium dialysis systems, or by preincubating AAG with the metal ions at the same concentrations used in the previous experiments (Fig. XIII-4 to 7) at 4° for 48 hrs and then adding EDTA to give a mM solution; testing for binding affinity followed immediately. Controls of the same batch of AAG were used. The results (Table XIII-8) show that the AAG preparations used were somewhat low in binding affinity, illustrating the point discussed in the last paragraph of Chapter XIII. 2. It is evident, however, that the protein can be protected from the inhibitory action of the four metal ions when a strongly chelating agent is present. The n k values in the presence of both metal ion and EDTA approximate those of the metal-free control in case of Cu^{+} and Fe^{++}, whereas the recovery is not as complete in the experiments with the more active ions Hg^{++} and Ag^{+}. EDTA itself was found to be without influence on the progesterone-binding affinity of AAG, when up to 100 moles EDTA per mole of the pure glycoprotein were added to the system. Table XIII-8 also shows that the inhibitory effect of these metal ions on the progesterone-AAG interaction is reversible, since the binding affinity returned approximately to the control level when EDTA was added after the metal ions had been incubated with the protein for 48 hr.

The results on metal ion effects given in Table XIII-7 have shown that the inhibition most likely is not competitive. This conclusion is corroborated by the more complete analysis given in Figs. XIII-4 to 7. In case of competitive inhibition, convergence of the lower lines at the intersection of the upper lines with the abscissa would be expected, indicating displacement of the metal ions by increasing amounts of progesterone. This is not the case except for the weak inhibition shown by ferrous ions (Fig. XIII-7).

In support of noncompetitive binding it should also be considered that the inhibiting metal ions and progesterone are quite dissimilar in their structure, making effective competition at the binding site improbable for steric reasons. In contrast, true competition for the progesterone binding site of AAG has been observed with several Δ^4-3-ketosteroids closely related to progesterone in their molecular structure (see Chapter XIII. 4).

The present findings are consistent with the assumption of an allosteric mechanism of the metal ion inhibition. Interaction of the ion inhibitors at a site in AAG other than the progesterone-binding site may lead to a conformational change of the AAG molecule and concomitant decrease of the apparent affinity of the primary binding site for the steroid. The resulting AAG, possessing the same primary binding site but of reduced affinity, might well interact with progesterone according to the lower lines in Figs. XIII-4 to 7.

The inhibition of progesterone-binding affinity by the four metal ions might suggest the involvement of sulfhydryl groups since these ions are known to react with thiol structures with great affinity (Gurd and Wilcox, 1956; Cecil and McPhee, 1959). However, all attempts to demonstrate a sulfhydryl group in human AAG, even in the presence of 8 M urea, have been unsuccessful (KROTOSKI and WEIMER, 1967; GANGULY and WESTPHAL, 1968 b). Interaction with disulfide has to be considered since 6 half-cystine residues have been found in human AAG (see Chapter XIII. 6). It is known that especially Hg^{++} and Ag^{+} interact with disulfide structures (CECIL, 1963) and cuprous ions form strong complexes not only with sulfhydryl groups, but generally with the lone electron pairs of sulfur atoms (BRILL et al., 1963). In view of the ready reversibility of the metal ion effect by EDTA, a noncovalent interaction of the ions with a disulfide grouping is therefore a possible interpretation. However, no direct evidence for this assumption is presently available, and interaction of the metal ions with various groupings other than disulfide must be considered at least equally possible.

In view of some of the early results on fractional values for n, experiments were carried out to test the possibility that more than one AAG molecule interacted with one progesterone molecule. Studies of sedimentation rates in the ultracentrifuge and gel filtration experiments with Sephadex G-200 under various conditions failed to provide evidence for such mechanism of interaction. Details may be found in the original publication (KERKAY and WESTPHAL, 1969).

4. Interaction of Other Steroids with AAG

It has been discussed in a previous chapter (VI. 5 d) that the apparent association constant of the cortisol-HSA complex decreased with increasing protein concentration (Table VI-10). An effect in the opposite direction has been observed in the interaction of AAG with various steroids. This was determined by equilibrium dialysis at 4° and 37°; the binding affinity was measured as $\bar{\nu}/[S] = n\,k$. The AAG concentration was varied from 0.1 mg/ml (0.24×10^{-5} M) to 5.0 mg/ml (12.2×10^{-5} M) in the inside solution. The steroid concentration was held constant at 0.02 μg/ml outside solution (about 6×10^{-8} to 7×10^{-8} M) in one set of dialysis vessels, while in another set the molar ratio of steroid to AAG was kept constant at 1 : 100, calculated on the basis of total steroid and protein present.

Table XIII-9 shows the results obtained with six steroids over a 50-fold range of AAG concentration. At 4° and 37°, an increase in the apparent binding affinity was generally observed with protein level rising to approximately 1 mg AAG/ml; the n k values then leveled off or decreased except for the estradiol-AAG complex. The change of apparent steroid-binding affinity with rising AAG level was found at a constant steroid level as well as at a constant steroid: AAG ratio of 1 : 100. Adsorption of AAG to the dialysis membrane or to other surfaces was excluded as an explanation of the results by sensitive protein analysis before and after equilibrium dialysis. VAN BAELEN and DE MOOR (1970) recently reported a similar dependency of progesterone affinity on the AAG concentration.

A likely explanation of the concentration effect is based on the finding that steroid-binding affinity of AAG can be inhibited by extremely small quantities of lipid-soluble contaminants (Chapter XIII. 2) or of heavy metal ions (Chapter XIII. 3). Evidently, a given small amount of contaminant which, for example, would affect only 7—8% of the AAG molecules at 1.0 mg/ml, would inhibit approximately 70—80% of the AAG molecules at a concentration of 0.1 mg/ml. The glycoprotein would thus be proportionally less affected at higher concentrations, and the apparent binding affinity would increase as seen in Table XIII-9. It has been discussed in Chapter XIII. 2 that unexplained inhibition of steroid binding was occasionally observed with our AAG preparations in spite of all careful precautions to eliminate contamination.

Earlier studies had shown that AAG interacts with a number of Δ^4-3-ketosteroid hormones in accordance with the polarity rule (WESTPHAL, 1964; WESTPHAL and CARNIGHAN, 1964). For an accurate determination of the number of binding sites and the apparent association constants of the complexes, binding data obtained by equilibrium dialysis at 4° and 37° were evaluated according to the Scatchard procedure (Chapter V. 4. d). An IBM 1130 computer was used to obtain a best fit to the experimental points of a theoretical curve given by the first two terms of Eq. (V-18). The quantities n_1, k_1, n_2, k_2 were varied until optimal coincidence with the experimental data was achieved. In most cases, however, the solubility of the steroids did not permit the use of sufficiently high concentrations for $\bar{\nu}$ values greater than one. In these instances, only the first term of Eq. (V-18) was used.

The results of these studies are illustrated in Figs. XIII-8 to 13 for the AAG complexes of progesterone, cortexone, corticosterone, cortisol, testosterone and estradiol, respectively. With the exception

Table XIII-9. *Apparent binding affinities of steroid-AAG complexes at increasing concentrations. In 0.05 M phosphate buffer, pH 7.4.* From KERKAY and WESTPHAL (1968, 1969)

AAG concn. (mg/ml)	Steroid: AAG molar ratio[a]	$n\,k\ M^{-1} \times 10^{-5}$										
		Progesterone	Cortexone		Corticosterone		Cortisol		Testosterone		Estradiol	
		4°	4°	37°	4°	37°	4°	37°	4°	37°	4°	37°
0.1	1 : 20	2.1	1.46	0.51	0.34	0.29	0.18	0.18	3.7	1.31	0.38	0.33
	1 : 100	2.6	1.25	0.57	0.19	0.21	0.13	0.22	3.1	1.31	0.38	0.27
0.5	1 : 100	6.7	2.14	0.81	0.81	0.35	0.21	0.12	6.0	1.94	0.74	0.46
	1 : 100	6.4	2.14	0.81	0.81	0.35	0.21	0.12	6.0	1.98	0.74	0.46
1.0	1 : 200	8.6	2.26	0.82	1.01	0.36	0.26	0.15	6.0	2.16	0.99	0.58
	1 : 100	10.4	2.08	0.80	0.99	0.37	0.27	0.15	5.8	2.12	1.00	0.56
2.0	1 : 400	9.2	2.05	0.88	1.01	0.38	0.26	0.14	5.1	2.21	1.21	0.67
	1 : 100	8.7	1.98	0.85	1.04	0.38	0.26	0.15	5.1	1.88	1.15	0.65
5.0	1 : 1000	9.5	1.47	0.69	0.97	0.38	0.25	0.13	3.5	1.59	1.39	0.70
	1 : 100	9.1	1.45	0.65	0.93	0.36	0.24	0.13	2.6	1.41	1.31	0.74

[a] Calculated on the basis of total steroid and total protein present in the dialysis system.

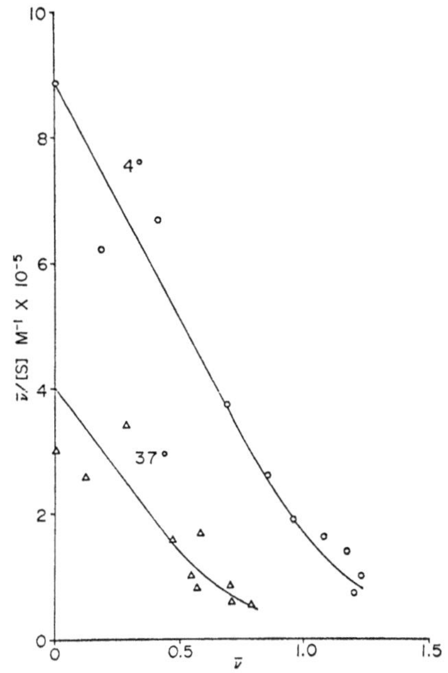

Fig. XIII-8. Binding of progesterone to AAG in 0.05 M phosphate buffer, pH 7.4. Protein concentration: 0.5 mg/ml (1.22×10^{-5} M). Circles: 4°; triangles: 37°. Lines represent theoretical curves for Eq. (V-18) obtained as described in the text. From KERKAY and WESTPHAL (1969)

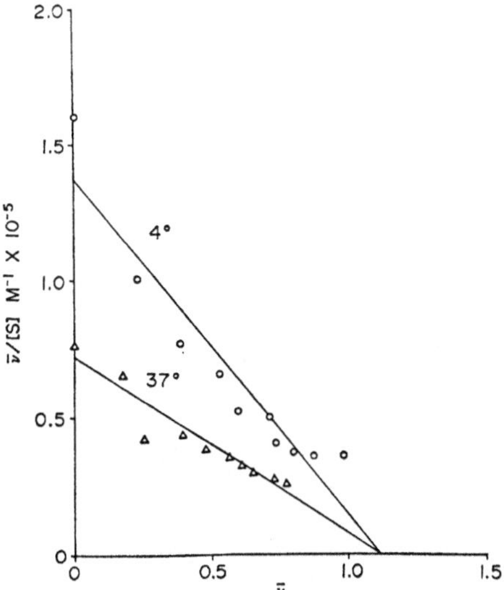

Fig. XIII-9. Binding of cortexone to AAG at a protein concentration of 1.0 mg/ml ($2.44 \cdot 10^{-5}$ M) and steroid concentrations of 0.006—$5.45 \cdot 10^{-5}$ M. Other explanations as in Fig. XIII-8. From KERKAY and WESTPHAL (1968)

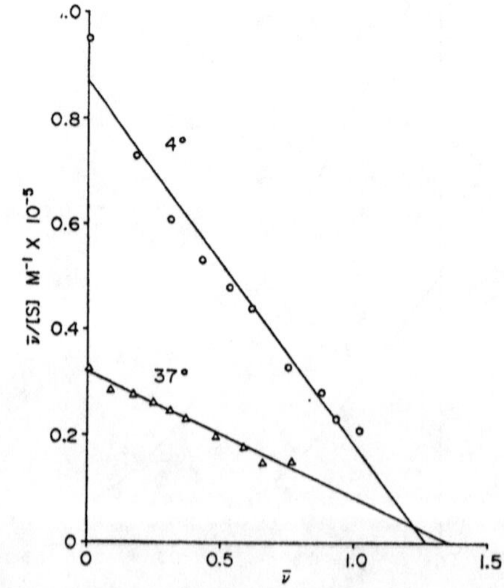

Fig. XIII-10. Binding of corticosterone to AAG at a protein concentration of 1.0 mg/ml ($2.44 \cdot 10^{-5}$ M) and steroid concentrations of 0.006—$8.66 \cdot 10^{-5}$ M. Other explanations as in Fig. XIII-8. From Kerkay and Westphal (1968)

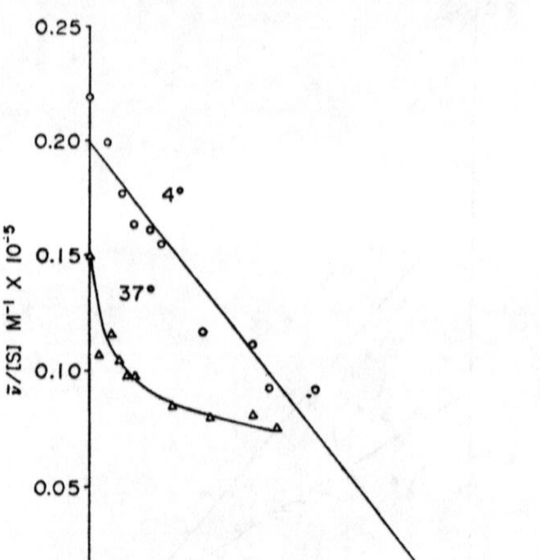

Fig. XIII-11. Binding of cortisol to AAG at a protein concentration of 1.0 mg/ml ($2.44 \cdot 10^{-5}$ M) and steroid concentrations of 0.006—$13.79 \cdot 10^{-5}$ M. Other explanations as in Fig. XIII-8. From Kerkay and Westphal (1968)

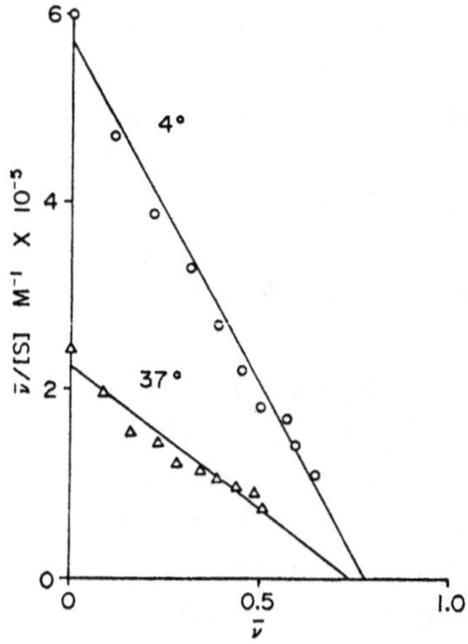

Fig. XIII-12. Binding of testosterone to AAG at a protein concentration of 2.0 mg/ml ($4.88 \cdot 10^{-5}$ M) and steroid concentrations of 0.007—$1.73 \cdot 10^{-5}$ M. Other explanations as in Fig. XIII-8. From KERKAY and WESTPHAL (1968)

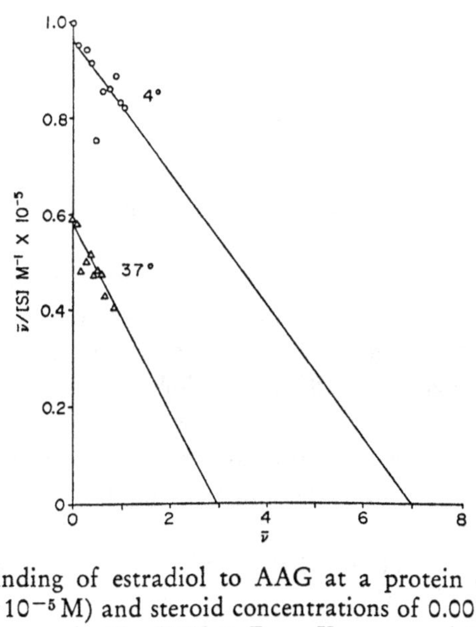

Fig. XIII-13. Binding of estradiol to AAG at a protein concentration of 2.0 mg/ml ($4.88 \cdot 10^{-5}$ M) and steroid concentrations of 0.007—$3.30 \cdot 10^{-5}$ M. Other explanations as in Fig. XIII-8. From KERKAY and WESTPHAL (1968)

of testosterone (Fig. XIII-12), the values for the primary binding sites, n_1, are close to unity for all Δ^4-3-ketosteroids analyzed. Most of them are somewhat greater than one. A contributing factor to this discrepancy from a value of $n_1 = 1.0$ may be the use of a mol. wt. of 41,000 for AAG (Table XIII-1) which may be too high; more recent publications give values of 37,000 (Kalous and Poncová, 1965), 39,000 (Li and Li, 1970), and 39,600 (Schmid et al., 1968 b) for this glycoprotein.

The values for n_1 and k_1 are compiled in Table XIII-10. AAG has one primary, high-affinity binding site for progesterone. It is apparent from the shape of the curves (Fig. XIII-8) that secondary sites, n_2, exist which were calculated to have values of about 10 and 5 at 4° and 20°, respectively. However, these values must be considered very approximate since they were obtained by extrapolation over a long range, due to the low solubility of progesterone. The apparent association constants for the secondary binding sites, k_2, were approximately $4—6 \times 10^3$ M^{-1}.

With few exceptions, the experimental points in Figs. XIII-9 to 13 form straight lines although some spreading is evident. Only the first term of Eq. (V-18) was therefore applied, even for the slightly curved line for cortexone interaction at 4° (Fig. XIII-9) which indicates secondary binding sites; computer calculation gave the very approximate value of $n_2 = 15$. Evaluation of the curve obtained for cortisol binding at 37° (Fig. XIII-11) according to Eq. (V-18) gave a fractional value of 0.33 for n_1. This finding together with some other observations prompted consideration of the heterogeneity of AAG as a possible cause for the appearance of abnormal n_1 values.

Polymorphic variants of AAG have been described by Schmid et al. (1962, 1964, 1965) and by Marshall (1966). A simple method for their identification in the desialized form was developed (Hunziker and Schmid, 1967) which facilitated chemical investigation of the variants (Schmid et al., 1968 b). Isolation of pure AAG from Caucasian patients with various diseases showed that the relative distribution of the three types of AAG variant patterns was very similar to that of normal adults (Tokita and Schmid, 1967). Constancy of the types of AAG variants was further observed in patients after major operations (Tokita et al., 1966), after hypophysial stalk section (Schmid et al., 1968 a), and after complete hysterectomy and subsequent irradiation (Yoshizaki et al., 1969)—conditions of severe

Table XIII-10. *Numbers of binding sites, n_1, apparent association constants, k_1, and thermodynamic parameters of steroid-AAG complexes. In 0.05 M phosphate buffer, pH 7.4.* From KERKAY and WESTPHAL (1968, 1969)

Steroid	Temp.	n_1	$k_1 \times 10^{-5}$ M^{-1}	$n k \times 10^{-5}$ M^{-1}	$\Delta F°$ kcal/mole	$\Delta H°$ kcal/mole	$\Delta S°$ cal mole^{-1} degree^{-1}
Progesterone	4	0.98 [a]	9.0	8.9	—7.5	—4.8	10
	37	1.05	3.5	3.5	—7.9		
Testosterone	4	0.78	7.3	6.0	—7.3	—4.0	12
	37	0.74	3.0	2.4	—7.8		
Cortexone	4	1.12	2.2	2.3	—6.7	—3.4	12
	37	1.11	1.0	0.88	—7.1		
Corticosterone	4	1.27	0.87	1.0	—6.0	—4.8	5
	37	1.35	0.32	0.37	—6.2		
Cortisol	4	1.24	0.16	0.20	—5.5	—1.5	14
	37	—	—	0.15	—5.9		
Estradiol	4	7	0.14	1.0	—6.2	—2.7	13
	37	3	0.19	0.59	—6.8		
Progesterone-HSA [b]	4	2.1	1.0	2.1	—6.3 [c]	—4.9	9
	37	2.6	0.46	1.2	—6.6 [c]		

[a] A value of $n_1 = 1.00$ was found at 20°.
[b] Values for the progesterone-HSA complex from Tables VI-9 and VI-11.
[c] Value for k_1.

stress in which considerable increase and variation in the blood concentration of AAG occur. These observations were interpreted by SCHMID and his coworkers to be consistent with the concept of genetically controlled transmission of the AAG variants; recent immunofixation studies by JOHNSON et al. (1969) indicate inheritance as autosomal traits with codominant expression. YAMAUCHI and YAMASHINA (1969) assume that the AAG polymorphism is related to variations in the polysaccharide units.

Recent work in our laboratory (RYAN, 1970) resulted in the separation of 7 variants of AAG, characterized by 5 isoelectric points (see Table XIII-1, footnote d). Investigation of the variants of pI 3.0, 3.2, and 3.4 showed the first two to be indistinguishable from the pure pooled AAG preparations in physicochemical and progesterone-binding properties. The variant of pI 3.4 contained less glutamic acid (30.2%) and more alanine (12.7%) than the unfractionated AAG (34% and 10.4%, respectively). The progesterone-binding affinity of AAG, pI 3.4 ($k = 6.7 \times 10^5 M^{-1}$; $\Delta\varepsilon = -10.8\%$) was about 2/3 that of the standard AAG.

The relatively high solubility of cortisol at 37° permitted the use of greater concentrations to test for additional binding sites, clearly indicated by the shape of the curve in Fig. XIII-11. The theoretical curve [Eq. (V-18)] best fitting the experimental data was obtained for $n_2 = 30$; the corresponding apparent association constant, k_2, was 60 and thus approximately 250 times smaller than k_1.

Comparatively large numbers of primary binding sites were observed in the interaction of AAG with estradiol (Fig. XIII-13, Table XIII-10). The dilemma of low steroid solubility is particularly evident in this case; the Scatchard graph leaves no doubt, however, that n_1 is considerably larger than unity even though the numerical value is uncertain because of the extended extrapolation. An average value of $n_1 \cong 5$ gives the order of magnitude.

According to the apparent affinity constants shown in Table XIII-10, the interactions of AAG with the Δ^4-3-ketosteroids investigated follow the polarity rule. The thermodynamic parameters show negative enthalpy changes for the complexes. They form spontaneously as the negative ΔF° values indicate. All entropy changes are positive. This binding behavior is similar to that of steroid-HSA interaction; data for the progesterone-HSA complex are included in Table XIII-10 for comparison. The interpretation of the interaction on the basis of the thermodynamic parameters has been elaborated in Chapter VI. 5. e.

The number of binding sites in AAG for all Δ^4-3-ketosteroids measured is $n_1 = 1$ (Table XIII-10), whereas estradiol was bound to approximately 5 sites. It was of interest to determine if there was competition of the steroids for the one binding site that interacts strongly with progesterone. This was done by double-isotope experiments in which tritiated corticosterone, testosterone and estradiol

Table XIII-11. *Steroid competition for the progesterone-binding site of AAG. In 0.05 M phosphate buffer, pH 7.4, 4°. Concentration of all steroids, 2.0 µg/ml outside solution; AAG concentration, 0.5 mg/ml.* From KERKAY and WESTPHAL (1968)

Steroid tested	$\bar{\nu}$ obs.	$\bar{\nu}$ calc. by Eq. (V-29)	Competition
Corticosterone [a]	0.34	0.30	Yes
Testosterone [a]	0.36	0.31	Yes
Estradiol [a]	0.40	0.10	No
Progesterone [b]	0.54	—	—
Progesterone [c]	0.53	—	No

[a] In presence of progesterone.
[b] In absence of estradiol.
[c] In presence of estradiol.

were combined with progesterone-4-^{14}C and AAG in equilibrium-dialysis systems at 4° (KERKAY and WESTPHAL, 1968). Eq. (V-29) was used to calculate the theoretical $\bar{\nu}$ values for the three tritiated steroids in the presence of progesterone, and they were compared with the observed data. The results (Table XIII-11) show satisfactory agreement between theoretical and experimental $\bar{\nu}$ values for corticosterone and testosterone, indicating competition for the progesterone-binding site in the AAG molecule. In contrast, the difference obtained for the estradiol complex suggests that this phenolic steroid does not interact at the binding site for progesterone. This conclusion was corroborated by the finding that the $\bar{\nu}$ values observed for progesterone in the presence and absence of estradiol were virtually identical (Table XIII-11).

These observations would suggest an important role of the Δ^4-3-keto group for interaction of the steroids with AAG. This grouping is present in both corticosterone and testosterone as well as in progesterone although the two competing steroids are different in other parts of the molecule. If the binding of testosterone to AAG would occur at the 17β-hydroxy group one might expect a similar binding behavior of estradiol which has the same 17β-hydroxy structure in ring D; such a mechanism is assumed for the steroid-binding β-globulin as discussed in Chapter XII. 3. Competition at the single progesterone-binding site of AAG was not considered likely for estradiol because this phenolic steroid interacts with clearly more than one binding site of the AAG molecule (Fig. XIII-13 and Table XIII-10). The results confirm the significance of the Δ^4-3-keto group for steroid-protein interaction as previously concluded from spectral studies of the association of Δ^4-3-ketosteroids with serum albumin and other proteins (Chapter VII). Similar spectral observations for the progesterone-AAG complex, which will be discussed in the next section of this chapter (see Table XIII-12), corroborate this mechanism also for steroid interaction with AAG.

5. Influence of Perturbants on Steroid-AAG Interactions

Basic observations. When the progesterone-AAG complex is exposed to increasing concentrations of sodium chloride, the apparent association constant rises in proportion to the salt concentration (Westphal et al., 1966). This is shown in Fig. XIII-14 which summarizes the original observations obtained (Ganguly and Westphal, 1968 a) with a chromatographically pure AAG preparation, Type C (see Chapter XIII. 2). The upper line indicates a logarithmic increase of n k ($\bar{\nu}$/[S]) in direct proportion to the molarity of the NaCl solutions; at 4 M NaCl, the apparent association constant is approximately 10 times greater than in phosphate buffer (0.05 M, pH 7.4).

In contrast, addition of lithium bromide to the progesterone-AAG complex resulted in a decrease of the apparent binding affinity with increasing concentration of the salt (Fig. XIII-14). This experiment was done in water instead of phosphate buffer as solvent, a variation which accounts for the lower level of binding affinity (see below Figs. XIII-15 and XIII-16). The decrease was not logarithmic with concentration. When urea was used as the perturbant, the n k value fell to less than 1/100 original, with urea levels rising up to 8 M; the reduction was again nearly logarithmic with molarity.

A comparison with the effect of NaCl on the apparent affinity constant of the progesterone-HSA complex (Fig. XIII-14, broken line) shows much less increase. The n k value at 4 M NaCl was only about twice that measured in the absence of sodium chloride.

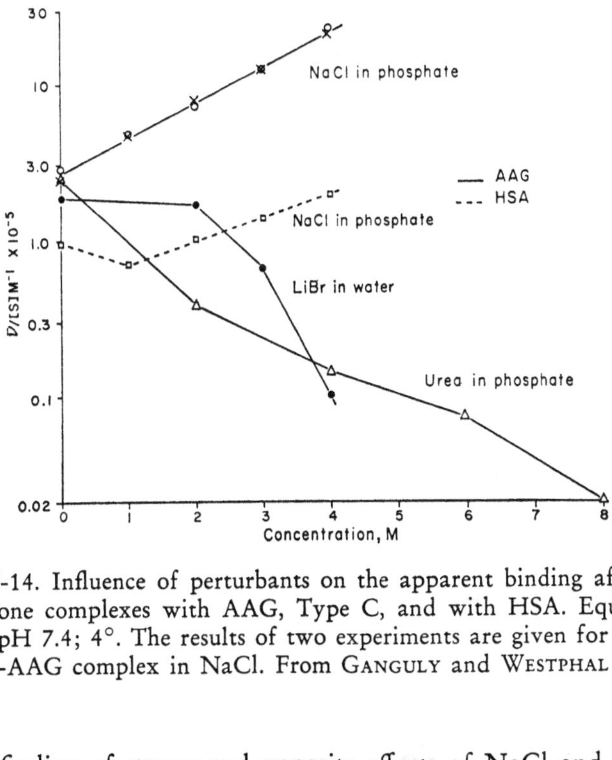

Fig. XIII-14. Influence of perturbants on the apparent binding affinity of progesterone complexes with AAG, Type C, and with HSA. Equilibrium dialysis; pH 7.4; 4°. The results of two experiments are given for the progesterone-AAG complex in NaCl. From GANGULY and WESTPHAL (1968 a)

The finding of strong and opposite effects of NaCl and LiBr on the binding affinity between progesterone and AAG prompted a study of the influence of a larger group of neutral salts. Urea was included in view of its known effects on protein conformation. In most cases, 4 M solutions of the perturbants were applied; some salts gave strong effects at lower concentrations. Figure XIII-15 summarizes the observations with AAG, Type C, in phosphate buffer and in distilled water. On the basis of the results with NaCl and urea (Fig. XIII-14), straight lines were drawn between the logarithmically plotted $\bar{\nu}$/[S] values at high and at zero perturbant concentrations although a logarithmic relationship may not obtain for all cases.

In a second series, similar measurements were made with the delipidated AAG preparation, Type P (Chapter XIII. 2), including some additional salts (Fig. XIII-16). Again, the level of the apparent

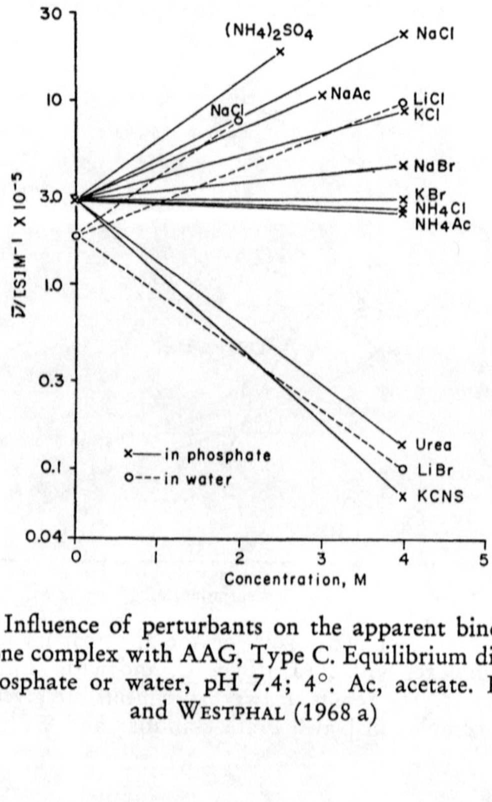

Fig. XIII-15. Influence of perturbants on the apparent binding affinity of the progesterone complex with AAG, Type C. Equilibrium dialysis was done in 0.05 M phosphate or water, pH 7.4; 4°. Ac, acetate. From GANGULY and WESTPHAL (1968 a)

binding affinity of AAG for progesterone is higher in phosphate, pH 7.4, than in distilled water adjusted to the same pH. This may be a specific influence of the phosphate ion as will be discussed below. The strongest affinity-enhancing effect is seen with sodium sulfate, at 1 M concentration exceeding that of 4 M NaCl. Potassium thiocyanate has the maximal influence in reducing the stability of the complex.

These studies indicate (Figs. XIII-14, 15, 16) that the apparent binding affinity between AAG and progesterone is greatly affected

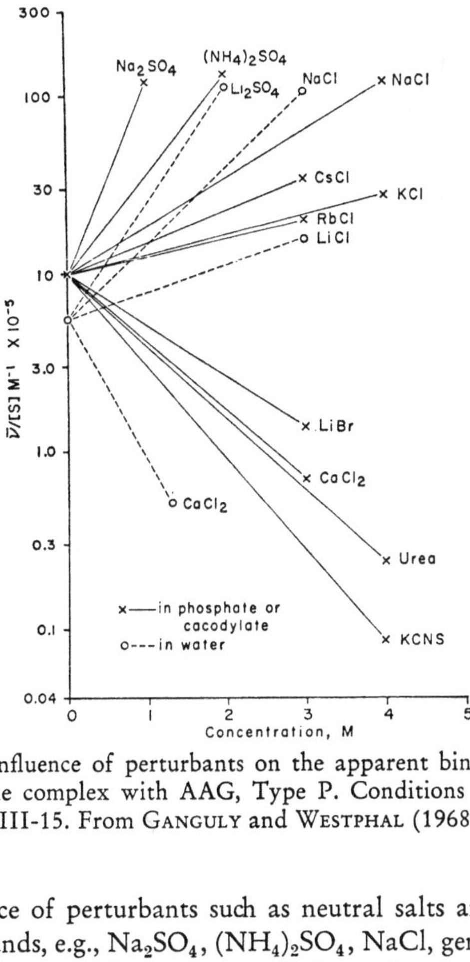

Fig. XIII-16. Influence of perturbants on the apparent binding affinity of the progesterone complex with AAG, Type P. Conditions were as in Fig. XIII-15. From GANGULY and WESTPHAL (1968 a)

by the presence of perturbants such as neutral salts and urea. Some of the compounds, e.g., Na_2SO_4, $(NH_4)_2SO_4$, NaCl, generally salting-out agents*, increase the apparent affinity; others, such as LiBr or urea, generally salting-in agents, decrease the apparent binding strength; and some salts are without effect. Starting with the salt most effective in stabilizing the complex, a perturbant series may be derived which resembles the Hofmeister series:

$Na_2SO_4 > Li_2SO_4 > (NH_4)_2SO_4 > NaCl > CsCl > LiCl >$ sodium acetate $> KCl > RbCl > NaBr > KBr \cong NH_4Cl \cong NH_4$-acetate $> LiBr > CaCl_2 >$ urea $> KCNS$.

* Although NaCl is usually considered a salting-out agent (SCHULTZE and HEREMANS, 1966), a salting-in effect has been observed with carbon monoxide-hemoglobin up to an ionic strength greater than 4 (TAYLOR, 1953).

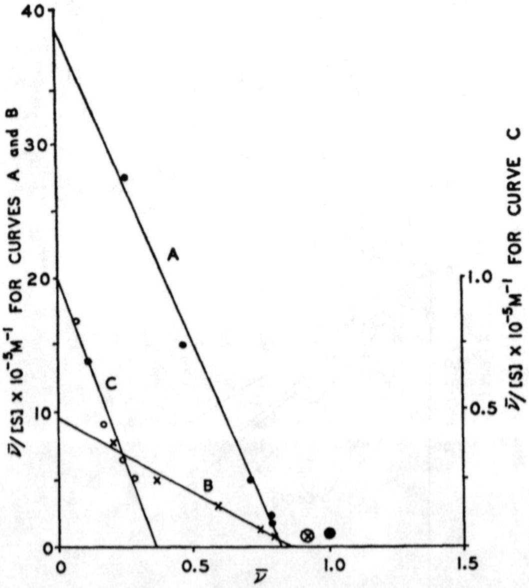

Fig. XIII-17. Scatchard plots of binding data for the complex of progesterone with AAG, Type P, in the presence and absence of perturbants. All solutions were in 0.05 M phosphate, pH 7.4; 4°. A, in presence of 2 M NaCl; n = 0.82; k = 46.8×10^5 M^{-1}. B, control, in 0.05 M phosphate; n = 0.86; k = 10.9×10^5 M^{-1}. C, in presence of 3 M LiBr; n = 0.36; k = 2.8×10^5 M^{-1}. From GANGULY and WESTPHAL (1968 a)

The turning point between affinity-enhancing and affinity-lowering influence is reached with KBr; this salt, as well as NH_4Cl and NH_4-acetate, does not change the apparent binding affinity to any measurable extent.

Scatchard analysis. In an effort to elucidate the mechanism of the perturbant effects on the progesterone-AAG complex, the binding systems were analyzed by the Scatchard method. Figure XIII-17 shows this evaluation for the interaction in the presence of 2 M NaCl (line A), of 3 M LiBr (line C), and in the standard phosphate buffer as control in the absence of perturbants (line B). The apparent association constant, k_1, of the progesterone-AAG complex in the presence of 2 M NaCl is 46.8×10^5 M^{-1}, increased more than 4-fold from the control value of 10.9×10^5 M^{-1} for the same AAG preparation in phosphate buffer. The number of binding sites is approximately n = 1 in both cases. This result shows that the increase of the n k value observed in the presence of a relatively high NaCl concentration is caused by enhancement of the apparent affinity constant and not by a larger number of binding sites interacting with the steroid.

Scatchard analysis of the progesterone-AAG complex in the presence of 3 M LiBr (Fig. XIII-17, line C; notice different scale at right hand ordinate) reveals a reduction of the apparent affinity constant to $k = 2.8 \times 10^5 M^{-1}$. In addition, the number of binding sites is reduced to $n = 0.36$, so that n k becomes about 1/10 of the control value. To explain the observation of a fractional binding site it is assumed that LiBr interferes with the free access of the steroid to a major portion of the specific primary sites, or inactivates the sites by other mechanisms. Only a fraction of the molecules, corresponding to n, remains able to associate with progesterone, although with reduced affinity; the remaining 64% of the glycoprotein molecules have no or very little binding affinity. This interpretation is in accordance with the general experience that the destabilizing, salting-in perturbants such as LiBr are more firmly bound to the proteins than the stabilizing, salting-out compounds (Robinson and Jencks, 1965 a, 1965 b; Bianchi et al., 1967). Furthermore, the polypeptide chain in its unfolded form may provide less affinity to steroids, presumably because of a poorer fit of groups adjacent to the binding site.

Spectral determination of binding. The results discussed so far on the influence of perturbants on the progesterone-AAG interaction were obtained by equilibrium dialysis. Since it was desirable to examine the effects by a second, independent method, spectrophotometric measurements (Chapter IV. 13) were performed at the wavelength of maximal progesterone absorption in the ultraviolet. The $\Delta\varepsilon$ values for the progesterone-AAG complex are given in Table XIII-12 for different ratios of steroid to protein, in the presence and absence of 4 M NaCl and 3 M $CaCl_2$. The cacodylate buffer used in these experiments has been found to provide the same binding parameters as the standard phosphate buffer. In the perturbant-free solutions, maximum depression of about 16% was observed at a progesterone to AAG ratio of approximately unity or lower. The $\Delta\varepsilon$ values decreased when more than 1 mole of progesterone was present per mole of glycoprotein, as was to be expected when part of the steroid remained unbound. In 4 M NaCl, the depression of ε was increased to 19 or 20%, whereas $\Delta\varepsilon$ was reduced in 3 M $CaCl_2$ to about 11 to 12%. Generally, lower $\Delta\varepsilon$ values were obtained with Type C AAG. These observations clearly corroborate by an independent method the findings obtained by equilibrium dialysis.

Reversibility. For an interpretation of the perturbant effects, it is important to know whether or not the changes in the apparent binding affinities are reversible. The results of pertinent studies (Table XIII-13) show that the increase of n k produced by 1 M NaCl is

Table XIII-12. *Influence of perturbants on the ultraviolet absorption of the progesterone-AAG complex. $\Delta\varepsilon$ indicates depression of the extinction coefficient at 249 nm. The buffer used was 0.15 M NaCl-0.013 M sodium cacodylate, pH 7.4, at 22° ± 1°.* From Ganguly and Westphal (1968 a)

AAG type	Concentration of Progesterone	AAG	Perturbant	$\Delta\varepsilon$
	$M \times 10^5$			%
P	1.47	2.51	None	15.3
	1.47	1.25	None	15.7
	1.30	1.18	None	16.4
	0.39	0.39	None	16.6
	2.34	1.01	None	8.3
	0.76	1.09	4 M NaCl	19.0
	0.39	0.39	4 M NaCl	19.7
	0.94	1.09	3 M $CaCl_2$	11.3
	0.39	0.39	3 M $CaCl_2$	12.1
C	1.53	1.32	None	10.2
	2.50	1.02	None	6.5

reversible; the value of 17.1×10^5 M^{-1} decreased to 3.9×10^5 M^{-1} when the salt was removed by dialysis. Subsequent precipitation with alcohol-acetone raised n k to 7.0×10^5 M^{-1}, i. e., fairly close to that of the untreated control.

The effect of urea was also reversible (Table XIII-13, Experiment 3). The low n k value of <0.3 in the presence of 4 M urea was restored to nearly normal by removal of the perturbant. Schmid and Kamiyama (1963) have concluded from optical rotatory dispersion studies on AAG that the unfolding which occurred in the presence of 4 M urea or 5 M LiBr was completely reversible; removal of the perturbants by dialysis resulted in AAG preparations with the optical properties of the untreated glycoprotein. The present finding of virtually complete reversibility of the effect of 4 M urea on the apparent binding affinity for progesterone is in accordance with these optical rotatory dispersion measurements.

The stabilizing effect of 4 M NaCl on the progesterone-AAG complex was also observed in ammediol at pH 10 and in sodium

Table XIII-13. *Reversibility of perturbant effects on the apparent binding affinity of the progesterone-AAG complex. AAG concentration was 0.5 mg per ml; equilibrium dialysis was carried out in 0.05 M phosphate, pH 7.4, at 4°.* From GANGULY and WESTPHAL (1968 a)

Experiment No. and treatment	n k
	$M^{-1} \times 10^{-5}$
1-1. None	10.2
1-2. None	9.4
1-3. 0.05 M phosphate, pH 7.4 [a]	9.2
2-1. Equilibrium dialysis in 0.1 M Tris-1.0 M NaCl, pH 8.0	17.1
2-2. 0.1 M Tris-1.0 M NaCl, pH 8.0 [b]	3.9
2-3. As in experiment 2-2, followed by precipitation with alcohol-acetone	7.0
3-1. 4 M urea [c]	7.3
3-2. Lyophilized AAG from experiment 3-1, precipitated with alcohol-acetone	8.3

[a] Exposed for 48 hours at 4°.
[b] Exposed for 48 hours at 4°, then adjusted to 0.05 M phosphate, pH 7.4, by dialysis.
[c] Exposed for 48 hours at 4°, then dialyzed exhaustively against water and lyophilized.

acetate at pH 4.1 (Table XIII-3, Experiments 3-4 and 5-4). The size of the increase of the apparent binding affinity by the neutral salt appeared to be pH dependent: factors of approximately 5, 10, and 19 were found at pH 10.0, 7.4, and 4.1, respectively. This relationship should be remembered in the further analysis of the salt effect on steroid-AAG interaction.

Conformational interpretation. To arrive at a possible interpretation of the perturbant effects described, the physicochemical properties of AAG must be considered. The general structure of the native molecule appears to be best represented by a combination of β conformation and random coil as concluded from studies by SCHMID and KAMIYAMA (1963), JIRGENSONS (1965), SARKAR and DOTY (1966), and YAMAGAMI and SCHMID (1967). Optical measurements have revealed (SCHMID and KAMIYAMA, 1963) a large increase in the negative specific optical rotation of AAG at high concentrations of guanidine hydrochloride, urea, or LiBr. It has been suggested that at, respectively, 5 M, 8 M, and 4 M concentrations of these compounds complete unfolding of the glycoprotein molecule takes place. No change of the optical rotation of AAG was observed in NaCl solutions up to 5 M, indicating that NaCl does not have the unfolding effect of the above agents, which are known to destabilize globular proteins (VON HIPPEL and WONG, 1965).

Extensive studies have been reported on the effects of neutral salts and other perturbants on the stability of macromolecular conformations (HAMAGUCHI and GEIDUSCHEK, 1962; VON HIPPEL and WONG, 1964), in particular on thermal transition and conformational structure of RNase (MANDELKERN and ROBERTS, 1961; GINSBURG and CARROLL, 1965; SHERWOOD and POTTS, 1965; VON HIPPEL and WONG, 1965). Some salts, e. g., potassium phosphate or ammonium sulfate, increase the transition temperature of RNase, indicating a stabilization of the conformational structure toward an ordered, presumably helical, arrangement. Other perturbants, e. g., LiBr, $CaCl_2$, KCNS, and urea, decrease the transition temperature, suggesting a change from an ordered structure to a random coil. The effects are proportional to the concentrations of the perturbants. The order of their relative molar effectiveness follows approximately the classical Hofmeister series (EDSALL and WYMAN, 1958, p. 277).

WARREN and CHEATUM (1966) and WARREN, STOWRING, and MORALES (1966) studied the influence of neutral salts on activity and structure of myosin and other enzymes. Anions and cations in high concentrations inhibited the activity in the same order of effectiveness in which they disrupted the organized structure of proteins and other macromolecules. These results, obtained by various authors pursuing different objectives, bear on our findings in that the order of effectiveness of the perturbants toward the conformational structure of ribonuclease and other macromolecules is in general agreement with their influence on the apparent binding affinity of AAG for progesterone. It is suggested, therefore, that this affinity is directly or indirectly related to the effects on the conformational structure of the glycoprotein. The following paragraphs describe efforts to obtain evidence for this hypothesis.

Viscosity. Since the viscosity, in conjunction with other properties, is a useful parameter to assess conformational changes in proteins (YANG, 1961), the intrinsic viscosity of AAG was determined in the presence of different perturbants. Fig. XIII-18 shows that the value

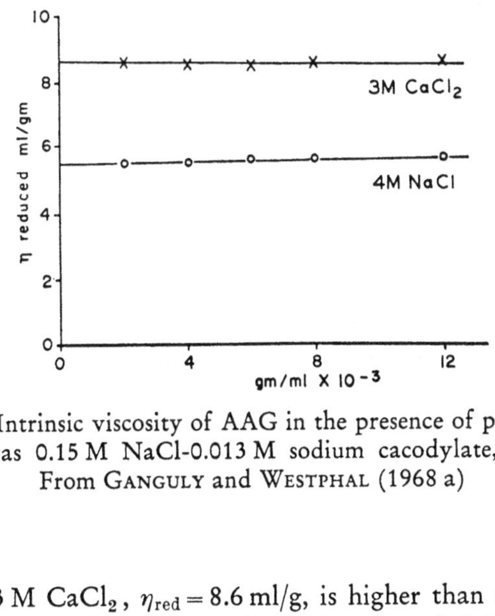

Fig. XIII-18. Intrinsic viscosity of AAG in the presence of perturbants. The buffer used was 0.15 M NaCl-0.013 M sodium cacodylate, pH 7.4; 30°. From GANGULY and WESTPHAL (1968 a)

obtained in 3 M $CaCl_2$, $\eta_{red} = 8.6$ ml/g, is higher than that measured in 4 M NaCl, 5.5 ml/g. The intrinsic viscosity of AAG in distilled water has been reported to be 6.9 ml/g (SMITH et al., 1950), i. e., between those found in the two salt solutions. This result would agree (HOLCOMB and VAN HOLDE, 1962) with the assumption of an unfolded AAG molecule in $CaCl_2$, and a more compact structure in 4 M NaCl. The apparent binding affinity between progesterone and AAG would then be decreased in the less ordered AAG structure, and increased when the ordered state was reinforced.

A similar relationship between viscosity and effect on the apparent progesterone-binding affinity has been found for other perturbants. Table XIII-14 shows that AAG in solutions of those salts which stabilize the progesterone-AAG complex (NaCl and Na_2SO_4) has a lower relative viscosity than in the perturbant solutions which destabilize the complex (LiBr, $CaCl_2$, KCNS, and urea). This result corresponds to that of VON HIPPEL and WONG (1965) who found a higher viscosity for ribonuclease in 3 M $CaCl_2$ than in 3 M NaCl.

It will be noted (Table XIII-14) that the relative viscosity of AAG in 4 M NaCl and 1 M Na_2SO_4 is slightly higher than in the 0.15 M NaCl—0.013 M cacodylate buffer. A similarly elevated value is seen in the presence of 4 M ammonium acetate, a perturbant which

Table XIII-14. *Relative viscosities of AAG. The buffer used was 0.15 M NaCl-0.013 M sodium cacodylate, pH 7.4; 30°.* From GANGULY and WESTPHAL (1968 a)

Perturbant	$\eta_{1\%}$ relative	
None	1.054;	1.053 [a]
3 M LiBr	1.072	—
3 M $CaCl_2$	1.092;	1.084 [a]
3 M KCNS	—	1.080 [a]
4 M urea	1.077	—
4 M NaCl	1.060;	1.058 [a]
1 M Na_2SO_4	1.060	—
4 M NH_4-acetate	—	1.059 [a]

[a] These data were obtained with the viscometer constructed according to SCHACHMAN (1957); all other values were determined in a Cannon-Ubbelohde semimicroviscometer.

does not affect the binding affinity of the progesterone-AAG complex (Fig. XIII-15).

Ultraviolet absorption. Since the effects of neutral salts and other perturbants on protein conformation have been thoroughly investigated for ribonuclease, this enzyme was utilized in ultraviolet absorption measurements as a standard for comparison with AAG. Absorbances of AAG and ribonuclease were determined at 278 nm in different perturbant solutions and compared with control values obtained in buffer. The percentage differences between absorbance in the presence and absence of perturbant were calculated (Table XIII-15). Average absorbance values from four independent experiments each, measured in the standard cells, showed a slight decrease in 3.3 M $CaCl_2$ for both proteins, and about 2% increase for AAG in

4 M NaCl. For greater precision, the split compartment cells of YANKEELOV (1963) were used, which generally confirmed and established the values obtained with the standard cells. Consistent, if slight, increase of the absorbance was obtained in 2.5 M NaCl in the split compartment cell for AAG and ribonuclease; the decrease of absorption in 3 M $CaCl_2$, approximately 2 to 3%, was equally con-

Table XIII-15. *Ultraviolet absorbance of AAG and RNase in different perturbant solutions containing 0.15 M NaCl-0.013 M sodium cacodylate, pH 7.4, as buffer. Absorbance at 278 nm has been corrected to a protein concentration of 1.0 mg per ml; experiments were done at 22°.* From GANGULY and WESTPHAL (1968 a)

Protein	Standard cells			Split cells		
	Solvent	Absorbance [a]	Difference from control	Solvent	Absorbance	Difference from control
			%			%
AAG	Buffer	0.893		Buffer	0.893	
	4 M NaCl	0.913	+2.2	2.5 M NaCl	0.904	+1.2
	3.3 M $CaCl_2$	0.883	—1.1	3 M $CaCl_2$	0.872	—2.3
	4 M urea	0.887	—0.6	4 M urea	0.904	+1.2
RNase	Buffer	0.647		Buffer	0.647	
	4 M NaCl	0.648	+0.1	2.5 M NaCl	0.655	+1.2
	3.3 M $CaCl_2$	0.621	—4.0	3 M $CaCl_2$	0.628	—2.9
	4 M urea	0.655	+1.2	4 M urea	0.658	+1.7

[a] Average values from four experiments each.

sistent for the two proteins. Urea produced a slight increase of absorption in AAG as well as in ribonuclease. Measurement of the ultraviolet absorption spectra of AAG and ribonuclease revealed no significant differences at other wavelengths between the solutions in the absence and in the presence of the above perturbants.

The effects of the individual perturbants measured in the split cell, which are outside experimental error, are closely similar for AAG and ribonuclease, i. e., positive for NaCl and urea and negative for $CaCl_2$. WHITE and ANFINSEN (1959) found no difference in maximal ultraviolet absorbance for ribonuclease solutions in phos-

Table XIII-16. *Effect of NaCl on apparent binding affinity between AAG and several steroids. AAG concentration was 0.5 mg per ml, steroid concentration 0.02 μg per ml; the buffer used was 0.05 M phosphate, pH 7.4.* From GANGULY and WESTPHAL (1968 a)

Steroid and perturbant	$\bar{\nu}/[S] = n\,k$ at 4°	37°
	$M^{-1} \times 10^{-5}$	$M^{-1} \times 10^{-5}$
Progesterone		
None	10.8	3.7
4 M NaCl	117.0	14.3
Testosterone		
None	5.7	2.4
4 M NaCl	62.4	13.6
Cortexone		
None	2.5	1.4
4 M NaCl	29.9	10.2
Corticosterone		
None	0.92	0.37
4 M NaCl	14.9	3.4
Cortisol		
None	0.29	0.18
4 M NaCl	5.1	1.5
Estrone		
None	1.2	—
4 M NaCl	33.1	—
Estradiol		
None	0.71	—
4 M NaCl	42.1	—

phate buffer in the presence and in the absence of 8 M urea, presumably measured in standard cells; the clearly established conformational change of ribonuclease in urea (VON HIPPEL and WONG, 1965) thus is difficult to demonstrate by ultraviolet absorption, and cannot be conveniently measured except by the refined differential procedure (YANKEELOV, 1963). The results in Table XIII-15 are therefore in agreement with the assumption that NaCl, $CaCl_2$, and urea produce in AAG conformational changes similar to those established for ribonuclease by various independent physicochemical procedures.

Steroid-AAG complexes. The investigation of the perturbant effect was extended to the interaction of AAG with steroids other than progesterone. Apparent affinity constants were determined in the absence and presence of 4 M NaCl since this salt in the case of progesterone affected only the apparent association constant and not the number of binding sites (Fig. XIII-17). Table XIII-16 shows the results obtained with five Δ^4-3-ketosteroids of increasing polarity, and with the phenolic steroids, estrone and estradiol. The n k values for the different steroids in salt-free buffer are generally in good agreement with the apparent association constants obtained by the Scatchard procedure (Table XIII-10). This result provides experimental evidence that the use of $\bar{r}/[S]$ as an expression of n k, first deduced from theoretical considerations (GANGULY et al., 1967), is justified (see Chapter V. 4). The apparent affinity constants are greatly increased in the presence of 4 M NaCl. It is evident that within a comparable series of steroid structures, such as progesterone and its hydroxylated derivatives, the affinity-enhancing effect of 4 M NaCl becomes stronger with increasing polarity of the steroids; testosterone fits into this sequence if one assumes a polarity between progesterone and cortexone. These relationships are illustrated more clearly in Figs. XIII-19 and XIII-20 for 4° and 37°, respectively: the numbers given on top of the dotted bars indicate the factors by which n k of the perturbant-free control increases in the presence of the salt. The effects at 37° are approximately half those observed at 4°.

The data in Table XIII-16 and Figs. XIII-19 and XIII-20 might seem to imply that the salt effect has some inverse proportionality to the n k value of the complex. Generalization of such relationship, however, is ruled out by the findings with estrone and estradiol (Fig. XIII-19); a comparison with the results on the Δ^4-3-ketosteroids shows that the absolute value of n k bears no relationship to the perturbant effect. It is not known whether the relatively greater increase of binding affinity of the estradiol-AAG complex in 4 M NaCl is related to the difference in the number of binding sites for the estrogenic hormone; it has been shown (Table XIII-10) that AAG has approximately n = 5 binding sites for estradiol, whereas all Δ^4-3-ketosteroid complexes are characterized by n = 1.

Thermodynamic parameters. In an attempt to further analyze the effects of the perturbants on the steroid-AAG interaction, thermodynamic data on the binding equilibria were obtained. The quantities

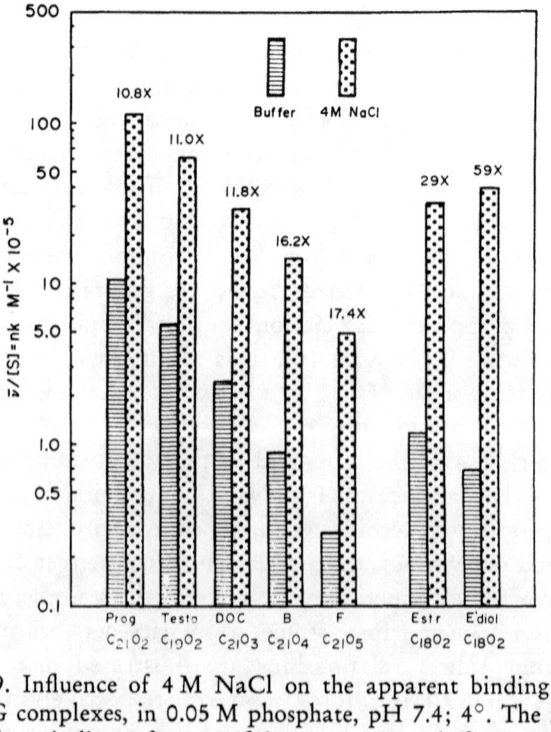

Fig. XIII-19. Influence of 4 M NaCl on the apparent binding affinity of steroid-AAG complexes, in 0.05 M phosphate, pH 7.4; 4°. The figures over the dotted bars indicate factors of increase over salt-free controls. Prog, progesterone; Testo, testosterone; DOC, cortexone; B, corticosterone; F, cortisol; Estr, estrone; E'diol, estradiol-17β

$\Delta F°$, the approximate change of free energy, $\Delta H°$, of enthalpy, and $\Delta S°$, of entropy, for the complexes with the five Δ^4-3-ketosteroids in the absence and in the presence of 4 M NaCl are shown in Table XIII-17. The calculations have been made with the use of $\bar{\nu}/[S]$ instead of k since $\bar{\nu}/[S]$ is indistinguishable from n k, and $n=1$ for all Δ^4-3-ketosteroid-AAG complexes studied. In confirmation of the validity of this procedure, the values for $\Delta F°$, $\Delta H°$, and $\Delta S°$ for the perturbant-free systems in Table XIII-17 are practically identical with those calculated on the basis of association constants, k, obtained by Scatchard analysis (Table XIII-10). In the binding affinity-reducing system, i. e., 3 M LiBr, n was found to be a fraction of unity (Fig. XIII-17). For this reason, and because of the limited accuracy of the differences between the low apparent binding affinity values measured at 4° and 37° in the presence of the destabilizing perturbants, the thermodynamic data in Table XIII-17 do not include perturbant systems which decrease the apparent steroid-AAG affinity. Thermodynamic data for the estrogen complexes with AAG were not obtained since the number of primary binding sites is uncertain.

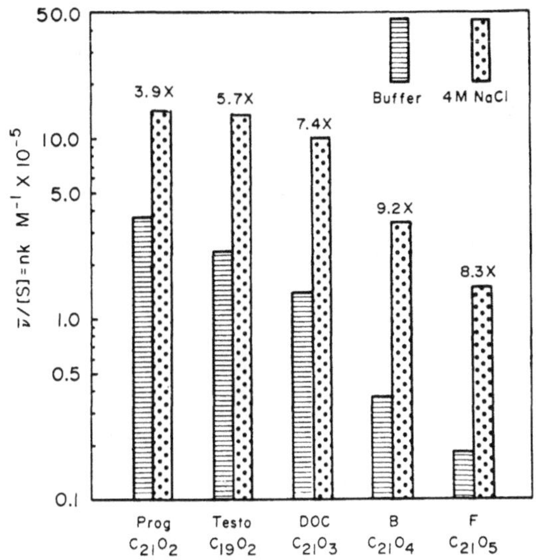

Fig. XIII-20. Influence of 4 M NaCl on the apparent binding affinity of steroid-AAG complexes, at 37°. Other conditions and explanations as in Fig. XIII-19

The negative ΔF° values of the steroid-AAG complexes (Table XIII-17) indicate spontaneous associations. The apparent negative free energies in 4 M NaCl are higher than in buffer by about 1.3 to 1.5 kcal per mole at 4° and by 0.8 to 1.4 kcal per mole at 37°. This difference seems to rise with increasing polarity of the steroid, reflecting the stronger perturbant effects shown as magnification factors in Figs. XIII-19 and XIII-20. The entropy changes are positive in all cases, except for the stabilized progesterone complex in 4 M NaCl solution. The apparent enthalpy change in the presence of 4 M NaCl is markedly greater than in its absence, possibly specifying more closely the effect of the perturbant which increases the apparent k of the steroid-AAG complex.

Conclusions. Although the thermodynamic parameters cannot be fully evaluated before the salt effects on AAG and its conformational structure are better understood, the association between progesterone and AAG may be interpreted in the following manner. The steroid interacts at one primary binding site of the glycoprotein to form a complex of considerably lower heat content. The negative

Table XIII-17. *Approximate free energy, enthalpy and entropy of steroid-AAG complexes in absence and in presence of 4 M NaCl. The buffer used was 0.05 M phosphate, pH 7.4.* From GANGULY and WESTPHAL (1968 a)

Steroid and perturbant	ΔF° at 4°	ΔF° at 37°	ΔH°	ΔS°
	kcal/mole		kcal/mole	cal mole^{-1} degree^{-1}
Progesterone				
None	−7.7	−7.9	−5.5	+8
4 M NaCl	−9.0	−8.7	−10.9	−7
Testosterone				
None	−7.3	−7.6	−4.4	+10
4 M NaCl	−8.6	−8.7	−7.9	+3
Cortexone				
None	−6.9	−7.3	−3.1	+14
4 M NaCl	−8.2	−8.5	−5.6	+10
Corticosterone				
None	−6.3	−6.5	−4.7	+6
4 M NaCl	−7.8	−7.9	−7.6	+1
Cortisol				
None	−5.7	−6.0	−2.5	+12
4 M NaCl	−7.2	−7.4	−6.3	+3

enthalpy change is highest for the steroid with the fewest polar groups, and increases greatly in the presence of 4 M NaCl. The interaction between steroid and protein is presumably accompanied by displacement of structured hydration water from the protein and the steroid to form randomized water molecules, causing a positive entropy change of the system. These thermodynamic reactions may be complicated by conformational changes of the AAG molecule in the presence of the perturbants. Interaction with progesterone and other steroids appears to be energetically favored by salts known to produce a more ordered conformation of the protein, whereas solvent conditions which cause unfolding to a random coil lead to a decrease in the apparent binding affinity, possibly by perturbant interference and reduced availability of hydrophobic residues near the binding site.

6. Chemical Modification of AAG in Relation to Steroid Binding

Removal of sialic acid. It has been discussed in Chapter XI. 3 that the sialic acid content of CBG and TBG is not essential for the hormone-binding qualities of these glycoproteins. Complete removal of sialic acid from AAG by mild enzymatic hydrolysis using *vibrio cholerae* neuraminidase over a period of 14 days (GANGULY and WESTPHAL, 1968 a) did not diminish the ability of AAG to interact with progesterone (Table XIII-18); the n k value may be even somewhat higher in the neuraminidase-treated AAG. The influence of 4 M

Table XIII-18. *Significance of sialic acid in progesterone-AAG interaction. AAG concentration was 0.5 mg per ml; equilibrium dialysis was carried out in 0.15 M NaCl-0.013 M sodium cacodylate, pH 7.4, at 4°.* From GANGULY and WESTPHAL (1968 a)

AAG and perturbant	n k	Effect [a]
	$M^{-1} \times 10^{-5}$	
Type P		
None	10.5	
4 M NaCl	117.0	11.1
3 M $CaCl_2$	0.69	0.07
Sialic acid-free [b]		
None	12.5	
4 M NaCl	151.6	12.4
3 M $CaCl_2$	0.37	0.03
Sialic acid-free [c]		—
None	10.5	

[a] Factor by which n k of perturbant-free control changes in the presence of perturbant.

[b] A molecular weight of 36,000 was assumed for the sialic acid-free AAG.

[c] Preparation obtained from Dr. K. SCHMID.

NaCl and of 3 M $CaCl_2$ on the apparent binding affinity of the sialic acid-free glycoprotein was similar to values observed with the intact AAG. These results add another example to the general conclusion that sialic acid is not indispensable for the steroid-binding ability of serum glycoproteins. It is of interest in this context that the sialic acid and other oligosaccharide units have been found not to contribute significantly to the conformational specificity of AAG (SCHMID and KAMIYAMA, 1963).

Treatment of AAG with acid and alkali. Whereas exposure of AAG to 0.05 M H_2SO_4 for 24 hrs at 22° did not affect the ultracentrifugal pattern, the same treatment at 28° resulted in the appearance of two peaks (see Fig. 1 in the original publication by GANGULY and WESTPHAL, 1968 b). The major peak had the sedimentation coefficient of the untreated glycoprotein, whereas the second one moved more rapidly ($s_{20,w} = 10.9$ in a 1% solution of the acid-treated AAG). This heavier component may correspond to the faster-moving peak which SCHMID et al. (1967) observed in the ultracentrifugal analysis of AAG at pH 2.5 and 3.5.

Gel filtration of the acid-treated glycoprotein over Sephadex G-200 indicated at least two major components each of which was subjected to a second Sephadex G-200 filtration. The fractionation of the first component yielded a fairly symmetrical main peak (Fraction I) preceded by a small amount of material of presumably larger molecular size which was not further investigated. Similar fractionation of the second component gave a symmetrical main peak (Fraction II), preceded by a small shoulder. The amino acid compositions of Fractions I and II did not differ significantly from that of untreated AAG (see Table I in the original publication by GANGULY and WESTPHAL, 1968 b). As to be expected from the acid treatment, the sialic acid content was decreased as shown in Table XIII-19. The sedimentation coefficients, $s_{20,w}^{1\%}$, of Fractions I and II were the same as that of untreated AAG (SCHMID et al., 1968 b). The immunoprecipitation reaction (see Fig. 2 in the original publication by GANGULY and WESTPHAL, 1968 b) was not affected by the exposure to the dilute acid, in confirmation of earlier results by KROTOSKI and WEIMER (1966). Table XIII-19 furthermore indicates that the progesterone-binding affinity is not altered by the acid treatment of AAG as shown for Fraction I after precipitation with alcohol-acetone. Fraction II had a lower binding affinity for progesterone which may indicate some alteration of the glycoprotein molecule; the possibility of an increased n k value after delipidation has not been examined.

Treatment of AAG with 0.1 M NaOH for 24 hrs at 28° yielded a product which appeared homogeneous on gel filtration over Sephadex G-200. Ultracentrifugation showed a single peak of $s^{1\%}_{20,w} = 4.41$ (see Fig. 3 in original publication by GANGULY and WESTPHAL, 1968 b).

Table XIII-19. *Physicochemical properties of AAG, treated with acid or alkali.* From GANGULY and WESTPHAL (1968 b)

Treatment	Sialic acid %	Half cystine [a]	$s^{1\%}_{20,w}$	n k $M^{-1} \times 10^{-5}$	Immuno-reaction [b]
None	11.6	11.8	2.76	10.3	+ + +
0.1 equiv/l acid; Fraction I	6.2	11.1	2.73	5.0	+ + +
0.1 equiv/l acid; Fraction I [c]	—	—	—	10.8	+ + +
0.1 equiv/l acid; Fraction II	8.0	10.7	2.68	1.6 [d]	+ + +
0.1 equiv/l NaOH, 28°, 24 hr	11.3	0.8	4.41	0.06	Neg.

[a] Residues/100,000 g.

[b] Immunoelectrophoresis; symbols in this and subsequent tables indicate strongly positive (+ + +), weak (+) or negative (Neg.) reaction with antiserum against human AAG (rabbit).

[c] Precipitated with alcohol-acetone (9 : 1, v/v).

[d] Sufficient material for reprecipitation with alcohol-acetone was not available.

As Table XIII-19 shows, the progesterone-binding affinity was practically lost. The amino acid composition of this product (see Table I in original publication by GANGULY and WESTPHAL, 1968 b) was in fair agreement with the untreated AAG, except for the low half-cystine content; cysteic acid was relatively high. It is possible that the decline of half-cystine followed reactions similar to those described by BOHAK (1964) for alkali treatment of ribonuclease. The alkali-treated AAG failed to show antigenic properties in immunoelectrophoresis (see Fig. 2 in original publication by GANGULY and WESTPHAL, 1968 b).

Modification of cystine residues. Titrations of AAG were carried out with p-hydroxymercuribenzoate (BOYER, 1954), N-ethylmaleimide (BENESCH and BENESCH, 1962), and 5,5′-dithiobis-(2-nitrobenzoic acid) (DTNB) (ELLMAN, 1959) in an attempt to demonstrate free sulfhydryl groups. No thiol was found by any of these reactions. Efforts to obtain SH reactions with iodoacetic acid or iodoacetamide,

Table XIII-20. *Treatment of AAG with iodoacetic acid compounds.* From GANGULY and WESTPHAL (1968 b)

Treatment	n k [a] $M^{-1} \times 10^{-5}$	Immuno-reaction [b]
None	10.6	+ + +
Iodoacetic acid	9.4	Not tested
Iodoacetic acid in 8 M urea	9.0	+ + +
Iodoacetamide in 8 M urea	9.1	Not tested

[a] After dialysis, lyophilization and precipitation with alcohol-acetone.
[b] For explanation see Table XIII-19.

even in the presence of 8 M urea, were equally unsuccessful (for experimental details see GANGULY and WESTPHAL, 1968 b). Treatment with radiolabeled iodoacetic acid or iodoacetamide resulted in uptake of an insignificant amount of ^{14}C indicating virtual absence of non-specific alkylation. These results are in agreement with studies by KROTOSKI and WEIMER (1967) who have in addition discussed similarly negative observations made in other laboratories. Table XIII-20 shows that the progesterone-binding affinity of AAG was not changed after reaction with the iodoacetic acid compounds, even in the presence of urea. Similarly, the antigenicity as tested by immunoelectrophoresis was retained.

In an effort to elucidate the role of the disulfide groups in regard to the steroid-binding ability, AAG was treated with reducing agents (for experimental conditions see GANGULY and WESTPHAL, 1968 b). In the presence of 2-mercaptoethanol or dithiothreitol, the progesterone-binding affinity was somewhat decreased (Table XIII-21). When the reducing agents were removed by dialysis, the original n k value was restored.

Table XIII-21. *Treatment of AAG with reducing and alkylating agents.* From GANGULY and WESTPHAL (1968 b)

Treatment	n k $M^{-1} \times 10^{-5}$	Immuno-reaction [a]
Mercaptoethanol present during equilibrium dialysis	6.5	Not tested
Mercaptoethanol [b]	10.1	Not tested
Dithiothreitol present during equilibrium dialysis	6.8	Not tested
Dithiothreitol [c]	8.0	Not tested
Dithiothreitol in 8 M urea [d], equilibrium dialysis at pH 4.1 under N_2	0.24	Not tested
AAG, untreated control at pH 4.1	4.0	Not tested
Dithiothreitol in 8 M urea [b], air not excluded	9.7	+++
Dithiothreitol in 4 M urea followed by iodoacetic acid [b]	0.19	+++
Dithiothreitol in 8 M urea followed by iodoacetamide [b]	0.35	+++
Sialic acid-free AAG, reduced and alkylated	0.14 [e]	+++

[a] For explanation see Table XIII-19.
[b] Followed by dialysis, lyophilization and precipitation with alcohol-acetone.
[c] Followed by dialysis and lyophilization.
[d] Followed by gel filtration over Sephadex G-25.
[e] Assumed MW, 36,000.

Reduction of AAG by dithiothreitol in the presence of 8 M urea, separation over Sephadex G-25, and equilibrium dialysis at pH 4.1 under N_2 gave a very low n k value, indicating virtual inactivation for progesterone binding. This is seen in Table XIII-21 by comparison with the control value at the same pH. When this reduced AAG without exclusion of oxygen was dialyzed, lyophilized and precipitated with alcohol-acetone, full progesterone-binding affinity was found.

These results show that the reduction of disulfide linkages in AAG by dithiothreitol in the presence of urea causes loss of progesterone-binding activity. Air oxidation of this reduced glycoprotein restores full binding affinity, presumably by reformation of the original disulfide groups. This assumption was corroborated by blocking the sulfhydryl groups of the reduced AAG with iodoacetic acid or iodoacetamide (Table XIII-21). The progesterone-binding affinity was virtually eliminated in both cases. However, the antigenic properties of the carboxymethylated preparations were about the same as those of untreated AAG (Fig. XIII-21, A, B).

The amino acid analysis (Table XIII-22) of AAG, reduced and alkylated with iodoacetic acid or its amide, showed that almost 80% of the half-cystine residues were modified. Radioactivity measurement of the product which was alkylated with labeled iodoacetic acid gave 4.8 thiol groups per mole AAG (for experimental details see Ganguly and Westphal, 1968 b). Titration of the same reduced product with DTNB indicated the presence of 4.6 sulfhydryl groups per mole of AAG. Similar results were obtained when the glycoprotein was first reduced and alkylated with iodoacetic acid or its amide and then separated by gel filtration; 5.1 and 3.6 thiol groups, respectively, were found by radiocounting, whereas the amino acid analysis of the iodoacetic acid-treated product gave 3.5 moles carboxymethylcysteine per 41,000 g of AAG.

Loss of progesterone-binding activity was also observed with a sialic acid-free AAG preparation obtained from Dr. K. Schmid, which had been reduced with mercaptoethanol and alkylated with iodoacetamide (Table XIII-21). The antigenic properties were retained as shown by immunoelectrophoresis (Fig. XIII-21, F). Since removal of sialic acid by itself does not affect the progesterone-binding affinity of AAG, these results are in accordance with the interpretations given above.

Krotoski and Weimer (1967) have reported iodoacetamide alkylation of reduced AAG under a variety of conditions and have discussed the mechanism of the reduction. They conclude that the three disulfide groups in the AAG molecule are responsible for the tertiary structure of the glycoprotein; their quantitative studies of the antigenic properties of the alkylated products indicated a slight loss of activity. More recent observations by Ishiguro et al. (1969) led to the conclusion that two of the three disulfide bridges can be reduced readily and without apparent changes in the three-dimensional structure of AAG. The authors assume that these two disulfide groupings are located at the surface of the molecule; the third one is considered "buried" and appears to be essential for maintaining the conformational structure of AAG.

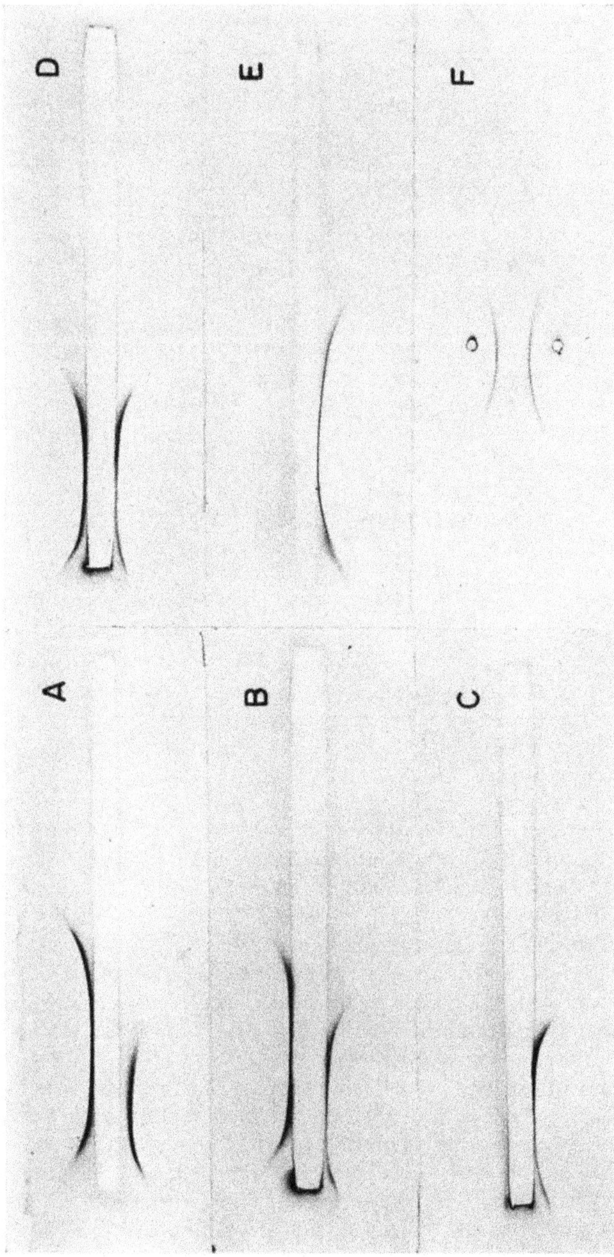

Fig. XIII-21. Immunoelectrophoretic analyses of modified AAG preparations in barbital buffer (pH 8.6, μ 0.1) using antihuman AAG serum produced in rabbit. Antigen concentrations, 5%. Top wells contained AAG treated with (A) dithiothreitol, urea, iodoacetic acid; (B) dithiothreitol, urea, iodoacetamide; (C) ethyl thioltrifluoroacetate; (D) ethyl thioltrifluoroacetate followed by piperidine; (E) sialic acid-free AAG, trifluoroacetylated; (F) sialic acid-free AAG, treated with mercaptoethanol and iodoacetamide. Bottom wells contained (A—E) untreated AAG; (F) the same antigen as top well. Selected from GANGULY and WESTPHAL (1968 b)

Table XIII-22. *Amino acid composition*[a] *of AAG before and after chemical modifications.* From GANGULY and WESTPHAL (1968 b)

Amino acid	Native	Dithio-threitol, urea, iodoacetic acid	Dithio-threitol, urea, iodoacet-amide	KCNO, urea	Biacetyl trimer	KCNO, urea, biacetyl trimer
Asp	63.1	64.7	62.8	63.7	61.5	61.8
Thr[b]	47.6	48.9	47.8	46.5	45.4	44.3
Ser[b]	19.7	21.2	21.3	21.1	21.0	20.6
Glu	86.5	85.1	92.7	90.1	91.9	89.7
Pro	22.6	21.7	20.5	20.3	21.1	20.6
Gly	20.2	20.2	22.1	21.4	22.4	21.8
Ala	26.0	26.6	26.2	25.9	26.0	25.9
Val[c]	—	—	—	—	—	—
Cys[b]	11.8	2.6	3.0	7.9	2.3	5.0
Met	3.7	3.5	3.5	3.2	3.2	2.9
Ile	25.8	24.3	26.3	25.5	26.7	26.2
Leu	42.4	42.4	41.5	42.3	42.7	42.6
Tyr[b]	31.8	30.0	32.2	31.0	26.6	25.2
Phe	27.8	27.6	28.5	27.9	28.5	28.7
Lys	40.1	39.2	40.4	12.0	29.7	10.7
His	8.6	8.8	8.6	8.5	8.6	8.2
Arg	25.1	24.8	25.5	25.5	5.8	2.9
Cysteic acid	—	Trace	Trace	1.2	7.1	3.2
CM-cysteine	—	8.4	7.0	—	—	—
Homo-citrulline	—	—	—	d	—	d

[a] Residues/100,000 g, corrected for water content. In order to compensate for possible minor deviations in sample quantity application and in the level of ninhydrin color development, corrections were applied by a method analogous to the internal reference procedure used by YANKEELOV and KOSHLAND (1965). The basis for correction was the sum of μmoles of the five stable amino acids, Asp, Ala, Ileu, Leu and Phe determined in a pure AAG preparation (standard) which was found to have an amino acid composition in close agreement with the values reported in the literature. The sum of μmoles of the same amino acids was calculated from the results obtained for each of the other runs. None of these amino acid residues is modified by the reactions applied. These sums are then compared with the sum obtained with the AAG standard; any deviation is compensated by a factor. This factor is calculated for each run by dividing the total μmoles of the five amino acids obtained with standard AAG, by the total μmoles of the five amino acids of a given run. In this way, internal consistency was assured for all amino acid analyses.

[b] Results for labile amino acids corrected approximately for decomposition during 24 hr of acid hydrolysis of deaerated samples by applying the factors: for threonine, half-cystine, tyrosine, 1.05; for serine, 1.10 (MOORE and STEIN, 1963).

[c] Valine was not determined since it separated only partially from glucosamine.

[d] Homocitrulline was formed.

Table XIII-23. *Treatment of AAG with modifying reagents.* From GANGULY and WESTPHAL (1968 b)

Treatment [a]	n k $M^{-1} \times 10^{-5}$	Immuno-reaction [b]
Acetic anhydride	3.7; 3.5	+ + +
Partial deacetylation at pH 10.5	6.3	+ + +
Partial deacetylation with hydroxylamine	6.8	Not tested
Tetranitromethane	1.5	+ + +
Ethyl thioltrifluoroacetate	0.39	Neg.
Trifluoroacetylated AAG treated with piperidine	1.3	+ + +
Trifluoroacetylation of sialic acid-free AAG	0.21	Neg.
O-Methyl isourea	0.18	+
O-Methyl isourea in 8 M urea	Not tested	+
KCNO in 4 M urea	0.32	+
KCNO in 8 M urea	Not tested	+
Biacetyl trimer	1.8	+ + +
KCNO in 8 M urea, followed by biacetyl trimer	0.14	Neg.

[a] Followed by dialysis, lyophilization and precipitation with alcohol-acetone.

[b] For explanation see Table XIII-19.

Modification of other residues. Reaction of AAG with acetic anhydride according to FRAENKEL-CONRAT (1957) and FRAENKEL-CONRAT and COLLOMS (1967) resulted in decreased progesterone-binding affinity (Table XIII-23). The antigenic properties were retained (Fig. XIII-22, B). In acetylation studies with radiolabeled acetic anhydride, only six acetyl groups were introduced into the molecule which contains approximately 17 lysine and 12 tyrosine residues (SCHMID et al., 1968 b). Partial deacetylation (2 acetyl groups removed) by exposure to pH 10.5 or by treatment with hydroxylamine restored some of the original steroid-binding activity.

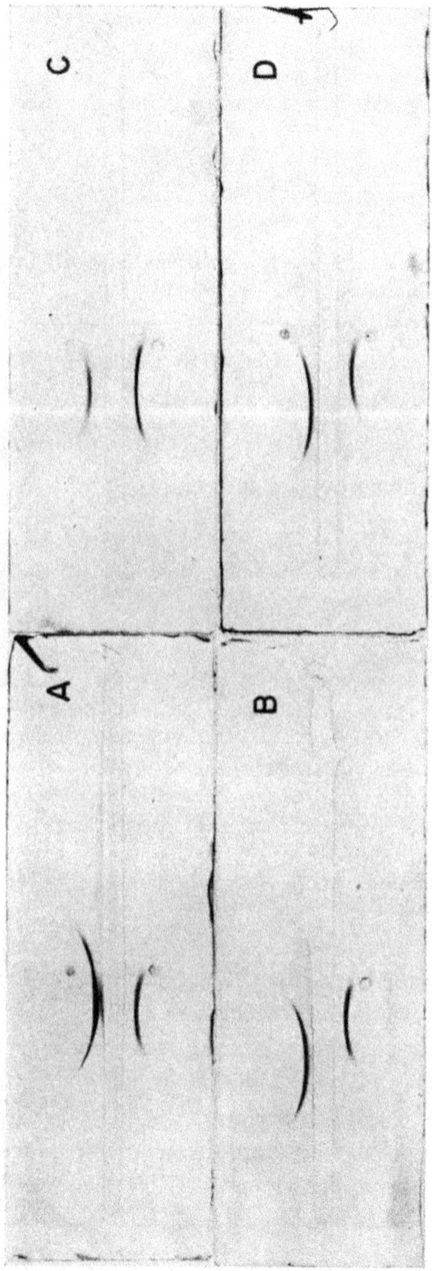

Fig. XIII-22. Immunoelectrophoretic analyses of modified AAG preparations. Conditions as in Fig. XIII-21. Top wells contained (A) untreated AAG (control), and AAG treated with (B) acetic anhydride; (C) tetranitromethane; (D) bi-acetyl trimer. Bottom wells contained whole human serum. Selected from GANGULY and WESTPHAL (1968 b)

The progesterone-binding potency of AAG was also reduced by reaction with tetranitromethane according to the procedure of SOKOLOVSKY et al. (1966) as shown in Table XIII-23. The product was heterogeneous in polyacrylamide electrophoresis. Immunoreactivity was maintained, even after repeated reaction with tetranitromethane in the presence of 8 M urea (Fig. XIII-22, C).

In order to assess the significance of lysine for the progesterone-binding activity, AAG was reacted with ethyl thioltrifluoroacetate under conditions which have been found (GOLDBERGER and ANFINSEN, 1962) to mask all available amino groups in ribonuclease. The trifluoroacetylated AAG had lost its steroid-binding ability almost completely (Table XIII-23). No immunoreaction was detected (Fig. XIII-21, C). Treatment with 1 M aqueous piperidine to regenerate the amino groups resulted in partial recovery of the affinity for progesterone. At the same time, the immunoreaction was restored to the approximate level of the untreated glycoprotein (Fig. XIIII-21, D). A similarly trifluoroacetylated sialic acid-free AAG preparation was also devoid of steroid-binding and antigenic activity (Table XIII-23 and Fig. XIII-21, E).

The method of KLEE and RICHARDS (1957) was used (GANGULY and WESTPHAL, 1968 b) for repeated treatment with O-methylisourea under conditions of presumably complete guanidation of the ε-amino groups of lysine. Another lysine reagent, potassium cyanate, was employed according to the technique of STARK et al. (1960). Either modification of AAG resulted in virtually complete suppression of the progesterone-binding affinity (Table XIII-23). The immunoreaction was greatly reduced, but not eliminated even after repeated treatment with the reagents in the presence of 8 M urea (see Fig. 4 in GANGULY and WESTPHAL, 1968 b). The amino acid analysis of the AAG preparation that had reacted with KCNO in 8 M urea indicated a loss of about one third of the half-cystine residues originally present (Table XIII-22). For discussion of the lysine residues see GANGULY and WESTPHAL (1968 b).

When about 80% of the arginine residues in AAG were modified by the trimer of biacetyl (YANKEELOV et al., 1968), the progesterone-binding affinity was decreased whereas the immunoreaction was similar to that of untreated AAG (Tables XIII-22 and XIII-23; Fig. XIII-22, D). The low lysine value is in accordance with the observations of YANKEELOV et al. (1968). When the biacetyl reaction was preceded by treatment with KCNO in 8 M urea, the modification of arginine involved almost 90% of this amino acid; lysine reacted to a slightly greater extent than to be expected from the treatment with KCNO alone. The product was inactivated with regard to progesterone binding (Table XIII-23) and to immunoprecipitation (see Fig. 4 in GANGULY and WESTPHAL, 1968 b). It should be noted that the amino acid analysis of the biacetyl-treated preparations gave increased cysteic acid values at the expense of half-cystine (Table XIII-22).

Conclusions. The results obtained with the modified AAG preparations lead to the following conclusions. The disulfide group, present at a ratio of three per AAG molecule, is essential for the progesterone-binding activity. Reductive cleavage of the S-S bond(s) reduces the binding affinity; this inactivation is reversible by air oxidation. When the sulfhydryl groups formed by reduction are covalently blocked by alkylation with iodoacetic acid or its amide, the ability to interact with progesterone is lost irreversibly. These findings would allow an interpretation of direct participation of a disulfide grouping in the steroid association at the binding site. Perhaps a more likely assumption would be that disulfide groups in the AAG molecule are responsible for securing a conformational structure optimal for interaction with progesterone; reductive cleavage would result in changes of the tertiary structure of the glycoprotein and thus disturb the particular conformation which is most suited for a close fit between steroid and AAG at the binding site. The studies performed so far do not permit to decide between these and other possibilities for the role of disulfide in the binding mechanism.

Reagents that are known to modify predominantly the ε-amino groups of lysine virtually abolish the binding activity for progesterone. Partial removal of the modifying group restores part of the steroid affinity. Diminished interaction after treatment with YANKEELOV's reagent may be the result of the extensive modification of arginine as well as that of lysine; YANKEELOV (1970) has recently discussed the modifications of arginine and lysine in proteins by the butanedione reagent. For an understanding of the mechanism of the steroid-AAG association it would be important to elucidate the structural basis of the metal ion inhibition discussed in Section 3 of this chapter, especially with regard to the site of the metal interaction in the AAG molecule.

The antigenicity of AAG as determined by immunoelectrophoresis is independent of its ability to form a progesterone complex. The immunoreaction after reduction and alkylation was essentially that of untreated AAG although the progesterone-binding activity was virtually absent. This independence of antigenicity and steroid affinity was not observed in the reversed sense: whenever the antigenic activity was eliminated by the chemical modification, the ability to associate with progesterone was lost concomitantly.

It is evident from the preceding sections that we are far from an understanding of the chemical basis for the affinity between AAG and the steroids. But inroads have been made, and methodical tools are available for further exploration of the interactions involved. The AAG molecule promises to continue to be useful for the investigation of the structural requirements of the protein in relation to steroid binding, particularly when supplemented by advances in the chemical elucidation of this glycoprotein.

XIV. Effect of Protein Binding on Hormonal Activity of Steroids

1. Corticosteroid Hormones

The concept of Bennhold of a vehicle function for the plasma proteins (Chapter III) had led to early interpretations which emphasize the solubilization of the steroid hormones as an essential function of their protein interactions. However, a closer look at steroid solubilities in water and aqueous buffer systems showed that such aid for solubilization would not be required for most steroid hormones which are water-soluble at their physiological concentrations. It was then suggested (Westphal, 1955) that the unbound form of the steroid would be the species available to the reactive tissue. This implies that the protein-bound hormone is in an inactive form, and that protein binding serves a buffer and storage function; the steroid-protein complex forms a biologically inert reservoir in which the hormone circulates relatively protected from chemical and enzymatic attack which might result in structural alteration and excretion. The energy of the protein binding, however, is of such magnitude that by dissociation the steroid hormone becomes readily available as a physiologically active entity.

Indications and evidence for suppression of hormonal activity in the protein complex are available from numerous laboratories. As Sandberg et al. (1966) have discussed, plasma cortisol levels have been known for some time to be elevated in pregnancy, or following estrogen therapy, without the appearance of the usual symptoms of hypercorticism. This has been explained by inactivation of cortisol bound to an increased amount of CBG. The concentration of the binding protein rises before the cortisol level increases (Sandberg et al., 1960); the lower amount of unbound corticoid results in increased ACTH production through the negative feedback mechanism (Kawai and Yates, 1966). The general effects of the corticosteroid hormones as well as the specific feedback inhibition of ACTH release thus appear to be given by the unbound form. Studies in our laboratory (see Chapters IX and X) showed that the thymus weights were clearly influenced by the unbound, not the total concentration of corticosterone. This was observed in the rat (GALA and WESTPHAL, 1965 b) and in the mouse (GALA and WESTPHAL, 1967).

The difference in functional capacity between protein-bound and unbound cortisol may also be inferred from its metabolic fate. WALLACE et al. (1957) observed a distinct delay in the rate of clearance of exogenous cortisol in patients whose CBG level was elevated by estrogen treatment. Indications were obtained that the delayed clearance was related to an alteration in hepatic handling of the steroid. In studies by MILLS et al. (1960), the protein-bound and unbound fractions of cortisol were determined in human subjects under estrogen administration, and the influence on cortisol metabolism was followed. Whereas total corticosteroid increased, the percentage of unbound cortisol was reduced; the absolute level of unbound corticoid remained fairly constant. The elevated plasma corticosteroid levels resulting from the estrogen therapy were not associated with depression of eosinophils, suggesting that only the unbound form was biologically active. The half-time of disappearance of plasma cortisol was again considerably prolonged by the administration of estrogen; this was interpreted as protection from destruction by the liver through increased binding to protein. The alterations in cortisol metabolism under estrogen therapy were comparable to those occurring in pregnancy.

Alternately, a low CBG level should be accompanied by a short half-time of disappearance of cortisol from an equivalent plasma pool. This has been shown by PLAGER et al. (1963) for the dog. This species has a much lower CBG activity than the human as may be seen from ultrafiltration experiments that show 88—90% cortisol bound in both a human plasma containing 9 μg and dog plasma having 2 μg total cortisol per 100 ml. After maximal ACTH stimulation, the plasma cortisol pool in the dog will be turning over with a half-time of about 25 min., whereas the equivalent pool in the human has a half-time of about 90 min. In a similar way, the sheep with its extremely low CBG capacity excretes intravenously injected cortisol at a faster rate than man does (LINDNER, 1964).

After the isolation of human CBG by SEAL and DOE (1962 b), it became possible to investigate directly the effect of the binding protein on the hormonal and other properties of the steroid. SLAUNWHITE et al. (1962) studied the influence of purified CBG in a liver glycogen deposition test for corticosteroids. Table XIV-1 shows the results obtained with adrenalectomized mice. It is evident that transcortin by itself does not affect the level of liver glycogen. Subcutaneous injection of 10 to 15 μg cortisol in seven divided doses at hourly

Table XIV-1. *The effect of injected transcortin and/or cortisol on the deposition of glycogen in the liver of fasting adrenalectomized mice (T, transcortin; F, cortisol).* From SLAUNWHITE et al. (1962)

Amount injected T (mg)	F (μg)	Expt. No.	Number of Mice	Mean glycogen ± S.E.M. (mg/10 g of body wt.)
0[a]	0[a]	1	5	1.7 ± 1.1
25	0	1	6	1.7 ± 0.7
25	10	1	4	1.0 ± 0.6
25	10	2	4	1.8 ± 1.0
5.6	15	3	6	1.1 ± 0.4
0	10	1	7	6.2 ± 1.6 [b]
0	10	2	9	20.5 ± 2.3 [c]
0	15	3	7	7.1 ± 1.2 [c]

[a] Control injection of corn oil only.

[b] $P = 0.05$. This and the following P-value give the probability that the group receiving only cortisol is not different from the group receiving transcortin plus cortisol.

[c] $P < 0.01$.

intervals produced consistently a significant increase of the glycogen value. Intravenous injection of CBG at the start of the assay completely nullified the cortisol effect. There was no significant difference between the results for transcortin and those for transcortin plus cortisol. The authors conclude that the biologically effective level of cortisol in the organism is not related to the total plasma concentration of the steroid, but rather to the portion that is not bound to transcortin.

Results obtained in the same laboratory (SANDBERG and SLAUNWHITE, 1963) seemed to provide evidence for inhibition by transcortin of metabolic alterations of cortisol. However, DE MOOR et al. (1963 a) showed that the inhibitory action of the transcortin preparations could not be due to CBG alone. TROOP and BIGGS (1965) reported subsequently that heparin may be responsible for the effect.

MATSUI and PLAGER (1966) studied the oxidation of glucose to CO_2 in mouse ear slices, a reaction which is inhibited by cortisol. When BSA or an enriched CBG preparation was added to the incubation medium, the inhibitory activity of cortisol was found to be given by the unbound portion and not by the total or the protein-bound cortisol. The authors concluded that in the mouse ear strip system the unbound form of cortisol determines the action on cellular function (PLAGER and MATSUI, 1966). An analogous conclusion was reported by BLECHER (1964) who investigated the inhibition by desoxycorticosterone of the influx of α-aminoisobutyrate into rat thymic lymphocytes. Binding of the steroid to BSA was assumed to eliminate its inhibitory effect on the amino acid uptake. KAWAI and YATES (1966), in studying corticosterone inhibition of ACTH release by histamine in the rat, injected corticosterone in combination with rat serum proteins rich in CBG. The association of corticosterone with the protein suppressed the feedback inhibition, again suggesting that the unbound form of the corticoid is the biologically active one. According to a recent hypothesis by KELLER et al. (1969), the corticosteroid-binding proteins may increase the specificity of the adrenocortical system by determining the distribution of the corticosteroid effects. REIF-LEHRER and CHADER (1969) have suggested that CBG plays a role in mediating the cortisol induction of the glutamic synthetase in chick embryo retina.

2. Progesterone

HOOKER and FORBES (1949) reported more than 20 years ago that progesterone when bound to unfractionated plasma protein was biologically inactive; activity was restored when the hormone was released by acid treatment and extracted. Similarly, exogenous progesterone seemed to be inactivated in the mouse by protein binding (FORBES and HOOKER, 1949). After identification of the blood serum proteins that interact with the progestational hormone, studies were undertaken in collaboration with Dr. T. R. FORBES to examine in more detail the influence of protein association on the biological activity of progesterone (WESTPHAL and FORBES, 1963; HOFFMANN et al., 1969).

As test procedure for progestational activity, the bioassay of HOOKER and FORBES (1947) was applied. Protein-progesterone mixtures or protein-free solutions of progesterone were injected into the ligated uterine horns of young adult, previously castrated mice. In a positive response, some of the stromal nuclei of the endometrium develop an elongated oval outline, a conspicuous nucleolus, and fine, evenly scattered chromatin.

The main advantage of the use of the intrauterine test procedure in the present problem is the confined localization of the injected material in the ligated segment of the uterine horn. The steroid-protein mixture is not injected into the general circulation, so that various possibilities of systemic interference, e. g., metabolism or excretion of the steroid, are avoided; the topical test therefore permits a simpler interpretation than a systemic assay technique. An additional advantage is the small quantity of material required.

It has been shown in previous work (ZANDER et al., 1957, 1958; FORBES, 1964) that the intrauterine test procedure in the mouse is not completely specific for progesterone. This limitation does not appear to be pertinent for the objectives of the present investigation in which the established effect of pure progesterone is compared with that of the same progesterone solution in the presence of binding protein. The question of steroid specificity in the assay therefore does not enter the interpretation.

The results of our studies on the influence of AAG on the progestational activity of progesterone are summarized in Table XIV-2. The minimal effective dose of progesterone in the uterine test procedure with the strain of mice used (CHI) is 0.2 ng. In accordance, administration of this or greater amounts of progesterone in glycoprotein-free solution produced positive responses in all uterine segments injected. Injection of the test volume of the standard progesterone-AAG solution (molar ratio 1 : 80), providing 925 pg progesterone and 10.2 μg glycoprotein, resulted in an all-positive response, whereas the $^1/_3$ and $^1/_4$ diluted solutions, providing between 231 and 321 pg progesterone and between 2.55 and 3.54 μg glycoprotein, were negative. Administration of a relatively high amount of AAG failed to elicit a progestational response.

Table XIV-2. *Progestational effect of progesterone in the presence of AAG.* From WESTPHAL and FORBES (1963)

Volume, μl	Injection of Progesterone, pg	AAG, μg	Response [a]
0.74	—	20.3	− − − − −
0.74	925	—	+ + + + +
0.74	231	—	+ + + + +
0.74	925	10.2	+ + + + +
0.74	231	2.55	− − − − + [b]
0.77	241	—	+ + + +
0.77	321	3.54	− − − − − − [c]
0.77	241	2.66	− − − −

[a] Each — or + sign indicates negative or positive response in a ligated uterine horn of one mouse.

[b] Positive response of one horn doubtful.

[c] One negative response doubtful.

Table XIV-3. *Bound and unbound progesterone (P) in the presence of AAG at 37°.* From WESTPHAL and FORBES (1963)

Assumed dilution at reactive site	Concentrations of injected solutions						
	Standard 1.25 μg P/ml 13.8 mg AAG/ml		Standard 1/3 0.417 μg P/ml 4.60 mg AAG/ml		Standard 1/4 0.313 μg P/ml 3.45 mg AAG/ml		
	Unbound P in 0.74 μl		Unbound P in 0.77 μl			Unbound P in 0.74 μl	Unbound P in 0.77 μl
	%	pg	%	pg	%	pg	pg
(1)	2.93	27	8.3	27	10.8	25	26
1/10	23.2	215	47.5	152	54.7	126	132
1/20	37.6	348	64.4	207	70.7	163	170
1/30			73.1	235	78.4	181	189
1/40					82.9	191	200
1/50					85.8	198	207
1/60					87.9	203	212

On the basis of the observed combining affinity of the progesterone-AAG complex at 37° (C = 2.4), the dissociation into the two components has been calculated. The data of the first line in Table XIV-3 show that about 2.9% of the progesterone in the standard mixture is in the unbound state; this percentage is higher in the 1/3 and 1/4 diluted standard solutions. Calculation for the actual quantities injected indicates that between 25 and 27 pg progesterone is present as free hormone. This is far less than the minimal effective dose of the test procedure.

The injected mixture of progesterone and AAG may be assumed to be subject mainly to two effects: dilution by intrauterine fluid, and adsorption to dissolved substances or structural elements, including reactive sites at the target tissue. The net effect of these two factors on the dissociation equilibrium of the steroid-protein complex would be qualitatively the same and additive, resulting in increased dissociation of the complex. The combined influence of all factors which would thus affect the dissociation has been comprised as "assumed dilution at reactive site", and treated as a dilution effect. The calculations show (Table XIV-3) that at an assumed dilution of 1/10, 215 pg progesterone would be in the unbound state in the standard solution, and between 126 and 152 pg in the diluted standard solutions. One would expect therefore that the standard solution would give a positive result in the intrauterine test with its minimal effective dose of 0.2 ng whereas the 1/3 and 1/4 diluted standard solutions would not. This has been actually found (Table XIV-2). If the assumed dilution were approximately 1/20 or higher, a positive effect would be expected also from the 1/3 or 1/4 diluted standard solutions. Table XIV-2 shows that this is not the case. Fig. XIV-1 illustrates the essential results of these calculations, and indicates more clearly the unbound progesterone fractions in relation to the threshold quantity of 200 pg.

The results are in accordance with the assumption that the progesterone-AAG complex is inactive as a progestational agent, and that the injected solution, under the conditions of the test procedure applied, is subject to dilution and adsorption effects which result in a net assumed dilution of not less than 1/10 and not more than 1/20. This "dilution" would effect dissociation of the progesterone-glycoprotein complex, as indicated in Table XIV-3, resulting in a positive test response with the solution of the standard concentrations.

There is another possible interpretation of the results obtained. If the undissociated progesterone-AAG complex as such should have a definite progestational activity lower than that of progesterone, results as those of Table XIV-2 might be obtained. This explanation, however, is considered less likely because it is generally accepted that the Δ^4-3-keto grouping in the progesterone molecule is essential for biological activity. Evidence has been discussed in Chapter XIII that the interaction between AAG and progesterone involves the Δ^4-3-keto

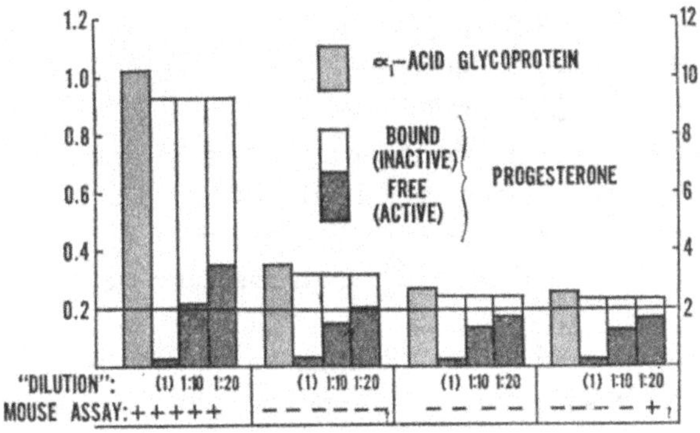

Fig. XIV-1. Bound and unbound progesterone after injection of complex with AAG in the amounts shown in Table XIV-2. Left ordinate: ng progesterone; right ordinate: μg AAG. Progestational effect in relation to threshold of 0.2 ng progesterone (horizontal line)

structure, so that the biologically active site would be blocked in the complex.

The molar ratio, AAG : progesterone in human blood varies between approximately 20 and several hundred, depending mainly on the progesterone concentration; in the present experiments (Tables XIV-2 and -3), this ratio is about 80 : 1. Although in all these cases the number of AAG molecules greatly exceeds those of progesterone, it has to be considered that under physiological conditions in the blood the binding of progesterone to AAG involves only a small fraction of this hormone. It has been shown in Table VIII-15 that by far the major portion of progesterone is associated with CBG and HSA, and only about 1% with AAG. However, the present results add another example to the observations of biological inactivation by protein binding and are considered significant for that reason.

In view of the strong binding affinity between progesterone and CBG (Table VIII-21), and of the high albumin concentration in relation to progesterone, the effect of association with these proteins on the hormonal activity of progesterone was studied (HOFFMANN et al., 1969). This information is necessary for the interpretation of steroid-CBG interaction observed under specific physiological conditions such as pregnancy (ROSENTHAL et al., 1969 a).

The bioassay procedure of HOOKER and FORBES (1947) was applied in the manner described above. The strain of mice used, Jackson Laboratory B6D2F1, was less sensitive; the smallest amount of progesterone giving a consistently positive response was 0.5 ng per uterine horn. The CBG preparation was obtained by paper-curtain electrophoresis of rat serum; it corresponded to Fraction No. 27 in Fig. VIII-19. The albumin used in the bioassay was also prepared from rat serum by continuous paper electrophoresis (Fig. VIII-19), chromatography on DEAE-cellulose and repeated gel filtrations over Sephadex G-200.

The solutions used to test the influence of CBG were adjusted to contain a 1 : 1 molar ratio of progesterone : CBG, on the basis of the physicochemical parameters given in Chapter IX. In the solutions containing albumin, the molar ratios of progesterone : albumin were 1 : 200, 1 : 100, and 1 : 10; a molecular weight of 65,000 was assumed for rat serum albumin (JUNGBLUT and TURBA, 1963).

These ratios of progesterone to the two binding proteins may be related to the following physiological data. The CBG concentration in the serum of normal male rats has been found to be about 1.1×10^{-6} M (Table VIII-21), and that in pregnant rats about twice that value (Table IX-10). This concentration would equal that of progesterone-binding sites assuming rat CBG to have the same number of binding sites for progesterone as for corticosterone, i. e., $n = 1$ (Table XI-7). The albumin concentration observed in rat serum is 3.7×10^{-4} M (WESTPHAL et al., 1963); the serum progesterone level in pregnant rats has been reported to be 3.6×10^{-7} M (WIEST, 1968). These values show that the progesterone-protein ratios used in the present study are of the same order of magnitude or exceed those found under physiological conditions.

Table XIV-4 shows the responses obtained in the uterine horns of the test mice after injection of the progesterone solutions in the absence and presence of CBG or albumin. Each + or − symbol represents the result in one horn. It is evident that the positive responses given by 2.41 and 0.60 ng progesterone without protein are abolished by the presence of CBG; the same result was obtained with 0.80 ng. The molar ratio of CBG to progesterone was the same in the three mixtures tested.

The results obtained with the highest albumin concentration (Table XIV-4) indicate complete inactivation of progesterone. A calculation of bound and unbound progesterone for this solution, providing 29.8 μg per horn, under the assumption of a 1 : 10 dilution (see above) gives an equilibrium concentration of 0.48 ng bound and 0.12 ng unbound progesterone per uterine horn. A similar calculation for the solution containing the lowest albumin concentration (molar ratio of progesterone to albumin of 1 : 10), assuming a dilution of 1 : 5, gives 0.17 ng bound and 0.43 ng unbound progesterone per uterine horn. The latter amount of hormone is not sufficient to give a positive response in all tests. The results obtained (Table XIV-4) are thus in agreement with the assumption that dilutions of this magnitude have taken place in the bioassay and that the progesterone-albumin complex has no or very little biological activity.

Table XIV-4. *Effect of CBG or albumin on the biological activity of progesterone.* From HOFFMANN et al. (1969)

Progesterone ng/horn	Rat serum preparation		Response in uterine horn
	CBG μg/horn	Albumin μg/horn	
2.41	none	none	+ + + −
0.60	none	none	+[a] +[b] + + + + (+) −
2.41	59.5	none	− − (−)
0.80	19.8	none	− − − − −
0.60	14.9	none	−[a] −[b] − − − − − − + +
0.60	none	29.8	− − − − − − − − − −[c]
0.60	none	14.9	+ + + + + + + − − − − − − − − − −
0.60	none	1.5	+ + + + + − − − − −[d]

[a, b] Identical letters indicate the 2 horns of the same animal.
[c] Two responses doubtful.
[d] One positive response doubtful.

An *in-vitro* system was used by BILLIAR et al. (1969) to study the influence of CBG and HSA on the rate of reduction of progesterone by human placental 20α-hydroxysteroid dehydrogenase. The purified enzyme was incubated with the steroid in the presence and absence of purified human CBG or HSA. Convincing evidence was obtained that the transcortin inhibited the enzymatic conversion in direct correlation with the binding of progesterone by CBG. Cortisol which by itself had no effect on the enzymatic reduction, counteracted the inhibitory action of transcortin, presumably by displacing progesterone from the binding sites on CBG. It was concluded that progesterone when associated with CBG is not available for the enzymatic reaction. Relatively high concentrations of HSA also inhibited the 20-keto reduction by the 20α-hydroxysteroid dehydrogenase, again in good correlation with the progesterone-HSA interaction. These well-controlled *in-vitro* studies supplement the bioassay results discussed above and corroborate the concept of hormonal inactivation by protein binding.

The experimental evidence presented in this chapter warrants, in the author's opinion, the conclusion that the unbound form of the steroid hormone is biologically active, and that the hormonal potency is suppressed in the protein complex. It has to be realized, however, that the steroid-protein complex is dissociable, that a dynamic equilibrium exists in which continuous dissociation and association take place, regulated primarily by the association constant of the complex. Depending on the size of this constant, a certain percentage will be present, as a statistical average, in unbound form. Biological inactivation by protein binding then means reduction of the statistical average of steroid molecules present as unbound, i.e., biologically active species.

Association of the steroid with the binding protein may alternately be understood as a reduction of the chemical activity of the ligand. In this more general interpretation, various consequences of the complex formation become evident (WESTPHAL, 1961). Protection of the steroid hormone against chemical attack in the circulating blood, against enzymatic reactions in the liver leading to formation of inactive derivatives, to catabolism and/or excretion. Reduction of chemical activity would result in protection from adsorption to blood vessel walls or other structural surfaces. At the same time, the steroid-protein complex would represent an innocuous storage form in which relatively large quantities can be maintained without exerting biological effects.

The hormonal action at the target tissue in this interpretation becomes a competition between the specific receptor sites and the plasma protein carrier. A similar competition in the reverse sense may occur at the steroid-producing tissue. The binding affinity of the serum protein may facilitate the transfer from the site of biosynthesis into the bloodstream, possibly through the lymph system (WESTPHAL, 1970 a). The question of a specific carrier function of the steroid-binding serum proteins in the sense of a directed transport (BENNHOLD, 1932), in which the protein portion of the complex would participate in the interaction with the receptors, has not yet been answered. DE MOOR et al. (1968 a) in a thoughtful discussion have considered the various possible roles protein binding may play in the hormonal function of the steroids.

XV. Steroid-Binding Immunoglobulins: Preparation, Properties and Application for Radioimmunoassay of Steroid Hormones

The interactions between steroids and proteins discussed so far in this monograph, and the reactions between steroids and enzymes to be dealt with in Chapter XVI, involve proteins which occur in every normal individual of a given species in about the same quantities. Normal biosynthesis, metabolism and degradation applies to these proteins which possess more or less specific binding sites for steroid molecules. No experimental procedure is known to influence the biosynthesis of the binding proteins with respect to steroid specificity, with the possible exception of bacterial enzymes.

A unique class of serum proteins, the immunoglobulins, can interact with high specificity and great affinity with those very compounds that are responsible for their formation, *viz.*, the antigens. The steroids, as compounds of low molecular weight, are not immunogenic (for nomenclature and molecular aspects of immunological phenomena see SELA, 1969). However, LANDSTEINER (1933, 1945) has shown that certain small molecules (haptens) when covalently linked to proteins become antigenic, i.e., they provoke the production of unique antibodies. These antibodies reflect in their combining sites the antigenic specificity of the particular hapten that was responsible for their formation.

If a steroid were covalently coupled to a protein, an artificial antigen would be produced capable of eliciting the formation of antibodies with specificity not only for the protein carrier but also for the particular haptenic steroid. The result would be an *in-vivo* production of tailor-made binding sites for any steroid that can be attached as a hapten to a protein by covalent bonds and thus become antigenic.

Preparation of steroid-binding immunoproteins according to these principles was the objective of studies begun in 1946 in the author's laboratory with E. MEINZINGER. Estradiol was selected as the haptenic steroid. In anticipation of possible relationships to estrogen interaction at the target tissue, at sites which are now known to be provided by receptor proteins, diethylstilbestrol was selected as a second hapten, so that a comparison could be made between the properties of the antibodies against the natural and the artificial estrogen. These ideas and aims have been discussed in the doctoral dissertation of E. MEINZINGER (1949)—the work itself was discontinued when I moved to the United States in 1949.

In contrast to steroid-protein conjugates prepared in subsequent years in other laboratories (see below), our aim was to attach the steroid hormone to the protein in such a way that the functional groups, essential for the biological activity, remained intact in the coupled product. For this reason, estradiol (1) was substituted in ring A by coupling with phenyldiazonium chloride to form two isomeric phenylazoderivatives (MEINZINGER, 1949) which were assumed (from analogous reactions with *ar*-2-tetralol) to be 4-phenylazoestradiol [(3), m. 167°, main product] and 2-phenylazoestradiol [(2), m. 177—178°]. Catalytic hydrogenation of the phenylazoestradiol (m. 167°) gave a crystalline product (m. 268—270°) of the composition $C_{18}H_{25}O_2N$. This compound is assumed to be 4-amino-estradiol (4) although the 2-position for the amino group is not excluded. The following formulas illustrate the reactions:

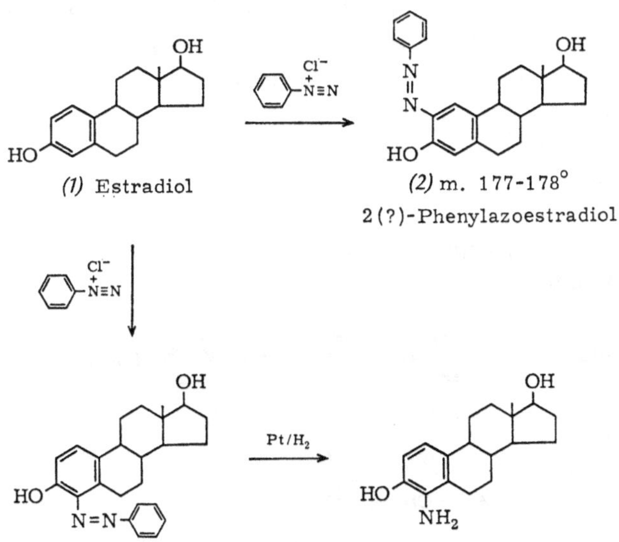

(1) Estradiol

(2) m. 177-178°
2(?)-Phenylazoestradiol

(3) m. 167°
4 (?)-Phenylazoestradiol

(4) m. 268-270°
4 (?)-Aminoestradiol

Clearly, the aminoestradiol (4) obtained fulfills the above requirements in having an amino group which may be coupled to proteins without blocking the functional groups at C-3 and C-17. In bioassays for estrogenic activity (Allen-Doisy test), 1 μg of the aminoestradiol had the same activity as 0.2 μg of diethylstilbestrol dipropionate.

In earlier studies, KING and FRANKS (1941) have covalently linked estrone to casein through the p-aminophenylether. Since the biologically important phenolic hydroxyl was blocked the estrogenic activity of the p-aminophenylether and of the casein conjugate was very low. Formation of antibodies has not been reported. MOOSER and GRILICHESS (1941) esterified androst-5-ene-3β,17β-diol with p-amino-o-sulfobenzoic acid and performed diazo coupling with several proteins. No antibodies against the androstenediol derivatives were formed, possibly because of hydrolysis of the esters *in vivo*. The same reason may have been responsible for the failure of GRAB and GOLDBERG (1949) to produce antibodies against protein-coupled esters of androstenediol and estradiol. Antisera were obtained by SPRUNT et al. (1951) against alpha-estradiol-albumin conjugates; no details were given on the mode of linkage.

In all these studies, the steroid molecules were attached to the proteins by covalent bonds. SACHS and KLOPSTOCK (1925) originally used mixtures of cholesterol or lecithin with animal sera to provoke in rabbits antisera with specific affinities for the lipids; these earlier experiments have been discussed by LANDSTEINER (1933).

The following methods for covalent bonding of haptens to proteins have been applied in the preparation of protein conjugates with steroids.

1. Diazonium coupling method. Based on the reactions described by PAULY (1904), LANDSTEINER and LAMPL (1918) coupled diazonium group-containing haptens (6) to proteins according to the following general scheme:

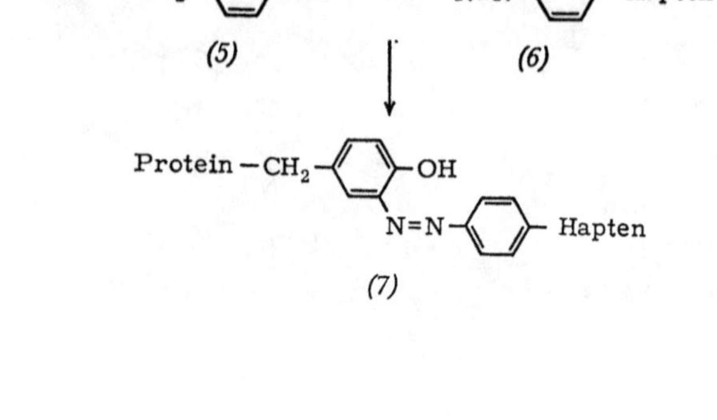

(7)

The coupling reaction occurs either at tyrosine residues of the protein (7) or with other reactive residues such as histidine, tryptophan, proline, hydroxyproline and certain free amino groups (EAGLE and VICKERS, 1936). The procedure can be applied to various types of haptens as the work of LANDSTEINER (1933, 1945) has shown; its use for the preparation of steroid-protein conjugates has been discussed above.

2. *Isocyanate method.* In an endeavor to eliminate the non-physiological azo grouping from the hapten-protein conjugates, HOPKINS and WORMALL (1933) reacted proteins with phenylisocyanate (9) and obtained phenylureidoproteins (10):

(8) *(9)* *(10)*

Phenylisocyanate

This principle was applied by GOODFRIEND and SEHON (1958) to couple estrone to rabbit serum albumin, according to the following reactions:

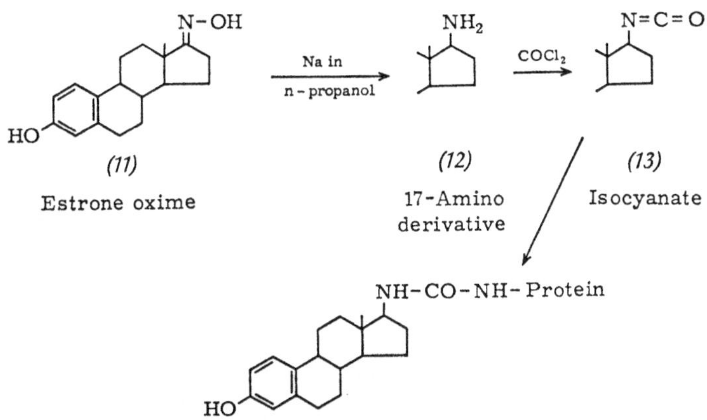

(11) Estrone oxime

(12) 17-Amino derivative

(13) Isocyanate

(14) Conjugate at C-17

The conjugate (14) contained about 25 estrone residues per molecule albumin. The coupling occurred essentially at the ε-amino groups of lysine. GOODFRIEND and SEHON (1960) prepared similar estrone-17-carbamido conjugates with HSA and BSA. They observed formation in the rabbit of specific antibodies to the estrone moiety as well as to the protein carrier. A detailed account of the immunochemical studies with these and additional conjugates has been given by GOODFRIEND and SEHON (1961 a). The isocyanate method was later extended by GOODFRIEND et al. (1961) to the preparation of HSA conjugates of estrone (20), coupled by a carbamide bond at C-2, according to the following reactions:

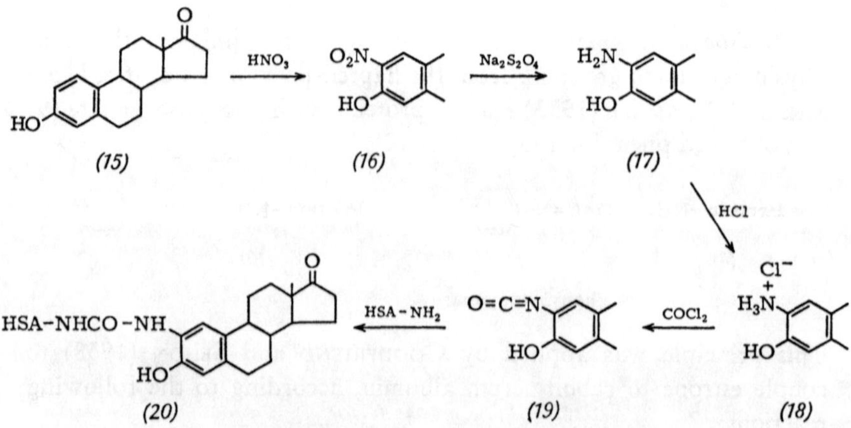

(15) (16) (17) (18) (19) (20)

Estrone-2-carbamido-HSA Isocyanate

3. Schotten-Baumann method. The classical acylation technique of SCHOTTEN and BAUMANN, in which an acid chloride reacts with an alcoholic group or another reactive hydrogen, was applied by ERLANGER et al. (1957) for covalent linkage of testosterone to BSA (22). A peptide bond is formed between an amino group of the protein and the chloroformate of the steroid (21) which is prepared by reaction of the hydroxyl with phosgene:

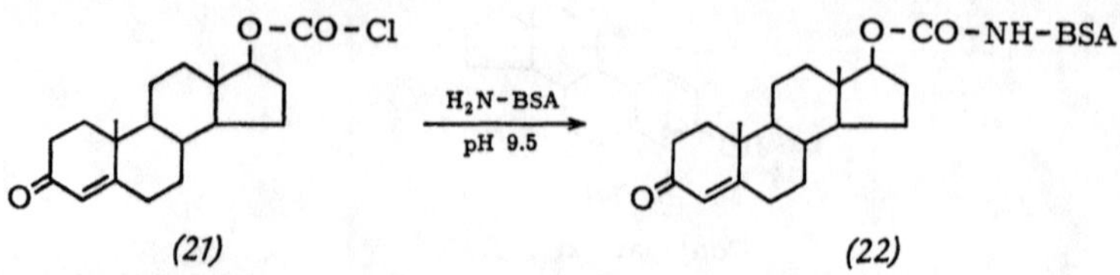

(21) Testosterone-17-chloroformate

(22) Testosterone-17-conjugate with BSA

KLOPSTOCK et al. (1964) have coupled cholanic acid and cholesterol to HSA by use of the Schotten-Baumann reaction in pyridine, starting from cholanyl chloride (23) and cholesteryl-chloroformate (25), respectively:

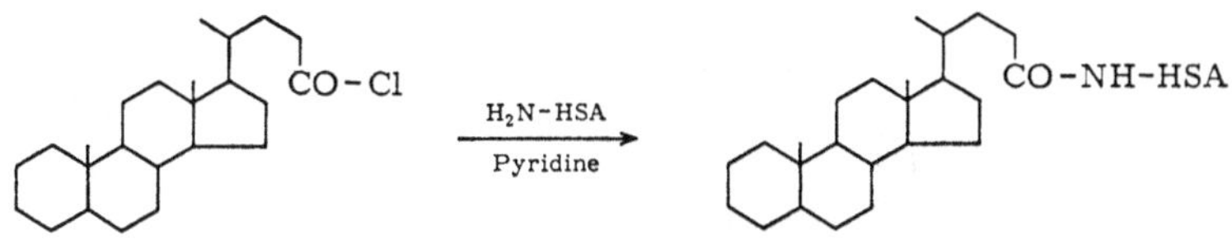

(23) Cholanylchloride

(24) Cholanyl-24-conjugate with HSA

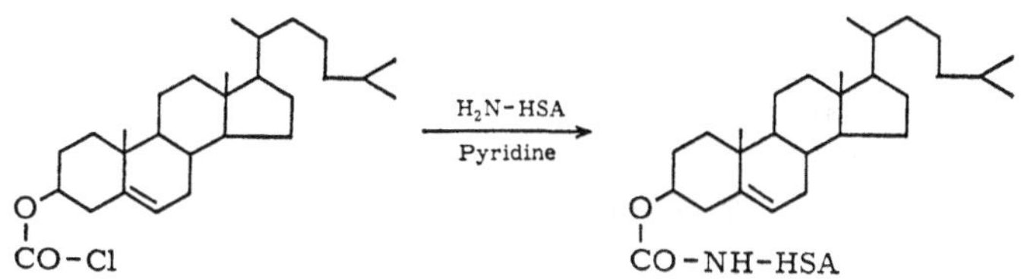

(25) Cholesteryl chloroformate

(26) Cholesterol-3-conjugate with HSA

4. *Mixed anhydride method.* An alternate technique of producing peptide bonds between carboxy groups of the steroid molecule and amino groups of the protein was utilized by ERLANGER et al. (1957). A mixed anhydride (29) was formed between the carboxy group of the steroid derivative (27) and isobutyl chloroformate (28), followed by acylation of a reactive amino group of the protein (30):

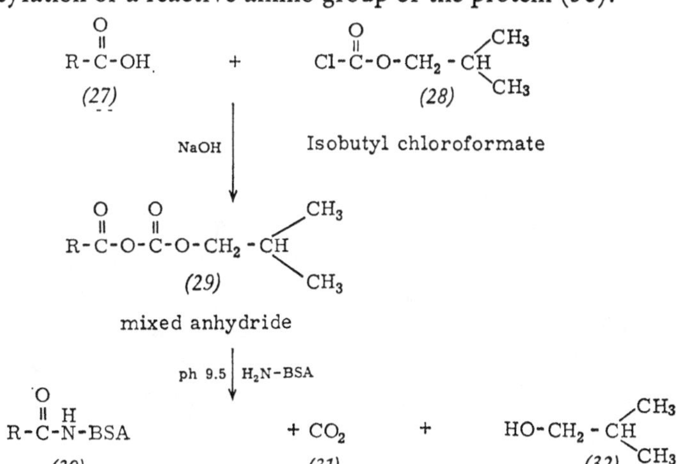

Steroid-BSA conjugate

The steroid compounds initially employed by ERLANGER et al. (1957) were the (O-carboxymethyl)oxime of testosterone (33) and the 21-hemisuccinate of cortisone (34).

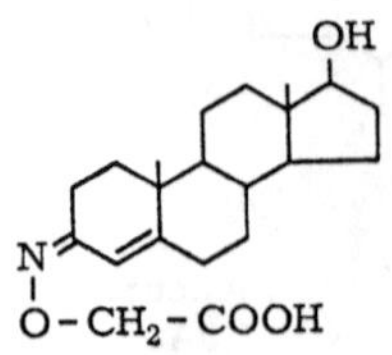

(33)

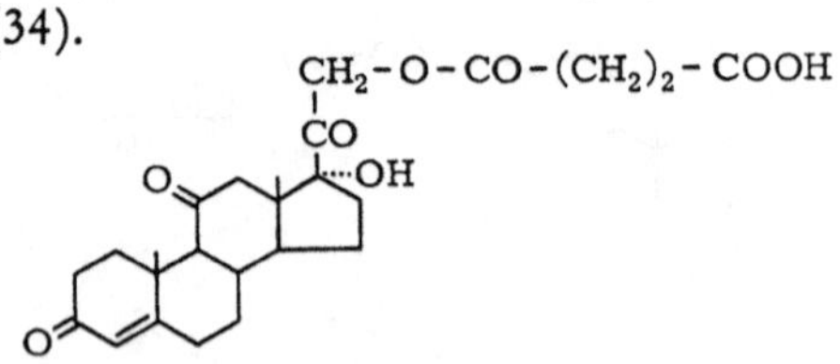

(34)

LIEBERMAN et al. (1959) have described additional carboxyl-containing steroid derivatives which can be coupled to proteins, *i.e.*, the 20-(O-carboxymethyl)oxime of progesterone (35), cortexone hemisuccinate (36), and the (O-carboxymethyl)oxime of estrone (37):

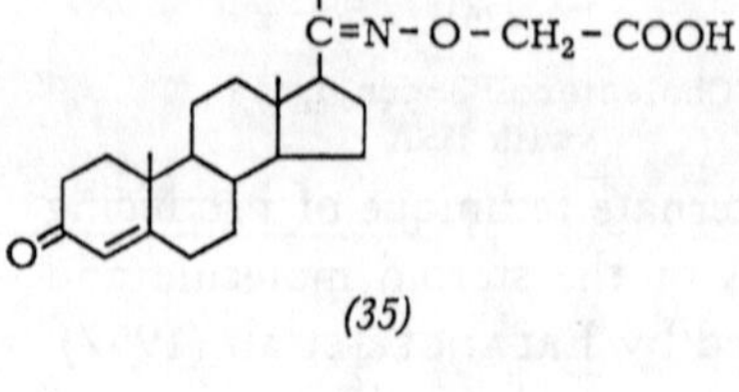

(35)

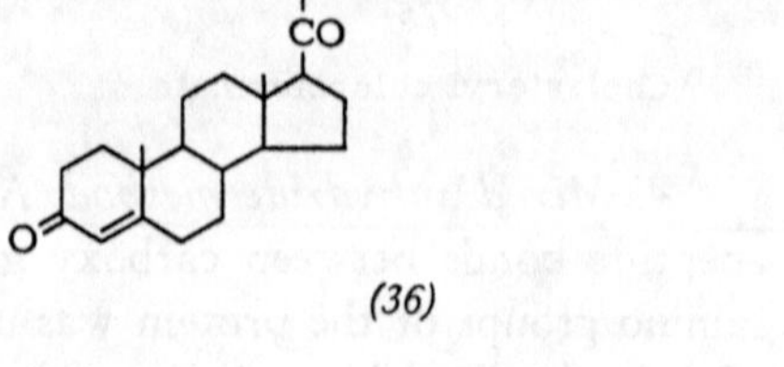

(36)

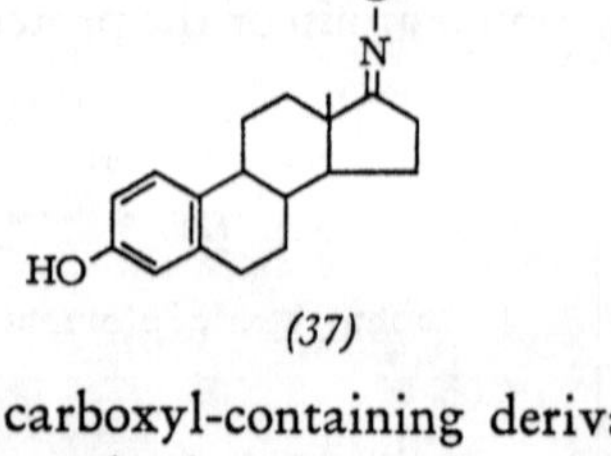

(37)

For each of these carboxyl-containing derivatives, formation of the protein conjugate may be formulated by replacing the carboxy group by —CO—NH-Protein. Clearly, numerous possibilities exist to attach the protein at different locations of the steroid molecule.

5. Carbodiimide method. A combination of diazo coupling and peptide bond linkage has been designed by GROSS et al. (1968). In this procedure, the condensation between the carboxy and the amino group was facilitated by 1-cyclohexyl-3-[2-morpholinyl-(4)-ethyl] carbodiimide (41). The following formulas explain the reactions which result in the formation of an estradiol-protein conjugate (42) coupled at C-4 or C-2 through an azo grouping:

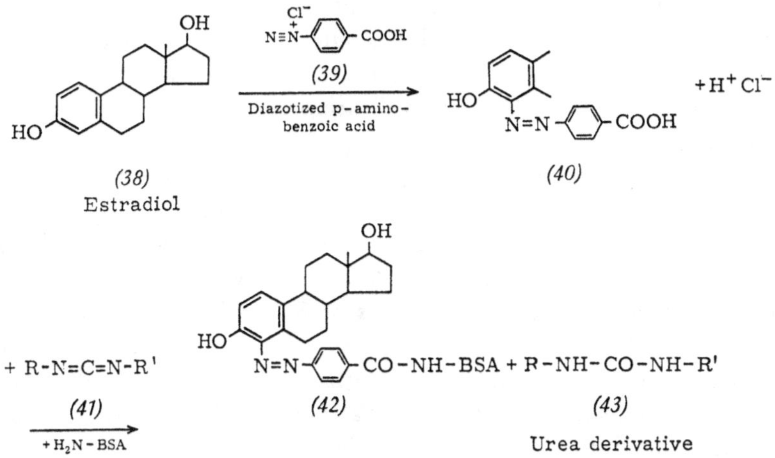

The proteins used for bonding were BSA and the keyhole limpet hemocyanin. An analysis of the substituted proteins revealed 5 to 29 steroid residues per molecule of protein.

LINDNER et al. (1970) applied dicyclohexyl carbodiimide to couple estrogens to proteins or synthetic polypeptides (BAUMINGER et al., 1969). For considerations similar to those outlined above, the linkage was accomplished at steroid positions remote from the functional groups at C-3 and C-17. The following formulas show the preparation of a 6(α?)-aminoestradiol (46) from 6-oxo-estradiol-diacetate (44) which can be obtained by chromic acid oxidation of estradiol diacetate. Note that the aminosteroid becomes bonded to a free carboxy group of the protein (47):

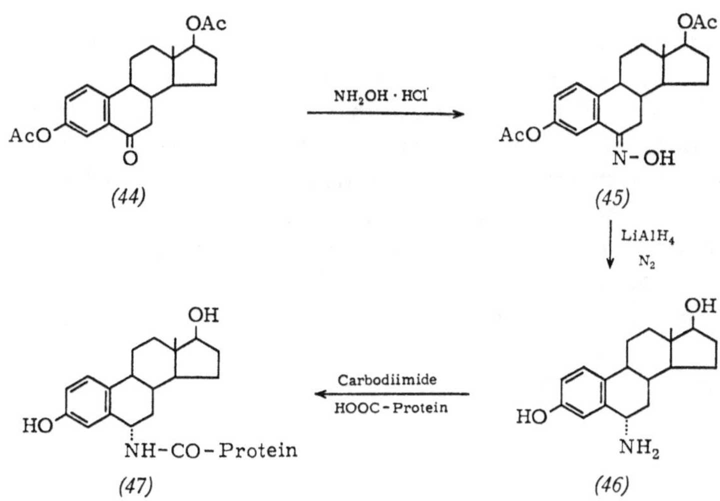

To increase the distance between haptenic steroid and the macromolecule, an ε-aminocaproic acid was interposed between the two components. The 6-oxo-estradiol could also be coupled directly through the (O-carboxymethyl)oxime to lysine residues, or through the p-aminobenzoylhydrazone to dicarboxylic residues of the carrier protein.

In 1958, LIEBERMAN et al. (1959) have presented an excellent review of the chemical, immunochemical and endocrinological properties of the steroid-protein conjugates—results that were almost totally achieved in Dr. LIEBERMAN's laboratory. The conjugates discussed include the coupling products of BSA with steroids (22) and (33) through (37) formulated above. Experimental details on the preparation of BSA conjugates with progesterone, cortexone and estrone have been published by ERLANGER et al. (1959). The number of steroid residues linked to BSA was determined by an ultraviolet-spectrophotometric method, by determination of the ratio of amino nitrogen to total nitrogen, and by a dinitrophenylation technique. The attachment occurred mainly at the ε-amino groups of lysine; BSA possesses 59 of them. The various conjugates were found to have about 20 to 30 steroid molecules coupled per molecule BSA.

Bioassays showed hormonal activity for the BSA conjugates of testosterone, estrone and cortisone if ring A was not involved in the covalent bonding. The attachment of testosterone to BSA through the oxime derivative at C-3 (formula 33) resulted in loss of androgenic and myogenic activity.

Immunization of rabbits with the alum-precipitates of the BSA conjugates of testosterone, cortisone, cortexone, estrone, and progesterone elicited the formation of antibodies specific for the haptenic steroids injected (BEISER et al., 1959). All antisera precipitated with BSA as well as with the conjugate used for immunization. After absorption with BSA, it was found that, in addition to the specific precipitation, cross-reactions occurred with other steroid-protein conjugates. For example, at the respective points of maximal precipitation, the BSA conjugates of cortexone (36) and of progesterone (35) precipitated 67%, and the BSA conjugate of testosterone coupled at C-3 (33) precipitated 51% of the antibody from an anti-testosterone C-17-BSA serum precipitable by the homologous antigen (22).

The highest specificity for the homologous antigen was also demonstrated in hapten inhibition tests. Characteristically, an antiserum to the testosterone-C-17-BSA conjugate (22), which was inhibited most effectively by testosterone hemisuccinate (48) in its precipitation with the homologous antigen, was almost equally inhibited by compound S hemisuccinate (49).

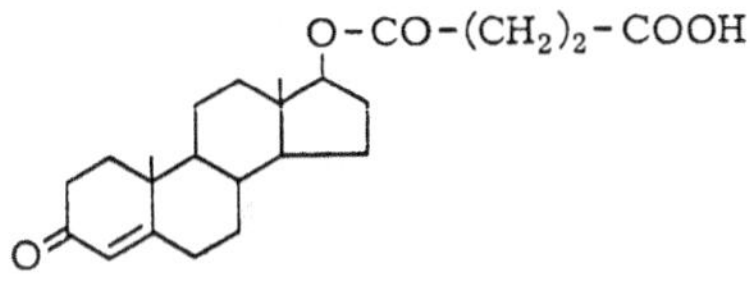

(48)
Testosterone hemisuccinate

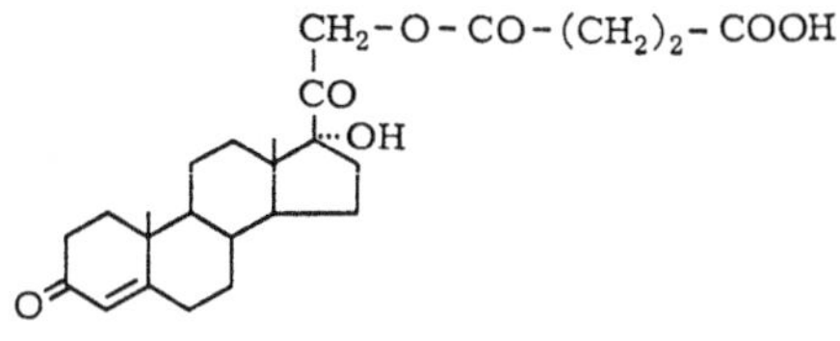

(49)
Compound S 21-hemisuccinate

Cortisone-21-hemisuccinate inhibited slightly less, cortisol-21-hemisuccinate much less. In all these heterologous haptens, ring A of the steroid structure was not changed from that in the original antigen (22) or the homologous hapten (48). Alteration of ring A in the haptens to a structure given by compound (33) reduced the inhibiting power further; aromatization of ring A [estradiol-17-hemisuccinate or estrone-(O-carboxymethyl) oxime (37)] abolished hapten inhibition completely. The immune sera prepared by GROSS et al. (1968) against the C-4-(or C-2)-coupled conjugate between estradiol and BSA (42) showed slight cross reactivity between estradiol, estriol, and estrone.

The important question of antihormonal properties of the antisera to steroid hormone-protein conjugates has been investigated in numerous laboratories. GOODFRIEND and SEHON (1961 b) found that immune sera to estrone-17-carbamido-HSA neutralized the uterotropic activity of exogenous estrone in immature rats. NERI et al. (1964) administered to appropriate test animals antibodies that were elicited by the BSA conjugates of testosterone, estrone, cortisol, and aldosterone. They could demonstrate that these antisera neutralized the biological effects ordinarily caused by the corresponding exogenous steroid hormones.

The effects of antibodies to steroid hormone-BSA conjugates on hormonal activities were studied extensively in the laboratory of VANDE WIELE. Antisera produced in sheep against the BSA conjugate of estradiol-17β-hemisuccinate inhibited the biological effects of estradiol administered to immature or mature ovariectomized rats and mice (FERIN et al., 1968). They also prevented uterine weight increases, endometrial stimulation and vaginal cornification due to endogenous estrogens following treatment with human chorionic gonadotropin. Pretreatment of intact animals with antisera inhibited the uptake of subcutaneously administered tritiated estradiol by various target tissues. Uterine stimulation produced by diethylstilbestrol was not inhibited, even by large amounts of antiserum, indicating that the antibodies to the estradiol conjugate do not cross-react with the synthetic estrogen. Since the specific antibodies presumably neutralize endogenous hormone in the peripheral blood, a means is provided of controlling endogenous hormone outside the target organ, contrary to other anti-estrogens which act at the target site.

FERIN et al. (1969 a) utilized the antibodies to estradiol-17β-succinyl-BSA in studies with immature rats and obtained direct evidence for the existence of a positive feedback role of estrogens in the ovulatory release of luteinizing hormone (LH) from the pituitary. Again it was found that the estrogenic activity of diethylstilbestrol was not inhibited by the antiserum to the natural hormone. The antibodies to estradiol had no effects other than inhibition of the biological activity of the hormone in the circulation; this property renders them uniquely useful as a tool to study the role of estrogens and other hormones in reproductive processes.

Additional studies by FERIN et al. (1969 b), in which antibodies to the BSA conjugate of progesterone-20-(0-carboxymethyl) oxime (35) were applied, corroborated the conclusion that in the adult rat the initiation of LH release is under the control of estrogens but not of progesterone. The significance of a high level of estradiol for induction of a release of LH from the hypothalamic-pituitary axis was illustrated by SCARAMUZZI et al. (1970) in active immunization experiments in which estradiol-17β-BSA conjugate was used as antigen in ovariectomized adult ewes. Effective neutralization of exogenous estradiol was observed in the immunized animals. VANDE WIELE et al. (1970) have reviewed the various applications of steroid hormone-specific immunoproteins. Physiological studies involving the use of antibodies to steroids have been discussed by CALDWELL et al. (1970 b) as well as by VANDE WIELE and FERIN (1970).

Purification studies were performed by ZIMMERING et al. (1965) with antibodies prepared in sheep against the BSA conjugate of testosterone-17-hemisuccinate. The antibody was separated from the sera by means of sodium sulfate which precipitated the globulin fraction, by specific precipitation using a hapten analogue, and by chromatography on DEAE-cellulose. The preferred method of purification of the anti-testosterone antibodies was precipitation by the corresponding conjugate of testosterone-17-hemisuccinate with HSA at the equivalence point and subsequent dissolution by the hapten analogue, N,N-dimethylaminoethyl-17β-(3-oxoandrost-4-enyl)-carbonate. The antibodies were then separated by gel filtration over Sephadex G-25. Chromatography on DEAE-cellulose produced globulin solutions which contained 33 to 59% specifically precipitable antibody.

In later studies by ZIMMERING et al. (1967), the binding properties of steroid-specific antibodies were investigated in more detail. Antisera were produced in different unbred ewes against BSA conjugates of testosterone-17-hemisuccinate, testosterone-3-(O-carboxymethyl)-oxime, and progesterone-20-(O-carboxymethyl)oxime. The immune sera were absorbed with BSA to remove antibody to carrier protein, and were then purified sufficiently to remove all albumin and all other steroid-binding proteins.

The antibodies were found to have a wide spectrum of binding affinities. Two main types of binding sites could be characterized, one with high strength of binding and the other with a relatively weak affinity. At steroid concentrations below 3×10^{-6} M, a strong steroid specificity to the high-affinity sites was observed, whereas considerable cross-reactivity was present at the low-affinity sites. It is interesting to note that ZIMMERING et al. (1967) reported, for the weak binding sites in the three antibodies studied, augmentation effects for additional binding after a first steroid molecule was bound. The high-affinity binding sites of the antibodies to the testosterone-17-conjugate of BSA were also in part available to the cross-reacting haptens, i.e., cortexone-21-hemisuccinate, cortisone-21-hemisuccinate, and 11-dehydrocorticosterone-21-hemisuccinate; these sites had no affinity for the 21-hemisuccinates of cortisol or corticosterone, nor for a series of other hydroxylated steroids. Antibodies to the testosterone-BSA conjugate, bonded through the O-carboxymethyloxime group at C-3, had tight binding sites available only to the homologous hapten and to some extent to 11β-hydroxytestosterone.

The properties of the antibodies depended on the mode of preparation. Salt fractionation and DEAE-cellulose chromatography gave preparations containing high-affinity binding sites. Antibodies against the testosterone-17-BSA conjugate, purified by the two procedures, bound the hapten, testosterone hemisuccinate, with apparent affinity constants of $k = 3.9 \times 10^5$ and $3.5 = 10^5$ M^{-1}, respectively. The specific precipitation technique yielded preparations with lower binding affinities, $k = 0.9 \times 10^5$ M^{-1}. All these affinity constants are small compared to the values known for the steroid complexes of other serum proteins such as CBG, SBP, AAG, and even albumin. MIDGLEY (1970) has observed association constants at 4° in the range of 10^9 to 10^{10} M^{-1} for antibodies elicited to steroid conjugates with bovine or rabbit serum albumin. Definitive values for apparent affinity constants of steroid-antibody complexes cannot be expected before the complete—and difficult—purification of the specific immunoglobulins has been achieved.

As an important approach to a specific purification of antibodies, the method reported by LINDNER (1970) deserves mention. The hapten-specific immunoglobulins are adsorbed to an immunoadsorbent consisting of an insoluble matrix such as bromoacetylcellulose or -sepharose, to which the steroid, or its conjugate with a heterologous carrier, has been covalently attached. Elution is carried out with 0.1 M acetic acid. Another promising adsorbent for this purpose is prepared from the acyl-azide derivative of cross-linked polyacrylamide beads (INMAN and DINTZIS, 1969) which will quantitatively bind amino steroids (LINDNER, 1970). Application of affinity chromatography for the isolation of steroid-binding proteins is in its beginning, and various difficulties, particularly that of elution from the specific adsorbents, will have to be overcome (BURSTEIN, 1969, 1970).

The immunoproteins produced in various animals against steroid-protein conjugates have become particularly important recently for the determination of extremely small quantities of steroid hormones by protein binding methods. An immunochemical principle was first utilized by BEISER and ERLANGER (1967) who determined testosterone, estradiol and progesterone by a haemagglutination-inhibition technique. The steroid hormone-HSA conjugates were covalently linked to erythrocytes by way of bis-diazotized benzidine. These erythrocytes can then be agglutinated by the appropriate antiserum, and the effectiveness of various concentrations of haptens in inhibiting agglutination can be measured. The minimal amounts for complete inhibition of agglutination were found to be 0.05 μg testosterone, and 0.06 μg estradiol per ml.

The sensitivity of steroid quantification utilizing specific immunoproteins was increased by several orders of magnitude when the principle of competitive binding was applied. The antibodies are employed as the binding protein, to which the tritium-labeled steroid of high specific radioactivity is non-covalently bound. The unlabeled steroid of the same structure then displaces the radioactive molecules in proportion to the amount added, and the unbound and bound radiosteroid fractions are separated. A series of assays with known amounts of unlabeled steroid provides the calibration curve from which the results of the unknown test samples are read. Modifications of the basic method have been developed in several laboratories.

The acute interest in this highly sensitive technique, known as radioimmunoassay of steroid hormones, is reflected in several recent symposia that were totally or partly devoted to this topic. In November, 1969, the Worcester Foundation Symposium on Immunological Methods in Steroid Determination was held; the proceedings are being published in book form, edited by PÉRON and CALDWELL (1970). At the IVth Meeting of the International Study Group for Steroid Hormones, in Rome, in December of 1969, a good portion of the program was concerned with the immunological aspects of steroid-hormones. The transactions, edited by CONTI (1970), will appear in "Research on Steroids", volume 4. The second of the Karolinska Symposia on Research Methods in Reproductive Endocrinology, organized by the Karolinska Institutet, Stockholm, with the collaboration of the World Health Organization, was arranged in Geneva, March 1970. The topic "Steroid Assay by Competitive Protein Binding" included various problems of radioimmunoassays. The transactions of this symposium have been published under the editorship of DICZFALUSY (1970). Individual papers of these three symposia are cited in this monograph.

Important theoretical aspects of the general type of analysis discussed here and known as competitive protein binding, saturation analysis, or radioligand binding assay, have been presented by EKINS and NEWMAN (1970). Reliability criteria have been defined by REEVES and CALHOUN (1970). RODBARD and LEWALD (1970) have written a computer program for analysis of radioimmunoassay and competitive protein binding assays.

A general introduction into the field of steroid-specific antisera has been given by GOODFRIEND and SEHON (1970), and GROSS and LAI (1970) have discussed the antibody specificities. Practical aspects of preparation and purification of antibodies to steroids have been presented by THORNEYCROFT et al. (1970). MIDGLEY et al. (1969) have described a radioimmunoassay procedure applicable for any small molecular weight substance which can act as a hapten and thus become immunogenic. In this technique, the protein portion of the conjugate is radioiodinated and used in combination with antibodies specific for the hapten, for assay of free hapten. A number of radioimmunoassay systems have been developed for quantification of various steroid hormones, and the advantages and problems associated with the use of radioiodinated conjugates for labeled steroids have been discussed (NISWENDER and MIDGLEY, 1970). According to MIDGLEY and NISWENDER (1970), as little as one pg of some steroids could be reliably determined in a single assay tube.

A preliminary communication by JIANG and RYAN (1969) gave some details on a radioimmunoassay for estrogens, using antisera produced in rabbits against the BSA conjugate with estradiol-17β-hemisuccinate. Binding of tritiated estradiol to the diluted antiserum was measured by gel filtration over Sephadex G-25, or the bound and free fractions of the radiolabeled estradiol were separated by dextran-coated charcoal or Dowex-1 resin. Addition of unlabeled estradiol, estrone and estriol reduced, in decreasing order, the percentage of bound tritiated estradiol. A quantity of 0.2 ng of unlabeled estradiol could be readily detected by this technique.

Greater sensitivity in the radioimmunoassay of estradiol was achieved by ABRAHAM (1969) who used a solid-phase technique. Polystyrene disposable test tubes were coated with a highly diluted antiserum, which had been produced in ewes against the estradiol-17β-succinyl-BSA conjugate. A standard amount of tritiated estradiol was added, in mixture with the unknown quantity of unlabeled estradiol. After incubation, measurement of radioactivity in the solution gave the amount of free radiolabeled estradiol which was dependent on the quantity of unlabeled steroid to be assayed. At a confidence limit of 95%, 10 pg of estradiol were found to be significantly different from zero pg. By prolonging the incubation time at room temperature from 2 to 16 hrs (ABRAHAM and ODELL, 1970), the sensitivity could be further increased to a minimal amount of 5 pg (Fig. XV-1). Estrone has also been determined by the solid-phase technique (CALDWELL et al., 1970 a). The solid-phase radioimmunoassay of steroids has been reviewed by TILLSON et al. (1970), and several modifications of the general method have been described by ABRAHAM et al. (1970).

A sensitivity of 10 pg of estradiol in a radioimmunoassay has also been reported by VANDE WIELE et al. (1970). The method used by MIKHAIL et al. (1970 c) includes ether extraction, separation of estrone and estradiol by Sephadex LH-20 column chromatography, and radioimmunoassay using a mixture of polymerized estradiol antibodies and tracer ^{3}H-estradiol. The free and protein-bound portion of the radiolabeled steroid is separated by centrifugation. The water blank of the

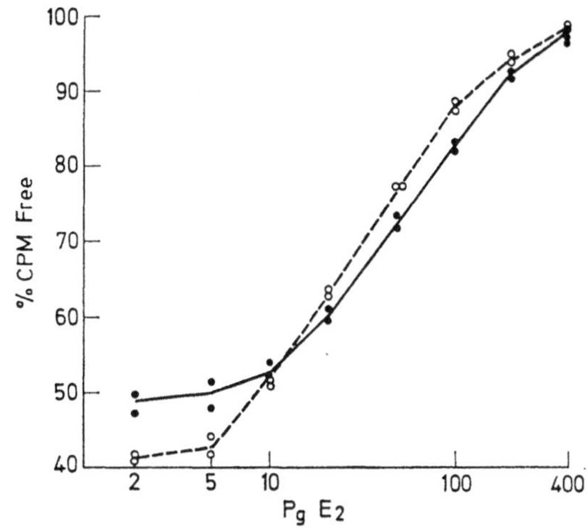

Fig. XV-1. Radioimmunoassay for estradiol (E_2). Solid-phase technique; antiserum dilution 1 : 20,000, 40 pg tritiated estradiol. Room temperature. Full circles, solid line: 2 hr incubation; open circles, broken line: 16 hr incubation. From ABRAHAM et al. (1970)

method was 11 ± 1 pg, the lower limit of sensitivity 50 pg (MIKHAIL et al., 1970 a, 1970 b).

The preceding review of the radioimmunoassay procedures applied for the determination of extremely small quantities of steroid hormones gives the present state of development in a rapidly moving field of investigation. The techniques and results are sketched only briefly since the author of this monograph has not contributed to them; for details, the original literature should be consulted. Although the principle of the radioimmunoassay appears simple, the actual performance of the techniques is difficult and requires both meticulous attention to experimental details and, very likely, a certain "incubation time" before enough experience is gained to obtain the amazingly high sensitivities reported from several laboratories. One of the more important factors for successful application is the rigorous exclusion of contaminations from the assay systems. The discussions, held after the presentation of the papers at the meetings mentioned, convey most vividly the happy as well as unhappy experiences of the investigators in this newest area of steroid hormone research.

XVI. Notes on Steroid Interactions with Enzymes

A unique type of steroid-protein interaction is the complex formation between a steroid and its specific enzyme. Here the steroid plays the role of a substrate to an enzyme, and at the same time may be considered a ligand to a protein which has a high and rather specific affinity for the particular steroid. Clearly, the study of steroid-enzyme complexes is of interest for more than one reason, and work carried out in recent years on several purified enzyme systems has demonstrated their usefulness for the investigation of basic problems of steroid-protein interaction. The present brief chapter will summarize the more important results obtained, and indicate trends of research in an area to which this author has not contributed. An excellent review article by TALALAY (1965) emphasizes the molecular aspects of enzymatic mechanisms in steroid biochemistry, including considerations similar to those concerned with non-enzymic steroid-protein complexes.

The most important steroid enzyme to be discussed in this context is the *Δ^5-3-ketosteroid isomerase of Pseudomonas testosteroni*, an inducible enzyme, which was discovered by TALALAY and WANG (1955). The function of this isomerase, which has also been found widely distributed in the mammalian organism, is to catalyze the migration of the Δ^5 double bond to the Δ^4 position in conjugation to the 3-keto group as was shown for androst-5-ene-3, 17-dione and for pregn-5-ene-3,20-dione. The enzyme thus accelerates *in vivo* an isomerization reaction which had been performed *in vitro* about 20 years earlier, when pregn-5-ene-3,20-dione was obtained as an oxidation product of pregn-5-en-3β-ol-20-one, and was readily isomerized in acidic solution to progesterone (WESTPHAL and SCHMIDT-THOMÉ, 1936).

The Δ^5-3-ketosteroid isomerase from *Pseudomonas testosteroni* has been obtained in crystalline form (KAWAHARA and TALALAY, 1960) and has been extensively purified (KAWAHARA et al., 1962; KAWAHARA, 1962; TALALAY and BOYER, 1965; JARABAK et al., 1969).

At present it is the most refined steroid enzyme system available for study; it is also one of the most active catalytic proteins known today (KAWAHARA et al., 1962). The molecular weight of the homogeneous, crystalline enzyme was determined by the approach to sedimentation equilibrium method (pH 6.8, 20°) to be 40,800 ± 5%. Some physicochemical parameters are given in Table XVI-1.

Table XVI-1. *Properties of the Δ^5-3-ketosteroid isomerases from Pseudomonas testosteroni* (JARABAK et al., 1969) *and from beef adrenal* (ALFSEN et al., 1967)

	Bacterial enzyme	Mammalian enzyme
Molecular Weight	40,800	112,000 [a]
$s_{20,w}$	3.3	5.1
Frictional ratio, f/fo	1.3	
$E^{1\%}_{1cm}$	4.13	10.0
$A_{280} : A_{260}$	1.33	
pH_i		5
Tyr (spectrometrically) [b]	9	11
SH, free [b]	0	2
SH, masked [b]	0	4
S-S [b]	0	6
Rate } for androst-5-ene-3,17-dione	1.7×10^7 min^{-1}	
Km } (25°, pH 7)	320 μM	
n	3	

[a] Gel filtration of pure preparations of the bovine enzyme over Sephadex G-200 and G-100 gave subunit or polymer fractions of 60,000; 250,000; and 500,000 molecular weight.

[b] Residues per mole.

The amino acid composition calculated from analyses obtained after various times of hydrolysis (BOYER and TALALAY, 1966) corresponded to the approximate formula: Asx_{38}, Thr_{21}, Ser_{16}, Glx_{38}, Pro_{15}, Gly_{29}, Ala_{68}, Val_{45}, Met_8, Ile_{12}, Leu_{26}, Tyr_9, Phe_{25}, Lys_{13}, His_9, Arg_{22}, for a total of 394 residues; Asx and Glx represent the sums of amidated and unamidated aspartic and glutamic acid residues, respectively (JARABAK et al., 1969). Tryptophan and cyst(e)ine are absent. Several lines of evidence suggest that the enzyme molecule consists of three identical chains; methionine is the N-terminal residue of each chain.

The crystalline Δ^5-3-ketosteroid isomerase has a very high catalytic activity. The maximum rate of isomerization of androst-5-ene-3,17-dione is 17.3×10^6 min.$^{-1}$ at 25° and pH 7.0, and the Michaelis constant for this substrate is 320 μM. The bacterial enzyme also isomerizes pregn-5-ene-3,20-dione, but does not affect cholest-5-en-3-one. Isomerization of various $\Delta^{5(10)}$-19-norsteroids occurs at rates about 0.25—0.75% of that of androst-5-ene-3,17-dione (TALALAY, 1965).

Further insight into the mechanism of the isomerization reaction has been acquired by WANG et al. (1963) from ultraviolet absorption and fluorescence spectra of steroid-enzyme complexes. Several steroids were found to be powerful competitive inhibitors of the isomerase reaction: 19-nortestosterone ($K_i = 5.2$ μM; K_i values determined in phosphate buffer, pH 7.0 at 25°), 17β-estradiol ($K_i =$ 10 μM), and 17β-dihydroequilenin (1,3,5(10),6,8-estrapentaene-3,17β-diol, $K_i = 6.3$ μM); they undergo marked spectral changes upon interaction with stoichiometric quantities of the isomerase. For example, the principal absorption band of 19-nortestosterone, which is due to the α,β-unsaturated keto group, is shifted from 248 to 258 nm. Spectral titrations indicate that the enzyme at saturation binds 3 moles of this steroid per unit weight of 40,800; these three binding sites have an average association constant of 0.6×10^6 M^{-1} at pH 7.0. It is not known if more than one of these sites is catalytically active.

The aromatic steroids 17β-estradiol and 17β-dihydroequilenin also undergo profound spectral shifts when they interact with the bacterial isomerase. These changes in ultraviolet absorption, and those of 19-nortestosterone are most reasonably interpreted as resulting from the formation on the protein of structures closely resembling the phenolate and enol ions of the respective steroids.

The fluorescence activation and emission spectra of the Δ^5-3-ketosteroid isomerase cannot be distinguished qualitatively from those of tyrosine, in accordance with the fact that tryptophan is absent, and that the contribution of phenylalanine to protein fluorescence is generally negligible. WANG et al. (1963) conclude from comparative fluorescence studies with tyrosine that four tyrosine residues in the enzyme, which titrate normally, are fully fluorescent, whereas the remaining five tyrosine residues, which are "blocked", contribute little to the fluorescence.

The characteristic tyrosine fluorescence of the native isomerase is quenched progressively by the addition of increasing quantities of 19-nortestosterone. This would suggest that the enzyme in the excited state is capable of a specific transfer to the steroid of as much as three quarters of the energy normally emitted as fluorescence. TALALAY (1965) offered the plausible suggestion that the normally titratable tyrosine residues are located at or near the steroid-binding sites, and that they may be concerned with the binding of the steroid molecule at the active site.

Studies with hydrogen isotopes have provided valuable insight into the mechanism and the stereochemistry of the isomerization reaction. Experiments with highly purified enzyme preparations failed to show entrance of isotope from deuterium- or tritium-labeled water during the isomerization from Δ^5 to Δ^4-androstenedione, in contrast to acid and base-promoted exchange reactions with the same steroids. WANG et al. (1963) concluded that the isomerization involved a "direct" transfer of hydrogen from C-4 to C-6. The question of whether the hydrogen shift is intramolecular, or whether an intermolecular transfer of hydrogen occurs, involving two steroid molecules, has been answered by MALHOTRA and RINGOLD (1965) in isomerization experiments using mixtures of a labeled and an unlabeled Δ^5-steroid; the authors conclude that the isomerization reaction occurs largely or exclusively by intramolecular rather than intermolecular hydrogen transfer. MALHOTRA and RINGOLD (1965) have also established the stereochemical course of the reaction. Using steroid substrates appropriately labeled with deuterium at C-4 and C-6, they found that the 4β-proton of the Δ^5-3-ketone is removed and bonded at the 6β-position. The enzymatic reaction of the Δ^5-3-ketosteroid isomerase of *P. testosteroni* is thus completely stereospecific and involves a diaxial proton transfer from the 4β to the 6β-position.

On the basis of these findings, TALALAY (1965) has proposed the following mechanism of the isomerization reaction (Fig. XVI-1). An acidic group on the enzyme donates a proton to the carbonyl oxygen (1) and a basic group accepts the 4β-proton, thus forming the $\Delta^{3,5}$-dienol structure (2). The basic group then reprotonates at C-6β, the proton from the carbonyl oxygen is removed and the Δ^4-3-ketone

(3) is formed. Histidine is assumed to be involved in the reaction: it is known that photooxidation of Δ^5-3-ketosteroid isomerase in the presence of methylene blue results in progressive loss of enzymatic activity which parallels the destruction of histidine, under conditions essentially innocuous to other amino acids (TALALAY, 1965). The two nitrogen atoms of the imidazole ring of histidine are assumed to function alternately as proton donors and acceptors by tautomerization (Fig. XVI-1, (1) to (3)). Scale models indicate that the spatial configuration of the atoms involved are in agreement with this mechanism.

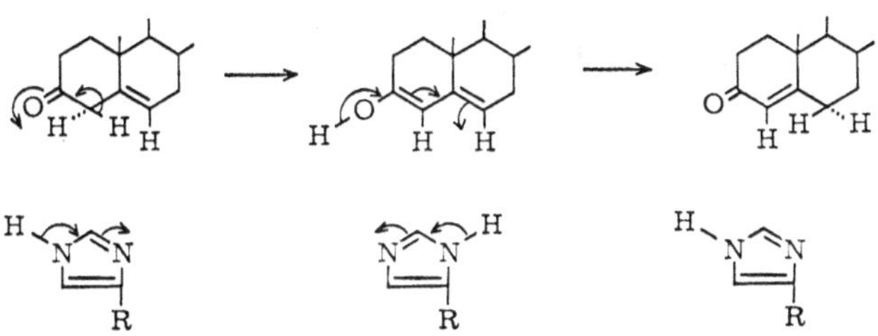

(1) (2) (3)

Fig. XVI-1. Proposed mechanism for imidazole ring catalysis of steroid isomerization by Δ^5-3-ketosteroid isomerase of *Pseudomonas testosteroni*. From TALALAY (1965)

The high activity of the bacterial isomerase is evident from the fact that the rates of acid-, base-, and enzyme-catalyzed isomerizations of androst-5-ene-3,17-dione are in the ratio of $1:700:17\times 10^6$. This has been analyzed by JONES and WIGFIELD (1967) who found that the advantage for the enzymic isomerization over those catalyzed by acid and base derives mainly from an extremely low enthalpy of activation. It should be mentioned that the $\Delta^{3,5}$-dienol mechanism (Fig. XVI-1), demonstrated for the *Pseudomonas* isomerase, differs from that found for a mammalian Δ^5-3-ketosteroid isomerase. OLEINICK and KORITZ (1966) studied the mechanism of the enzyme from rat adrenal and concluded that the isomerization takes place with an enolate anion as intermediate.

FALCOZ-KELLY et al. (1968) have utilized the Δ^5-3-ketosteroid isomerase of *P. testosteroni* for a study of affinity problems in steroid-protein complexes. A close relationship was observed between the partition coefficient and the affinity constants of substrates or of inhibitors obtained from enzyme kinetics data. The results are interpreted to show that the enzyme-substrate or inhibitor complex is formed by an extraction process of steroid from the aqueous medium by the hydrophobic part of the protein. In addition to this "hydrophobic extraction", there is another factor which seems related to the hydrophilic groups of the substrate.

Enzyme purification and physicochemical characterization, elucidation of the isomerization mechanism and concepts about the steroid-protein interactions are farther advanced for the bacterial Δ^5-3-ketosteroid isomerase than for any other steroid enzyme, mainly owing to the work of TALALAY and his coworkers. The corresponding Δ^5-3-ketosteroid isomerases of animal tissues have been studied by EWALD et al. (1964); KRÜSKEMPER et al. (1964); ALFSEN et al. (1965) and EWALD and CHAIKOFF (1968); some properties are given in Table XVI-1. ALFSEN et al. (1967) have compared certain parameters of the bacterial and mammalian isomerases in relation to their specificity and various interactions. Fractionation studies with 3-oxosteroid Δ^5-isomerase from human term placenta by KOIDE and TORRES (1966) have led to the conclusion that different enzymes are responsible for the isomerization of androst-5-ene-3,17-dione and for pregn-5-ene-3,20-dione. On the basis of activity ratios during purification and other criteria, WARREN and his coworkers assume a single active site for both C_{19} and C_{21} steroids in the Δ^5-3-ketosteroid isomerase of bovine corpora lutea (CHEATUM and WARREN, 1966) and of beef adrenals (CHEATUM et al., 1967). However, the enzymes have not yet been purified to the state of homogeneous proteins.

Other steroid enzymes that can be induced in *Pseudomonas testosteroni* catalyze dehydrogenation reactions. MARCUS and TALALAY (1956) induced and purified a 3α-hydroxysteroid dehydrogenase and a β-hydroxysteroid dehydrogenase which oxidizes 3β- and 17β-hydroxysteroids to the corresponding oxosteroids. Both enzymes are DPN-dependent. Steroid specificity, kinetics and other properties of these bacterial dehydrogenases have been studied by TALALAY and MARCUS (1956). TALALAY (1957) reported a lowering of binding affinity between steroid and enzyme, and a decrease of the oxidation velocity, when hydroxyl groups were introduced into the steroid.

Extensive purification of the two enzymes has been carried out by DELIN et al. (1964) and by TALALAY (1965), resulting in products that appeared essentially homogeneous by several criteria (SQUIRE et al., 1964). The molecular weight of the 3α-hydroxysteroid dehydrogenase was found to be 47,100, and that of the β-hydroxysteroid dehydrogenase 100,000; their turnover numbers were 9.3×10^3 min^{-1} (androsterone) and 18.6×10^3 min^{-1} (testosterone), respectively. BOYER et al. (1965) have provided additional details for the 3α-hydroxy dehydrogenase on pyridine nucleotide specificity, pH-activity relationship, and use in microassay of steroids.

RINGOLD et al. (1967) carried out an extensive investigation on substrate recognition and specificity with the 3α-hydroxysteroid dehydrogenase. Starting with cyclohexanone or cyclohexanol, which are extremely poor substrates, hydroxy or keto steroid analogs of increasing complexity were employed, and the relative significance of various portions of the steroid molecule and their stereochemical influence were assessed.

The 3α- and the β-hydroxysteroid dehydrogenases described have been obtained from the wild-type *Pseudomonas testosteroni*. ROE and KAPLAN (1969) searched for more specific hydroxysteroid dehydrogenases, particularly at the 3α-position, and examined three organisms that were closely related to *P. testosteroni* (STDH-m; strain 28; M_3). By extensive fractionation, four highly selective enzymes were obtained. Substrate specificity studies revealed them to be 1) ring A/B *trans* 3β-hydroxysteroid dehydrogenase; 2) 3α-hydroxysteroid dehydrogenase, unable to oxidize 3α-hydroxysteroids, ring A/B *cis*, with an 11-keto group; 3) ring A/B *trans* 3α-hydroxysteroid dehydrogenase; and 4) an axial 3-hydroxysteroid dehydrogenase. Their use for stereospecific quantitative assays appears possible.

Pseudomonas testosteroni also biosynthesizes enzymes which catalyze the introduction of certain double bonds into ring A of the steroid molecule. Three such dehydrogenases have been found by LEVY and TALALAY (1959 a) in this microorganism; one introduces a Δ^4 double bond into 5α-steroids, another does the same for 5β-steroids, and the third one catalyzes the formation of a Δ^1 double bond in either type of steroid. The enzymes have been partly purified by LEVY and TALALAY (1959 b), and their mechanisms have been studied; additional purification has been described by DAVIDSON and TALALAY (1966). The preparation of the highest specific activity, obtained recently by DAVIDSON (1969), catalyzed the dehydrogenation of 6,600 micromoles of 5β-androstane-3,17-dione per mg enzyme per minute under standard assay conditions. A crystalline 20β-hydroxysteroid dehydrogenase has been prepared from *streptomyces hydrogenans* by HÜBENER et al. (1959); see also HÜBENER and SAHRHOLZ (1960).

In comparison with the bacterial steroid dehydrogenases, purification and characterization of *steroid-transforming enzymes of mammalian origin* are far less advanced. In accordance, little use has been made of these enzyme proteins to elucidate general problems of steroid-protein interaction. A few recent publications will be cited dealing with those enzymes that may become relevant to the study of molecular aspects of steroid binding by proteins; these references form a convenient guide to additional articles in the literature. A comprehensive treatment of the multitude of enzymatic reactions involving steroids has been given by DORFMAN and UNGAR (1965); they are also discussed by HEFTMANN (1970). Preparation and properties of various steroid-metabolizing enzymes are described in volume 5 of Methods in Enzymology, edited by COLOWICK and KAPLAN (1962), and in volume 15 of the same series, edited by CLAYTON (1969).

WIEST (1969) reported recently on the purification of the 20α-hydroxysteroid dehydrogenase from rat ovaries, and KERSEY and WILCOX (1970) studied the stereochemistry of hydrogen transfer from reduced triphosphopyridine nucleotide (TPNH), catalyzed by this enzyme. The 3α-hydroxysteroid dehydrogenase of rat liver, which reduces DPN or TPN in the presence of a 3α-hydroxysteroid, has been purified by KOIDE (1969); the enzyme possesses a pyridine nucleotide transhydrogenase activity.

THEORELL et al. (1966) separated a steroid-active dehydrogenase ($LADH_S$) from liver alcohol dehydrogenase prepared from fresh horse liver. The enzyme, which catalyzes the reduction of DPN by 3β-hydroxy-5β-cholanic acid, has been obtained in crystalline form; it is characterized by an $s^{\circ}_{20,w}$ value of 4.9, a molecular weight of 83,300 and an absorption at 280 nm of $E^{1\%}_{1cm} = 4.60$. Its amino acid composition resembles closely that of the horse liver alcohol dehydrogenase ($LADH_E$) whose main substrate is ethanol. The steroid-active $LADH_S$ was found to be about half as active as $LADH_E$

toward ethanol per coenzyme-binding site; this activity relationship is also valid for C_3-C_5 aliphatic normal alcohols and cyclohexanol. The binding sites for ethanol and for the 3β-hydroxycholanic acid in $LADH_S$ are different and independent of each other. THEORELL et al. (1966) obtained indications that the binding sites for ethanol in $LADH_S$ and $LADH_E$ are similar. The stereospecificity of hydrogen transfer by a crude preparation of 3β-hydroxysteroid dehydrogenase from bovine adrenals has been studied by WARREN and CHEATUM (1968).

MONDER and FURFINE (1969) have described the purification of a DPN^+—dependent 21-hydroxysteroid dehydrogenase from lamb liver, a TPNH-dependent 21-hydroxysteroid dehydrogenase from the same source, and a DPNH-dependent 21-hydroxysteroid dehydrogenase from bovine adrenals. A broad pH maximum at pH 7.4, and a molecular weight of 129,000 (sucrose density technique) was found for the latter enzyme. The development of liver 21-hydroxysteroid dehydrogenases was followed in the growing rat and calf by MONDER and MARTINSON (1969 a). A TPNH-dependent 21-hydroxysteroid dehydrogenase was partially purified by MONDER and MARTINSON (1969 b) from human placenta. It catalyzes the reduction of 21-dehydrocorticosteroids to the corresponding 21-hydroxy compounds. It contains sulfhydryl groups which appear to be necessary for the enzymatic activity. Gel filtration over Sephadex G-100 indicated a molecular weight between 33,000 and 35,000. The same technique gave an estimated MW of 41,000 to 43,000 for the TPNH-dependent 21-hydroxysteroid dehydrogenase of lamb liver which has been mentioned above.

The 16α-hydroxysteroid dehydrogenase of rat kidney is an adaptive enzyme regulated by sex hormones. Its purification has been described by MEIGS and RYAN (1969). It is inactivated by sulfhydryl reagents. Other sex-dependent steroid enzymes, mostly observed in rat liver, have not been purified to any extent (MEIGS and RYAN, 1969). TALALAY (1962) has reported a partial purification of male rat liver microsomes applicable to the 11β-hydroxysteroid dehydrogenase.

A recent article by BREUER and KNUPPEN (1969) on the enzymes of estrogen metabolism, containing almost 200 references, gives a comprehensive review of the subject. It shows that none of the large number of enzymes known have been purified to any extent with the exception of the 17β-hydroxysteroid dehydrogenase. Only this enzyme has been investigated sufficiently to warrant discussion in relationship to steroid-protein interaction.

The 17β-hydroxysteroid dehydrogenase, or 17β-estradiol dehydrogenase, of human placenta is a pyridine nucleotide-linked enzyme that catalyzes the interconversion of estradiol and estrone. This oxidoreductase has played a role in early attempts to interpret the mechanism of estrogen action by enzymatic transhydrogenation reactions, *i. e.*, by reversible transfer of hydrogen from DPNH to TPN^+ (TALALAY et al., 1958; HAGERMAN and VILLEE, 1959; LANGER et al., 1959; TALALAY and WILLIAMS-ASHMAN, 1960; WILLIAMS-ASHMAN and LIAO, 1964). The dehydrogenase has been purified (JARABAK et al., 1962; JARABAK, 1969). DESCOMPS et al. (1968) by chromatography (DEAE-Sephadex; Sephadex G-150) obtained an enzyme which was homogeneous on polyacrylamide electrophoresis, and had a MW of 92,000 ± 1500 (gel filtration). This preparation was used by MOUSSERON-CANET et al. (1969) to determine oxidation rates and K_m values for numerous 17β-hydroxysteroids, including fluorinated derivatives. DESCOMPS and CRASTES DE PAULET (1969) have studied the geometry of the enzymatic site using about fifty 17β-hydroxysteroids. It is of interest that the enzyme is protected from cold inactivation (formation of enzymatically inactive high molecular weight components following conformational changes) by interaction with estradiol and testosterone (JARABAK et al., 1966).

The dehydrogenase activity has been separated from the transhydrogenase action (HAGERMAN and VILLEE, 1965; KARAVOLAS and ENGEL, 1966; KARAVOLAS et al., 1969). Working with a dehydrogenase preparation which was free of the unique transhydrogenase, HAGERMAN (1969) observed a concentration-dependent, reversible monomer-trimer transition. The kinetic properties of the enzyme show that the monomeric form can catalyze both net steroid dehydrogenation and substrate-mediated transhydrogenation. In contrast, the trimer catalyzes only substrate-mediated transhydrogenation. A molecular weight of 47,000 was found for the monomer on Sephadex columns G-150 and G-100; the elution volume of the trimer corresponded to a MW of 135,000. A monomer-dimer-trimer equilibrium is assumed. JARABAK and SACK (1969) reported a MW of 62,000—65,000 and a single binding site for 17β-estradiol and estrone.

The properties of the 17β-estradiol dehydrogenase obtained from ovaries are quite different from those of the placental enzyme. KAUTSKY and HAGERMAN (1970) have purified this dehydrogenase from sheep ovaries more than 800-fold. The preparations approached homogeneity by sedimentation analysis and by discontinuous polyacrylamide gel electrophoresis at pH 9.5 and 4.3. A MW of 104,000 was derived from sucrose density gradient centrifugation; the Stokes radius was calculated from gel filtration studies to be 50.5 Å. The purified enzyme was unstable in simple buffers; it could be stabilized by 20% glycerol and 0.05 M 2-mercaptoethanol. No evidence of monomer-polymer interconversion was obtained.

The problem of influence of steroid hormones on enzymatic activity became of acute interest when YIELDING and TOMKINS (1959) found that glutamic dehydrogenase was inhibited by interaction with steroid hormones and diethylstilbestrol. YIELDING and TOMKINS (1960, 1962) studied the interaction as a model for a mechanism of the biological action of these compounds, and included a large number of steroid hormones of various structures. The steroids, by interacting at allosteric sites, produced disaggregation of the enzyme molecules into smaller species, accompanied by inhibition of the glutamic dehydrogenase and stimulation of the alanine dehydrogenase activity. It was later recognized (TOMKINS and YIELDING, 1964) that the changes effected by the steroid may not be physiologically important, although they may play a certain role in the regulation of glutamate dehydrogenase activity *in vivo*. VILLEE (1963) observed an increase of the glutamic dehydrogenase activity by addition of corticosterone or progesterone.

The glutamate oxidation by the crystalline enzyme from beef liver was inhibited also by guanosine triphosphate; at the same time, increased monocarboxylic L-amino acid dehydrogenase activity with alanine, valine and other alpha-amino acids was observed (TOMKINS et al., 1965). These various catalytic functions were explained by an equilibrium between different molecular forms of the enzyme with different relative substrate specificities. One of these forms is the oligomeric species of an average MW of 313,000 at the limit of zero concentration (EISENBERG and TOMKINS, 1968). Light scattering experiments in guanidine hydrochloride in the presence of mercaptoethanol revealed that the oligomer consists of 6 subunits of MW 53,000 or 52,000. At increasing enzyme concentration, the oligomers aggregate to form polymers of MW$>2\times10^6$. These results have been confirmed by MOSEBACH and KEMPFLE (1969). The conclusion is that dissociation to the oligomer units inhibits the glutamate dehydrogenase activity and increases alanine dehydrogenase function; this is promoted by the steroid hormones, guanosine triphosphate and diethylstilbestrol as described above.

The interaction of steroids with glutamate dehydrogenase, glucose 6-phosphate dehydrogenase from bovine corpus luteum, and rabbit liver aldehyde dehydrogenase, has been studied by DOUVILLE and WARREN (1968) who found that the activity of the three enzymes is inhibited by a variety of steroids. This inhibition is effective immediately on addition of steroid, is reversible on dilution, and not competitive in regard to substrate and cofactor. The association constants for the interaction at those (allosteric) sites which influence catalytic activity were in the range of 10^5 to $10^3\,M^{-1}$. Different steroids effect inhibition by binding at identical sensitive sites. For the glutamate dehydrogenase of bovine liver, the data are consistent with a mechanism which requires the combination with two steroid molecules for complete inactivation. The steroids that bind with highest affinity to the sensitive sites are 17*β*-estradiol for the glutamate dehydrogenase, Δ^5-3*β*-hydroxysteroids for the glucose 6-phosphate dehydrogenase, and 3-ketosteroids for the aldehyde dehydrogenase.

A noteworthy advance to a characterization of macromolecular steroid-binding sites by affinity labeling has been made by WARREN and his coworkers. CHIN and WARREN (1968) synthesized 4-mercuri-17*β*-estradiol, a compound analogous to p-mercuribenzoate with the mercury bonded to the aromatic ring A of estradiol. The mercuristeroid forms covalent bonds with the sulfhydryl groups of cysteine, 2-mercaptoethanol, and reduced glutathione. One mole of egg albumin reacts with 3 moles of 4-mercuri-17*β*-estradiol, indicating three reactive sulfhydryl groups. The rate of the mercaptide formation between 4-mercuri-17*β*-estradiol and glutathione or egg albumin is not affected by 17*β*-estradiol.

When the analogous reaction is carried out with the mercuriestradiol and glutamate dehydrogenase or pyruvate kinase, both of which have allosteric binding sites for 17*β*-estradiol, the rate of mercaptide formation is significantly decreased by 17*β*-estradiol. Similar concentrations of cortisol or estriol have no effect; alternately, the mercaptide formation between p-hydroxymercuribenzoate and pyruvate kinase is not inhibited by 17β-estradiol.

It is assumed that the steroid moiety of 4-mercuri-17β-estradiol delivers mercury to the allosteric binding site for 17β-estradiol. The resulting increase in local concentration favors mercaptide formation with a reactive residue at or near the binding site. Inhibition of mercaptide formation by 17β-estradiol occurs because it can occupy the binding site and thus exclude the mercuristeroid. The method can be expected to prove useful for identification of amino acid residues at the steroid-binding sites of high affinity, present in steroid-converting enzymes or in other proteins.

This was shown by MULDOON and WARREN (1969) for the receptor protein of rat uterus. Using the induction of glucose 6-phosphate dehydrogenase and 6-phosphogluconate dehydrogenase as biological response, the investigators found that 4-mercuri-17β-estradiol and its mercaptide with 2-mercaptoethanol were active on intrauterine administration to ovariectomized rats. The uterine tissue showed the classic histological effects seen after similar application of estradiol. The possibility of *in vivo* cleavage of the carbon-mercury bond was excluded. Binding of 4-mercuri-17β-estradiol to the cytoplasmic receptor protein of immature rat uterus could be demonstrated by density gradient centrifugation; pretreatment with 17β-estradiol abolished the binding.

The usefulness of 4-mercuri-17β-estradiol is restricted to cases where a sulfhydryl group is present at or near the binding site. CHIN and WARREN (1970) have recently extended their approach to characterization of steroid-binding sites through affinity labeling by the synthesis of 2-diazoestrone sulfate. This derivative couples under physiological conditions with tryptophan, cysteine, and histidine, and with their various esters and peptides. A single coupling product is formed in each case. Cysteine reacted more rapidly than the other compounds mentioned.

Upon reaction of small amounts of 2-diazoestrone sulfate with egg albumin, coupling takes place mainly with cysteine residues; the presence of 17β-estradiol has no effect. If the egg albumin is first treated with excess p-mercuribenzoate, subsequent reaction with 2-diazoestrone sulfate results in predominant coupling with tryptophan and histidine. When the diazosteroid derivative reacts with glutamate dehydrogenase, coupling occurs with tryptophan, cysteine and histidine. The presence of 17β-estradiol or estrone sulfate in the reaction mixture significantly diminishes the coupling to cysteine, but has little or no effect on that with tryptophan or histidine. These observations are in accord with a mechanism whereby the steroid moiety of 2-diazoestrone sulfate delivers the reactive diazonium group to the steroid-binding site of glutamate dehydrogenase, favoring coupling with a cysteine residue present at that site.

This short chapter on steroid-enzyme interactions is to call the attention of the reader to this field, and to emphasize its promise for future research. The few cases in which pure enzymes have been used in the interaction studies with steroids attest to the importance of the objectives. The promising approach of WARREN and his coworkers clearly indicates the significance of the affinity-labeling techniques for structural elucidation of steroid-binding sites in general, beyond the area of enzyme proteins.

XVII. Synopsis

More than 35 years have passed since BRUNELLI first observed binding of steroid hormones to plasma proteins, and thus opened a field of research which has branched into various areas of biological and chemical interest. Table XVII-1 summarizes the types of proteins known to be involved in interactions with steroid hormones. Best studied by far are the numerous representatives of the serum proteins which cover a wide range of steroid-binding affinities. They vary in their concentration over five orders of magnitude—and it is noteworthy that the most abundant serum protein, albumin, stands at the bottom in steroid affinity, with association constants about 4 to 5 powers of ten smaller than those of the high-affinity binders.

The transport function of the serum proteins at first seemed most important. It has been de-emphasized over the years in favor of a regulatory effect on hormonal activity because the formation of the hormone-protein complex suppresses the biological activity. However, the question of the binding proteins acting as steroid carriers for specific biological functions is not yet closed. The possibility remains that the serum binders facilitate transition of the newly biosynthesized steroid hormones from the secretory tissues into the blood capillaries, or, *vice versa,* that they penetrate the capillary walls and transport the bound hormones to the reactive tissues.

And here is where additional possibilities present themselves—all speculative at this time. We do not know the mode of transfer of the steroid hormones to the receptor proteins in cytosol and nucleus of the target cells. Does the serum carrier protein contribute to the specificity of the uptake by interacting with specific loci of the target cell surface, and delivering it at these points by dissociation? Is the steroid binder itself perhaps able to pass through the cell membrane and thus provide a true transport of the hormone into the cell?

The relationship between the high-affinity steroid binders of the blood serum and the cellular receptor proteins is of particular interest. It has been established for a number of steroid target tissues that reversible transformation occurs between cytoplasmic receptors having sedimentation coefficients of approximately 8 S and smaller entities of 4 S. These in turn seem to be closely related to 5 S molecules, a size typical of nuclear receptors. This phenomenon of deaggregation and aggregation is reminescent of the reversible polymerization observed with CBG of rat serum. It is noteworthy in this context that the characteristics of the progesterone receptor in rat uterus cytosol closely resemble those of serum CBG. The high concentration in the

Table XVII-1. *Areas of research in steroid-protein interactions*

Type of protein	Biological significance for	Application for
Serum proteins, normally present	Transport Regulation of hormonal activity and distribution Protection of steroids	Competitive binding methods
Serum immunoglobulins, generated	Neutralization of specific hormones	Competitive binding methods
Receptor proteins	Mechanism of action	Competitive binding methods
Enzyme proteins	Biosynthesis Metabolic transformations	Analytical and preparative methods

pregnant guinea pig of a PBG, which does not interact with cortisol, invites the speculation that this serum protein with the very high affinity for progesterone of $k > 10^{10}\,M^{-1}$ may originally be a uterine component and play a role as progesterone receptor. A number of approaches can be taken to explore these intriguing possibilities.

The steroid-binding serum proteins are important for another reason: the elucidation of the relation of chemical structure to binding properties. Which amino acids form the core of the binding site? What is the influence of identity and conformation of the neighboring residues? In which way do the inhibitors of steroid binding interfere? Clearly, the best objects for such study would be the high-affinity binders that can be isolated from mammalian blood sera. But their low concentration and the difficulty of complete purification limit this analysis. It may be mainly for this reason that albumin has been extensively studied with respect to steroid interaction, although the apparent conformational adaptability, resulting in low specificity of binding, would seem to decrease the significance of the findings as far as the problem of steroid-protein association is concerned. Nevertheless, meaningful conclusions have been drawn from such studies, especially with respect to the hydrophobic contributions to steroid binding.

It was the multifarious nature of ligand interaction with serum albumin that led us to investigate the glycoprotein AAG in its steroid-binding properties. Although we have no evidence for any biological significance of steroid affinity to AAG, numerous similarities with the high-affinity steroid binders in the serum make the AAG a worthwhile object to investigate the chemical basis for steroid binding. Recent progress in protein chemistry has paved the way for such studies, which would seem to be limited by manpower and time rather than by the given possibilities.

Progress is a problem of priorities. The investigation of structure-function relationship in steroid-protein interaction will receive new impetus from the discovery of the specific PBG in the pregnant guinea pig—a protein of the highest steroid affinity known for any serum component, and occurring in concentrations that make isolation in adequate quantities practicable. The possible relationship to cellular receptor proteins will be additional stimulus for this research.

There are fringe benefits arising from the knowledge of steroid-binding proteins. Ingenious application of steroid hormone binders in competitive systems has made possible the quantitative determination of the biologically important steroids at the picogram level, thus increasing the sensitivity of analysis by several orders of magnitude. Steroid-binding serum proteins as well as cellular receptor proteins are being used in these methods.

A third group of proteins is being employed in these steroid assays by protein binding: the immunoglobulins, tailor-made in a responsive animal by immunochemical techniques. This latest addition to the family may be considered, at least by design, the most specific of the steroid-binding macromolecules. Clearly, anti-steroid immunoglobulins offer important possibilities for biological as well as chemical studies. The unique capability of neutralizing *in vivo* the effects of a single steroid hormone will aid the endocrinologist in his search for hormonal pathways. And the biochemist may be able to attack fundamental problems of steroid-binding specificity once the difficult task of complete purification of the anti-steroid globulins has been accomplished.

It is perhaps an indication for the different viewpoint of the investigators involved, that a wealth of information is available on the function of steroid enzymes, on the pathways they catalyze and their biological significance, and that, with a few notable exceptions, very little is known about the enzyme proteins themselves. These molecules appear ideally suited for the study of steroid-protein interaction from a chemical approach, as has been recognized in a few laboratories. The dual function of these enzymes, i.e., both binding and structurally transforming the steroid, augments the experimental possibilities; it may supplement in important ways the results obtained with other types of steroid-binding proteins.

Many avenues are open to the curious mind.

Appendix

A. List of Abbreviations and Symbols

(a)	Axial position on steroid molecule
$A_{280} : A_{260}$	Ratio of absorbances at 280 and 260 nm
AAG	α_1-acid glycoprotein, orosomucoid
ACTH	Adrenocorticotropic hormone
BLG	Beta-lactoglobulin
BSA	Bovine serum albumin
BW	Body weight
C	Combining affinity or C-value
cal	Calories
CBG	Corticosteroid-binding globulin, transcortin
CBG-Hu	Human CBG
CBG-Rb	Rabbit CBG
CBG-Rt	Rat CBG
CM	Carboxymethyl
cpm	Counts per minute
$D_{20,w}$	Diffusion coefficient at 20° in water ($cm^2 sec^{-1}$)
DEAE	Diethylaminoethyl
DNA	Desoxyribonucleic acid
(e)	Equatorial position on steroid molecule
$E^{1\%}_{1cm}$	Absorbance of 1% solution at 1 cm light path
EBI	Estradiol binding index
F	Gibbs free energy
f/fo	Frictional ratio
HMB	p-Hydroxymercuribenzoate
HSA	Human serum albumin
$H\Phi$	Hydrophobicity of amino acid side chains
$H\Phi_{avg}$	Average hydrophobicity per amino acid residue
k	Association constant
K	Dissociation constant
kcal	Kilocalories

K_i	Dissociation constant of enzyme-inhibitor complex
K_m	Michaelis-Menten constant
MW	Molecular weight
mμ	Millimicrons or nanometers
n	Number of binding sites in a protein molecule
ng	Nanogram $= 10^{-9}$ g
nm	Nanometers or millimicrons
nM	Nanomolar, 10^{-9} moles/liter
NPS	Non-polar side chain frequency
17-OHCS	17-hydroxycorticosteroid(s)
[P]	Concentration of unbound protein
$[P_{bd}]$	Concentration of bound protein
$[P_t]$	Concentration of total protein
PBG	Progesterone-binding globulin
pg	Picogram $= 10^{-12}$ g
R	Molar gas constant = 1.987 calories per degree per mole
r	Dispersion or solubility increase by binding, in moles of steroid per mole protein
RNA	Ribonucleic acid
S	Svedberg unit of sedimentation (10^{-13} cm/sec/dyne/g)
$\mathring{s}_{20,w}$	Sedimentation coefficient extrapolated to zero concentration, 20°, water
[S]	Concentration of unbound steroid (ligand)
SBβG, SBP	Sex steroid hormone-binding β-globulin
$[S_{bd}]$	Concentration of bound steroid (ligand)
sc	Subcutaneous
S.D.	Standard deviation
S.E.M.	Standard error of the mean
S_f	Svedberg unit of flotation (10^{-13} cm/sec/dyne/g)
SHBG	Sex steroid hormone-binding β-globulin
$[S_t]$	Concentration of total steroid (ligand)
T	Absolute temperature in °K
TBG	Thyroxine-binding globulin
TSH	Thyroid-stimulating hormone
u	Electrophoretic mobility ($cm^2\ volt^{-1}\ sec^{-1}$)
$\bar{v}$	Partial specific volume (ml/gm)
α or α_1	Unbound fraction of steroid (ligand)
$\Delta F°$	Change of free energy

ΔH°	Change of enthalpy
ΔS°	Change of entropy
Δ^1	Double bond from C-1 to C-2
=	Double bond
$\Delta\varepsilon$	Depression of molar extinction coefficient of steroid by protein interaction; in % of steroid absorbance; value negative
ε	Molecular extinction coefficient
λ_{max}	Wavelength of maximal light absorption
μ	Ionic strength
$\bar{\nu}$	Average number of steroid (ligand) molecules bound per molecule of protein

B. Calculation of Average Hydrophobicities and of Non-polar Side Chain Frequencies

Hydrophobicities, $H\Phi$, of amino acid side chains, in kcal/residue. From original data of free energy changes for transfer from ethanol to water, determined by TANFORD (1962). For calculation of $H\Phi_{avg}$ according to BIGELOW (1967), the total $H\Phi$ for all residues is divided by the number of residues. Residue weights are included to aid calculations of amino acid compositions. For calculation of the fraction of non-polar side chain residues, NPS, according to WAUGH (1954), the number of non-polar residues, marked [a], is divided by the total number of amino acid residues.

Amino Acid	Residue weight	$H\Phi$ kcal/res.	Amino Acid	Residue weight	$H\Phi$ kcal/res.
Lys	128.17	1.50	Cys/2	102.13	1.00
His	137.14	0	Val [a]	99.13	1.69
Arg	156.18	0.73	Met	131.19	1.30
Asp	115.08	0	Ile [a]	113.15	2.97
Thr	101.10	0.44	Leu [a]	113.15	2.42
Ser	87.07	0.04	Tyr [a]	163.17	2.87
Glu	129.11	0	Phe [a]	147.17	2.65
Pro [a]	97.11	2.60	Try [a]	186.20	3.00
Gly	57.05	0	Asn	114.08	−0.01
Ala	71.07	0.73	Gln	128.11	−0.10

[a] Non-polar residues used for calculation of NPS.

Bibliography

ABELSON, D., DEPATIE, C., CRADDOCK, V.: Interactions of testosterone with amino acids. Arch. Biochem. **91**, 71 (1960).

ABRAHAM, G. E.: Solid-phase radioimmunoassay of estradiol-17β. J. clin. Endocr. **29**, 866 (1969).

— ODELL, W.: Solid phase radioimmunoassay of estrogens. In: Immunological methods in steroid determination. Eds.: F. G. PÉRON and B. V. CALDWELL. New York: Appleton-Century Crofts 1970 (in press).

— — EDWARDS, R., PURDY, J. M.: Solid-phase radioimmunoassay of estrogens in biological fluids. Transactions of 2nd Karolinska Symposium on Research Methods in Reproductive Endocrinology. Steroid assay by protein binding, Geneva, March, 1970. Acta endocr. (Kbh.) Suppl. **147**, 332 (1970).

ÁCS, Z., STARK, E., CSAKI, L.: The effect of long-term corticotrophin treatment on the corticosteroid-binding capacity of transcortin. J. Endocr. **39**, 565 (1967).

ALFSEN, A.: Étude de la liaison testostérone-sérumalbumine et action de l'heptane sur la molécule de sérumalbumine. C. R. Carlsberg **33**, 415 (1963).

— BAULIEU, E. E., CLAQUIN, M. J.: Isolation of an adrenal Δ^{5-4}-3-ketocholestene isomerase. Biochim. biophys. Res. Commun. **20**, 251 (1965).

— — — FALCOZ-KELLY, F.: Protein conformation and enzymatic specificity. A study of $\Delta^5 \rightarrow \Delta^4$-3-oxosteroid isomerases. Proc. Second Intern. Congr. on Hormonal Steroids, Milan, 1966. Excerpta Med. Intern. Congr. Series No. 132, 1967, p. 697.

ALLEN, C., KENDALL, J. W.: Maturation of the circadian rhythm of plasma corticosterone in the rat. Endocrinology **80**, 926 (1967).

ALLEN, W. S., BERNSTEIN, S.: Steroidal cyclic ketals. XII. The preparation of Δ^{16}-steroids. J. Amer. chem. Soc. **77**, 1028 (1955).

ALLINSON, R., COLEBY, B., WEISS, J.: Chemical effects of X-rays upon aqueous solutions of cortisone and deoxycorticosterone. Nature **175**, 720 (1955).

ALY, F. W., BRAUN, H. J., HADAM, W. R., KALLEE, E.: Zur Bedeutung der Eiweißvehikelfunktion. Arzneimittel-Forsch. **18**, 1467 (1968).

ANGHILERI, L. J.: *In vitro* binding of 3,4-benzpyrene to albumin. European J. Pharmacol. **3**, 153 (1968).

ANTONIADES, H. N., DAUGHADAY, W. H., SLAUNWHITE, W. R., JR.: The binding of steroid hormones and their metabolites by plasma proteins. In: Hormones in Human Plasma, chapter XIV. Ed.: H. N. ANTONIADES. Boston (Mass.): Little, Brown and Co. 1960.

ANTONIADES, H. N., MCARTHUR, J. W., PENNELL, R. B., INGERSOLL, F. M., ULFELDER, U., ONCLEY, J. L.: Distribution of infused estrone in human plasma. Amer. J. Physiol. **189,** 455 (1957 a).

— PENNELL, R. B., SLAUNWHITE, W. R., JR., SANDBERG, A. A.: Binding and distribution of intravenously injected C^{14}-steroids and their metabolites in human plasma and its fractions. J. Biol. Chem. **229,** 1071 (1957 b).

ANTONICHEV, A. V., ROZEN, V. B.: Significance of the hypophysis in corticosteroid binding with transcortin in guinea pigs. Probl. Endocr. (Moscow) **13,** No. 6, 85 (1967).

APPLEZWEIG, N.: Steroid drugs. New York-Toronto-London: McGraw-Hill 1962.

ARIENS, E. J.: Molecular pharmacology. The mode of action of biologically active compounds, vol. 1. New York-London: Academic Press 1964.

ATHINEOS, E., KUKRAL, J. C., WINZLER, R. J.: Biosynthesis of glycoproteins. II. The site of glucosamine incorporation into canine plasma α_1-acid glycoprotein. Arch. Biochem. **106,** 338 (1964).

ATKINSON, W. B., LEATHEM, J. H.: The day to day level of estrogen and progestin during lactation in the mouse. Anat. Rec. **95,** 147 (1946).

ATTALLAH, N. A., LATA, G. F.: Steroid-protein interactions studied by fluorescence quenching. Biochim. biophys. Acta (Amst.) **168,** 321 (1968).

AUGUST, G. P., TKACHUK, M., GRUMBACH, M. M.: Plasma testosterone-binding affinity and testosterone in umbilical cord plasma, late pregnancy, prepubertal children, and adults. J. clin. Endocr. **29,** 891 (1969).

AVIGAN, J.: A method for incorporating cholesterol and other lipids into serum lipoproteins *in vitro*. J. biol. Chem. **234,** 787 (1959).

BAIRD, D. T., HORTON, R., LONGCOPE, C., TAIT, J. F.: Steroid dynamics under steady-state conditions. Rec. Progr. Hormone Res. **25,** 611 (1969).

BALLOU, G. A., BOYER, P. D., LUCK, J. M.: The electrophoretic mobility of human serum albumin as affected by lower fatty acid salts. J. biol. Chem. **159,** 111 (1945).

BANG, I.: Über Lipämie. Biochem. Z. **90,** 383 (1918).

BARDIN, C. W., HEMBREE, W. C., CHRAMBACH, A.: Studies on a testosterone-estradiol-binding protein (TEBP) in human plasma. 3rd Internatl. Congr. Endocrinol. Excerpta Med. Intern. Congr. Ser. No. 157, 1968, p. 160.

BARLOW, J. J., MACLAREN, J. A., POTHIER, L.: *In vivo* observations on testosterone and estradiol-17β protein binding in women. J. clin. Endocr. **29,** 767 (1969).

BARRON, E. S. G.: The role of free radicals and oxygen in reactions produced by ionizing radiations. Radiation Res. **1,** 109 (1954).

BASSETT, J. M., HINKS, N. T.: Micro-determination of corticosteroids in ovine peripheral plasma: effects of venipuncture, corticotrophin, insulin and glucose. J. Endocr. **44,** 387 (1969).

BAULIEU, E. E.: Les récepteurs hormonaux. Mise en évidence de la liaison spécifique de l'oestradiol à des protéines utérines. Nouv. Rev. franç. Hémat. **7,** 589 (1967).

BAULIEU, E. E.: Les "récepteurs hormonaux". Mise en évidence de la liaison spécifique de l'oestradiol à des protéines utérines. Ann. Endocr. (Paris) **29**, 131 (1968).

— Personal communication (1969).

— ALBERGA, A., JUNG, I.: Récepteurs hormonaux. Liaison spécifique de l'oestradiol à des protéines utérines. C. R. Acad. Sci. (Paris) **265**, 354 (1967 a).

— — — Récepteurs hormonaux. Liaison de divers oestrogènes à des protéines utérines. Une méthode de dosage de l'oestradiol. C. R. Acad. Sci. (Paris) **265**, 501 (1967 b).

— CORPECHOT, C., DRAY, F., EMILIOZZI, R., LEBEAU, M. C., MAUVAIS-JARVIS, P., ROBEL, P.: An adrenal-secreted "androgen": Dehydroisoandrosterone sulfate. Its metabolism and a tentative generalization on the metabolism of other steroid conjugates in man. Rec. Progr. Hormone Res. **21**, 411 (1965).

— RAYNAUD, J. P.: The use of radioactive steroids for the study of "transport" and "receptor" proteins. A new graphical methodology. Progr. Biochem. Pharm. **5**, 46 (1969).

— — A "proportion graph" method for measuring binding systems. Europ. J. Biochem. **13**, 293 (1970).

— — MILGROM, E.: Measurement of steroid binding proteins. Transactions of 2nd Karolinska Symposium on Research Methods in Reproductive Endocrinology, Steroid assay by protein binding, Geneva, March, 1970. Acta endocr. (Kbh.), Suppl. No. **147**, 104 (1970).

BAUMINGER, S., LINDNER, H. R., PEREL, E., ARNON, R. J.: J. Endocr. **44**, 567 (1969).

BAYLY, R. J., EVANS, E. A.: Stability and storage of compounds labelled with radioisotopes. J. Labelled Compounds **2**, 1 (1966).

— — Storage and stability of compounds labelled with radioisotopes. Part II. J. Labelled Compounds **3**, 349 (1967).

BEARDWELL, C. G., BURKE, C. W., COPE, C. L.: Urinary free cortisol measured by competitive protein binding. J. Endocr. **42**, 79 (1968).

BECHHOLD, H.: Ultrafiltration. Biochem. Z. **6**, 379 (1907 a).

— Kolloidstudien mit der Filtrationsmethode. Z. Physiol. Chem. **60**, 257 (1907 b).

BEIER, F. R., LAHEY, M. E., HEINER, D. C.: Serum transcortin levels in acute leukemia. Amer. J. Dis. Child. **106**, 381 (1963).

BEIERWALTES, W. H., ROBBINS, J.: Familial increase in the thyroxine-binding sites in serum alpha globulin. J. clin. Invest. **38**, 1683 (1959).

BEIGELMAN, P. M., SLUSHER, M. A., SLATER, G. G., ROBERTS, S.: Effect of anesthetics and collection time on corticosteroid secretion by rat adrenal. Proc. Soc. exp. Biol. (N. Y.) **93**, 608 (1956).

BEISEL, W. R., DIRAIMONDO, V. C., CHAO, P. Y., ROSNER, J. M., FORSHAM, P. H.: The influence of plasma protein binding on the extra-adrenal metabolism of cortisol in normal, hyperthyroid and hypothyroid subjects. Metabolism **13**, 942 (1964 a).

Beisel, W. R., Diraimondo, V. C., Forsham, P. H.: Cortisol transport and disappearance. Ann. intern. Med. **60**, 641 (1964 b).

Beiser, S. M., Erlanger, B. F.: Estimation of steroid hormones by an immunochemical technique. Nature **214**, 1044 (1967).

— — Agate, F. J., Lieberman, S.: Antigenicity of steroid-protein conjugates. Science **129**, 564 (1959).

Bencze, W. L., Schmid, K.: Determination of tyrosine and tryptophan in proteins. Anal. Chem. **29**, 1193 (1957).

Benesch, R., Benesch, R. E.: Heterogeneity of serum albumin. Fed. Proc. **15**, 218 (1956).

— — Determination of —SH groups in proteins. Meth. biochem. Anal. **10**, 43 (1962).

Bennett, J. V., Kirby, W. M. M.: A rapid, modified ultrafiltration method for determining serum protein binding and its application to new penicillins. J. Lab. clin. Med. **66**, 721 (1965).

Bennhold, H.: Congo red test. Klin. Wschr. **3**, 1711 (1924).

— Über die Adsorptionsfähigkeit der Serumkolloide tubulärer Nierenkranker gegenüber Farbstoffen. Z. ges. exp. Med. **49**, 71 (1926).

— Über die Bindung des Cholesterins an die Globuline; zugleich ein weiterer Beitrag zur Frage der Funktion der Serumeiweißkörper. Verh. dtsch. Ges. inn. Med. **43**, 211 (1931).

— Über die Vehikelfunktion der Serumeiweißkörper. Ergebn. inn. Med. Kinderheilk. **42**, 273 (1932).

— Die Vehikelfunktion der Bluteiweißkörper. In: Die Eiweißkörper des Blutplasmas. Hrsg.: H. Bennhold, E. Kylin und S. Rusznyak. Dresden: Steinkopf 1938, p. 220.

— Über die Bedeutung der Serumeiweißkörper für Transportvorgänge im lebenden Organismus. Antibiot. et Chemother. (Basel) **12**, 1 (1964).

— Transport function of the serum proteins: historical review and report on recent investigations on the transport of dyestuffs and of iron. In: Transport function of plasma proteins. Eds.: P. Desgrez and P. M. de Traverse. Amsterdam: Elsevier 1966, p. 1.

— Ott, H.: Der Stofftransport. Handbuch der allgemeinen Pathologie, Bd. 5, Kapit. 1, S. 166. Hrsg.: F. Buchner, E. Letterer und F. Roulet. Berlin: Springer 1961.

— — Scheurlen, G.: Beiträge zur Frage der genbedingten Bluteiweißstörungen. Verh. dtsch. Ges. inn. Med. **64**, 279 (1958).

— Peters, H., Roth, E.: Über einen Fall von kompletter Analbuminämie ohne wesentliche klinische Krankheitszeichen. Verh. dtsch. Ges. inn. Med. **60**, 630 (1954).

Bernstein, S., Cantrall, E. W., Dusza, J. P., Joseph, J. P.: Steroid conjugates. A bibliography. Washington D. C.: American Chemical Society 1966.

Beutner, R.: The binding power of serum for drugs tested by a new *in vitro* method. J. Pharmacol. exp. Ther. **25**, 365 (1925).

— The reaction between serum and alkaloids. J. Pharmacol. exp. Ther. **29**, 95 (1926).

BEZKOROVAINY, A.: Some hydrodynamic properties of native and modified serum orosomucoids. Biochim. Biophys. Acta (Amst.) **101**, 336 (1965 a).
— Comparative study of the acid glycoproteins isolated from bovine serum, colostrum, and milk whey. Arch. Biochem. **110**, 558 (1965 b).
— DOHERTY, D. G.: Comparison of bovine and human orosomucoids. Nature **195**, 1003 (1962).
— WINZLER, R. J.: Isolation and characterization of an acidic α_2-glycoprotein. Biochim. Biophys. Acta (Amst.) **49**, 559 (1961).
— ZSCHOCKE, R., GROHLICH, D.: Some physical-chemical properties of succinylated transferrin, conalbumin, and orosomucoid. Biochim. Biophys. Acta (Amst.) **181**, 295 (1969).
BIANCHI, E., CONIO, G., CIFERRI, A., PUETT, D., RAJAGH, L.: The role of pH, temperature, salt type, and salt concentration on the stability of the crystalline, helical, and randomly coiled forms of collagen. J. biol. Chem. **242**, 1361 (1967).
BIGELOW, C. C.: On the average hydrophobicity of proteins and the relation between it and protein structure. J. theor. Biol. **16**, 187 (1967).
BILLIAR, R. B., TANAKA, Y., KNAPPENBERGER, M., HERNANDEZ, R., LITTLE, B.: Influence of transcortin and albumin on the rate of reduction of progesterone by human placental 20α-hydroxysteroid dehydrogenase. Endocrinology **84**, 1152 (1969).
BISCHOFF, F.: Carcinogenic effects of steroids. Advan. Lipid Res. **7**, 165 (1969).
— KATHERMAN, R. E.: Distribution of estradiol between serum and red cells. Amer. J. Physiol. **152**, 189 (1948).
— — Solubility of estrone in biologic systems. Fed. Proc. **11**, 188 (1952).
— — YEE, Y. S., MORAN, J. J.: Solubilities of testosterone and estradiol esters in biologic systems. Fed. Proc. **11**, 189 (1952).
— PILHORN, H. R.: The state and distribution of steroid hormones in biological systems. III. Solubilities of testosterone, progesterone and alpha estradiol in aqueous salt and protein solution and in serum. J. biol. Chem. **172**, 663 (1948).
— STAUFFER, R. D.: The dispersion of testosterone in aqueous bovine serum albumin solution. J. Amer. chem. Soc. **76**, 1962 (1954).
— — Orientation of circulating human estrogens by albumin. Amer. J. Physiol. **191**, 313 (1957).
— — GRAY, C. L.: Physio-chemical state of the circulating steroids. Amer. J. Physiol. **177**, 65 (1954).
— TURNER, J. G., BRYSON, G.: Competition between human plasma albumin and globulins for estradiol and sodium estrone sulfate. Amer. J. Physiol. **195**, 81 (1958).
BLAQUIER, J. A., SEGRE DE CAMEO, M.: Determination of the binding capacity and affinity of proteins from human pregnancy plasma for androstanediol. Steroids **15**, 219 (1970).
BLATT, W. F., ROBINSON, S. M., BIXLER, H. J.: Membrane ultrafiltration: The diafiltration technique and its application to microsolute exchange and binding phenomena. Anal. Biochem. **26**, 151 (1968).

BLECHER, M.: Serum protein-steroid hormone interactions. I. Binding of deoxycorticosterone to serum proteins assayed in a metabolic system *in vitro*. Biochim. Biophys. Acta (Amst.) **93**, 158 (1964).

— Serum protein-steroid hormone interactions. Effects of glucocorticoids on glucose metabolism in rat adipose isolated cells, and the influence of human plasma corticosteroid binding protein. Endocrinology **79**, 541 (1966).

BLUMBERG, B. S., WARREN, L.: The effect of sialidase on transferrins and other serum proteins. Biochim. Biophys. Acta (Amst.) **50**, 90 (1961).

BOETTIGER, E.: The adsorption of estriol by the plasma proteins. J. cell. comp. Physiol. **28**, 139 (1946).

BOHAK, Z.: N^{ε}-(DL-2-Amino-2-carboxyethyl)-L-lysine, a new amino acid formed on alkaline treatment of proteins. J. biol. Chem. **239**, 2878 (1964).

BOHUS, R., ENDRÖCZI, E., LISSÁK, K.: Further data concerning the sex differences of the pituitary-adrenal system in the rat. Acta physiol. Acad. Sci. hung. **24**, 85 (1963).

BONGIOVANNI, A. M., EBERLEIN, W. R., WESTPHAL, M., BOGGS, T.: Prolonged turnover rate of hydrocortisone in the newborn infant. J. clin. Endocr. **18**, 1127 (1958).

BOOTH, M., DIXON, P. F., GRAY, C. H., GREENAWAY, J. M., HOLNESS, H. J.: Protein binding of cortisol in health and in pregnancy. J. Endocr. **23**, 25 (1961).

BOULOUARD, R., FONTAINE, Y. A.: Sur l'état de la corticostérone plasmatique chez le rat. Étude par filtration sur gels Sephadex G-100 et G-200. C. R. Acad. Sci. (Paris) **257**, 1379 (1963).

BOWMAN, R. E.: Ultramicro method for nonconjugated 17-hydroxycorticosteroids in one-half milliliter of monkey or human plasma. Anal. Biochem. **19**, 166 (1967).

— DELUNA, R. F.: Assessment of a protein-binding method for cortisol determination. Anal. Biochem. **26**, 465 (1969).

BOYER, J., BARON, D. N., TALALAY, P.: Purification and properties of a 3α-hydroxysteroid dehydrogenase from *Pseudomonas testosteroni*. Biochemistry **4**, 1825 (1965).

— TALALAY, P.: The amino acid composition of crystalline Δ^5-3-ketosteroid isomerase and its tryptic peptides. J. biol. Chem. **241**, 180 (1966).

BOYER, P. D.: Spectrophotometric study of the reaction of protein sulfhydryl groups with organic mercurials. J. Amer. chem. Soc. **76**, 4331 (1954).

— BALLOU, G. A., LUCK, J. M.: The combination of fatty acids and related compounds with serum albumin. III. The nature and extent of the combination. J. biol. Chem. **167**, 407 (1947).

BRADLEY, E. M., WATERHOUSE, C.: Effect of estrogen administration on extravascular cortisol. J. clin. Endocr. **26**, 705 (1966).

BRAUDE, E. A.: Ultraviolet and visible light absorption. In: Determination of organic structures by physical methods, vol. I, chapter 4. Eds.: E. A. BRAUDE and F. C. NACHOD. New York: Academic Press 1955.

BRAUNSBERG, H., JAMES, V. H. T.: The determination of cortisol and corticosterone in blood: A review. J. clin. Endocr. **21,** 1146 (1961).

BRAVERMAN, L. E., INGBAR, S. H.: Effects of norethandrolone on the transport in serum and peripheral turnover of thyroxine. J. clin. Endocr. **27,** 389 (1967).

BRAY, G. A.: A simple efficient liquid scintillator for counting aqueous solutions in a liquid scintillation counter. Anal. Biochem. **1,** 279 (1960).

BREUER, H., KNUPPEN, R.: Enzymes of estrogen metabolism. Methods Enzymol. **15,** 691 (1969).

BRILL, A. S., MARTIN, R. B., WILLIAMS, R. J. P.: Copper in biological systems. In: Electronic aspects in biochemistry. Ed.: B. PULLMAN. New York: Academic Press 1963, p. 519.

BRODIE, B. B.: Displacement of one drug by another from carrier or receptor sites. Proc. Roy. Soc. Med. **58,** 946 (1965).

BRODISH, A., LONG, C. N. H.: Changes in blood ACTH under various experimental conditions studied by means of a cross-circulation technique. Endocrinology **59,** 666 (1956).

BROOKS, R. V., DUPRE, J., GOGATE, A. N., MILLS, I. H., PRUNTY, F. T. G.: Appraisal of adrenocortical hyperfunction: patients with Cushing's syndrome or "non-endocrine" tumors. J. clin. Endocr. **23,** 725 (1963 a).

BROOKS, S. C., HORN, L., JACKSON, J., LOUD, A. V., HORWITZ, J. P.: On the formation of estroprotein. Biochim. Biophys. Acta (Amst.) **74,** 569 (1963 b).

BROWN, E. A., MUIR, R. D., CELLA, J. A.: Steroidal aldosterone blockers. III. J. Org. Chem. **25,** 96 (1960).

BROWN-GRANT, K.: Gonadal function and thyroid activity. J. Physiol. (Lond.) **131,** 70 (1956).

— Changes in thyroid gland activity during the oestrous cycle in rats. J. Physiol. (Lond.) **161,** 557 (1962).

BRUNELLI, B.: Sulla funzione veicolante delle proteine plasmatiche per l'ormone follicolare. Arch. int. Pharmacodyn. **49,** 262 (1934).

BRUNKHORST, W. K., HESS, E. L.: Interaction of cortisol with serum albumins. Arch. Biochem. **111,** 54 (1965).

BULL, H. B., BREESE, K.: Protein Hydration. I. Binding Sites. Arch. Biochem. **128,** 488 (1968 a).

— — Protein hydration. II. Specific heat of egg albumin. Arch. Biochem. **128,** 497 (1968 b).

BURKE, C. W.: Accurate measurement of steroid-protein binding by steady-state gel filtration. Biochim. Biophys. Acta (Amst.) **176,** 403 (1969 a).

— Steroid adsorption to gel filtration media, and its prevention by serum. Biochim. Biophys. Acta (Amst.) **187,** 564 (1969 b).

— Uncertainties in the determination of the "cortisol-binding capacity" of plasma and their removal. J. Clin. Path. **22,** 683 (1969 c).

BURLINGAME, A. L., HAUG, P., BELSKY, T., CALVIN, M.: Occurrence of biogenic steranes and pentacyclic triterpanes in an eocene shale (52 million years) and in an early pre-cambrian shale (2.7 billion years): a preliminary report. Proc. Nat. Acad. Sci. (Wash.) **54,** 1406 (1965).

BURSTEIN, S. H.: The use of affinity chromatography in the purification of steroid-binding proteins. In: Immunological methods in steroid determination. Eds.: F. G. PÉRON and B. V. CALDWELL. New York: Appleton-Century Crofts 1970 (in press).

— The removal of testosterone-binding globulin from plasma by affinity chromatography. Steroids **14**, 263 (1969).

BURTON, R. M., HARDING, G. B., RUST, N., WESTPHAL, U.: Steroid-protein interactions. XXIII. Nonidentity of cortisol-binding globulin and progesterone-binding protein in guinea pig serum. (1970) Steroids (in press).

BUSCK, G.: Die photobiologischen Sensibilisatoren und ihre Eiweißverbindungen. Biochem. Z. **1**, 425 (1906).

BUSH, I. E.: Species differences and other factors influencing adrenocortical secretion. Ciba Found. Colloq. Endocrinol., vol. 7. Eds.: G. E. W. WOLSTENHOLME and E. C. P. MILLAR. Boston: Little, Brown and Co. 1953 a, p. 210.

— Species differences in adrenocortical secretion. J. Endocr. **9**, 95 (1953 b).

— The physicochemical state of cortisol in blood. In: Hormones in blood, Ciba Found. Colloq. Endocrinol., vol. 11. Eds.: G. E. W. WOLSTENHOLME and E. C. P. MILLAR. Boston: Little, Brown and Co. 1957, p. 263.

— Chemical and biological factors in the activity of adrenocortical steroids. Pharmacol. Rev. **14**, 317 (1962).

BUTENANDT, A., WESTPHAL, U.: Beiträge zur Konstitutionsermittlung des Follikelhormons. II. Über den Sättigungsgrad und den aromatischen Charakter des Follikelhormons. Untersuchungen über das weibliche Sexualhormon. Hoppe-Seylers Z. physiol. Chem. **223**, 147 (1934).

BUTLER, J., DIXON, P. F.: The binding of cortisol to CBG and albumin in plasma from women during the menstrual cycle and in pregnancy. Ann. Endocr. (Paris) **30**, 217 (1969).

CABIB, E., ALGRANATI, I. D.: A rapid dialyzer for small samples. Nature **188**, 409 (1960).

CALDWELL, B. V., ABRAHAM, G. E., THORNEYCROFT, I. H., TILLSON, S. A., SCARAMUZZI, R. J.: Solid phase radioimmunoassay of estradiol-17-beta and estrone. Transactions of the 4th Meeting of the International Study Group for Steroid Hormones, Rome, Dec. 1969; Research on Steroids **4** (1970 a), in press.

— SCARAMUZZI, R. J., TILLSON, S., THORNEYCROFT, I. H.: Physiological studies using antibodies to steroids. In: Immunological methods in steroid determination. Eds.: F. G. PÉRON and B. V. CALDWELL. New York: Appleton-Century Crofts 1970 b (in press).

CAMPBELL, B. J., SCHLUETER, R. J., WEBER, G. F., WHITE, W. F.: Erythropoietin. III. Chemical characterization of a highly purified fraction from sheep erythropoietin concentrates. Biochim. Biophys. Acta (Amst.) **46**, 279 (1961).

CASPI, E., SCHMID, W., WITTSTRUCK, T. A.: Reaction of lead tetra-acetate with steroidal α,α'-dihydroxy ketones in alcoholic media. Tetrahedron **16**, 271 (1961).

CATANESE, B., LISCIANI, R., PICCINELLI, D.: Erythrocyte membrane stabilization and protein binding of some anti-inflammatory drugs and of deoxycholic acid. Biochem. Pharmacol. **18**, 1707 (1969).

CECIL, R.: Intramolecular bonds in proteins. I. The role of sulfur in proteins. In: The proteins, vol. I, chapter 5. Ed.: H. NEURATH. New York-London: Academic Press 1963, p. 379.

— MCPHEE, J. R.: The sulfur chemistry of proteins. Advanc. Protein Chem. **14**, 255 (1959).

CHADER, G. J.: Isolation and characterization of the corticosteroid-binding globulin of the rabbit. University of Louisville: Doctoral dissertation 1966.

— WESTPHAL, U.: Unpublished observations.

— — Isolation of rat corticosteroid-binding globulin (CBG) and observations on CBG polymerism. Fed. Proc. **27**, 807 (1968 a).

— — Steroid-protein interactions. XVI. Isolation and characterization of the corticosteroid-binding globulin of the rabbit. J. biol. Chem. **243**, 928 (1968 b).

— — Steroid-Protein Interactions. XVIII. Isolation and observations on the polymeric nature of the corticosteroid-binding globulin of the rat. Biochemistry **7**, 4272 (1968 c).

CHANUTIN, A., LUDEWIG, S., MASKET, A. V.: Studies on the calcium-protein relationship with the aid of the ultracentrifuge. I. Observations on calcium caseinate solutions. J. biol. Chem. **143**, 737 (1942).

CHEATUM, S. G., DOUVILLE, A. W., WARREN, J. C.: Site specificity of bovine adrenal 3β-hydroxysteroid dehydrogenase and Δ^5-3-ketosteroid isomerase. Biochim. Biophys. Acta (Amst.) **137**, 172 (1967).

— WARREN, J. C.: Purification and properties of 3β-hydroxysteroid dehydrogenase and Δ^5-3-ketosteroid isomerase from bovine corpora lutea. Biochim. Biophys. Acta (Amst.) **122**, 1 (1966).

CHEN, P. S., JR., MILLS, I. H., BARTTER, F. C.: Ultrafiltration studies of steroid-protein binding. J. Endocr. **23**, 129 (1961).

CHEN, R. F.: Removal of fatty acids from serum albumin by charcoal treatment. J. biol. Chem. **242**, 173 (1967).

CHIN, C. C., WARREN, J. C.: Synthesis of 4-mercuri-17β-estradiol and study of its reaction with sulfhydryl compounds and proteins. J. biol. Chem. **243**, 5056 (1968).

— — Synthesis of 2-diazoestrone sulfate and use for affinity labeling of steroid binding sites. Biochemistry **9**, 1917 (1970).

CHRISTY, N. P., WALLACE, E. Z., GORDON, W. E. L., JAILER, J. W.: On the rate of hydrocortisone clearance from plasma in pregnant women and in patients with Laennec's cirrhosis. J. clin. Invest. **38**, 299 (1959).

CHULSKI, T., FORIST, A. A.: The effects of some solid buffering agents in aqueous suspension on prednisolone. J. Amer. pharm. Ass., sci. Ed. **47**, 553 (1958).

CLARK, F., GRANT, J. K.: The biosynthesis and secretion of adrenocortical steroids. Cambridge: Cambridge University Press 1960.

CLAYTON, R. B. (Ed.): Enzyme systems of steroid hormone metabolism. Methods Enzymol. **15**, section V (1969).

COHEN, M., STIEFEL, M., REDDY, W. J., LAIDLAW, J. C.: The secretion and disposition of cortisol during pregnancy. J. clin. Endocr. **18**, 1076 (1958).

COHEN, P., CHIN, R. C., KIDSON, C.: Interactions of hormonal steroids with nucleic acids. II. Structural and thermodynamic aspects of binding. Biochemistry **8**, 3603 (1969).

— KIDSON, C.: Interactions of hormonal steroids with nucleic acids. I. A specific requirement for guanine. Proc. Natl. Acad. Sci. (Wash.) **63**, 458 (1969).

COHEN, S. C., HUSEBY, R. A.: The effects of estrogen on the serum glucuronidase activity of patients with breast cancer. Cancer Res. **11**, 52 (1951).

COHN, E. J.: Chemical, physiological and immunological properties and clinical uses of blood derivatives. Experientia **3**, 125 (1947).

— The chemical specificity of the interaction of diverse human plasma proteins. Blood **3**, 471 (1948).

— GURD, F. R. N., SURGENOR, D. M., BARNES, B. A., BROWN, R. K., DEROUAUX, G., GILLESPIE, J. M., KAHNT, F. W., LEVER, W. F., LIN, C. H., MITTELMAN, C. H., MOUTON, R. F., SCHMID, K., UROMA, E.: A system for the separation of the components of human blood: quantitative procedures for the separation of the protein components of human plasma. J. Amer. chem. Soc. **72**, 465 (1950).

— HUGHES, W. L., JR., WEARE, J. H.: Preparation and properties of serum and plasma proteins. XIII. Crystallization of serum albumins from ethanol-water mixtures. J. Amer. chem. Soc. **69**, 1753 (1947).

— STRONG, L. E., HUGHES, W. L., JR., MULFORD, D. J., ASHWORTH, J. N., MELVIN, M., TAYLOR, H. L.: Preparation and properties of serum and plasma proteins. IV. A system for the separation into fractions of the protein and lipoprotein components of biological tissues and fluids. J. Amer. chem. Soc. **68**, 459 (1946).

COLOWICK, S. P., KAPLAN, N. O. (Eds.): Enzymes of lipid metabolism. Methods Enzymol. **5**, section III (1962).

CONTI, C. (Ed.): Transactions of the 4th Meeting of the International Study Group for Steroid Hormones, Rome, Dec. 1969. Research on Steroids **4**, (1970), in press.

COOLIDGE, T. B.: A simple ultrafiltration apparatus. J. biol. Chem. **135**, 541 (1940).

COOPER, A., DUAX, W. L.: Structure and conformation of the cortisone side chain. J. pharm. Sci. **58**, 1159 (1969).

— GOPALAKRISHNA, E. M., NORTON, D. A.: The crystal structure and absolute configuration of the 2 : 1 complex between testosterone and mercuric chloride. Acta Cryst. **B 24**, 935 (1968).

— NORTON, D. A.: Estrogenic steroids. III. The crystal and molecular structure of estriol. Acta Cryst. **B 25**, 814 (1969).

COOPER, G. R.: Electrophoretic and ultracentrifugal analysis of normal human serum. In: The plasma proteins, vol. I. Ed.: F. W. PUTNAM. New York-London: Academic Press 1960, p. 51.

COPE, C. L., SEWELL, C. F.: The fate of hydrocortisone-4-^{14}C in the tissues of the mouse. J. Endocr. **13,** 417 (1956).

CORKER, C. S., EXLEY, D., NAFTOLIN, F.: Assay of 17β-oestradiol by competitive protein binding methods. Transactions of 2nd Karolinska Symposium on Research Methods in Reproductive Endocrinology, Steroid assay by protein binding, Geneva, March, 1970. Acta endocr. (Kbh.), Suppl. No. **147,** 305 (1970).

CORNWELL, D. G.: Lipoproteins. Lipids and lipidoses. Ed.: G. SCHETTLER. New York: Springer 1967, p. 168.

COWIE, A. T., FOLLEY, S. J.: The mammary gland and lactation. In: Sex and internal secretions, vol. I. Ed.: W. C. YOUNG. Baltimore: Williams and Wilkins 1961, p. 590.

CRAIG, L. C.: Differential dialysis. Science **144,** 1093 (1964).

— KONIGSBERG, W.: Dialysis studies. III. Modification of pore size and shape in cellophane membranes. J. Phys. Chem. **65,** 166 (1961).

CRÉPY, O., DRAY, F., SEBAOUN, J.: Rôle des hormones thyroïdiennes dans les interactions entre la testostérone et les protéines sériques. C. R. Acad. Sci. (Paris) **264,** 2651 (1967).

— GUÉRIGUIAN, J.: Transport of steroids by proteins. Transactions of the 4th Meeting of the International Study Group for Steroid Hormones, Rome, Dec. 1969. Research on Steroids **4** (1970), in press.

— SCHWOB, A., DUCRET, M. A., GUÉRIGUIAN, J. L., MOWSZOWICZ, I., DRAY, F.: Étude critique de quelques méthodes de dosage de la testostérone plasmatique. Ann. Endocr. (Paris) **30,** 165 (1969).

CRITCHLOW, V., LIEBELT, R. A., BAR-SELA, M., MOUNTCASTLE, W., LIPSCOMB, H. S.: Sex difference in resting pituitary-adrenal function in the rat. Amer. J. Physiol. **205,** 807 (1963).

CUSHMAN, P., JR., ALTER, S., HILTON, J. G.: Effects of clomiphene on oestradiol alterations in thyroxine and cortisol transport in man. Acta Endocr. **50,** 329 (1965).

DALY, J. R., SPENCER-PEET, J.: Fluorimetric determination of adrenal corticosteroids: Observations on interfering fluorogens in human plasma. J. Endocr. **30,** 255 (1964).

D'ANGELO, S. A.: Simultaneous effects of estradiol on TSH secretion and adrenocortical function in male and female rats. Endocrinology **82,** 1035 (1968).

— FISHER, J. S.: Influence of estrogen on the pituitary-thyroid system of the female rat: Mechanisms and loci of action. Endocrinology **84,** 117 (1969).

— GRODIN, J. M.: Experimental hyperthyroidism and adrenocortical function in the rat. Endocrinology **74,** 509 (1964).

DAS, B. R.: Isolation and partial characterization of sheep plasma α_1-mucoprotein. Biochim. Biophys. Acta (Amst.) **58,** 52 (1962).

DAUBEN, W. G., PAYOT, P. H.: Radiation induced oxidation of cholesterol. J. Amer. chem. Soc. **78**, 5657 (1956).

DAUGHADAY, W. H.: Evidence for two corticosteroid binding systems in human plasma. J. Lab. clin. Med. **48**, 799 (1956 a).

— Binding of corticosteroids by plasma proteins. I. Dialysis equilibrium and renal clearance studies. J. clin. Invest. **35**, 1428 (1956 b).

— Binding of corticosteroids by plasma proteins. II. Paper electrophoresis and equilibrium paper electrophoresis. J. clin. Invest. **35**, 1434 (1956 c).

— Binding of corticosteroids by plasma proteins. V. Corticosteroid-binding globulin activity in normal human beings and in certain disease states. Arch. intern. Med. **101**, 286 (1958 a).

— Binding of corticosteroids by plasma proteins. III. The binding of corticosteroid and related hormones by human plasma and plasma protein fractions as measured by equilibrium dialysis. J. clin. Invest. **37**, 511 (1958 b).

— Binding of corticosteroids by plasma proteins. IV. The electrophoretic demonstration of corticosteroid binding globulin. J. clin. Invest. **37**, 519 (1958 c).

— Steroid protein interactions. Physiol. Rev. **39**, 885 (1959).

— The binding of corticosteroids by serum proteins. In: Hormones in human plasma. Ed.: H. N. ANTONIADES. Boston: Little, Brown and Co. 1960, p. 495.

— The binding of corticosteroids by plasma protein. In: The adrenal cortex. Ed.: A. B. EISENSTEIN. Boston: Little, Brown and Co. 1967, p. 385.

— ADLER, R. E., MARIZ, I. K., RASINSKI, D. C.: Measurement of the binding capacity of corticosteroid-binding globulin in human plasma. J. clin. Endocr. **22**, 704 (1962).

— BREMER, R.: The association of 17-OH corticosteroids and 17-OH corticosteroid glucuronides with plasma proteins and the mechanism of corticosteroid excretion. J. Lab. clin. Med. **46**, 807 (1955).

— HOLLOSZY, J., MARIZ, I. K.: Binding of corticosteroids by plasma proteins. VI. The binding of cortisol and aldosterone by corticosteroid-binding globulin and by the estrogen-induced binding system of plasma. J. clin. Endocr. **21**, 53 (1961).

— KOZAK, I., BIEDERMAN, O.: Corticosteroid-binding globulin activity in body fluids and in fetal and maternal plasma. J. clin. Invest. **38**, 998 (1959).

— MARIZ, I. K.: The binding of steroid hormones by plasma proteins. In: Biological activities of steroids in relation to cancer. Ed.: G. PINCUS and I. P. VOLLMER. New York: Academic Press 1960, p. 61.

— — Corticosteroid-binding globulin: Its properties and quantitation. Metabolism **10**, 936 (1961).

DAVIDSON, E. T., DE VENUTO, F., WESTPHAL, U.: Steroid-protein interactions. IX. Interaction of aldosterone with human serum proteins in competition with spirolactones and other steroids. Endocrinology **71**, 893 (1962).

DAVIDSON, E. T., DE VENUTO, F., WESTPHAL, U.: Interaction of aldosterone with liver and kidney cell membranes and subcellular fractions of kidney. Proc. Soc. exp. Biol. (N. Y.) **113**, 387 (1963).

DAVIDSON, S. J.: The Δ^4-5β-steroid dehydrogenase of *Pseudomonas testosteroni*. Methods Enzymol. **15**, 656 (1969).

— TALALAY, P.: Purification and mechanism of action of a steroid Δ^4-5β-dehydrogenase. J. biol. Chem. **241**, 906 (1966).

DAVIS, B. J.: Disc electrophoresis—II. Method and application to human serum proteins. Ann. N. Y. Acad. Sci. **121**, 404 (1964).

DAVIS, J. W., LIU, T. M. Y.: The adrenal gland and lactogenesis. Endocrinology **85**, 155 (1969).

DAY, C. E., LEVY, R. S.: Molecular structure of serum lipoproteins. J. theor. Biol. **23**, 387 (1969).

DEAKINS, S., ROSNER, W.: Electrophoretic studies of testosterone-binding proteins in human plasma. 3rd Intern. Congr. Endocr. Excerpta Med. Intern. Congr. Ser. No. 157, p. 159 (1968).

DECKX, R., RAUS, J., DENEF, C., DE MOOR, P.: Sex difference in the metabolism of cortisol by rat liver. Steroids **6**, 129 (1965).

DE HAAN, J., VAN CREVELD, S.: Über die Wechselbeziehungen zwischen Blutplasma und Gewebeflüssigkeiten, insbesondere Kammerwasser und Cerebrospinalflüssigkeit. Biochem. Z. **124**, 172 (1921).

DE LALLA, O. F., GOFMAN, J. W.: Ultracentrifugal analysis of serum lipoproteins. In: Methods of biochemical analysis, vol. I. Ed.: D. GLICK. New York: Interscience 1954, p. 459.

DELIN, S., SQUIRE, P. G., PORATH, J.: Purification of steroid-induced enzymes from *Pseudomonas testosteroni*. Biochim. Biophys. Acta (Amst.) **89**, 398 (1964).

DEMISCH, K., GRANT, J. K., BLACK, W.: Plasma testosterone in women in late pregnancy and after delivery. J. Endocr. **42**, 477 (1968).

DE MOOR, P.: The binding of radiocortisol on transcortin as measured by gel filtration. Pflügers Arch. ges. Physiol. **285**, 349 (1965).

— DECKX, R., RAUS, J., DENEF, C.: About the possible influence of transcortin on the catabolism of cortisol in the liver. Metabolism **12**, 592 (1963 a).

— HEIRWEGH, K., HEREMANS, J., DECLERCK-RASKIN, M.: Protein binding of corticoids studied by gel filtration. J. clin. Invest. **41**, 816 (1962).

— — STEENO, O.: Protein binding of progesterone studied by gel filtration. Arch. Biochem. **103**, 506 (1963 b).

— HENDRIKX, A., STEENO, O.: Idiopathic lowering of the cortisol binding capacity of human plasma transcortin. Ann. Endocr. (Paris) **26**, 488 (1965 a).

— HEYNS, W.: Cortisol binding capacity of plasma transcortin in diabetic subjects. J. clin. Endocr. **27**, 706 (1967).

— — Cortisol binding affinity of plasma transcortin (CBA) as studied by competitive adsorption. J. clin. Endocr. **28**, 1281 (1968).

DE MOOR, P., HEYNS, W., STEENO, O.: Clinical and biochemical features associated with extreme CBG values; possible significance of these associations. Ann. Endocr. (Paris) **29,** 181 (1968 a).

— — VAN BAELEN, H., STEENO, O.: Binding of steroids by plasma globulins. Abstract, 3rd Intern. Congr. Endocrinology. Excerpta Med. Intern. Congr. Ser. No. 157, p. 159 (1968 b).

— HINNEKENS, M.: The disappearance rate of plasma corticoids in patients with suppressed adrenocortical production. Acta Endocr. **37,** 1 (1961).

— MEULEPAS, E., HENDRIKX, A., HEYNS, W., VANDENSCHRIECK, H. G.: Cortisol-binding capacity of plasma transcortin: A sex-linked trait? J. clin. Endocr. **27,** 959 (1967).

— — STEENO, O.: Idiopathic lowering of the cortisol binding capacity of plasma transcortin. V. Familial evidence, sex distribution and proposed mode of inheritance. Ann. Endocr. (Paris) **27,** 234 (1966 a).

— ROELS, H., DELAERE, K., CRABBÉ, J.: Unusual case of adrenocortical hyperfunction. J. clin. Endocr. **25,** 612 (1965 b).

— STEENO, O.: About the existence in or on erythrocytes of a specific cortisol binding agent. J. Endocr. **26,** 301 (1963).

— — The biological role of the cortisol-binding globulin (CBG) in plasma. Koninkl. Vlaam. Acad. Geneeskunde Belg. **27,** 675 (1965).

— — BROSENS, I., HENDRIKX, A.: Data on transcortin activity in human plasma as studied by gel filtration. J. clin. Endocr. **26,** 71 (1966 b).

— — DECKX, R.: Factors affecting the binding of cortisol by transcortin. Acta Endocr. **44,** 107 (1963 c).

— — HENDRIKX, A.: Clinical features observed in patients with an unexplained low cortisol binding capacity. Acta Endocr. **48,** 272 (1965 c).

— — HEYNS, W.: Possible role or significance of protein-steroid binding in plasma. Ann. Endocr. (Paris) **29,** 119 (1968 c).

— — — VAN BAELEN, H.: The steroid binding β-globulin in plasma: pathophysiological data. Ann. Endocr. (Paris) **30,** 233 (1969).

— — RASKIN, M., HENDRIKX, A.: Fluorometric determination of free plasma 11-hydroxycorticosteroids in man. Acta Endocr. **33,** 297 (1960).

— VERBERCKMOES, V., HEYNS, W.: Unbound plasma cortisol as measured by ultrafiltration *in vivo:* Effect of estrogen treatment. J. clin. Endocr. **28,** 313 (1968 d).

DESCOMPS, B., CRASTES DE PAULET, A.: Étude du site actif de la 17β-hydroxystéroide: NAD(P) oxydoréductase soluble du placenta humain. Bull. Soc. Chim. biol. (Paris) **51,** 1591 (1969).

— NICOLAS, J. C., CRASTES DE PAULET, A.: Isolement de la 17β-hydroxystéroide-NAD oxydoréductase du placenta humain. Bull. Soc. Chim. biol. (Paris) **50,** 1681 (1968).

DESGREZ, P.: Role des protéines plasmatiques dans le transport des corticostéroides: la transcortine ou CBG. Ann. Biol. clin. **25,** 931 (1967).

— DE TRAVERSE, P. M. (Eds.): Transport function of plasma proteins. Amsterdam-London-New York: Elsevier 1966.

DE SOUZA, M. L. A., WILLIAMSON, H. O., MOODY, L. O., DICZFALUSY, E.: Further assessment of the reliability of progesterone assays by competitive protein binding. Transactions of the 2nd Karolinska Symposium on Research Methods in Reproductive Endocrinology, Steroid assay by protein binding, Geneva, March, 1970. Acta Endocr. (Kbh.), Suppl. No. **147**, 171 (1970).

DESSYPRIS, A., LUUKKAINEN, T., ADLERCREUTZ, H.: Testosterone binding in plasma of twin pregnancies. Scand. J. clin. Lab. Invest. **23**, Suppl. **108**, 39 (1969).

DE VENUTO, F.: Osmotic fragility and spontaneous lysis of human red cells preserved with addition of progesterone. Proc. Soc. exp. Biol. (N. Y.) **128**, 997 (1968 a).

— Swelling of human red blood cells, preserved with progesterone, as determined by spectrophotometric measurements. Proc. Soc. exp. Biol. (N. Y.) **129**, 106 (1968 b).

— ASHLEY, B. D., WILLIAMS, W. C., JR., WESTPHAL, U.: Corticosteroid-binding activity of pooled versus individual sera. U. S. Army Med. Res. Lab. Rep. 614, Fort Knox, Ky. December, 1964.

— KELLEHER, P. C., WESTPHAL, U.: Interactions between corticosteroids and fractions of rat liver and muscle cells as determined by "equilibrium fractionation" and equilibrium dialysis. Biochim. Biophys. Acta (Amst.) **63**, 434 (1962).

— LIGON, D. F., FRIEDRICHSEN, D. H., WILSON, H. L.: Human erythrocyte membrane. Uptake of progesterone and chemical alterations. Biochim. Biophys. Acta (Amst.) **193**, 36 (1969).

DIAMOND, M.: Androgen-induced masculinization in the ovariectomized and hysterectomized guinea pig. Anat. Rec. **157**, 47 (1967).

— RUST, N., WESTPHAL, U.: High-affinity binding of progesterone, testosterone and cortisol in normal and androgen-treated guinea pigs during various reproductive stages; relationship to masculinization. Endocrinology **84**, 1143 (1969).

— YOUNG, W. C.: Differential responsiveness of pregnant and nonpregnant guinea pigs to the masculinizing action of testosterone propionate. Endocrinology **72**, 429 (1963).

DICKERSON, R. E.: X-ray analysis and protein structure. In: The proteins, vol. II. Ed.: H. NEURATH. New York and London: Academic Press 1964, p. 603.

DICZFALUSY, E. (Ed.): Steroid assay by protein binding. Karolinska Symposia on Research Methods in Reproductive Endocrinology. Transactions of the 2nd Symposium, Geneva, March, 1970. Acta endocr. (Kbh.), Suppl. No. **147** (1970).

DISCHE, Z.: Reciprocal relation between fucose and sialic acid in mammalian glycoproteins. Ann. N. Y. Acad. Sci. **106**, 259 (1963).

DIXON, P. F.: The kinetics of the exchange between transcortin-bound and unbound cortisol in plasma. J. Endocr. **40**, 457 (1968).

— BOOTH, M., BUTLER, J.: The corticosteroids. Hormones in blood, 2nd ed., vol. 2. Eds.: C. H. GRAY and A. L. BACHARACH. New York-London: Academic Press 1967, p. 305.

DJERASSI, C.: Steroid reactions. San Francisco: Holden-Day, Inc. 1963.

Doe, R. P., Fernandez, R. N., Seal, U.S.: Measurement of corticosteroid-binding globulin in man. J. clin. Endocr. **24**, 1029 (1964).
— Lohrenz, F. N., Seal, U. S.: Familial decrease in corticosteroid-binding globulin. Metabolism **14**, 940 (1965).
— Mellinger, G. T., Swaim, W. R., Seal, U. S.: Estrogen dosage effects on serum proteins: a longitudinal study. J. clin. Endocr. **27**, 1081 (1967).
— Seal, U. S.: Protein binding of cortisol in man. J. clin. Invest. **42**, 929 (1963).
— Zinneman, H. H., Flink, E. B., Ulstrom, R. A.: Significance of the concentration of nonprotein-bound plasma cortisol in normal subjects, Cushing's syndrome, pregnancy and during estrogen therapy. J. clin. Endocr. **20**, 1484 (1960).
Donohue, J.: Selected topics in hydrogen bonding. In: Structural chemistry and molecular biology. Eds.: A. Rich and N. Davidson. San Francisco-London: Freeman and Co. 1968, p. 443.
Dorfman, L.: Ultraviolet absorption of steroids. Chem. Rev, **53**, 47 (1953).
Dorfman, R. I.: Corticoids. In: Methods in hormone research, vol. II. Ed.: R. I. Dorfman. New York-London: Academic Press 1962 a, pp. 325, 355.
— Methods in hormone research, vols. I and II. New York-London: Academic Press 1962 b.
— Methods in hormone research, vols. I and II, 2nd ed. New York-London: Academic Press 1969.
— Dorfman, A. S.: Thymolytic activity of corticoids in combination with other steroids. Endocrinology **71**, 271 (1962).
— Shipley, R. A.: Androgens. New York: Wiley 1956.
— Ungar, F.: Metabolism of steroid hormones. New York-London: Academic Press 1965.
Dounce, A. L., Witter, R. F., Monty, K. J., Pate, S., Cottone, M. A.: A method for isolating intact mitochondria and nuclei from the same homogenate and the influence of mitochondrial destruction on the properties of cell nuclei. J. biophys. biochem. Cytol. **1**, 139 (1955).
Douville, A. W., Warren, J. C.: Steroid-protein interaction at sites which influence catalytic activity. Biochemistry **7**, 4052 (1968).
Dowling, J. T., Freinkel, N., Ingbar, S. H.: Thyroxine-binding by sera of pregnant women. J. clin. Endocr. **16**, 280 (1956).
Dray, F.: Physiopathologie de la liaison de la testostérone, d l'oestradiol et du cortisol avec les protéines sériques. Rapports de la X⁰ Réunion des Endocrinologistes de Langue Française, Paris, Sept. 1969, p. 159.
— Ledru, M. J., Modigliani, E., Delzant, G.: Effects of clomiphene citrate on binding affinity of testosterone, estradiol and cortisol in serum of normal adult men. Transactions 4th Meeting Intern. Study Group for Steroid Hormones. Research on Steroids **4** (1970), in press.
— Mowszowicz, I., Ledru, M. J., Crépy, O., Delzant, G., Sebaoun, J.: Anomalies de l'affinité de liaison de testostérone dans le sérum des

sujets thyrotoxicosiques et dans le virilisme pilaire idiopathique. Augmentation de l'affinité de liaison après administration d'hormones thyroidiennes, chez le sujet dont l'affinité est normale ou basse. Ann. Endocr. (Paris) **30**, 223 (1969).

— SEBAOUN, J., DELZANT, G., LEDRU, M. J., MOWSZOWICZ, I.: Activité de liaison de la testostérone dans le sérum des femmes présentant un virilisme pilaire idiopathique. Rev. franç. Étud. clin. biol. **13**, 622 (1968).

— — MOWSZOWICZ, I., DELZANT, G., DESGREZ, P., DREYFUS, G.: Facteurs influençant le taux de la testostérone plasmatique chez l'homme: rôle des hormones thyroidiennes. C. R. Acad. Sci. (Paris) **264**, 2578 (1967).

DUSZA, J. P., HELLER, M., BERNSTEIN, S.: Ultraviolet absorption. In: Physical properties of the steroid hormones. Ed.: L. L. ENGEL. New York: MacMillan Company 1963, p. 69.

DZIUK, P. J., COOK, B.: Passage of steroids through silicone rubber. Endocrinology **78**, 208 (1966).

EAGLE, H., VICKERS, P.: The nature of the reaction between diazotized sulfanilic acid and proteins. J. biol. Chem. **114**, 193 (1936).

EDELMAN, I. S., FIMOGNARI, G. M.: On the biochemical mechanism of action of aldosterone. Rec. Progr. Hormone Res. **24**, 1 (1968).

EDSALL, J. T., WYMAN, J.: Some general aspects of molecular interactions. In: Biophysical chemistry, vol. I, chapter 11. New York: Academic Press 1958, p. 591.

EECHAUTE, W.: A simple method for the separate determination of cortisol and corticosterone in plasma. Clin. chim. Acta **13**, 785 (1966 a).

— A simple chromatographic technique for the determination of cortisol and corticosterone in plasma. Steroids **8**, 633 (1966 b).

— DEMEESTER, G., LACROIX, E., LEUSEN, I.: The adrenal cortex activity during experimental renal hypertension in rats. Arch. int. Pharmacodyn. **136**, 161 (1962).

EHRENPREIS, S., FLEISCH, J. H., MITTAG, T. W.: Approaches to the molecular nature of pharmacological receptors. Pharmacol. Rev. **21**, 131 (1969).

EHRLICH, P.: Über provozierte Fluorescenzerscheinungen am Auge. Dtsch. med. Wschr. **8**, 21, 35, 54 (1882).

— Beiträge zur Theorie und Praxis der histologischen Färbung. Universität Leipzig: Inaugural Dissertation 1878. The collected papers of PAUL EHRLICH, vol. I. Ed.: F. HIMMELWEIT. London-New York: Pergamon Press 1956, p. 29.

— On immunity with special reference to cell life. Croonian Lecture, Proc. Roy. Soc. 1900. The collected papers of PAUL EHRLICH, vol. II. Ed.: F. HIMMELWEIT. London-New York: Pergamon Press 1956, p. 178.

EIK-NES, K. B., SCHELLMAN, J. A., LUMRY, R., SAMUELS, L. T.: The binding of steroids to protein. I. Solubility determinations. J. biol. Chem. **206**, 411 (1954).

EISENBERG, H., TOMKINS, G. M.: Molecular weight of the subunits, oligomeric and associated forms of bovine liver glutamate dehydrogenase. J. molec. Biol. **31**, 37 (1968).

EKINS, R., NEWMAN, B.: Theoretical aspects of saturation analysis. Transactions of the 2nd Karolinska Symposium on Research Methods in Reproductive Endocrinology, Steroid assay by protein binding, Geneva, March, 1970. Acta endocr. (Kbh.), Suppl. No. **147,** 122 (1970).

ELLMAN, G. L.: Tissue sulfhydryl groups. Arch. Biochem. **82,** 70 (1959).

ENGEL, L. L.: Physical properties of the steroid hormones. New York: MacMillan 1963.

— CARTER, P.: Partition coefficients. In: Physical properties of the steroid hormones. Ed.: L. L. ENGEL. New York: MacMillan Company 1963, p. 1.

ENSINCK, J., VOLWILER, W.: Characterization of "transcortin", the corticosteroid-binding protein of human plasma. Clin. Res. **10,** 115 (1962).

ERLANGER, B. F., BOREK, F., BEISER, S. M., LIEBERMAN, S.: Steroid-protein conjugates. I. Preparation and characterization of conjugates of bovine serum albumin with testosterone and with cortisone. J. biol. Chem. **228,** 713 (1957).

— — — — Steroid-protein conjugates. II. Preparation and characterization of conjugates of bovine serum albumin with progesterone, deoxycorticosterone and estrone. J. biol. Chem. **234,** 1090 (1959).

EVANS, E. A.: Control of self-irradiation decomposition of tritium-labeled compounds at high specific activity. Nature **209,** 169 (1966).

— STANFORD, F. G.: Decomposition of tritium-labeled organic compounds. Nature **197,** 551 (1963).

EWALD, W., CHAIKOFF, I. L.: Δ^5-3-ketosteroid isomerase activities in guinea-pig adrenal glands: the role of copper ions. Biochim. Biophys. Acta (Amst.) **167,** 444 (1968).

— WERBIN, H., CHAIKOFF, I. L.: Evidence for two substrate-specific Δ^5-3-ketosteroid isomerases in beef adrenal glands, and their separation from 3β-hydroxysteroid dehydrogenase. Biochim. Biophys. Acta (Amst.) **81,** 199 (1964).

EYLAR, E. H.: On the biological role of glycoproteins. J. theor. Biol. **10,** 89 (1965).

FAHEY, J. L., MCCOY, P. F., GOULIAN, M.: Chromatography of serum proteins in normal and pathological sera: The distribution of protein-bound carbohydrate and cholesterol, siderophilin, thyroxin-binding protein, B_{12}-binding protein, alkaline and acid phosphatases, radioiodinated albumin and myeloma proteins. J. clin. Invest. **37,** 272 (1958).

FAIRCLOUGH, G. F., JR., FRUTON, J. S.: Peptide-protein interaction as studied by gel filtration. Biochemistry **5,** 673 (1966).

FAJER, A. B., VOGT, M.: Adrenocortical secretion in the guinea pig. J. Physiol. **169,** 373 (1963).

FALCOZ-KELLY, F., BAULIEU, E. E., ALFSEN, A.: Studies on $\Delta^{5\to4}$-3-oxo steroid isomerases. I. An extraction model for enzymatic activity. Biochemistry **7,** 4119 (1968).

FARAH, A.: On the combination of some cardio-active glycosides with serum proteins. J. Pharmacol. exp. Ther. **83,** 143 (1945).

FARESE, R. V., PLAGER, J. E.: The *in vitro* red blood cell uptake of C_{14}-cortisol; studies of plasma protein binding of cortisol in normal and abnormal states. J. clin. Invest. **41,** 53 (1962).

FAWAZ, G., FARAH, A.: A study of the digitoxin binding power of serum and other soluble tissue proteins of the rabbit. J. Pharmacol. exp. Ther. **80,** 193 (1944).

FAZEKAS, A. G., WEBB, J. L.: A fluorescent complex between steroids and nucleotide coenzymes. Canad. J. Biochem. **45,** 1479 (1967).

FAZEKAS, I. G., FAZEKAS, A. T.: Papierchromatographischer Nachweis der Corticosteroid-Fraktionen in Kaninchenorganen und Geweben. Endokrinologie **50,** 130 (1966).

FEDER, H. H., RESKO, J. A., GOY, R. W.: Progesterone concentrations in the arterial plasma of guinea pigs during the oestrous cycle. J. Endocr. **40,** 505 (1968).

FEDERMAN, D. D., ROBBINS, J., RALL, J. E.: Effects of methyl testosterone on thyroid function, thyroxine metabolism and thyroxine-binding protein. J. clin. Invest. **37,** 1024 (1958).

FELDMAN, I., DANLEY, R. A., O'LEARY, J. F.: Improved centrifuge type of ultrafiltration apparatus. Anal. Chem. **22,** 837 (1950).

FELDMAN, J. D.: Iodine metabolism in pregnancy. Amer. J. Physiol. **192,** 273 (1958).

— DANOWSKI, T. S.: Effect of estrogen on the metabolism of protein-bound iodine. Endocrinology **59,** 463 (1956).

FERIN, M., TEMPONE, A., ZIMMERING, P. E., VANDE WIELE, R.: Effect of antibodies to 17β-estradiol and progesterone on the estrous cycle of the rat. Endocrinology **85,** 1070 (1969 b).

— ZIMMERING, P. E., LIEBERMAN, S., VANDE WIELE, R. L.: Inactivation of the biological effects of exogenous and endogenous estrogens by antibodies to 17β-estradiol. Endocrinology **83,** 565 (1968).

— — VANDE WIELE, R. L.: Effects of antibodies to estradiol-17β on PMS-induced ovulation in immature rats. Endocrinology **84,** 893 (1969 a).

FIESER, L. F., FIESER, M.: Natural products related to phenanthrene, 3rd. ed. New York: Reinhold Publishing Corp. 1949.

— — Steroids. New York: Reinhold 1959.

— — Steroids. New York: Reinhold 1959, p. 15.

FINEAN, J. B., ROBERTSON, J. D.: Lipids and the structure of myelin. Brit. med. Bull. **14,** 267 (1958).

FINLAYSON, J. S., BAUMANN, C. A.: Protein-bound sterols in rodent urine. Amer. J. Physiol. **190,** 297 (1957).

FIRSCHEIN, H. E., DE VENUTO, F., FITCH, W. M., PEARCE, E. M., WESTPHAL, U.: Distribution of injected cortisol-4-C^{14} in normal and shocked rats. Endocrinology **60,** 347 (1957).

FISHER, H. F.: A limiting law relating the size and shape of protein molecules to their composition. Proc. nat. Acad. Sci. (Wash.) **51,** 1285 (1964).

FISHER, W. R., GURIN, S.: Structure of lipoproteins: covalently bound fatty acids. Science **143,** 362 (1964).

FLORINI, J. R., BUYSKE, D. A.: Plasma protein binding of triamcinolone-H^3 and hydrocortisone-4-C^{14}. J. biol. Chem **236**, 247 (1961).

FLORSHEIM, W. H., MIRISE, R. T., FAIRCLOTH, M. A., WILLIAMS, A. D.: Effects of adrenalectomy on thyroid function in starved rats. Metabolism **18**, 311 (1969).

FOLLEY, S. J.: The physiology and biochemistry of lactation. London: Oliver and Boyd 1956.

— YOUNG, F. G.: Prolactin as a specific lactogenic hormone. Lancet **240**, 380 (1941).

FORBES, T. R.: Antagonisms and synergisms in progestin: progestin mixtures. Endocrinology **75**, 799 (1964).

— HOOKER, C. W.: Inactivity of bound plasma progesterone. Proc. Soc. exp. Biol. (N. Y.) **70**, 682 (1949).

FOREST, M. G., RIVAROLA, M. A., MIGEON, C. J.: Percentage binding of testosterone, androstenedione and dehydroisoandrosterone in human plasma. Steroids **12**, 323 (1968).

FORTIER, C.: Pituitary ACTH and plasma free corticosteroids following bilateral adrenalectomy in the rat. Proc. Soc. exp. Biol. (N. Y.) **100**, 13 (1959).

FOSTER, J. F., SOGAMI, M., PETERSEN, H. A., LEONARD, W. J., JR.: The microheterogeneity of plasma albumins. I. Critical evidence for and description of the microheterogeneity model. J. biol. Chem. **240**, 2495 (1965).

FRAENKEL-CONRAT, H.: Methods for investigating the essential groups for enzyme activity. Methods in Enzymol. **4**, 247 (1957).

— COLLOMS, M.: Reactivity of tobacco mosaic virus and its protein toward acetic anhydride. Biochemistry **6**, 2740 (1967).

FRANCIS, T., JR., LING, N. S., KRASTEFF, T.: Transcholesterin, a cholesterol-binding globulin: serological demonstration of a specific interaction between cholesterol and serum globulin. Proc. nat. Acad. Sci. (Wash.) **53**, 1061 (1965).

FRANK, H. S.: The structure of water. Fed. Proc. **24**, Suppl. No. 15, S-1 (1965).

FRANKE, R.: Die hydrophobe Wechselwirkung von polycyclischen aromatischen Kohlenwasserstoffen mit Humanserumalbumin. Biochim. Biophys. Acta (Amst.) **160**, 378 (1968).

— The possible role of hydrophobic interactions of polycyclic aromatic hydrocarbons with protein in chemical carcinogenesis. Mol. Pharmacol. **5**, 640 (1969).

FRANKEL, A. I., COOK, B., GRABER, J. W., NALBANDOV, A.: Determination of corticosterone in plasma by fluorometric techniques. Endocrinology **80**, 181 (1967).

— NALBANDOV, A. V.: Effect of the evaporation of various solvents upon the stability of corticosterone-1,2-^{3}H and corticosterone-4-^{14}C. Steroids **8**, 749 (1966).

FREDRICKSON, D. S., LEVY, R. I., LEES, R. S.: Fat transport in lipoproteins—an integrated approach to mechanisms and disorders. New Engl. J. Med. **276**, 34, 94, 148, 215, 273 (1967).

FREEMAN, H. C., IDLER, D. R.: Transcortin binding of cortisol in Atlantic salmon *(salmo salar)* plasma. Gen. comp. Endocr. **7**, 37 (1966).

— — Sex hormone binding proteins. II. Isolation from serum of an elasmobranch *(Raja radiata)*. Gen. comp. Endocr. **13**, 83 (1969).

FRICK, J., KINCL, F. A.: The measurement of plasma testosterone by competitive protein-binding assay. Steroids **13**, 495 (1969).

FRIEDEMANN, U.: In: Handbuch der Biochemie des Menschen und der Tiere, vol. III. Ed.: C. OPPENHEIMER. Jena: Fischer 1909, p. 282.

FRIESEN, H. G.: Lactation induced by human placental lactogen and cortisone acetate in rabbits. Endocrinology **79**, 212 (1966).

FRITZ, G. R., KNOBIL, E.: The measurement of testosterone in plasma by competitive protein-binding analysis. Fed. Proc. **26**, 757 (1967).

FRÖHLICH, C.: Arbeitserfahrungen mit einem trägerfreien kontinuierlichen Elektrophoreseverfahren und seine Anwendung auf Bindungsuntersuchungen. Klin. Wschr. **45**, 461 (1967).

FURMAN, R. H., ALAUPOVIC, P., HOWARD, R. P.: Effects of androgens and estrogens on serum lipids and the composition and concentration of serum liproproteins in normolipemic and hyperlipidemic states. Progr. Biochem. Pharmacol. **2**, 215 (1967).

GABRILOVE, J. L., SOFFER, L. J.: The relation of the adrenal to thyroid function. J. clin. Invest. **29**, 814 (1950).

GALA, R. R., WESTPHAL, U.: Corticosteroid-binding globulin in the rat and other species. Fed. Proc. **23**, 357 (1964).

— — Corticosteroid-binding globulin in the rat: possible role in the initiation of lactation. Endocrinology **76**, 1079 (1965 a).

— — Corticosteroid-binding globulin in the rat: studies on the sex difference. Endocrinology **77**, 841 (1965 b).

— — Relationship between the pituitary gland and the corticosteroid-binding globulin in the rat. Endocrinology **78**, 277 (1966 a).

— — Influence of anterior pituitary hormones on the corticosteroid-binding globulin in the rat. Endocrinology **79**, 55 (1966 b).

— — Further studies on the corticosteroid-binding globulin in the rat: proposed endocrine control. Endocrinology **79**, 67 (1966 c).

— — Corticosteroid-binding activity in serum of mouse, rabbit and guinea pig during pregnancy and lactation: possible involvement in the initiation of lactation. Acta endocr. **55**, 47 (1967).

GANGULY, M., CARNIGHAN, R. H., WESTPHAL, U.: Steroid-protein interactions. XIV. Interaction between human α_1-acid glycoprotein and progesterone. Biochemistry **6**, 2803 (1967).

— WESTPHAL, U.: Steroid-protein interactions. XVII. Influence of solvent environment on interaction between human α_1-acid glycoprotein and progesterone. J. biol. Chem. **243**, 6130 (1968 a).

— — Unpublished.

— — Steroid-protein interactions. XX. Effect of chemical modifications on progesterone-binding affinity of α_1-acid glycoprotein. Biochim. Biophys. Acta (Amst.) **170**, 309 (1968 b).

GANONG, W. F.: The effect of ACTH on adrenal size in hypophysectomized rats after removal of one adrenal. Endocrinology **55**, 117 (1954).

GARCIA, J., HARRIS, G. W., SCHINDLER, W. J.: Vasopressin and thyroid function in the rabbit. J. Physiol. (Lond.) **170**, 487 (1964).

GARDNER, J. A., GAINSBOROUGH, H.: Studies on the cholesterol content of normal human plasma. Part II. The attraction of the proteins of plasma for sterols. Biochem. J. **21**, 141 (1927).

GARDNER, L. I.: Role of human plasma protein fractions IV and V in 17-ketosteroid transport. Bull. Johns Hopk. Hosp. **94**, 105 (1954).

GELLER, L. E., SILBERMAN, N.: The effect of chemical impurities on the stability of labelled steroids. Steroids **9**, 157 (1967).

GELOTTE, B.: Studies on gel filtration sorption properties of the bed material Sephadex. J. Chromatog. **3**, 330 (1960).

GEMZELL, C. A.: Increase in the formation and the secretion of ACTH in rats following administration of oestradiol-monobenzoate. Acta Endocr. **11**, 221 (1952).

— Blood levels of 17-hydroxycorticosteroids in normal pregnancy. J. clin. Endocr. **13**, 898 (1953).

GENAZZANI, E., DI CARLO, F., PAGNINI, G., CORIGLIANO, G. G., SANTAMARIA, R.: Competizione tra digitossina ed acidi grassi nel legame con le sieroproteine. Atti Soc. ital. Sci. vet. **22**, 273 (1968).

GINSBURG, A., CARROLL, W. R.: Some specific ion effects on the conformation and thermal stability of ribonuclease. Biochemistry **4**, 2159 (1965).

GIORGI, E. P., CROSIGNANI, P. G.: The binding "*in vitro*" of oestradiol and oestradiol glucuronide to plasma proteins. 2nd Intern. Congr. Hormonal Steroids, Milan, Italy, 1966. Excerpta Med. Found. Intern. Congr. Ser. No. 111, p. 270 (1966).

— — Competitive binding of free and conjugated oestrogens to plasma proteins. J. Endocr. **44**, 219 (1969).

GIORGIO, N. A., JR., TABACHNICK, M.: Thyroxine-protein interactions. V. Isolation and characterization of a thyroxine-binding globulin from human plasma. J. biol. Chem. **243**, 2247 (1968).

GITLIN, D., JANEWAY, C. A.: Genetic alterations in plasma proteins of man. In: The plasma proteins, vol. 2. Ed.: F. W. PUTNAM. New York and London; Academic Press 1960, p. 407.

GLENISTER, D. W., YATES, F. E.: Sex difference in the rate of disappearance of corticosterone-4-C^{14} from plasma of intact rats: further evidence for the influence of hepatic Δ^4-steroid hydrogenase activity on adrenal cortical function. Endocrinology **68**, 747 (1961).

GLICK, D., VON REDLICH, D., LEVINE, S.: Fluorometric determination of corticosterone and cortisol in 0.02—0.05 milliliters of plasma or submilligram samples of adrenal tissue. Endocrinology **74**, 653 (1964).

GOLDBERGER, R. F., ANFINSEN, C. B.: The reversible masking of amino groups in ribonuclease and its possible usefulness in the synthesis of the protein. Biochemistry **1**, 401 (1962).

GOLDIE, D. J., HASHAM, N., KEANE, P. M., PEARSON, J., WALKER, W. H. C.: Cortisol binding characteristics measured by dialysis. J. Endocrinology **43,** 127 (1969).

— — — WALKER, W. H. C.: Temperature dependence of cortisol binding to plasma proteins. Nature **217,** 852 (1968).

GOLDSTEIN, A.: The interactions of drugs and plasma proteins. Pharmacol. Rev. **1,** 102 (1949).

GOLIKOV, P. P.: A rapid method for estimating transcortin on biogels. Clin. chim. Acta **26,** 41 (1969).

GOODFRIEND, L., LEZNOFF, A., SEHON, A. H.: Antibodies to estrone-protein conjugates. III. Tissue localization of estrogens. Canad. J. Biochem. **39,** 967 (1961).

— SEHON, A. H.: Preparation of an estrone-protein conjugate. Canad. J. Biochem. **36,** 1177 (1958).

— — Antigenicity of oestrone-protein conjugates. Nature **185,** 764 (1960).

— — Antibodies to estrone-protein conjugates. I. Immunochemical studies. Canad. J. Biochem. **39,** 941 (1961 a).

— — Antibodies to estrone-protein conjugates. II. Endocrinological studies. Canad. J. Biochem. **39,** 961 (1961 b).

— — Early approaches to production, analysis, and use of steroid antisera. In: Immunological methods in steroid determination. Eds.: F. G. PÉRON and B. V. CALDWELL. New York: Appleton-Century Crofts 1970 (in press).

GOODMAN, D. S.: Preparation of human serum albumin free of long-chain fatty acids. Science **125,** 1296 (1957).

— The interaction of human serum albumin with long-chain fatty acid anions. J. Amer. chem. Soc. **80,** 3892 (1958).

GOPALAKRISHNA, E. M., COOPER, A., NORTON, D. A.: Crystal and molecular structure of 6β-bromoprogesterone. Acta Cryst. **B 25,** 639 (1969).

GORNALL, A. G., BARDAWILL, C. J., DAVID, M. M.: Determination of serum proteins by means of the biuret reaction. J. biol. Chem. **177,** 751 (1949).

GORSKI, J., TOFT, D., SHYAMALA, G., SMITH, D., NOTIDES, A.: Hormone receptors: studies on the interaction of estrogen with the uterus. Rec. Progr. Hormone Res. **24,** 45 (1968).

GOTTO, A. M., LEVY, R. I., BIRNBAUMER, M. E., FREDRICKSON, D. S.: Human serum beta lipoprotein and beta apoprotein. Nature **223,** 835 (1969).

GOTTSCHALK, A., GRAHAM, E. R. B.: The basic structure of glycoproteins. In: The proteins, second ed., vol. IV. Ed.: H. NEURATH. New York-London: Academic Press 1966, p. 95.

GRAHAM, J. M., GREEN, C.: The binding of sterols in cellular membranes. Biochem. J. **103,** 16 c (1967).

GRANT, D. L., MARTIN, W. G., ANASTASSIADIS, P. A.: Isolation and characterization of M-1 and M-2 glycoproteins from bovine, porcine and avian plasma. J. biol. Chem. **242,** 3912 (1967).

GRAY, C. H., BACHARACH, A. L. (Eds.): Hormones in blood, second edition, vol. 2. London-New York: Academic Press 1967.

GRAY, C. H., GREENAWAY, J. M., HOLNESS, N. J.: The corticosteroids. In: Hormones in blood. Eds.: C. H. GRAY and A. L. BACHARACH. New York-London: Academic Press 1961, p. 439.

— — — SHAW, D. A.: The metabolism of [4-^{14}C] cortisol in a patient with Cushing's syndrome. J. Endocr. **24**, 199 (1962).

GREENAWAY, J. M., HOLNESS, N. J.: Decomposition of 4-^{14}C cortisol on storage. J. Endocr. **18**, xiv (abstract) (1959).

GRÉGOIRE, C.: Factors involved in maintaining involution of the thymus during suckling. J. Endocr. **5**, 68 (1947).

GROB, C. A., GOLDBERG, M. W.: Darstellung künstlicher Komplex-antigene der Steroidreihe. I. Diazotierbare Ester des Δ^5-Androsten-3β,17β-diols. Helv. chim. Acta **32**, 172 (1949).

GROSS, S.: Specificities of steroid antibodies. In: Immunological methods in steroid determination. Eds.: F. G. PÉRON and B. V. CALDWELL. New York: Appleton-Century Crofts 1970 (in press).

GROSS, S. J., CAMPBELL, D. H., WEETALL, H. H.: Production of antisera to steroids coupled to proteins directly through the phenolic A-ring. Immunochemistry **5**, 55 (1968).

GUÉRIGUIAN, J., PEARLMAN, W. H.: Separation of testosterone-binding protein of human pregnancy serum from CBG. Fed. Proc. **26**, 757 (1967).

GUÉRIGUIAN, J. L., CRÉPY, O.: Méthode de dosage de la protéine sérique liant la testostérone. C. R. Acad. Sci. (Paris) **269**, 1457 (1969).

— PAVARD, J., CRÉPY, O.: Affinité de liaison du sérum de femme enceinte pour l'aldostérone. Ann. Endocr. (Paris) **30**, 211 (1969).

— PEARLMAN, W. H.: Some properties of a testosterone-binding component of human pregnancy serum. J. biol. Chem. **243**, 5226 (1968).

GUIDOLLET, J., LOUISOT, P.: Rôle de l'oestradiol et de la thyroxine dans l'augmentation du nombre de sites subcellulaires fixateurs d'hormones corticostéroides en l'absence de biosynthèse glycoprotéinique. C. R. Acad. Sci. (Paris) **269**, 659 (1969 a).

— — Biosynthèse de la transcortine. II. Rôle des oestrogènes dans l'activation de sites moléculaires préalablement masqués. Acta Endocr. **62**, 468 (1969 b).

— — Biosynthèse de la transcortine. Localisation des sites fixateurs de cortisol et de corticostérone dans les fractions sub-cytoplasmiques des cellules hépatiques du rat. Clin. chim. Acta **23**, 121 (1969 c).

GUILLEMANT, S., CORVOL, P., CRÉPY, O., MAUVAIS-JARVIS, P.: Action du cyprotérone sur la liaison plasmatique et le métabolisme de la testostérone. C. R. Acad. Sci. (Paris) **268**, 2595 (1969).

GUILLEMIN, R., CLAYTON, G. W., LIPSCOMB, H. S., SMITH, J. D.: Fluorometric measurement of rat plasma and adrenal corticosterone concentration. J. Lab. clin. Med. **53**, 830 (1959 a).

— — SMITH, J. D., LIPSCOMB, H. S.: Measurement of free corticosteroids in rat plasma: physiological validation of a method. Endocrinology **63**, 349 (1958).

GUILLEMIN, R., DEAR, W. E., LIEBELT, R. A.: Nychthemeral variations in plasma free corticosteroid levels of the rat. Proc. Soc. exp. Biol. (N. Y.) **101**, 394 (1959 b).

GULYASSY, P., DE VENUTO, F., WESTPHAL, U.: Effect of route of injection on distribution and excretion of cortisol-4-^{14}C in the rat. Proc. Soc. exp. Biol. (N. Y.) **98**, 711 (1958).

GURD, F. R. N.: Association of lipids with proteins. In: Lipid chemistry. Ed.: D. J. HANAHAN. New York-London: Wiley 1960 a, p. 208.

— Some naturally occurring lipoprotein systems. In: Lipid chemistry. Ed.: D. J. HANAHAN. New York-London: Wiley 1960 b, p. 260.

— WILCOX, P. E.: Complex formation between metallic cations and proteins, peptides, and amino acids. Advanc. Protein Chem. **11**, 311 (1956).

HAAN, J., LANG, N.: Quantitative Bestimmung der Antigen-Bindung im Antigen-Antikörperkomplex mit der Überwanderungselektrophorese. Klin. Wschr. **35**, 1013 (1957).

HAARMANN, W., HAGEMEIER, A., LENDLE, L.: Über die Bindung von Digitalisglykosiden und Digitaloiden an die Eiweißstoffe des Blutserums. Arch. exp. Path. Pharmakol. **194**, 205 (1940).

HADD, H. E., BLICKENSTAFF, R. T.: Conjugates of steroid hormones. New York: Academic Press 1969.

HAFT, D. E., ROHEIM, P. S., WHITE, A., EDER, M.: Plasma lipoprotein metabolism in perfused rat livers. I. Protein synthesis and entry into plasma. J. clin. Invest. **41**, 842 (1962).

HAGERMAN, D. D.: Monomer-trimer transition of human placenta estradiol dehydrogenase. Arch. Biochem. **134**, 196 (1969).

— VILLEE, C. A.: Separation of human placental estrogen-sensitive transhydrogenase from estradiol-17β dehydrogenase. J. biol. Chem. **234**, 2031 (1959).

— — Kinetics of nicotinamide nucleotide transhydrogenation by human placenta. Biochem. J. **95**, 150 (1965).

HAGGARD, H. W.: An accurate method of determining small amounts of ethyl ether in air, blood and other fluids, together with a determination of the coefficient of distribution of ether between air and blood at various temperatures. J. biol. Chem. **55**, 131 (1923).

HAHN, D. W., TURNER, C. W.: Effect of corticosterone and aldosterone upon milk yield in the rat. Proc. Soc. exp. Biol. (N. Y.) **121**, 1056 (1966).

HALBERG, F., ALBRECHT, P. G., BITTNER, J. J.: Corticosterone rhythm of mouse adrenal in relation to serum corticosterone and sampling. Amer. J. Physiol. **197**, 1083 (1959 a).

— PETERSON, R. E., SILBER, R. H.: Phase relations of 24-hour periodicities in blood corticosterone, mitoses in cortical adrenal parenchyma and total body activity. Endocrinology **64**, 222 (1959 b).

HALLBERG, H. C., ZORN, E. M., WIELAND, R. G.: A sensitive testosterone assay by protein binding. Steroids **12**, 241 (1968).

HALTMEYER, G. C., DENENBERG, V. H., THATCHER, J., ZARROW, M. X.: Response of the adrenal cortex of the neonatal rat after subjection to stress. Nature **212**, 1371 (1966).

HAMAGUCHI, K., GEIDUSCHEK, E. P.: The effect of electrolytes on the stability of the deoxyribonucleate helix. J. Amer. chem. Soc. **84**, 1329 (1962).

HANDOVSKY, H.: Untersuchungen über die Zusammensetzung des Blutserums und ihre Bedeutung für Giftwirkungen. Pflügers Arch. ges. Physiol. **210**, 35 (1925).

HARRIS, G. W., LEVINE, S., SCHINDLER, W. J.: Vasopressin and thyroid function in the rat: the effect of oestrogens. J. Physiol. (Lond.) **170**, 516 (1964).

HARTE, R. A., RUPLEY, J. A.: Three-dimensional pictures of molecular models — lysozyme. J. biol. Chem. **243**, 1663 (1968).

HASKINS, A. L., JR., TAUBERT, H. D.: Progesterone transportation in blood. Obstet. and Gynec. **21**, 395 (1963).

HAUROWITZ, F.: Hemoglobin, anhydrohemoglobin, and oxyhemoglobin. J. biol. Chem. **193**, 443 (1951).

— Immunochemistry and the biosynthesis of antibodies. New York-London-Sidney: Interscience Publishers 1968.

HAYES, F. N.: Solutes and solvents for liquid scintillation counting. Technical Bulletin No. 1. La Grange (Ill.): Packard Instrument Co. January, 1962.

HAYES, J. E., JR., VELICK, S. F.: Yeast alcohol dehydrogenase: molecular weight, coenzyme binding, and reaction equilibria. J. biol. Chem. **207**, 225 (1954).

HEAP, R. B., DEANESLY, R.: Progesterone in systemic blood and placentae of intact and ovariectomized pregnant guinea-pigs. J. Endocr. **34**, 417 (1966).

— — The increase in plasma progesterone levels in the pregnant guinea-pig and its possible significance. J. Reprod. Fertil. **14**, 339 (1967).

HECHTER, O., PINCUS, G.: Genesis of adrenocortical secretion. Physiol. Rev. **34**, 459 (1954).

HEFTMANN, E.: Steroid biochemistry. New York-London: Academic Press 1970.

— MOSETTIG, E.: Biochemistry of steroids. New York: Reinhold 1960.

HEIDELBERGER, C.: The relation of protein binding to hydrocarbon carcinogenesis. Ciba Found. Symp. Carcinogenesis. Eds.: G. E. W. WOLSTENHOLME and M. O'CONNER. Boston: Little, Brown 1959, p. 179.

— MOLDENHAUER, M. G.: The interaction of carcinogenic hydrocarbons with tissue constituents. IV. A quantitative study of the binding to skin proteins of several C^{14}-labeled hydrocarbons. Cancer Res. **16**, 442 (1956).

HEIDENHAIN, M.: Neue Versuche über die chemischen Umsetzungen zwischen Eiweißkörpern und Anilinfarben, insbesondere unter Benutzung der Dialyse. Arch. ges. Physiol. **96**, 440 (1903).

HEIMBURGER, N., HEIDE, K., HAUPT, H., SCHULTZE, H. E.: Bausteinanalysen von Humanserumproteinen. Clin. chim. Acta **10**, 293 (1964).

HELINSKI, D. R., YANOFSKY, C.: Genetic control of protein structure. In: The proteins, 2nd ed., vol. IV. Ed.: H. NEURATH. New York-London: Academic Press 1966, p. 1.

Herrmann, W. L., Schindl, I. K., Bondy, P. K.: Effect of estrogen on steroid levels in plasma and urine. Proc. Soc. exp. Biol. (N. Y.) **103**, 103 (1960).

Heyns, W., van Baelen, H., de Moor, P.: Study of steroid-protein binding by means of competitive adsorption: application to cortisol binding in plasma. Clin. chim. Acta **18**, 361 (1967).

— — — Study of steroid-protein binding by means of competitive adsorption: a theoretical and graphic analysis of the steroid distribution. Ann. Endocr. (Paris) **29**, 123 (1968).

— — — Study of the specificity of the steroid-binding β-globulin in human plasma. J. Endocr. **43**, 67 (1969 a).

— Verhoeven, G., van Baelen, H., de Moor, P.: Measurement of testosterone in plasma by competitive protein binding. Ann. Endocr. (Paris) **30**, 153 (1969 b).

Hirs, C. H. W., Stein, W. H., Moore, S.: The amino acid composition of ribonuclease. J. biol. Chem. **211**, 941 (1954).

Hochheuser, W., Marguth, F., Müller-Bardorff, M., Schwarz, K., Scriba, P. C., Thiele, H.: Diagnostische Bedeutung der Proteinbindung von Plasmacortisol, bestimmt durch Dextrangelfiltration. Klin. Wschr. **47**, 300 (1969).

— Müller-Bardorff, M., Thiele, H., Scriba, P. C., Schwarz, K.: Klinische Bedeutung der Trennung von proteingebundenem und sog. freiem Anteil des Plasmacortisols mit Hilfe der Dextrangelfiltration. Verh. dtsch. Ges. inn. Med. **74**, 1157 (1968).

Hoekstra, R. A.: Das Eindringen der Digitalisglykoside in die Organe. Arch. exp. Pathol. **162**, 663 (1931).

Hoffmann, W., Forbes, T. R., Westphal, U.: Biological inactivation of progesterone by interaction with corticosteroid-binding globulin and with albumin. Endocrinology **85**, 778 (1969).

— Westphal, U.: Steroid-protein interactions. XXII. Extrapolation technique for determination of corticosteroid-binding globulin by gel filtration. Analyt. Biochem. **32**, 48 (1969).

Holcomb, O. N., van Holde, R. W.: Ultracentrifugal and viscometric studies of the reversible thermal denaturation of ribonuclease. J. Phys. Chem. **66**, 1999 (1962).

Hollander, N., Chiu, Y. W.: Further purification of corticosteroid-binding protein from P 1798 lymphosarcoma. Abstract, 3rd Intern. Congr. Endocrinology, Excerpta Med. Intern. Congr. Ser. No. 157 (1968), p. 162.

— Stevens, Y. W.: Cortisol-binding protein from various tissues. Transactions 4th Meeting Internatl. Study Group for Steroid Hormones, Rome, Dec. 1969. Research on Steroids **4** (1970), in press.

Holt, P. G., Oliver, I. T.: Plasma corticosterone concentrations in the perinatal rat. Biochem. J. **108**, 339 (1968).

Hooker, C. W., Forbes, T. R.: A bio-assay for minute amounts of progesterone. Endocrinology **41**, 158 (1947).

HOOKER, C. W., FORBES, T. R.: The transport of progesterone in blood. Endocrinology **44**, 61 (1949).

HOPKINS, S. J., WORMALL, A.: Phenylisocyanate protein compounds and their immunological properties. Biochem. J. **27**, 740 (1933).

HORTON, R., KATO, T.: Studies of testosterone-binding globulin. 3rd. Internatl. Congr. Endocrinol. Excerpta Med. Intern. Congr. Ser. No. 157 (1968), p. 160.

— — SHERINS, R.: A rapid method for the estimation of testosterone in male plasma. Steroids **10**, 245 (1967).

HORWITZ, J. P., HORN, L., LOUD, A. V., BROOKS, S. C.: Liver catalysis in the association of estrogen to protein. Experientia **18**, 414 (1962).

HÜBENER, H. J., SAHRHOLZ, F. G.: 20β-Hydroxy-Steroid-Dehydrogenase. II. Darstellung und Kristallisation. Biochem. Z. **333**, 95 (1960).

— — SCHMIDT-THOMÉ, J., NESEMANN, G., JUNK, R.: 20β-Hydroxy-Steroid-Dehydrogenase, ein neues kristallines Enzym. Biochim. Biophys. Acta (Amst.) **35**, 270 (1959).

HUGGINS, C. B., PATAKI, J., HARVEY, R. G.: Geometry of carcinogenic polycyclic aromatic hydrocarbons. Proc. natl. Acad. Sci. (Wash.) **58**, 2253 (1967).

HUGHES, T. R., KLOTZ, I. M.: Analysis of metal-protein complexes. Meth. biochem. Anal. **3**, 265 (1956).

HUGHES, W. L., JR.: Protein mercaptides. Cold Spr. Harb. Symp. quant. Biol. **14**, 79 (1950).

HUMMEL, J. P., DREYER, W. J.: Measurement of protein-binding phenomena by gel filtration. Biochim. Biophys. Acta (Amst.) **63**, 530 (1962).

HUNZIKER, K., SCHMID, K.: A brief method for the identification of the α_1-acid glycoprotein variants. Anal. Biochem. **20**, 495 (1967).

IDLER, D. R., FREEMAN, H. C.: Binding of testosterone, 1α-hydroxycorticosterone and cortisol by plasma proteins of fish. Gen. comp. Endocr. **11**, 366 (1968).

— — Sex hormone binding proteins. I. Binding of steroids by serum of an elasmobranch (Raja radiata). Gen. comp. Endocr. **13**, 75 (1969).

— — TRUSCOTT, B.: Biological activity and protein binding of 1α-hydroxycorticosterone: an interrenal steroid in elasmobranch fish. Gen. comp. Endocr. **9**, 207 (1967).

IINO, S., GREER, M. A.: Thyroid function in the rat during pregnancy and lactation. Endocrinology **68**, 253 (1961).

INGBAR, S. H., FREINKEL, N.: Corticotropin, cortisone and the metabolism of iodine. Metabolism **5**, 652 (1956).

INMAN, J. K., DINTZIS, H. M.: The derivatization of cross-linked polyacrylamide beads. Controlled introduction of functional groups for the preparation of special-purpose, biochemical adsorbents. Biochemistry **8**, 4074 (1969).

ISHIGURO, M., YAMAMOTO, A., SCHMID, K.: The number and reactivity of the disulfide bonds and the amino acid sequences of some peptides containing half-cystine residues of α_1-acid glycoprotein (orosomucoid). Biochim. Biophys. Acta (Amst.) **194**, 91 (1969).

ITURZAETA, N. F., HILLMAN, D. A., COLLE, E.: Measurement of plasma cortisol in children and adults: A comparison of the double isotope derivative assay, competitive protein-binding analysis and the modified competitive protein-binding analysis. J. clin. Endocr. **30**, 185 (1970).

JACQUES, J., KAGAN, H., OURISSON, G.: Constantes sélectionnés. Pouvoir rotatoire naturel — optical rotatory poper. I a. Steroides-Steroids. Paris-New York: Pergamon Press 1965.

JANSON, J. C.: Absorption phenomena on Sephadex. J. Chromat. **28**, 12 (1967).

JARABAK, J.: Soluble 17β-hydroxysteroid dehydrogenase of human placenta. Methods Enzymol. **15**, 746 (1969).

— ADAMS, J. A., WILLIAMS-ASHMAN, H. G., TALALAY, P.: Purification of a 17β-hydroxysteroid dehydrogenase of human placenta and studies on its transhydrogenase function. J. biol. Chem. **237**, 345 (1962).

— SACK, G. H., JR.: A soluble 17β-hydroxysteroid dehydrogenase from human placenta. The binding of pyridine nucleotides and steroids. Biochemistry **8**, 2203 (1969).

— SEEDS, A. E., JR., TALALAY, P.: Reversible cold inactivation of a 17β-hydroxysteroid dehydrogenase of human placenta: protective effect of glycerol. Biochemistry **5**, 1269 (1966).

JARABAK, R., COLVIN, M., MOOLGAVKAR, S. H., TALALAY, P.: Δ^5-3-ketosteroid isomerase of *Pseudomonas testeroni*. Methods Enzymol. **15**, 642 (1969).

JEANLOZ, R. W.: α_1-Acid glycoprotein. In: Glycoproteins. Ed.: A. GOTTSCHALK. Amsterdam-London-New York: Elsevier 1966, p. 362.

JENSEN, E. V., DE SOMBRE, E. R.: Receptors and action of estradiol. Monograph in this series (1970), in preparation.

— DE SOMBRE, E. R., JUNGBLUT, P. W.: Interaction of estrogens with receptor sites *in vivo* and *in vitro*. Proc. Second Intern. Congr. Hormonal Steroids. Excerpta Med. Intern. Congr. Series, No. 132, Amsterdam-New York (1967), p. 492.

— SUZUKI, T., KAWASHIMA, T., STUMPF, W. E., JUNGBLUT, P. W., DE SOMBRE, E. R.: A two-step mechanism for the interaction of estradiol with rat uterus. Proc. natl. Acad. Sci. (Wash.) **59**, 632 (1968).

JIANG, N. S., RYAN, R. J.: Radioimmunoassay for estrogens: a preliminary communication. Mayo Clin. Proc. **44**, 461 (1969).

JIRGENSONS, B.: The Cotton effects in the optical rotatory dispersion of proteins as new criteria of conformation. J. biol. Chem. **240**, 1064 (1965).

JOHANSSON, E. D. B.: Progesterone levels in peripheral plasma during the luteal phase of the normal human menstrual cycle measured by a rapid competitive protein binding technique. Acta Endocr. **61**, 592 (1969 a).

— Plasma levels of progesterone in pregnancy measured by a rapid competitive protein binding technique. Acta Endocr. **61**, 607 (1969 b).

— A simplified procedure for the assay of progesterone. Transactions of the 2nd Karolinska Symposium on Research Methods in Reproductive Endocrinology, Steroid assay by protein binding, Geneva, March, 1970. Acta endocr. (Kbh.), Suppl. No. **147**, 188 (1970).

JOHANSSON, E. D. B., NEILL, J. D., KNOBIL, E.: Periovulatory progesterone concentration in the peripheral plasma of the rhesus monkey with a methodologic note on the detection of ovulation. Endocrinology **82**, 143 (1968).

JOHN, D. W., MILLER, L. L.: Regulation of net biosynthesis of serum albumin and acute phase plasma proteins. Induction of enhanced net synthesis of fibrinogen, α_1-acid glycoprotein, α_2 (acute phase)-globulin, and haptoglobin by amino acids and hormones during perfusion of the isolated normal rat liver. J. biol. Chem. **244**, 6134 (1969).

JOHNSON, A. M., SCHMID, K., ALPER, C. A.: Inheritance of human α_1-acid glycoprotein (orosomucoid) variants. J. clin. Invest. **48**, 2293 (1969).

JONES, I. C., PHILLIPS, J. G., HOLMES, W. N.: Comparative physiology of the adrenal cortex. In: Comparative endocrinology. Ed.: A. GORBMAN. New York: Wiley 1959, p. 582.

JONES, J. A., MASON, J. W.: A critical comparison of a chromatographic and an isotope-dialysis method of plasma 17-hydroxycorticosteroid measurement. J. clin. Endocr. **26**, 1010 (1966).

JONES, J. B., WIGFIELD, D. C.: Steroids and steroidases. V. On the mechanism of action of the Δ^5-3-ketoisomerase of *Pseudomonas testosteroni*. J. Amer. chem. Soc. **89**, 5294 (1967).

JONES, J. E., SEAL, U. S.: X-chromosome linked inheritance of elevated thyroxine-binding globulin. J. clin. Endocr. **27**, 1521 (1967).

JUNGBLUT, P. W., TURBA, F.: Vergleich von Darstellungsverfahren für Ratten-Serumalbumin und Charakterisierung eines reinen Präparates. Biochem. Z. **337**, 88 (1963).

KAGAWA, C. M.: Blocking the renal electrolyte effects of mineralocorticoids with an orally active steroidal spirolactone. Endocrinology **67**, 125 (1960).

KALOUS, V.: Some physicochemical properties of chromatographically isolated human orosomucoid. Biochim. Biophys. Acta (Amst.) **107**, 139 (1965).

— PONCOVÁ, M.: Chromatographic isolation of orosomucoid from serum and some of its physicochemical properties. Collection Czech. Chem. Commun. **30**, 737 (1965).

KARAVOLAS, H. J., ENGEL, L. L.: Human placental 17β-estradiol dehydrogenase. III. The separation of a 17β-estradiol-dependent transhydrogenase. J. biol. Chem. **241**, 3454 (1966).

— ORR, J. C., ENGEL, L. L.: Human placental 17β-estradiol dehydrogenase. IV. Differentiation of 17β-estradiol-activated transhydrogenase from the transhydrogenase function of 17β-estradiol dehydrogenase. J. biol. Chem. **244**, 4413 (1969).

KARUSH, F.: Heterogeneity of the binding sites of bovine serum albumin. J. Amer. chem. Soc. **72**, 2705 (1950).

— The study of protein binding by partition analysis. The effect of protein charge. J. Amer. chem. Soc. **73**, 1246 (1951).

KASS, E. H., HECHTER, O., MACCHI, I. A., MOU, T. W.: Changes in patterns of secretion of corticosteroids in rabbits after prolonged treatment with ACTH. Proc. Soc. exp. Biol. (N. Y.) **85**, 583 (1954).

KATO, T., HORTON, R.: Studies of testosterone binding globulin. J. clin. Endocr. **28**, 1160 (1968).

KATZ, F. H., KAPPAS, A.: The effects of estradiol and estriol on plasma levels of cortisol and thyroid hormone-binding globulins and on aldosterone and cortisol secretion rates in man. J. clin. Invest. **46**, 1768 (1967).

KAUTSKY, M. P., HAGERMAN, D. D.: 17β-Estradiol dehydrogenase of ovine ovaries. J. biol. Chem. **245**, 1978 (1970).

KAUZMANN, W.: Some factors in the interpretation of protein denaturation. Advanc. Protein Chem. **14**, 1 (1959).

KAWAHARA, F. S.: Δ^5-3-ketosteroid isomerase from *Pseudomonas testeroni*. Methods Enzymol. **5**, 527 (1962).

— TALALAY, P.: Crystalline Δ^5-3-ketosteroid isomerase. J. biol. Chem. **235**, PC 1 (1960).

— WANG, S. F., TALALAY, P.: The preparation and properties of crystalline Δ^5-3-ketosteroid isomerase. J. biol. Chem. **237**, 1500 (1962).

KAWAI, A., YATES, F. E.: Interference with feedback inhibition of adrenocorticotropin release by protein binding of corticosterone. Endocrinology **79**, 1040 (1966).

KAWASAKI, T., KOYAMA, J., YAMASHINA, I.: Isolation and characterization of α_1-acid glycoprotein from rat serum. J. Biochem. **60**, 554 (1966).

KELLER, N., RICHARDSON, U. I., YATES, F. E.: Protein binding and the biological activity of corticosteroids: *In vivo* induction of hepatic and pancreatic alanine aminotransferases by corticosteroids in normal and estrogen-treated rats. Endocrinology **84**, 49 (1969).

— SENDELBECK, L. R., RICHARDSON, U. I., MOORE, C., YATES, F. E.: Protein binding of corticosteroids in undiluted rat plasma. Endocrinology **79**, 884 (1966).

KENDALL, F. E.: Studies on human serum proteins. II. Crystallization of human serum albumin. J. biol. Chem. **138**, 97 (1941).

KENDALL, J. W., EGANS, M. L., STOTT, A. K.: Fluorometric determination of corticosteroids: An interfering substance in impure dichloromethane which fluoresces with benzyl alcohol preservative in heparin. J. clin. Endocr. **28**, 1373 (1968).

KERKAY, J.: Interaction of human α_1-acid glycoprotein with progesterone and related steroids. University of Louisville: Doctoral dissertation 1969.

— WESTPHAL, U.: Steroid-protein interactions. XIX. Complex formation between α_1-acid glycoprotein and steroid hormones. Biochim. Biophys. Acta (Amst.) **170**, 324 (1968).

— — Steroid protein interactions. XXI. Metal ion inhibition of association between progesterone and α_1-acid glycoprotein. Arch. Biochem. **129**, 480 (1969).

KERSEY, W. H., WILCOX, R. B.: Stereochemistry of hydrogen transfer by rat ovary 20α-hydroxysteroid dehydrogenase. Biochemistry **9**, 1284 (1970).

KIER, L. B.: The conformation of 20-oxopregnane hormones from molecular orbital calculations and a consideration of the cortisol receptor. J. med. Chem. **11**, 915 (1968).

KING, L. F., FRANKS, W. R.: The conjugation of estrogens with proteins. J. Amer. chem. Soc. **63**, 2042 (1941).

KIRDANI, R. Y., BURGETT, M. J.: The pK of five estrogens in aqueous solution. Arch. Biochem. **118**, 33 (1967).

KITAY, J. I.: Sex differences in adrenal cortical secretion in the rat. Endocrinology **68**, 818 (1961).

— Effects of estradiol on pituitary-adrenal function in male and female rats. Endocrinology **72**, 947 (1963 a).

— Pituitary-adrenal function in the rat after gonadectomy and gonadal hormone replacement. Endocrinology **73**, 253 (1963 b).

— Amelioration of cortisone-induced pituitary adrenal suppression by estradiol. J. clin. Endocr. **24**, 231 (1964).

— Depression of adrenal corticosterone production in oophorectomized rats. Endocrinology **77**, 1048 (1965 a).

— Effects of oophorectomy and various doses of estradiol-17β on corticosterone production by rat adrenal slices. Proc. Soc. exp. Biol. (N. Y.) **120**, 193 (1965 b).

— Effect of oestradiol in adrenal corticoidogenesis: An additional step in steroid biosynthesis. Nature **209**, 808 (1966).

— COYNE, M. D., NEWSON, W., NELSON, R.: Relation of the ovary to adrenal corticosterone production and adrenal enzyme activity in the rat. Endocrinology **77**, 902 (1965).

KLEE, W. A., RICHARDS, F. M.: The reaction of O-methylisourea with bovine pancreatic ribonuclease. J. biol. Chem. **229**, 489 (1957).

KLOPSTOCK, A., PINTO, M., RIMON, A.: Antibodies reacting with steroid haptens. J. Immunol. **92**, **515** (1964).

KLOTZ, I. M.: The application of the law of mass action to binding by proteins. Interactions with calcium. Arch. Biochem. **9**, 109 (1946 a).

— Spectrophotometric investigations of the interactions of proteins with organic anions. J. Amer. chem. Soc. **68**, 2299 (1946 b).

— The nature of some ion-protein complexes. Cold Spr. Harb. Symp. quant. Biol. **14**, 97 (1950).

— Protein interactions. In: The proteins, vol. I, B, chapter 8. Eds.: H. NEURATH and K. BAILEY. New York: Academic Press 1953, p. 727.

— Role of water structure in macromolecules. Fed. Proc. **24**, Suppl. 15, S-24 (1965).

— Energy changes in biochemical reactions. New York-London: Academic Press 1967, p. 99.

— AYERS, J.: Interactions of some neutral organic molecules with proteins. J. Amer. chem. Soc. **74**, 6178 (1952).

— BURKHARD, R. K., URQUHART, J. M.: The binding of organic ions by proteins. Comparison of bovine and human serum albumins. J. Phys. Chem. **56**, 77 (1952).

KLOTZ, I. M., LONGFELLOW, J. M.: Interactions between proteins and azosulfonamides. Science **104**, 264 (1946).

— URQUHART, J. M.: The binding of organic ions by proteins. Effect of temperature. J. Amer. chem. Soc. **71**, 847 (1949).

— WALKER, F. M.: Spectral changes in some dye ions and their relation to the protein error in indicators. J. Phys. Colloid Chem. **51**, 666 (1947 a).

— — The binding of organic ions by proteins. Charge and pH effects. J. Amer. chem. Soc. **69**, 1609 (1947 b).

— — PIVAN, R. B.: The binding of organic ions by proteins. J. Amer. chem. Soc. **68**, 1486 (1946).

KLYNE, W.: The chemistry of the steroids. London: Methuen and Co. 1957.

KNIGGE, K. M., HOAR, R. M.: A procedure for measuring plasma binding of adrenal corticoids; effect of acute and chronic stress. Proc. exp. Biol. (N. Y.) **113**, 623 (1963).

KOCH, B.: Liaisons de la corticostérone aux protéines plasmatiques et régulation du niveau du stéroide libre au cours de la gestation et de la lactation chez la ratte. Hormone Metab. Res. **1**, 129 (1969).

— LUTZ, B., MIALHE-VOLOSS, C., STUTINSKY, F.: Liaison corticostérone-albumine plasmatique et rôle des acides gras dans la détermination des paramètres de liaison de cette protéine chez le rat. Hormone Metab. Res. **1**, 183 (1969).

— MIALHE-VOLOSS, C., LUTZ, B., STUTINSKY, F.: Absence d'effet de la thyroxine sur le nombre de sites de transcortine en fin de gestation chez la ratte. C. R. Acad. Sci. (Paris) **267**, 1315 (1968 b).

— — STUTINSKY, F.: Fractions liées et libre de la corticostérone plasmatique au cours de la gestation chez la ratte. C. R. Acad. Sci. (Paris) **264**, 1183 (1967 a).

— — — Fractions liées et libre de la corticostérone plasmatique au cours du développement postnatal chez le rat. C. R. Acad. Sci. (Paris) **264**, 2497 (1967 b).

— — — Capacité de liaison de la transcortine et réponse à l'agression pendant la période post-natale chez le rat. C. R. Acad. Sci. (Paris) **266**, 808 (1968 a).

KOCHAKIAN, C. D., EVANS, W. W.: The iodine metabolism of the thyroid of androgen-treated rats. Endocrinology **58**, 279 (1956).

KOIDE, S. S.: 3α-Hydroxysteroid dehydrogenase of rat liver. Methods Enzymol. **15**, 651 (1969).

— TORRES, M. T.: Isolation of 3-oxosteroid Δ^4-Δ^5 isomerase of human term placenta. Biochim. Biophys. Acta (Amst.) **122**, 151 (1966).

KOLANOWSKI, J., PIZARRO, M. A.: Critical evaluation of competitive protein-binding radioassay for cortisol. Ann. Endocr. (Paris) **30**, 177 (1969).

KONO, T., YOSHIMI, T., MIYAKE, T.: Metabolic clearance of aldosterone, cortisol, and corticosterone in various clinical conditions. In: Steroid dynamics. Eds.: G. PINCUS, T. NAKAO and J. F. TAIT. New York-London: Academic Press 1966, p. 429.

KOPETZ, K., SCHWARZ, K., EYMER, K. P.: Das Verhalten des Plasmacorticosteroidspiegels unter der Einwirkung einer Oestrogen-androgen-Kombination. Klin. Wschr. **40**, 1181 (1962).

KORENMAN, S. G.: Radio-ligand binding assay of specific estrogens using a soluble uterine macromolecule. J. clin. Endocr. **28**, 127 (1968).

— Comparative binding affinity of estrogens and its relation to estrogenic potency. Steroids **13**, 163 (1969).

— PERRIN, L. E., MCCALLUM, T. P.: A radio-ligand binding assay system for estradiol measurement in human plasma. J. clin. Endocr. **29**, 879 (1969).

— TULCHINSKY, D., EATON, L. W., JR.: Radio-ligand procedures for estrogen assay in normal and pregnancy plasma. Transactions of 2nd Karolinska Symposium on Research Methods in Reproductive Endocrinology. Steroid assay by protein binding, Geneva, March, 1970. Acta endocrinol. (Kbh.), Suppl. No. **147**, 291 (1970).

KORNEL, L., MOORE, J. T., JR., NOYES, I.: Corticosteroids in human blood: IV. Distribution of cortisol and its metabolites between plasma and erythrocytes *in vivo*. J. clin. Endocr. **30**, 40 (1970).

KRAEMER, E., WISWELL, J. G.: Familial thyroxine-binding globulin deficiency. Metabolism **17**, 260 (1968).

KRIEGLSTEIN, J.: Zur Plasmaproteinbindung von Arzneimitteln. Klin. Wschr. **47**, 1125 (1969).

KRIPALANI, K. J., SORBY, D. L.: Binding of cortisol and its degradation products by human serum albumin. J. pharm. Sci. **56**, 687 (1967).

KROTOSKI, W. A., WEIMER, H. E.: Effects of dilute acids on the chemical and immunochemical properties of human orosomucoid. Clin. chim. Acta **13**, 329 (1966).

— — Reduction and alkylation of human plasma orosomucoid in aqueous solution and in 8 M urea: a physicochemical and immunochemical study. Human orosomucoid: reduction and alkylation. Canad. J. Biochem. **45**, 1577 (1967).

KRUM, A. A., GLENN, R. E.: Adrenal steroid secretion in rabbits following prolonged ACTH administration. Proc. Soc. exp. Biol. (N. Y.) **118**, 255 (1965).

KRÜSKEMPER, H. L., FORCHIELLI, E., RINGOLD, H. J.: Δ^5-3-keto isomerases. Specificity and inhibition studies in bovine adrenal homogenate fractions and distribution in rat tissues. Steroids **3**, 295 (1964).

KUČEROVÁ, L., HOENIG, V., JIRSA, M., FABIAN, E.: Die Bindung *in vitro* von Bromsulphthalein, Fettsäuren, Salzen der Gallensäuren und von Bilirubin an Albumin. Acta hepatosplenol. (Stuttg.) **13**, 282 (1966).

KUNKEL, H. G., TRAUTMAN, R.: α_2-Lipoproteins of human serum: correlation of ultracentrifugal and electrophoretic properties. J. clin. Invest. **35**, 641 (1956).

KUSCHINSKY, K.: Über die Bindungseigenschaften von Plasmaproteinen für Herzglykoside. Naunyn Schmiedebergs Arch. Pharmak. **262**, 388 (1969).

LABRIE, F.: Interactions hormonales et rôle de la transcortine dans l'ajustement de l'activité hypophysosurrénalienne. Sc.D. dissertation, Quebec (Canada), Laval University 1967.

Labrie, F., Pelletier, G., Fortier, C.: Role of the thyroid in estrogen-induced stimulation of pituitary-adrenocortical activity in the male rat. Fed. Proc. **26**, 484 (1967).

— — Labrie, R., Ho-Kim, M. A., Delgado, A., MacIntosh, B., Fortier, C.: Liaison transcortine-corticostérone et contrôle de l'activité hypophyso-surrénalienne chez le rat. Interactions hypophyse-thyroide-surrénales-gonades. Ann. Endocr. (Paris) **29**, 29 (1968).

— Raynaud, J. P., Fortier, C.: Relationship of thyroid activity to protein-binding and disappearance rate of corticosterone in the rat. Excerpta Med., Internatl. Congr. Series No. 99, VIth Pan American Congress of Endocrinology, Abstract No. 109 (1965).

Landsteiner, K.: Die Spezifität der serologischen Reaktionen. Berlin: Springer 1933.

— The specificity of serological reactions. Cambridge (Mass.): Harvard University Press 1945.

— Lampl, H.: Über die Abhängigkeit der serologischen Spezifität von der chemischen Struktur. (Darstellung von Antigenen mit bekannter chemischer Konstitution der spezifischen Gruppen.) XII. Mitteilung über Antigene. Biochem. Z. **86**, 343 (1918).

Lang, N.: Ein Verfahren zum Nachweis von Reaktionen vom Typ der Antigen-Antikörperbindung. Klin. Wschr. **33**, 29 (1955).

— Quantitative Untersuchungen mit Überwanderungselektrophorese über die Bindungskapazität von Plasmaproteinen. Protides of biological fluids, Proc. 6th Colloq. Bruges 1958. Amsterdam: Elsevier 1959.

Langer, L. J., Alexander, J. A., Engel, L. L.: Human placental estradiol-17β dehydrogenase. II. Kinetics and substrate specificities. J. biol. Chem. **234**, 2609 (1959).

Lata, G. F., Dac, L. K.: Steroid solubility studies with aqueous solutions of urea and ureides. Arch. Biochem. Biophys. (Amst.) **109**, 434 (1965).

Lavietes, P. H.: Anaerobic ultrafiltration. J. biol. Chem. **120**, 267 (1937).

Layne, D.: Unpublished results obtained in 1962; cited by Tait and Burstein (1964).

Layne, D. S., Meyer, C. J.: Effects of 19-norsteroids on plasma-protein binding of hormones. Metabolism **14**, 429 (1965).

— — Vaishwanar, P. S., Pincus, G.: The secretion and metabolism of cortisol and aldosterone in normal and in steroid-treated women. J. clin. Endocr. **22**, 107 (1962).

Lebeau, M. C., Baulieu, E. E.: Binding of steroid conjugates to human corticosteroid-binding globulin. J. clin. Endocr. **30**, 166 (1970).

— Mercier-Bodard, C., Olds, J., Bourquin, D., Brécy, T., Raynaud, J. P., Baulieu, E. E.: Mesure de la liaison de la corticostérone, du sulfate de corticostérone, de la testostérone et de l'oestradiol à les protéines du plasma. Ann. Endocr. (Paris) **30**, 183 (1969).

Leclercq, R., Copinschi, G., Franckson, J. R. M.: Le dosage par compétition du cortisol plasmatique. Modification de la méthode de Murphy. Rev. franç. Étud. clin. biol. **14**, 815 (1969).

LECOMTE DU NOUY, P.: Surface tension of serum. III. Recovery after lowering by surface-active substances. J. exp. Med. **36**, 115 (1922).

— Sur la capacité d'adsorption des protéines du serum vis-a-vis sels biliaires. C. R. Soc. Biol. (Paris) **99**, 1097 (1928).

LEFRESNE, P., JACQUEMIN-SABLON, A.: Interaction entre un stéroide diamine, l'Irehdiamine A, et des homopolyribonucléotides de synthèse. C. R. Acad. Sci. (Paris) **268**, 2368 (1969).

LEONARD, P. J., D'ARBELA, P. G.: The plasma concentrations of free and protein-bound cortisol in hypoproteinemia. J. Endocr. **34**, 265 (1966).

— MACWILLIAM, K. M.: Cortisol binding in the serum in Kwashiorkor. J. Endocr. **29**, 273 (1964).

LETTRÉ, H., INHOFFEN, H. H., TSCHESCHE, R.: Über Sterine, Gallensäuren und verwandte Naturstoffe. Stuttgart: Enke, vol. I 1954, vol. II 1959.

LEVEDAHL, B. H.: Testosterone binding to bovine serum. Arch. Biochem. **59**, 300 (1955).

— BERNSTEIN, H.: Testosterone binding by modified bovine serum albumin. Arch. Biochem. **52**, 353 (1954).

— PERLMUTTER, R.: Testosterone binding to bovine albumin: effect of pH. Arch. Biochem. **61**, 442 (1956).

LEVIN, Ö.: Column chromatography of proteins: calcium phosphate. Methods Enzymol. **5**, 27 (1962).

LEVINE, S., GLICK, D., NAKANE, P. K.: Adrenal and plasma corticosterone and vitamin A in rat adrenal glands during postnatal development. Endocrinology **80**, 910 (1967).

LEVY, H. R., TALALAY, P.: Bacterial oxidation of steroids. I. Ring A dehydrogenations by intact cells. J. biol. Chem. **234**, 2009 (1959 a).

— — Bacterial oxidation of steroids. II. The enzymic mechanism of ring A dehydrogenation. J. biol. Chem. **234**, 2014 (1959 b).

LEVY, R. I., FREDRICKSON, D. S.: Heterogeneity of plasma high density lipoproteins. J. clin. Invest. **44**, 426 (1965).

LEVY, R. S., LYNCH, A. C., MCGEE, E. D., MEHL, J. W.: Amino acid composition of the proteins from chylomicrons and human serum lipoproteins. J. Lipid Res. **8**, 463 (1967).

LEYBOLD, K., RIEPER, J., WEISSBECKER, L.: Enzymatisch-fluorometrische Bestimmung der Cortisolbindungskapazität des Plasmas. Acta Endocr. **56**, 99 (1967).

LI, Y. T., LI, S. C.: Isolation and characterization of α_1-acid glycoprotein from chimpanzee plasma. J. biol. Chem. **245**, 825 (1970).

LIEBERMAN, S., ERLANGER, B. F., BEISER, S. M., AGATE, F. J., JR.: Steroid-protein conjugates: their chemical, immunochemical, and endocrinological properties. Rec. Progr. Hormone Res. **15**, 165 (1959).

LINDGREN, F. T., JENSEN, L. C., WILLS, R. D., FREEMAN, N. K.: Flotation rates, molecular weights and hydrated densities of the low-density lipoproteins. Lipids **4**, 337 (1969).

— NICHOLS, A. V.: Structure and function of human serum lipoproteins. In: The plasma proteins, vol. 2. Ed.: F. W. PUTNAM. New York-London: Academic Press 1960, p. 1.

LINDNER, H. R.: Comparative aspects of cortisol transport: lack of firm binding to plasma proteins in domestic ruminants. J. Endocr. **28**, 301 (1964).

— Participation of lymph in the transport of gonadal hormones. Proc. Second Intern. Congr. on Hormonal Steroids. Excerpta Med. Found., Intern. Congr. Series No. 132, p. 821. Amsterdam-New York 1967.

— PEREL, E., FRIEDLANDER, A.: Antibodies to oestrogens. Transactions 4th Meeting of the International Study Group for Steroid Hormones, Rome, Dec. 1969. Research on Steroids **4** (1970), in press.

LING, N. S., KRASTEFF, T.: Characterization of cholesterol-binding globulin by modified zone electrophoresis and O-(diethylaminoethyl) cellulose chromatography. Proc. natl. Acad. Sci. (Wash.) **60**, 928 (1968).

— — FRANCIS, T., JR., NEFF, B. J., BRODY, G. L.: Transcholesterin titers and their biological significance in experimental atherogenesis in rats. Proc. Soc. exp. Biol. (N. Y.) Med. **128**, 300 (1968).

LIPSETT, M. B., DOERR, P., BERMUDEZ, J. A.: Saturation assays for plasma progesterone and 17-hydroxyprogesterone. Transactions of the 2nd Karolinska Symposium on Research Methods in Reproductive Endocrinology. Steroid assay by protein binding, Geneva, March, 1970. Acta endocr. (Kbh.), Suppl. No. **147**, 155 (1970).

LOHRENZ, F., DOE, R. P., SEAL, U. S.: Idiopathic or genetic elevation of corticosteroid-binding globulin? J. clin. Endocr. **28**, 1073 (1968).

— SEAL, U. S., DOE, R. P.: Adrenal function and serum concentrations in a kindred with decreased corticosteroid-binding globulin (CBG) concentration. J. clin. Endocr. **27**, 966 (1967).

LOKEN, H. F., HAVEL, R. J., GORDAN, G. S., WHITTINGTON, S. L.: Ultracentrifugal analysis of protein-bound and free calcium in human serum. J. biol. Chem. **235**, 3654 (1960).

LONGSWORTH, L. G., JACOBSEN, C. F.: An electrophoretic study of the binding of salt ions by β-lactoglobulin and bovine serum albumin. J. Phys. Colloid Chem. **53**, 126 (1949).

LOUISOT, P., GUIDOLLET, J.: Localisation des glycoprotéines fixatrices d'hormones corticostéroïdes dans les fractions subcytoplasmiques des hépatocytes. C. R. Acad. Sci. (Paris) **267**, 1063 (1968).

LOW, B. W.: The structure and configuration of amino acids, peptides and proteins. In: The proteins, vol. I, part A. Eds.: H. NEURATH and K. BAILEY. New York: Academic Press 1953, p. 235.

LOWRY, O. H., ROSEBROUGH, N. J., FARR, A. L., RANDALL, R. J.: Protein measurement with the Folin phenol reagent. J. biol. Chem. **193**, 265 (1951).

LUKAS, D. S., DE MARTINO, A. G.: Binding of digitoxin and some related cardenolides to human plasma proteins. J. clin. Invest. **48**, 1041 (1969).

MACEK, T. J., BAADE, W. H., BORNN, A., BACHER, F. A.: Observations on the solubility of some cortical hormones. Science **116**, 399 (1952).

MACHEBOEUF, M.: Phosphoaminolipins and sterols of blood serum and plasma. II. Physical chemistry of the protein fraction richest in phospholipins and sterols. Bull. Soc. Chim. biol. (Paris) **11**, 485 (1929 a); Chem. Abstr. **23**, 2743 (1929).

MACHEBOEUF, M.: Researches on the phosphoamino lipoids and the sterols of blood plasma and serum. C. R. Acad. Sci. (Paris) **188**, 109 (1929 b); Chem. Abstr. **24**, 1653 (1930).

MAEDA, R., OKAMOTO, M., WEGIENKA, L. C., FORSHAM, P. H.: A clinically useful method for plasma testosterone determination. Steroids **13**, 83 (1969).

MAHLER, H. R., BAYLOR, M. B.: Effects of steroidal diamines on DNA duplication and mutagenesis. Proc. natl. Acad. Sci. (Wash.) **58**, 256 (1967).

— GOUTAREL, R., KHUONG-HUU, Q., HO, M. T.: Nucleic acid interactions. VI. Effects of steroidal diamines. Biochemistry **5**, 2177 (1966).

MAICKEL, R. P., MILLER, F. P., BRODIE, B. B.: Interaction of non-steroidal antiinflammatory agents with corticosterone binding to plasma proteins in the rat. Arzneimittel-Forsch. **19**, 1803 (1969).

MALHOTRA, S. K., RINGOLD, H. J.: Chemistry of conjugate anions and enols. V. Stereochemistry, kinetics, and mechanism of the acid- and enzymatic-catalyzed isomerization of Δ^5-3-ketosteroids. J. Amer. chem. Soc. **87**, 3228 (1965).

MALINOW, M. R., KORZON, W.: An experimental method for obtaining an ultrafiltrate of the blood. J. Lab. clin. Med. **32**, 461 (1947).

MANDELKERN, L., ROBERTS, D. E.: The effect of lithium bromide on the structural transition of ribonuclease in solution. J. Amer. chem. Soc. **83**, 4292 (1961).

MANGOLD, H. K., SAND, D. M.: The stability of ^{14}C-labelled fatty acids during prolonged storage. Biochim. Biophys. Acta (Amst.) **164**, 124 (1968).

MARCUS, P. I., TALALAY, P.: On the molecular specificity of steroid-enzyme combinations: The kinetics of β-hydroxysteroid dehydrogenase. Proc. Roy. Soc. (Lond.) **144 B**, 116 (1955).

— — Induction and purification of α- and β-hydroxysteroid dehydrogenases. J. biol. Chem. **218**, 661 (1956).

MARINIS, S., OTT, H.: Natürliche und künstliche Polymeren von Plasmaproteinen (Untersuchungen in Acrylamid-gel). In: Protides of the biological fluids. Ed.: H. PEETERS. Amsterdam: Elsevier 1965, p. 420.

MARKS, L. J., BENJAMIN, G., DUNCAN, F. J., O'SULLIVAN, J. V. I.: Comparative effects of ethinyl estradiol, 17α-ethyl-19-nortestosterone and methyltestosterone on the plasma clearance of infused cortisol. J. clin. Endocr. **21**, 826 (1961 a).

— FRIEDMAN, G. R., DUNCAN, F. J.: Effect of estrogen administration on hydrocortisone metabolism in man. J. Lab. clin. Med. **57**, 47 (1961 b).

MARSDEN, N. V. B.: Solute behavior in tightly cross-linked dextran gels. Ann. N. Y. Acad. Sci. **125**, 428 (1965).

MARSHALL, W. E.: The structure of glycoproteins. II. The separation and characterization of the polymorphic variants of the α_1-acid glycoprotein of human serum. J. biol. Chem. **241**, 4731 (1966).

— PORATH, J.: The structure of glycoproteins. I. The nature and number of oligosaccharide side chains of the α_1-acid glycoprotein, ceruloplasmin, and α_2-globulin, and a β-globulin of human serum. J. biol. Chem. **240**, 209 (1965).

Martin, B. T., Cooke, B. A., Black, W. P.: Evaluation of a competitive protein-binding method for determining progesterone in peripheral plasma. J. Endocr. **45**, iv (1969).

Martini, L., Pecile, A. (Eds.): Hormonal steroids. Biochemistry, Pharmacology and Therapeutics, vols. 1 and 2. New York-London: Academic Press 1964.

Mathews, A. P.: Absorption. Physiol. Rev. **1**, 553 (1921).

Matsui, N., Plager, J. E.: *In vitro* physiological activity of protein-bound and unbound cortisol. Endocrinology **78**, 1159 (1966).

Mautner, H. G.: The molecular basis of drug action. Pharmacol. Rev. **19**, 107 (1967).

Mayes, D., Nugent, C. A.: Determination of plasma testosterone by the use of competitive protein binding. J. clin. Endocr. **28**, 1169 (1968).

— — Plasma estradiol determined with a competitive protein binding method. Steroids **15**, 389 (1970).

McCarthy, J. L., Corley, R. C., Zarrow, M. X.: Diurnal rhythm in plasma corticosterone and lack of diurnal rhythm in plasma compound F-like material in the rat. Proc. Soc. exp. Biol. (N. Y.) **104**, 787 (1960).

McDonagh, R. P., Jr.: An investigation of the effects of cold exposure on the corticosteroid binding capacity of rat serum. California State College at Los Angeles: Master thesis 1966.

McElroy, W. D.: The application of the theory of absolute reaction rates to the action of narcotics. J. Cellular Comp. Physiol. **21**, 95 (1943).

McLean, F. C., Hastings, A. B.: The state of calcium in the fluids of the body. I. The conditions affecting the ionization of calcium. J. biol. Chem. **108**, 285 (1935).

McMenamy, R. H.: Binding studies by dialysis equilibrium. A description of an accurate and rapid technique. Analyt. Biochem. **23**, 122 (1968).

— Madeja, M. I., Watson, F.: Effects of various salts on the binding of indole compounds with bovine albumin. J. biol. Chem. **243**, 2625 (1968).

Meade, R. C., Stiglitz, R. A.: Improved solvent systems for liquid scintillation counting of body fluids and tissues. Int. J. appl. Radiat. **13**, 11 (1962).

Meigs, R. A., Ryan, K. J.: 16α-Hydroxysteroid dehydrogenase from rat kidney. Methods Enzymol. **15**, 753 (1969).

Meinzinger, E.: Versuche zur Darstellung von Steroidderivaten, die zur Bindung an Eiweiß geeignet sind. Universität Tübingen: Doctoral Dissertation 1949.

Meites, J., Hopkins, T. F., Talwalker, P. K.: Induction of lactation in pregnant rabbits with prolactin, cortisol acetate or both. Endocrinology **73**, 261 (1963).

Mercier, C.: Specificity of plasma testosterone-binding globulin. 2nd Internatl. Congr. on Hormonal Steroids, Milan, Italy, 1966. Excerpta Med. Intern. Congr. Ser. No. 111, p. 269 (1966).

MERCIER, C., ALFSEN, A., BAULIEU, E. E.: A testosterone binding globulin. Proc. 2nd Symp. on Steroid Hormones, Ghent, Belgium, 1965, Excerpta Med. Intern. Congr. Ser. No. 101, p. 212 (1966).
— — — Protéines liant la testostérone dans le sérum de femme enceinte. C. R. Acad. Sci. (Paris) **264**, 122 (1967).
— BAULIEU, E. E.: Récentes études des protéines plasmatiques liant la testostérone. Ann. Endocr. (Paris) **29**, 159 (1968).
MERCIER-BODARD, C., ALFSEN, A., BAULIEU, E. E.: Sex steroid binding plasma protein (SBP). Transactions of the 2nd Karolinska Symposium on Research Methods in Reproductive Endocrinology. Steroid assay by protein binding, Geneva, March, 1970. Acta endocr. (Kbh.), Suppl. No. **147**, 204 (1970).
— BAULIEU, E. E.: Affinité pour l'oestradiol de la protéine plasmatique de liaison de la testostérone. C. R. Acad. Sci. (Paris) **267**, 804 (1968).
MEYER, C. J.: Increase of free cortisol in plasmas of patients with toxemia of pregnancy. Steroids Suppl. **I**, 199 (1965).
— LAYNE, D. S., TAIT, J. F., PINCUS, G.: The binding of aldosterone to plasma proteins in normal, pregnant and steroid-treated women. J. clin. Invest. **40**, 1663 (1961).
MEYER, M. C., GUTTMAN, D. E.: The binding of drugs by plasma proteins. J. pharm. Sci. **57**, 895 (1968).
MICHAELIS, L.: Elektrische Überführung von Fermenten. I. Das Invertin. Biochem. Z. **16**, 81 (1909).
MICHON, J.: Preparation of orosomucoid (α_1-acid seromucoid). Nature **193**, 1077 (1962).
MIDDO, R. T., ANDO, S., SIEGEL, L., GOLD, E. M.: Protein bound cortisol: estimation by Sephadex gel-filtration. Clin. Res. **11**, 86 (1963).
MIDGLEY, A. R., JR.: Discussion to paper by MIDGLEY and NISWENDER (1970).
— NISWENDER, G. D.: Radioimmunoassay of steroids. Transactions of the 2nd Karolinska Symposium on Research Methods in Reproductive Endocrinology. Steroid assay by protein binding, Geneva, March, 1970. Acta endocr. (Kbh.), Suppl. No. **147**, 320 (1970).
— — RAM, J. S.: Hapten-radioimmunoassay: a general procedure for the estimation of steroidal and other haptenic substances. Steroids **13**, 731 (1969).
MIGEON, C. J., BERTRAND, J., WALL, P. E.: Physiological disposition of 4-C^{14}-cortisol during late pregnancy. J. clin. Invest. **36**, 1350 (1957).
— LESCURE, O., ANTONIADES, H. N.: Further *in vitro* studies with 16-C^{14}-estrone: distribution between plasma protein fractions and red blood cells of man. Bull. Johns Hopk. Hosp. **106**, 317 (1960).
MIKHAIL, G., FERIN, M., VANDE WIELE, R. L.: Radioimmunoassay of estrogens by polymerization. In: Immunological methods in steroid determination. Eds.: F. G. PÉRON and B. V. CALDWELL. New York: Appleton-Century Crofts 1970 c (in press).
— WU, C. H., FERIN, M., VANDE WIELE, R. L.: Radioimmunoassay of estrone and estradiol. Transactions of the 2nd Karolinska Symposium on

Research Methods in Reproductive Endocrinology. Steroid assay by protein binding, Geneva, March, 1970. Acta endocr. (Kbh.), Suppl. No. **147,** 347 (1970 a).

— — — — Radioimmunoassay of plasma estrone and estradiol. Steroids **15,** 333 (1970 b).

MILGROM, E., BAULIEU, E. E.: Liaison spécifique de la progestérone à une protéine dans l'uterus. C. R. Acad. Sci. (Paris) **267,** 2005 (1968).

— — A method for studying binding proteins, based upon differential dissociation of small ligand. Biochim. Biophys. Acta (Amst.) **194,** 602 (1969).

MILLER, E. C.: Studies on the formation of protein-bound derivatives of 3,4-benzpyrene in the epidermal fraction of mouse skin. Cancer Res. **11,** 100 (1951).

— MILLER, J. A.: Mechanisms of chemical carcinogenesis: Nature of proximate carcinogens and interactions with macromolecules. Pharmacol. Rev. **18,** 805 (1966).

MILLS, G. L., TAYLAUR, C. E.: The distribution of cholesterol in the low-density lipoproteins of human serum. Clin. chim. Acta **22,** 251 (1968).

MILLS, I. H.: Discussion of PETERSON (1959), p. 261.

— The transport state of steroid hormones. Soc. Endocrinol., Memoir No. 11, p. 81 (1961).

— Transport and metabolism of steroids. Brit. med. Bull. **18,** 127 (1962).

— BARTTER, F. C.: The effect of oestrogen administration on the protein binding and metabolism of hydrocortisone. J. Endocr. **18,** V (abstract) (1959).

— BROOKS, R. V.: Interaction of steroid hormones. I. Estrogen effects on cortisol metabolism. In: Hormonal steroids, vol. 1. Eds.: L. MARTINI and A. PECILE. New York-London: Academic Press 1964, p. 441.

— CHEN, P. S., BARTTER, F. C.: The protein-binding of steroids as studied by ultrafiltration. J. Endocr. **18,** XXX (1959).

— SCHEDL, H. P., CHEN, P. S., BARTTER, F. C.: The effect of estrogen administration on the metabolism and protein binding of hydrocortisone. J. clin. Endocr. **20,** 515 (1960).

MOHAMMADZADEH-K., A., FEENEY, R. E., SAMUELS, R. B., SMITH, L. M.: Solubility of alkanes in protein solutions. Biochim. Biophys. Acta (Amst.) **147,** 583 (1967).

— — SMITH, L. M.: Hydrophobic binding of hydrocarbons by proteins. I. Relationship of hydrocarbon structure. Biochim. Biophys. Acta (Amst.) **194,** 246 (1969 a).

— SMITH, L. M., FEENEY, R. E.: Hydrophobic binding of hydrocarbons by proteins. II. Relationship of protein structure. Biochim. Biophys. Acta (Amst.) **194,** 256 (1969 b).

MOLINARI, G., LATA, G. F.: Interaction of steroids with some pyrimidine and purine derivatives. Arch. Biochem. **96,** 486 (1962).

MOLNAR, J., ROBINSON, G. B., WINZLER, R. J.: The biosynthesis of glycoproteins. III. Glucosamine intermediates in plasma glycoprotein synthesis in livers of puromycin-treated rats. J. biol. Chem. **239,** 3157 (1964).

Molnar, J., Robinson, G. B., Winzler, R. J.: Biosynthesis of glycoproteins. IV. The subcellular sites of incorporation of glucosamine-1-^{14}C into glycoprotein in rat liver. J. biol. Chem. **240**, 1882 (1965).

Monder, C.: Interactions of nicotinamide adenine dinucleotide and its derivatives with steroids. Biochemistry **2**, 684 (1963).

— Stability of corticosteroids in aqueous solutions. Endocrinology **82**, 318 (1968).

— Furfine, C. S.: 21-Hydroxysteroid dehydrogenases of liver and adrenal. Methods Enzymol. **15**, 667 (1969).

— Martinson, F. A. E.: Development of the 21-hydroxysteroid dehydrogenases of rat and bovine liver. Endocrinology **84**, 149 (1969 a).

— — Partial purification of a NADPH-dependent 21-hydroxysteroid dehydrogenase from human placenta. Biochim. Biophys. Acta (Amst.) **171**, 217 (1969 b).

— Walker, M. C.: Identification of 11β,17α-dihydroxy-3,20-dioxo-4-pregnen-21-al-1,2-^{3}H (21-dehydrocortisol-1,2-^{3}H) and 11β-hydroxy-4-androstene-3,17-dione-1,2-^{3}H as contaminants in preparations of cortisol-1,2-^{3}H. Steroids **15**, 1 (1970).

Money, W. L.: The interrelation of the thyroid and the adrenals. Brookhaven Symp. Biol. **7**, 137 (1955).

— Kraintz, L., Fager, J., Kirschner, L., Rawson, R. W.: The effects of various steroids on the collection of radioactive iodine by the thyroid gland of the rat. Endocrinology **48**, 682 (1951).

Moore, B., Roaf, H. E.: On certain physical and chemical properties of solutions of chloroform in water, saline, serum and hemoglobin. A contribution to the chemistry of anesthesia. Proc. Roy. Soc. B, **73**, 382 (1904).

— — On certain physical and chemical properties of solutions of chloroform and other anesthetics. A contribution to the chemistry of anesthesia. Proc. Roy. Soc. B, **77**, 86 (1905).

Moore, S., Stein, W. H.: Chromatographic determination of amino acids by the use of automatic recording equipment. Methods Enzymol. **6**, 819 (1963).

Moore, W. E., Foster, J. F., The microheterogeneity of plasma albumins. IV. Evidence from reversible denaturation that three-dimensional folding is not responsible for microheterogeneity. Biochemistry **7**, 3409 (1968).

Mooser, H., Grilichess, R. K.: Immunisierungsversuche mit Steroiden. Schweiz. Z. Path. **4**, 375 (1941).

Mosebach, K. O., Kempfle, M.: Einfluß von Effektoren auf Molekulargewicht und Aktivität der Glutamat-Dehydrogenase (GluDH) aus Rinderleber *in vitro*. I. Das Molekulargewicht der GluDH und ihrer Untereinheiten. Z. Naturforsch. **24 b**, 580 (1969).

Moshang, T., Jr., Rudd, B. T., Eberlein, W. R., Bongiovanni, A. M.: The measurement of plasma Δ^4-androstenedione by competitive protein binding. Steroids **15**, 195 (1970).

MOUSSERON-CANET, M., HOUGH, D., MANI, J. C., ARNAUD, M.: Réactivité de quelques stéroïdes fluorés ou insaturés vis-à-vis de la 17β-hydroxy-stéroïde NAD oxydoréductase de placenta humain. C. R. Acad. Sci. (Paris) **269**, 2445 (1969).

MOXHAM, A., NABARRO, J. D. N.: The measurement of oestradiol and testosterone in urine by protein-binding. J. Endocr. **46**, ii (1970).

MULDOON, T. G.: Isolation and characterization of corticosteroid-binding globulin (transcortin) from human plasma. University of Louisville: Doctoral dissertation 1967.

— WARREN, J. C.: Characterization of steroid-binding sites by affinity labeling. II. Biological activity of 4-mercuri-17β-estradiol. J. biol. Chem. **244**, 5430 (1969).

— WESTPHAL, U.: Steroid-protein interactions. XV. Isolation and characterization of corticosteroid-binding globulin from human plasma. J. biol. Chem. **242**, 5636 (1967).

MÜLLER-BARDORFF, M., HOCHHEUSER, W., SCHWARZ, K., SCRIBA, P. C.: Effect of ACTH infusion tests on cortisol-4-C^{14} binding to plasma proteins. Acta Endocr. (Kbh.) Suppl. **119**, 180 (1967).

MUNCK, A., SCOTT, J. F., ENGEL, L. L.: The interaction of steroid hormones and coenzyme components. Biochim. Biophys. Acta (Amst.) **26**, 397 (1957).

MURPHY, B. E. P.: Application of the property of protein-binding to the assay of minute quantities of hormones and other substances. Nature **201**, 679 (1964).

— The determination of plasma corticoids and their diagnostic significance. In: An introduction to clinical neuroendocrinology. Ed.: E. BAJUSZ. Basel-New York: Karger 1967 a.

— Studies of the protein-binding of testosterone and estrogens in human plasma. 49th Meeting Endocrine Soc. Program, p. 83, Abstract No. 110 (1967 b).

— Some studies of the protein-binding of steroids and their application to the routine micro and ultramicro measurement of various steroids in body fluids by competitive protein-binding radioassay. J. clin. Endocr. **27**, 973 (1967 c).

— Clinical evaluation of urinary cortisol determinations by competitive protein-binding radioassay. J. clin. Endocr. **28**, 343 (1968 a).

— Binding of testosterone and estradiol in plasma. Canad. J. Biochem. **46**, 299 (1968 b).

— Protein binding and the assay of nonantigenic hormones. Rec. Progr. Hormone Res. **25**, 563 (1969).

— Methodological problems in competitive protein-binding techniques. The use of Sephadex column chromatography to separate steroids. Transactions of the 2nd Karolinska Symposium on Research Methods in Reproductive Endocrinology. Steroid assay by protein binding, Geneva, March, 1970. Acta endocr. (Kbh.), Suppl. No. **147**, 39 (1970).

— ENGELBERG, W., PATTEE, C. J.: Simple method for the determination of plasma corticoids. J. clin. Endocr. **23**, 293 (1963).

MURPHY, B. E. P., HOOD, A. B., PATTEE, C. J.: Clinical studies utilizing a new method for the serial determination of plasma corticoids. Canad. Med. Ass. J. **90**, 775 (1964).

— PATTEE, C. J.: A study of the binding capacity of corticosteroid-binding globulin in plasma. J. clin. Endocr. **23**, 459 (1963).

— — Determination of plasma corticoids by competitive protein-binding analysis using gel filtration. J. clin. Endocr. **24**, 919 (1964).

MURRAY, D.: Cortisol binding to plasma proteins in man in health, stress and at death. J. Endocr. **39**, 571 (1967).

MUSA, B. U., DOE, R., SEAL, U. S.: Species variations in the plasma corticosteroid-binding globulin elevation during pregnancy and estrogen administration. Excerpta Med. Internatl. Congr. Ser. No. 99, VIth Pan American Congress of Endocrinology Abstract No. 119 (1965 a).

— — — Serum protein alterations produced in women by synthetic estrogens. J. clin. Endocr. **27**, 1463 (1967 b).

— SEAL, U. S., DOE, R. P.: Elevation of certain plasma proteins in man following estrogen administration: A dose-response relationship. J. clin. Endocr. **25**, 1163 (1965 b).

— — — Excretion of corticosteroid-binding globulin, thyroxine-binding globulin and total protein in adult males with nephrosis: Effects of sex hormones. J. clin. Endocr. **27**, 768 (1967 a).

NAKAJIMA, H., MURATA, M., NAKATA, M., NARUSE, T., KUBO, S.: Determination of blood corticoids using the *in vitro* resin uptake of ^{3}H-prednisolone. Acta endocr. **57**, 23 (1968).

NEILL, J. D., JOHANSSON, E. D. B., DATTA, J. K., KNOBIL, E.: Relationship between the plasma levels of luteinizing hormone and progesterone during the normal menstrual cycle. J. clin. Endocr. **27**, 1167 (1967 a).

— — KNOBIL, E.: Levels of progesterone in peripheral plasma during the menstrual cycle of the rhesus monkey. Endocrinology **81**, 1161 (1967 b).

NELSON, D. H., TANNEY, H., MESTMAN, G., GIESCHEN, V., WILSON, L. D.: Potentiation of the biologic effect of administered cortisol by estrogen treatment. J. clin. Endocr. **23**, 261 (1963).

NÉMETHY, G., SCHERAGA, H. A.: The structure of water and hydrophobic bonding in proteins. III. The thermodynamic properties of hydrophobic bonds in proteins. J. phys. Chem. **66**, 1773 (1962).

NERI, R. O., TOLKSDORF, S., BEISER, S. M., ERLANGER, B. F., AGATE, F. J., JR., LIEBERMAN, S.: Further studies on the biological effects of passive immunization with antibodies to steroid-protein conjugates. Endocrinology **74**, 593 (1964).

NEUDERT, W., RÖPKE, H.: Atlas of steroid spectra. New York: Springer 1965.

NICHOLS, A. V.: Human serum lipoproteins and their interrelations. Advanc. biol. med. Phys. **11**, 109 (1967).

NICOLOFF, J. T., DOWLING, J. T., PATTON, D. D.: Inheritance of decreased thyroxine-binding by the thyroxine-binding globulin. J. clin. Endocr. **24**, 294 (1964).

NIESCHLAG, E., ISMAIL, A. A. A.: Bestimmung der Tagesschwankungen des Plasma-Testosterons normaler Männer durch kompetitive Proteinbindung. Klin. Wschr. **48**, 53 (1970 a).

— — Diurnal variations of plasma testosterone in normal and pathological conditions as measured by the technique of competitive protein binding. J. Endocr. **46**, iii (1970 b).

NIKOLAI, T. F., SEAL, U. S.: X-chromosome linked familial decrease in thyroxine-binding globulin activity. J. clin. Endocr. **26**, 835 (1966).

— — X-chromosome linked inheritance of thyroxine-binding globulin deficiency. J. clin. Endocr. **27**, 1515 (1967).

NISWENDER, G. D., MIDGLEY, A. R., JR.: Hapten-radioimmunoassay of steroid hormones. In: Immunological methods in steroid determination. Eds.: F. G. PÉRON and B. V. CALDWELL. New York: Appleton-Century Crofts 1970 (in press).

NOMOTO, M., NARAHASHI, Y., MURAKAMI, M.: A proteolytic enzyme of streptomyces griseus. VII. Substrate specificity of streptomyces griseus protease. J. Biochem. (Tokyo) **48**, 906 (1960).

NUGENT, C. A., MAYES, D. M.: Plasma corticosteroids determined by use of corticosteroid-binding globulin and dextran-coated charcoal. J. clin. Endocr. **26**, 1116 (1966).

— — Reliability of plasma testosterone assays by competitive protein binding methods. Transactions of the 2nd Karolinska Symposium on Research Methods in Reproductive Endocrinology. Steroid assay by protein binding, Geneva, March, 1970. Acta endocrinol. (Kbh.), Suppl. No. **147**, 257 (1970).

O'CONNELL, M., WELSH, G. W.: Unbound plasma cortisol in pregnant and Enovid-E treated women as determined by ultrafiltration. J. clin. Endocr. **29**, 563, (1969).

OERTEL, G. W.: Über Steroidkonjugate in Plasma. I. Isolierung von Sulfat-, Phosphat- und Lipoid-haltigen 17-Ketosteroidkonjugaten aus menschlichem Blutplasma. Biochem. Z. **334**, 431 (1961).

— On steroid conjugates in plasma. VIII. Extraction of 17-ketosteroid conjugates. Clin. chim. Acta **8**, 154 (1963).

— On conjugation of corticosteroids in plasma. In: Structure and metabolism of corticosteroids. Eds.: J. R. PASQUALINI and M. F. JAYLE. London-New York: Academic Press 1964, p. 65.

— GROOT, K., BRÜHL, P.: Über Steroid-Konjugate in Plasma. XIV. Zur Bindung von sulfokonjugierten Steroiden an Plasmaproteine. Hoppe-Seylers Z. physiol. Chem. **341**, 10 (1965).

— KAISER, E.: Conjugated steroids in the plasma. IV. Chromatography of conjugated steroids on ion exchangers and their use in the isolation of conjugated 17-ketosteroids from peripheral plasma. Biochem. Z. **336**, 10 (1962).

— — BRÜHL, P.: Über Steroidconjugate in Plasma. V. Untersuchungen über die Bindung von 17-Ketosteroidconjugaten an spezifische Plasmaproteine. Biochem. Z. **336**, 154 (1962).

OERTEL, G. W., KNAPSTEIN, P., HEIDE, K.: Distribution of C_{19}-steroid conjugates in plasma protein fractions. Experientia **25**, 1192 (1969).

OHTSUKA, E., KOIDE, S. S.: Binding of cortisol and 21-dehydrocortisol with transcortin. Endocrinology **83**, 999 (1968).

OLEINICK, N. L., KORITZ, S. B.: Studies on the mechanism of action of the Δ^5-3-ketosteroid isomerase from rat adrenal small particles. Biochemistry **5**, 3400 (1966).

ONCLEY, J. L.: The lipoproteins of human plasma. In: The lipoproteins, methods and clinical significance. Eds.: F. HOMBURGER and P. BERNFELD. Basel: Karger 1958, p. 14.

— WALTON, K. W., CORNWELL, D. G.: A rapid method for the bulk isolation of β-lipoprotein from human plasma. J. Amer. chem. Soc. **79**, 4666 (1957).

OPIEŃSKA-BLAUTH, J., CHARĘZIŃSKI, M., BERBEĆ, H.: A new, rapid method of determining tryptophan. Analyt. Biochem. **6**, 69 (1963).

OPPENHEIMER, E.: Zur Frage der Fixation der Digitaliskörper im tierischen Organismus und besonders deren Verhalten zum Blut. Biochem. Z. **55**, 134 (1913).

ORCUTT, F. S., SEEVERS, M. H.: The solubility coefficients of cyclopropane for water, oils and human blood. J. Pharmacol. exp. Ther. **59**, 206 (1937).

OSBORNE, W. A.: Intracellular colloidal salts. J. Physiol. **34**, 84 (1906).

OSORIO, C., SCHATS, D. L.: Determination of the binding capacity of corticosteroid-binding globulin by paper electrophoresis. J. clin. Endocr. **24**, 1067 (1964).

OSUGA, D. T., FEENEY, R. E.: Electrophoretic studies of interactions between proteolytic enzymes and inhibitors. Arch. Biochem. **118**, 340 (1967).

OYAKAWA, E. K., LEVEDAHL, B. H.: Testosterone binding to bovine and human serum albumin: The role of tyrosine groups. Arch. Biochem. **74**, 17 (1958).

PAGNINI, G., CORIGLIANO, G. G., DI CARLO, R., SANTAMARIA, R., GENAZZANI, E.: Influenza dell' acido oleico sull' interazione di alcuni glucosidi cardioattivi con sieroalbumina, con actina e con miosina cardiache. Bull. Soc. ital. Biol. Sper. **44**, 2003 (1968).

PAOLETTI, J., PAOLETTI, C.: Étude de l'interaction d'une diamine stéroidique, l'Irehdiamine A, et des acides ribonucléiques de transfert. C. R. Acad. Sci. (Paris) **268**, 2372 (1969).

PARR, M., KOIDE, S. S.: Cortisol and melting of human placental DNA. Endocrinology **84**, 683 (1969).

PATERSON, J. Y. F., HILLS, F.: The binding of cortisol by ovine plasma proteins. J. Endocrinology **37**, 261 (1967).

PAULY, H.: Über die Konstitution des Histidins. I. Mitteilung. Z. Physiol. Chem. **42**, 508 (1904).

PEARCE, E. M., DE VENUTO, F., FITCH, W. M., FIRSCHEIN, H. E., WESTPHAL, U.: Rapid determination of radiocarbon in animal tissues. Anal. Chem. **28**, 1762 (1956).

PEARLMAN, W. H.: Measurement of testosterone binding sites. Transactions of 2nd Karolinska Symposium on Research Methods in Reproductive Endocrinology. Steroid assay by protein binding, Geneva, March, 1970. Acta endocr. (Kbh.), Suppl. No. **147,** 225 (1970).

— CRÉPY, O.: Steroid-protein interaction with particular reference to testosterone-binding by human serum. 2nd Internatl. Congr. Hormonal Steroids, Milan, May, 1966. Excerpta Med. Intern. Congr. Ser. No. 111, p. 48.

— — Steroid-protein interaction with particular reference to testosterone binding by human serum. Proc. 2nd Internatl. Congr. Hormonal Steroids, Milano 1966, p. 516 (1967 a).

— — Steroid-protein interaction with particular reference to testosterone binding by human serum. J. biol. Chem. **242,** 182 (1967 b).

— — MURPHY, M.: Testosterone-binding levels in the serum of women during the normal menstrual cycle, pregnancy, and the post-partum period. J. clin. Endocr. **27,** 1012 (1967).

— FONG, I. F. F.: A testosterone and 17β-estradiol binding factor in human pregnancy serum. Fed. Proc. **27,** 624 (1968).

— — TOU, J. H.: A. further study of a testosterone-binding component of human pregnancy serum. J. biol. Chem. **244,** 1373 (1969).

PEARSON, J., KEANE, P. M., WALKER, W. H. C.: Binding of cortisol to human albumin. Nature **216,** 1334 (1967).

PEETS, E. A., STAUB, M., SYMCHOWICZ, S.: Plasma binding of betamethasone-^{3}H, dexamethasone-^{3}H, and cortisol-^{14}C—a comparative study. Biochem. Pharmacol. **18,** 1655 (1969).

PEGG, P. J., KEANE, P. M.: The simultaneous estimation of plasma cortisol and transcortin binding characteristics by a competitive protein binding technique. Steroids **14,** 705 (1969).

PELLETIER, G., FORTIER, C., LABRIE, F.: Different relationship of thyroid activity to transcortin binding of corticosteroids in rat and man. Abstract, 3rd Intern. Congr. Endocrinology, Excerpta Med., Intern. Congr. Ser. No. 157, p. 4 (1968).

PÉRON, F. G.: Adrenocorticosteroids. In: Methods in hormone research. Ed.: R. I. DORFMAN. New York-London: Academic Press 1962, p. 199.

— CALDWELL, B. V. (Eds.): Worcester Foundation Symposium. Immunological methods in steroid determination. New York: Appleton-Century-Crofts 1970 (in press).

PETERSEN, H. A., FOSTER, J.: The microheterogeneity of plasma albumins. II. Preparation and solubility properties of subfractions. J. biol. Chem. **240,** 2503 (1965).

PETERSON, E. A., SOBER, H. A.: Column chromatography of proteins: substituted celluloses. Methods Enzymol. **5,** 3 (1962).

PETERSON, P. E., CHEVLI, D. M.: Group effect traverses eight-carbon chain. Report on 154th meeting of the Amer. chem. Soc., Chem. Eng. News, Sept. 18, 1967, p. 47.

PETERSON, R. E.: The miscible pool and turnover rate of adrenocortical steroids in man. Rec. Progr. Hormone Res. **15,** 231 (1959).

PETERSON, R. E., KARRER, A., GUERRA, S. L.: Evaluation of Silber-Porter procedure for determination of plasma hydrocortisone. Analyt. Chem. **29**, 144 (1957).

— NOKES, G., CHEN, P. S., BLACK, R. L.: Estrogens and adrenocortical function in man. J. clin. Endrocr. **20**, 495 (1960).

PHELPS, R. A., PUTNAM, F. W.: Chemical composition and molecular parameters of purified plasma proteins. In: The plasma proteins, vol. I. Ed.: F. W. PUTNAM. New York-London: Academic Press 1960, p. 143.

PINCUS, G., HIRAI, M.: Effects of oestrous cycle variations and exogenous steroid hormones on the production and secretion of corticosterone by the rat adrenal. Acta Endocr. **45**, Suppl. 90, 191 (1964).

— NAKAO, T., TAIT, J. F. (Eds.): Steroid dynamics. New York-London: Academic Press 1966.

PIZARRO, M. A.: Testosterone binding globulin: an attempt at its purification. Ann. Endocr. (Paris) **30**, 205 (1969).

— ARREDONDO, R.: Measurement of plasma testosterone by competitive protein binding. Ann. Endocr. (Paris) **30**, 159 (1969).

PLAGER, J. E.: The binding of androsterone sulfate, etiocholanolone sulfate, and dehydroisoandrosterone sulfate by human plasma protein. J. clin. Invest. **44**, 1234 (1965).

— KNOPP, R., SLAUNWHITE, W. R., JR., SANDBERG, A. A.: Cortisol binding by dog plasma. Endocrinology **73**, 353 (1963).

— MATSUI, N.: An *in vitro* demonstration of the anti-insulin action of cortisol on glucose metabolism. Endocrinology **78**, 1154 (1966).

— — The effect of serum protein binding on the physiological activity of cortisol. Proc. Second Intern. Congr. on Hormonal Steroids. Excepta Med. Intern. Congr. Series, No. 132, Amsterdam-New York 1967, p. 447.

— SCHMIDT, K. G., STAUBITZ, W. J.: Increased unbound cortisol in the plasma of estrogen-treated subjects. J. clin. Invest. **43**, 1066 (1964).

POLONOVSKI, J.: Role of plasma proteins in transport of lipids. In: Transport function of plasma proteins. Eds.: P. DESGREZ and P. M. DE TRAVERSE. Amsterdam-London-New York: Elsevier 1966, p. 45.

PORATH, J.: Gel filtration of proteins, peptides and amino acids. Biochim. Biophys. Acta (Amst.) **39**, 193 (1960).

— FLODIN, P.: Gel filtration: a method for desalting and group separation. Nature **183**, 1657 (1959).

POULIK, M. D., SMITHIES, O.: Comparison and combination of the starch-gel and filter paper electrophoretic methods applied to human sera: Two-dimensional electrophoresis. Biochem. J. **68**, 636 (1958).

POULTON, B. R., REECE, R. P.: The activity of the pituitary-adrenal cortex axis during pregnancy and lactation. Endocrinology **61**, 217 (1957).

PRESSMAN, D., GROSSBERG, A. L.: The structural basis of antibody specificity. New York-Amsterdam: Benjamin, Inc. 1968.

PUCHE, R. C., NES, W. R.: Binding of dehydroepiandrosterone sulfate to serum albumin. Endocrinology **70**, 857 (1962).

PURDY, R. H., ENGEL, L. L., ONCLEY, J. L.: The characterization of estrone sulfate from human plasma. J. biol. Chem. **236,** 1043 (1961).

QUINCEY, R. V., GRAY, C. H.: A comparison of protein binding of cortisol as measured by equilibrium dialysis and gel filtration. J. Endocr. **26,** 509 (1963).

RADDING, C. M., BRAGDON, J. H., STEINBERG, D.: Synthesis of low- and high-density lipoproteins by rat liver *in vitro.* Biochim. Biophys. Acta (Amst.) **30,** 443 (1958).

— STEINBERG, D.: Studies on synthesis and secretion of serum lipoproteins by rat liver slices. J. clin. Invest. **39,** 1560 (1960).

RAKOFF, A. E., PASCHKIS, K. E., CANTAROW, A.: Conjugated estrogens in human pregnancy serum. Amer. J. Obstet. Gynecol. **46,** 856 (1943).

RALL, J. E.: Discussion to article by SANDBERG et al. (1957).

RAMIKH, M. V., ROZEN, V. B., KERTSMAN, B. I., MAROVA, E. I.: Binding of hydrocortisone with transcortin in Itsenko-Cushing's disease and in tumors of the adrenal cortex. Probl. Endokr. (Moscow) **15,** No. 4, 35 (1969).

RANDALL, R. V., LORENZ, N., ALBERT, A.: The effect of hypophysectomy on the biologic decay of thyroidal radioiodine. Endocrinology **48,** 339 (1951).

RECK, D. C., FORTIER, C.: Adrenal ascorbic acid and corticosteroidogenesis following unilateral adrenalectomy in the rat. Proc. Soc. exp. Biol. (N. Y.) **104,** 610 (1960).

REEVES, B. D., CALHOUN, D. W.: Reliability criteria for saturation analysis of steroids by competitive protein binding. Transactions of the 2nd Karolinska Symposium on Research Methods in Reproductive Endocrinology. Steroid assay by protein binding, Geneva, March, 1970. Acta endocr. (Kbh.), Suppl. No. **147,** 61 (1970).

— DE SOUZA, M. L. A., THOMPSON, I. E., DICZFALUSY, E.: An improved method for the assay of progesterone by competitive protein binding. Acta endocr. (Kbh.) **63,** 225 (1970).

REICH, M.: Differences in association of corticosteroids with serum proteins of different animal types. Aust. J. exp. Biol. med. Sci. **38,** 175 (1960).

REIF-LEHRER, L., CHADER, G. J.: The steroid induction of glutamine synthetase in chick embryo retinas in culture. Biochim. Biophys. Acta (Amst.) **192,** 310 (1969).

RICHMOND, J. E.: Studies on the metabolism of plasma glycoproteins. Biochemistry **2,** 676 (1963).

RIEGEL, I. L., MUELLER, G. C.: Formation of a protein-bound metabolite of estradiol-16-C^{14} by rat liver homogenates. J. biol. Chem. **210,** 249 (1954).

RIEKSTINYA, G. Y., SAMOKHVALOVA, M. A., ROZEN, V. B.: The role played by hypophysis in the regulation of the binding capacity of transcortin in man. Probl. Endokr. (Moscow) **15,** No. 5, 33 (1969).

RINGOLD, H. J., BATRES, E., ROSENKRANZ, G.: Steroids. LXXXIV. Synthesis of 6-methyl hormone analogs. J. Org. Chem. **22,** 99 (1957).

RINGOLD, H. J., BOWERS, A.: The ultraviolet light absorption spectra of C-6 substituted Δ^4-3-ketones. Experientia **17**, 65 (1961).

— GRAVES, J. M. H., CLARK, A., BELLAS, T.: Studies with the 3α-hydroxysteroid dehydrogenase from *Pseudomonas testosteroni*—Enzyme-substrate complementarity as the basis of selectivity and steric specificity. Rec. Progr. Hormone Res. **23**, 349 (1967).

RIVAROLA, M. A., FOREST, M. G., MIGEON, C. J.: Testosterone, androstenedione and dehydroepiandrosterone in plasma during pregnancy and at delivery: Concentration and protein binding. J. clin. Endocr. **28**, 34 (1968).

RIVERA, E. M.: Interchangeability of adrenocortical hormones in initiating mammary secretion *in vitro*. Proc. Soc. exp. Biol. (N. Y.) **116**, 568 (1964).

ROBBINS, J., RALL, J. E.: The interaction of thyroid hormones and protein in biological fluids. Rec. Progr. Hormone Res. **13**, 161 (1957).

ROBERTS, S., SZEGO, C. M.: The nature of circulating estrogen: lipoprotein-bound estrogen in human plasma. Endocrinology **39**, 183 (1946).

— — Steroid interaction in the metabolism of reproductive target organs. Physiol. Rev. **33**, 593 (1953).

— — Biochemistry of the steroid hormones. Ann. Rev. Biochem. **24**, 543 (1955).

ROBERTSON, M. E., STIEFEL, M., LAIDLAW, J. C.: The influence of estrogen on the secretion, disposition and biologic activity of cortisol. J. clin. Endocr. **19**, 1381 (1959).

ROBINSON, D. R., JENCKS, W. P.: The effect of compounds of the urea-guanidinium class on the activity coefficient of acetyltetraglycine ethyl ester and related compounds. J. Amer. chem. Soc. **87**, 2462 (1965 a).

— — The effect of concentrated salt solutions on the activity coefficient of acetyltetraglycine ethyl ester. J. Amer. chem. Soc. **87**, 2470 (1965 b).

ROBINSON, G. B., MOLNAR, J., WINZLER, R. J.: Biosynthesis of glycoproteins. I. Incorporation of glucosamine-^{14}C into liver and plasma proteins of the rat. J. biol. Chem. **239**, 1134 (1964).

ROBINSON, H. J., BERNHARD, W. G., GRUBIN, H., WANNER, H., SEWEKOW, G. W., SILBER, R. H.: 17,21-Dihydroxy-20-ketosteroids in plasma during and after pregnancy. J. clin. Endocr. **15**, 317 (1955).

ROBINSON, R. G., FANESTIL, D. D.: Assay of aldosterone by competitive protein binding. Transactions of the 2nd Karolinska Symposium on Research Methods in Reproductive Endocrinology. Steroid assay by protein binding, Geneva, March, 1970. Acta endocr. (Kbh.), Suppl. No. **147**, 275 (1970).

ROCHEFORT, H., BAULIEU, E. E.: Récepteurs hormonaux: relations entre les „récepteurs" utérins de l'oestradiol, „8 S" cytoplasmique, et „4 S" cytoplasmique et nucléaire. C. R. Acad. Sci. (Paris) **267**, 662 (1968).

— — New *in vitro* studies of estradiol binding in castrated rat uterus. Endocrinology **84**, 108 (1969).

ROCHLIN, P.: Self-decomposition of carbon-14-labelled organic compounds. Chem. Rev. **65**, 685 (1965).

RODBARD, D., LEWALD, J. E.: Computer analysis of radioligand assay and radioimmunoassay data. Transactions of the 2nd Karolinska Symposium on Research Methods in Reproductive Endocrinology. Steroid assay by protein binding, Geneva, March, 1970, Acta endocr. (Kbh.), Suppl. No. **147,** 79 (1970).

ROE, C. R., KAPLAN, N. O.: Purification and substrate specificities of bacterial hydroxysteroid dehydrogenases. Biochemistry **8,** 5093 (1969).

ROSENBAUM, W., CHRISTY, N. P., KELLY, W. G.: Electrophoretic evidence for the presence of an estrogen binding β-globulin in human plasma. J. clin. Endocr. **26,** 1399 (1966 a).

— — — ROGINSKY, M. S., TAVERNETTI, R. R.: Augmented binding of estradiol in the plasma of male patients with cirrhosis of the liver. 2nd Intern. Congr. Hormonal Steroids, Milan, Italy, 1966. Excerpta Med. Intern. Congr. Ser. No. **111,** p. 270 (1966 b).

ROSENBERG, R. M., KLOTZ, I. M.: Dye binding methods. In: A laboratory manual of analytical methods of protein chemistry, vol. 2. Eds.: P. ALEXANDER and R. J. BLOCK. New York: Pergamon Press 1960, p. 133.

ROSENFELD, M.: Cascading ultrafiltration of proteins. Biochim. Biophys. Acta (Amst.) **107,** 148 (1965).

ROSENFIELD, R. L.: A competitive protein binding method for the measurement of plasma androstenedione. Steroids **14,** 251 (1969).

— EBERLEIN, W. E., BONGIOVANNI, A. M.: Measurement of plasma testosterone by means of competitive protein binding analysis. J. clin. Endocr. **29,** 854 (1969).

ROSENTHAL, H. E.: A graphic method for the determination and presentation of binding parameters in a complex system. Analyt. Biochem. **20,** 525 (1967).

— SLAUNWHITE, W. R., JR., SANDBERG, A. A.: Transcortin: A corticosteroid-binding protein of plasma. X. Cortisol and progesterone interplay and unbound levels of these steroids in pregnancy. J. clin. Endocr. **29,** 352 (1969 a).

— — — Transcortin: a corticosteroid-binding protein of plasma. XI. Effects of estrogens or pregnancy in guinea pigs. Endocrinology **85,** 825 (1969 b).

ROSENTHAL, S. M.: The liberation of adsorbed substances from proteins. A function of the bile salts. J. Pharmacol. exp. Ther. **25,** 449 (1925).

ROSNER, W., CHRISTY, N. P., KELLY, W. G.: Partial purification and preliminary characterization of estrogen-binding globulins from human plasma. Biochemistry **8,** 3100 (1969).

— DEAKINS, S. M.: Testosterone-binding globulin in human plasma: Studies on sex distribution and specificity. J. clin. Invest. **47,** 2109 (1968).

ROTHLIN, E.: Das Blut und seine Funktion für den Transport und die Verteilung von Wirkstoffen. Vierteljahresschr. Naturforsch. Ges. Zürich **90,** 195 (1945).

ROYCE, P. C.: Testosterone binding by a protein isolated from urine of adult male rats. Endocrinology **86**, 419 (1970).

ROZEN, V. B., ANTONICHEV, A. V.: The role of the liver in transcortin-binding of plasma hydrocortisone in guinea pigs. Probl. Endokr. Gormonoter. (Moscow) **6**, 76 (1966).

— RIEKSTINYA, G. Y., KLYACHKO, V. R.: Hydrocortisone binding with transcortin in primary myxedema. Probl. Endokr. (Moscow) **13**, No. 5, 30 (1967).

RUDD, B. T., ROSENFIELD, R. L., BONGIOVANNI, A. M., EBERLEIN, W. R.: The measurement of urinary testosterone glucuronide by competitive protein binding and validation by a double isotope derivative assay. Steroids **13**, 227 (1969).

RUDMAN, D., GARCIA, L. A., ABELL, L. L., AKGUN, S.: Observations on the protein components of human plasma high- and low-density lipoproteins. Biochemistry **7**, 3136 (1968).

— KENDALL, F. E.: Bile acid content of human serum. I. Serum bile acids in patients with hepatic disease. J. clin. Invest. **36**, 530 (1957 a).

— — Bile acid content of human serum. II. The binding of cholanic acids by human plasma proteins. J. clin. Invest. **36**, 538 (1957 b).

RUSS, E. M., RAYMUNT, J.: Influence of estrogens on total serum copper and ceruloplasmin. Proc. Soc. exp. Biol. (N. Y.) **92**, 465 (1956).

RYAN, M. F.: Isolation and characterization of genetic variants of human α_1-acid glycoprotein. University of Louisville: Doctoral dissertation 1970.

RYAN, M. T.: Examination of steroid protein interaction by ultraviolet difference spectrophotometry. Arch. Biochem. **126**, 407 (1968).

— GIBBS, G.: Analysis of ultraviolet spectral perturbations arising in the interaction of steroids and human serum albumin. Arch. Biochem. **136**, 65 (1970).

SACHS, H., KLOPSTOCK, A.: Serologische Differenzierung von Lecithin und Cholesterin. Biochem. Z. **159**, 491 (1925).

SAIFER, A., GOLDMAN, L.: The free fatty acids bound to human serum albumin. J. Lipid Res. **2**, 268 (1961).

— PALO, J.: Amino acid composition of monomeric and polymeric human serum albumin. Analyt. Biochem. **27**, 1 (1969).

SAKIZ, E.: Étude fonctionelle du dimorphisme sexuel surrénalien chez le rat. C. R. Soc. Biol. (Paris) **154**, 1191 (1960).

SAMUELS, L. T.: Discussion following the article by SANDBERG et al. (1957).

— The dynamics of steroid hormone distribution in the body, particularly the distribution of cortisol. In: Steroid dynamics. Eds.: G. PINCUS, T. NAKAO and J. F. TAIT. New York-London: Academic Press 1966, p. 385.

SANDBERG, A. A., ROSENTHAL, H., SCHNEIDER, S. L., SLAUNWHITE, W. R., JR.: Protein-steroid interactions and their role in the transport and metabolism of steroids. In: Steroid dynamics. Eds.: G. PINCUS, T. NAKAO and J. F. TAIT. New York-London: Academic Press 1966, p. 1.

SANDBERG, A. A., ROSENTHAL, H., SLAUNWHITE, W. R., JR.: Transcortin: A corticosteroid-binding protein of plasma. VIII. Parameters affecting cortisol metabolism. Proc. Sec. Intern. Congr. on Hormonal Steroids. Excerpta Med. Intern. Congr. Series, No. **132**, p. 707, 1967.

— — — Interaction of cortisol and progesterone with transcortin, albumin and unspecified proteins in pregnancy. Excerpta Med. Intern. Congr. Series, No. **157**, p. 161, 1968. Abstract No. 402.

— SLAUNWHITE, W. R., JR.: Transcortin: a corticosteroid-binding plasma protein. J. clin. Invest. **37**, 928 (1958).

— — Transcortin: a corticosteroid-binding protein of plasma. II. Levels in various conditions and the effects of estrogens. J. clin. Invest. **38**, 1290 (1959).

— — Transcortin: a corticosteroid-binding protein of plasma. V. *In vitro* inhibition of cortisol metabolism. J. clin. Invest. **42**, 51 (1963).

— — ANTONIADES, H. N.: The binding of steroids and steroid conjugates to human plasma proteins. Rec. Progr. Hormone Res. **13**, 209 (1957).

— — CARTER, A. C.: Transcortin: a corticosteroid-binding protein of plasma. III. The effects of various steroids. J. clin. Invest. **39**, 1914 (1960).

— WOODRUFF, M., ROSENTHAL, H., NIENHOUSE, S., SLAUNWHITE, W. R., JR.: Transcortin: a corticosteroid-binding protein of plasma. VII. Half-life in normal and estrogen-treated subjects. J. clin. Invest. **43**, 461 (1964).

SARCIONE, E. J.: Synthesis of α_1-acid glycoprotein by the isolated perfused rat liver. Arch. Biochem. **100**, 516 (1963).

SARKAR, P. K., DOTY, P.: The optical rotatory properties of the β-configuration in polypeptides and proteins. Proc. nat. Acad. Sci. (Wash.) **55**, 981 (1966).

SAVARD, K., KOLFF, W. J., CORCORAN, A. C.: Corticosteroids of peripheral blood. Endocrinology **50**, 366 (1952).

— WOTIZ, H. W., MARCUS, P., LEMON, H. M.: Effect of ultraviolet light on steroids during paper chromatography. J. Amer. chem. Soc. **75**, 6327 (1953).

SCANU, A.: Factors affecting lipoprotein metabolism. Advanc. Lipid Res. **3**, 63 (1965).

— TOTH, J., EDELSTEIN, C., KOGA, S., STILLER, E.: Fractionation of human serum high density lipoprotein in urea solutions. Evidence for polypeptide heterogeneity. Biochemistry **8**, 3309 (1969).

SCARAMUZZI, R. J., CALDWELL, B. V., TILLSON, S. A., THORNEYCROFT, I. H.: Neutralization of exogenous estradiol activity following active immunization procedures. Transactions of the 4th Meeting of the International Study Group for Steroid Hormones, Rome, Dec. 1969. Research on Steroids **4** (1970), in press.

SCATCHARD, G.: The attractions of proteins for small molecules and ions. Ann. N. Y. Acad. Sci. **51**, 660 (1949).

— HUGHES, W. L., JR., GURD, F. R. N., WILCOX, P. E.: The interaction of proteins with small molecules and ions. In: Chemical specificity in biological interactions. Ed.: F. R. N. GURD. New York: Academic Press 1954, p. 193.

SCHACHMAN, H. K.: Ultracentrifugation, diffusion and viscometry. Methods Enzymol. **4,** 32 (1957).

SCHELLMAN, J., LUMRY, A. R., SAMUELS, L. T.: The binding of uncharged molecules to proteins. II. Testosterone and bovine serum albumin. J. Amer. chem. Soc. **76,** 2808 (1954).

SCHERAGA, H. A.: Protein structure. New York-London: Academic Press 1961, p. 217.

— Intramolecular bonds in proteins. II. Noncovalent bounds. In: The proteins, vol. I, chapter 6. Ed.: H. NEURATH. New York-London: Academic Press 1963.

SCHEURLEN, P. G.: Über Serumeiweißveränderungen beim Diabetes mellitus. Klin. Wschr. **33,** 198 (1955).

SCHMID, K.: Preparation and properties of an acid glycoprotein prepared from human plasma. J. Amer. chem. Soc. **72,** 2816 (1950).

— Preparation and properties of serum and plasma proteins. XXIX. Separation from human plasma of polysaccharides, peptides and proteins of low molecular weight. Crystallization of an acid glycoprotein. J. Amer. chem. Soc. **75,** 60 (1953).

— Isolation, characterization and polymorphism of glycoproteins. In: Biochemistry of glycoproteins and related substances. Cystic fibrosis, Part II. Eds.: E. ROSSI and E. STOLL. Basel-New York: Karger 1968, p. 4.

— BINETTE, J. P., KAMIYAMA, S., PFISTER, V., TAKAHASHI, S.: Studies on the structure of α_1-acid glycoprotein. III. Polymorphism of α_1-acid glycoprotein and the partial resolution and characterization of its variants. Biochemistry **1,** 959 (1962).

— — TOKITA, K., MOROZ, L., YOSHIZAKI, H.: The polymorphic forms of α_1-acid glycoprotein of normal Caucasian individuals. J. clin. Invest. **43,** 2347 (1964).

— FIELD, R. A., YOSHIZAKI, H.: Constancy of α_1-acid glycoprotein variants of Caucasian patients under conditions of severe stress. J. Med. Genet. **5,** 36 (1968 a).

— KAMIYAMA, S.: Studies on the structure of α_1-acid glycoprotein. IV. Optical properties. Biochemistry **2,** 271 (1963).

— OKUYAMA, T., KAUFMANN, H.: Isolation and chemical composition of the human plasma α_1-acid glycoprotein variants. Biochim. Biophys. Acta (Amst.) **154,** 565 (1968 b).

— POLIS, A., HUNZIKER, K., FRICKE, R., YAYOSHI, M.: Partial characterization of the sialic acid-free forms of α_1-acid glycoprotein from human plasma. Biochem. J. **104,** 361 (1967).

— TOKITA, K., YOSHIZAKI, H.: The α_1-acid glycoprotein variants of normal Caucasian and Japanese individuals. J. clin. Invest. **44,** 1394 (1965).

SCHOLTAN, W.: Die Bindung der Sulfonamide an Eiweißkörper. Arzneimittel-Forsch. **15,** 1433 (1965).

SCHOLTAN, W.: Die hydrophobe Bindung der Pharmaka an Humanalbumin und Ribonukleinsäure. Intern. Congr. Chemotherapy, Vienna (Austria) B **20**/8, p. 223 (1967).

— Die hydrophobe Bindung der Pharmaka an Humanalbumin und Ribonucleinsäure. Arzneimittel-Forsch. **18**, 505 (1968).

— SCHLOSSMANN, K., ROSENKRANZ, H.: Bestimmung der Eiweißbindung von Digitalis-Präparaten mittels der Ultrazentrifuge. Arzneimittel-Forsch. **16**, 109 (1966).

— — — Die Bindung von Steroidhormonen und von Steroidguanylhydrazonen an Serumalbumin. Arzneimittel-Forsch. **18**, 767 (1968).

SCHULTZE, H. E., HEREMANS, J. F.: Molecular biology of human proteins. Nature and metabolism of extracellular proteins, vol. I. Amsterdam-London-New York: Elsevier 1966.

SCHUMAKER, V. N., ADAMS, G. H.: Circulating lipoproteins. Ann. Rev. Biochem. **38**, 113 (1969).

SCOTT, J. F., ENGEL, L. L.: Molecular interaction between purines and steroids. Biochim. Biophys. Acta (Amst.) **23**, 665 (1957).

SEAL, U. S., DOE, R. P.: Purification and properties of "transcortin", the corticosteroid-binding globulin. Fed. Proc. **20**, 179 (1961).

— — Purification and properties of transcortin, the cortisol-binding globulin from patients with cancer of the prostate. Cancer Chemother. Rept. **16**, 329 (1962 a).

— — Corticosteroid-binding globulin. I. Isolation from plasma of diethylstilbestrol-treated men. J. biol. Chem. **237**, 3136 (1962 b).

— — Corticosteroid-binding globulin: species distribution and small-scale purification. Endocrinology **73**, 371 (1963).

— — Purification, some properties and composition of the corticosteroid- and thyroxine-binding globulins from human serum. Proc. 2nd Intern. Congr. Endocrinol., Excerpta Med. Intern. Congr. Series No. **83**, 325 (1964).

— — Vertebrate distribution of corticosteroid-binding globulin and some endocrine effects on concentration. Steroids **5**, 827 (1965).

— — Corticosteroid-binding globulin: Biochemistry, physiology, and phylogeny. In: Steroid dynamics. Eds.: G. PINCUS, T. NAKAO and J. F. TAIT. New York-London: Academic Press 1966, p. 63.

— — The role of corticosteroid-binding globulin in mammalian pregnancy. Proc. 2nd Intern. Congr. on Hormonal Steroids. Excerpta Med. Intern. Congr. Series No. **132**, 697 (1967).

— — Nonprotein-bound cortisol (NPC) and corticosteroid-binding globulin (CBG) in mammalian pregnancy. Abstract, 3rd Intern. Congr. Endocr., Excerpta Med. Intern. Congr. Series No. **157**, 161 (1968).

— ERICKSON, A. W.: Hematology, blood chemistry and protein polymorphisms in the white-tailed deer *(odocoileus virginianus)*. Comp. Biochem. Physiol. **30**, 695 (1969).

SELA, M.: Antigenicity: some molecular aspects. Science **166**, 1365 (1969).

SELDEN, G. L., WESTPHAL, U.: An apparatus for continuous paper electrophoresis. J. Lab. clin. Med. **49**, 786 (1957).

SHANNON, J. A.: The excretion of phenol red by the dog. Amer. J. Physiol. **113**, 602 (1935).

SHAPIRO, B.: Lipid metabolism. Ann. Rev. Biochem. **36**, 247 (1967).

SHENKIN, H. A., FELDMAN, W.: Plasma cortisol binding globulin levels in gastro-intestinal bleeding of stress origin. J. clin. Endocr. **28**, 15 (1968).

SHERWOOD, L. M., POTTS, J. T., JR.: Conformational studies of pancreatic ribonuclease and its subtilisin-produced derivatives. J. biol. Chem. **240**, 3799 (1965).

SHOLITON, L. J., WERK, E. E., JR., MACGEE, J.: The extra-adrenal metabolism of cortisol in triparanol treated subjects. Metabolism **11**, 1265 (1962).

SHOPPEE, C. W.: Chemistry of the steroids. London: Butterworths Scientific Publications 1958.

SHORE, B., SHORE, V.: Isolation and characterization of polypeptides of human serum lipoproteins. Biochemistry **8**, 4510 (1969).

SHUTT, D. A.: Measurement of estradiol-17β in plasma by competitive protein-binding radioassay. Steroids **13**, 69 (1969).

SILBER, R. H.: Fluorimetric analysis of corticoids. Meth. biochem. Anal. **14**, 63 (1966).

— BUSCH, R. D., OSLAPAS, R.: Practical procedure for estimation of corticosterone or hydrocortisone. Clin. Chem. **4**, 278 (1958).

SINANOĞLU, O., ABDULNUR, S.: Effect of water and other solvents on the structure of biopolymers. Fed. Proc. **24**, Suppl. **15**, 12 (1965).

SINGER, B., LOSITO, C., SALMON, S.: Effect of progesterone on adrenocortical hormone secretion in normal and hypophysectomized rats. J. Endocr. **28**, 65 (1963).

SKEGGS, L. T., JR., LEONARDS, J. R.: Studies on an artificial kidney. I. Preliminary results with a new type of continuous dialyzer. Science **108**, 212 (1948).

SKELTON, F. R., BERNARDIS, L. L.: Effect of age, sex, hypophysectomy and gonadectomy on plasma corticosterone levels and adrenal weights following the administration of ACTH and stress. Experientia **22**, 551 (1966).

SKIPSKI, V. P., BARCLAY, M., BARCLAY, R. K., FETZER, V. A., GOOD, J. J., ARCHIBALD, F. M.: Lipid composition of human serum lipoproteins. Biochem. J. **104**, 340 (1967).

SLAUNWHITE, W. R., JR., LICHTMAN, M. A., SANDBERG, A. A.: Studies on phenolic steroids in human subjects: VI. Biosynthesis of estriolglucosiduronic acid-16-^{14}C by human liver. J. clin. Endocr. **24**, 638 (1964).

— LOCKIE, G. N., BACK, N., SANDBERG, A. A.: Inactivity *in vivo* of transcortin-bound cortisol. Science **135**, 1062 (1962).

— ROSENTHAL, H., SANDBERG, A. A.: Interactions of steroids with human plasma proteins. Arch. Biochem. **100**, 486 (1963).

— SANDBERG, A. A.: The binding of urinary conjugated steroids to serum albumin: a new method of extraction. Endocrinology **62**, 283 (1958).

SLAUNWHITE, W. R., JR., SANDBERG, A. A.: Transcortin: A corticosteroid-binding protein of plasma. J. clin. Invest. **38**, 384 (1959).

— — Analysis of corticosteroids. Transactions of the 2nd Karolinska Symposium on Research Methods in Reproductive Endocrinology. Steroid assay by protein binding, Geneva, March, 1970. Acta endocr. (Kbh.), Suppl. No. **147**, 144 (1970).

— SCHNEIDER, S., WISSLER, F. C., SANDBERG, A. A.: Transcortin: A corticosteroid-binding protein of plasma. IX. Isolation and characterization. Biochemistry **5**, 3527 (1966).

SLUYSER, M.: Interaction of carcinogens with histones *in vitro*. Biochim. Biophys. Acta (Amst.) **154**, 606 (1968).

SMITH, E. K., MUEHLBAECHER, C. A.: A fluorometric method for plasma cortisol and transcortin. Clin. Chem. **15**, 961 (1969).

SMITH, E. L., BROWN, D. M., WEIMER, H. E., WINZLER, R. J.: Sedimentation, diffusion, and molecular weight of a mucoprotein from human plasma. J. biol. Chem. **185**, 569 (1950).

SMITH, W. W., SMITH, H. W.: Protein binding of phenol red, diodrast and other substances in plasma. J. biol. Chem. **124**, 107 (1938).

SMYTH, D. G., ELLIOTT, D. F.: Some analytical problems involved in determining the structure of proteins and peptides. J. Soc. Anal. Chem. **89**, 81 (1964).

SOBER, H. A., PETERSON, E. A.: Protein chromatography on ion exchange cellulose. Fed. Proc. **17**, 1116 (1958).

SOFFER, L. J., DORFMAN, R. I., GABRILOVE, J. L.: The human adrenal gland. Philadelphia: Lea and Febiger 1961.

SOGAMI, M., FOSTER, J. F.: Isomerization reactions of charcoal-defatted bovine plasma albumin. The N-F transition and acid expansion. Biochemistry **7**, 2172 (1968).

SOKOLOVSKY, M., RIORDAN, J. F., VALLEE, B. L.: Tetranitromethane. A reagent for the nitration of tyrosyl residues in proteins. Biochemistry **5**, 3582 (1966).

SOLEM, J. H.: Plasma corticosteroids in mice. With special regard to sex difference in adrenocortical responsiveness to exogenous corticotrophin. Scand. J. Clin. Lab. Invest. **18**, Suppl. **93**, 1 (1966).

— Plasma corticosteroids in mice. With special regard to the adrenal cortical function in mice with spontaneous adrenocortical lipid depletion. Oslo (Norway): Universitetsforlagets Trykningssentral 1967.

— BRINCK-JOHNSEN, T.: An evaluation of a method for determination of free corticosteroids in minute quantities of mouse plasma. Scand. J. Clin. Lab. Invest. **17**, Suppl. **80**, 1 (1965).

SØRENSEN, S. P. L.: Enzymstudien. II. Mitteilung. Über die Messung und die Bedeutung der Wasserstoffionenkonzentration bei enzymatischen Prozessen. Biochem. Z. **21**, 131 (1909).

SOUTHCOTT, C. M., BANDY, H. E., NEWSOM, S. E., DARRACH, M.: Metabolism of adrenal steroids in the mouse. I. Observations on 20α-dihydrocorticosterone and corticosterone in the plasma of mice treated with corticotropin. Canad. J. Biochem. **34**, 913 (1956).

SOUTHREN, A. L., GORDON, G. G., TOCHIMOTO, S., OLIVO, J., SHERMAN, D. H., PINZON, G.: Testosterone and androstenedione metabolism in the polycystic ovary syndrome: Studies of the percentage binding of testosterone in plasma. J. clin. Endocr. **29**, 1356 (1969).

SPECTOR, A. A., JOHN, K., FLETCHER, J. E.: Binding of long-chain fatty acids to bovine serum albumin. J. Lipid Res. **10**, 56 (1969).

SPENCER-PEET, J., DALY, J. R., SMITH, V.: A simple method for improving the specificity of the fluorimetric determination of adrenal corticosteroids in human plasma. J. Endocr. **31**, 235 (1965).

SPRUNT, D. H., DULANEY, A. D., CONGER, R.: Methods of preparation and antigenic activity of estrogen-protein conjugates. Cancer Res. **11**, 282 (1951).

SQUIRE, P. G., DELIN, S., PORATH, J.: Physical and chemical characterization of hydroxysteroid dehydrogenase from *Pseudomonas testosteroni*. Biochim. Biophys. Acta (Amst.) **89**, 409 (1964).

STARK, G. R., STEIN, W. H., MOORE, S.: Reactions of the cyanate present in aqueous urea with amino acids and proteins. J. biol. Chem. **235**, 3177 (1960).

STEENO, O., DE MOOR, P.: The corticosteroid-binding capacity of plasma transcortin in mammals and aves. Bull. Soc. Roy. Zool. d'Anvers, Bulletin No. **38**, p. 3 (1966).

— — Absence of a specific protein-testosterone binding in human seminal plasma. Andrologie **1**, 49 (1969).

— HEYNS, W., VAN BAELEN, H., DE MOOR, P.: Testosterone binding in human plasma. Ann. Endocr. (Paris) **29**, 141 (1968).

— LLOYD, C. W., Effect of constant light and postnatal testosterone sterilization on plasma transcortin capacity in rats. Ann. Endocr. (Paris) **28**, 720 (1967).

— SCHIRREN, C., DE MOOR, P.: Les glucocorticoïdes dans le sperme humain. Ann. Endocr. (Paris) **26**, 208 (1965).

STEIN, H. H.: Studies of binding by macromolecules. A new dialysis technique for obtaining quantitative data. Analyt. Biochem. **13**, 305 (1965).

STEINBERG, I. Z., SCHACHMAN, H. K.: Ultracentrifugation studies with absorption optics. V. Analysis of interacting systems involving macromolecules and small molecules. Biochemistry **5**, 3728 (1966).

STEINETZ, B. G., BEACH, V. L.: Some influences of thyroid on the pituitary-adrenal axis. Endocrinology **72**, 45 (1963).

— — DIPASQUALE, G., BATTISTA, J. V., JR.: Effects of different gestagenic steroid types on plasma free corticosteroid levels in ACTH-treated rats. Steroids **5**, 93 (1965).

STEINHARDT, J., BEYCHOK, S.: Interaction of proteins with hydrogen ions and other small ions and molecules. In: The proteins, vol. II, chapter 8. Ed.: H. NEURATH. New York-London: Academic Press 1964.

— REYNOLDS, J. A.: Multiple equilibria in proteins. New York: Academic Press 1970.

STENLAKE, J. B., DAVIDSON, A. G., JASANI, M. K., WILLIAMS, W. D.: The effect of acetylsalicylic acid, phenylbutazone and indomethacin on the

binding of 11-hydroxysteroids to plasma proteins in patients with rheumatoid arthritis. J. Pharm. Pharmacol. **20**, Suppl., 248 S (1968).
— WILLIAMS, W. D., DAVIDSON, A. G., DOWNIE, W., WHALEY, K.: The effect of anti-inflammatory drugs on the protein binding of 11-hydroxysteroids in human plasma *in vitro*. J. Pharm. Pharmacol. **21**, 451 (1969).
STOFFEL, W.: The chemistry of mammalian lipids. In: Lipids and lipidoses. Ed.: G. SCHETTLER. New York: Springer 1967, p. 1.
STONE, M. J., METZGER, H.: Study of macromolecular interactions by equilibrium molecular sieving. J. biol. Chem. **243**, 5049 (1968).
STORM VAN LEEUWEN, W.: On the influence of colloids on the action of non-colloidal drugs. I. J. Pharmacol. exp. Ther. **17**, 1 (1920).
— On sensitiveness to drugs in animals and men. J. Pharmacol. exp. Ther. **24**, 13 (1924 a).
— On antagonism of drugs. J. Pharmacol. exp. Ther. **24**, 21 (1924 b).
— ZEIJDNER, J.: On the influence of colloids on the action of non-colloidal drugs. II. J. Pharmacol. exp. Ther. **17**, 121 (1920).
STROTT, C. A., LIPSETT, M. B.: Measurement of 17-hydroxyprogesterone in human plasma. J. clin. Endocr. **28**, 1426 (1968).
— WEST, C. D., NAKAGAWA, K., KONDO, T., TYLER, F. H.: Plasma 11-deoxycorticosteroid and ACTH response to metyrapone (plasma metyrapone test). J. clin. Endocr. **29**, 6 (1969).
STRUCK, H.: Über die Bindung von Östron, 17β-Östradiol und Östriol an Proteine *in vitro*. Hoppe-Seylers Z. physiol. Chem. **333**, 89 (1963).
SUNAGA, K., KOIDE, S. S.: Interaction of calf thymus histones and DNA with steroids. Steroids **9**, 451 (1967).
SUTER, E.: Über die Bindung herzwirksamer Glykoside im Blutserum. Helv. physiol. pharmacol. Acta **2**, 211 (1944).
SZEGO, C. M.: The influence of the liver upon estrogen-protein binding *in vitro*. Endocrinology **52**, 669 (1953).
— The loss of estrogen-protein binding capacity in rat hepatoma. Endocrinology **57**, 541 (1955).
— Steroid interaction in the *in vitro* biosynthesis of steroid-protein complexes. In: Hormones in blood. Ciba Found. Colloq. Endocrinol., vol. 11. Eds.: G. E. W. WOLSTENHOLME and E. C. P. MILLAR. Boston: Little, Brown and Co. 1957, p. 286.
— The interaction of aldosterone and of cholesterol in the *in vitro* biosynthesis of estroprotein. Endocrinology **67**, 526 (1960).
— ROBERTS, S.: The nature of circulating estrogen. Proc. Soc. exp. Biol. (N. Y.) **61**, 161 (1946).
— — The determination of protein-bound blood estrogen. Endocrinology **41**, 322 (1947).
— — Steroid action and interaction in uterine metabolism. Rec. Progr. Hormone Res. **8**, 419 (1953).
— — Paper electrophoretic studies of protein-bound estrogen formed *in vitro*. Fed. Proc. **14**, 150 (1955).
— — Hepatic intervention in the binding of estrogen to rat serum albumin *in vitro*. J. biol. Chem. **221**, 619 (1956).

TAIT, J. F., BURSTEIN, S.: *In vivo* studies of steroid dynamics in man. In: The hormones, vol. V, chapter IV. Eds.: G. PINCUS, K. V. THIMANN and E. B. ASTWOOD. New York-London: Academic Press 1964.

TALALAY, P.: Molecular specificity in the enzymatic oxidation of steroids. Record Chem. Progr. Kresge-Hooker Sci. Lib. **18**, 31 (1957).

— Hydroxysteroid dehydrogenases. Methods Enzymol. **5**, 512 (1962).

— Enzymatic mechanisms in steroid biochemistry. Ann. Rev. Biochem. **34**, 347 (1965).

— BOYER, J.: Preparation of crystalline Δ^5-3-ketosteroid isomerase from *Pseudomonas testosteroni*. Biochim. Biophys. Acta (Amst.) **105**, 389 (1965).

— HURLOCK, B., WILLIAMS-ASHMAN, H. G.: On a coenzymatic function of estradiol-17β. Proc. natl. Acad. Sci. (Wash.) **44**, 862 (1958).

— MARCUS, P. I.: Specificity, kinetics, and inhibition of α- and β-hydroxysteroid dehydrogenases. J. biol. Chem. **218**, 675 (1956).

— WANG, V. S.: Enzymic isomerization of Δ^5-3-ketosteroids. Biochim. Biophys. Acta (Amst.) **18**, 300 (1955).

— WILLIAMS-ASHMAN, H. G.: Participation of steroid hormones in the enzymatic transfer of hydrogen. Rec. Progr. Hormone Res. **16**, 1 (1960).

TALIAFERRO, I., COBEY, F., LEONE, L.: Effect of diethylstilbestrol on plasma 17-hydroxycorticosteroid levels in humans. Proc. Soc. exp. Biol. (N. Y.) **92**, 742 (1956).

TALWALKER, P. K., NICOLL, C. S., MEITES, J.: Induction of mammary secretion in pregnant rats and rabbits by hydrocortisone acetate. Endocrinology **69**, 802 (1961).

TANFORD, C.: Transport processes. Multiple equilibria. Physical chemistry of macromolecules, chapters 6 and 8. New York and London: Wiley 1961.

— Contribution of hydrophobic interactions to the stability of the globular conformation of proteins. J. Amer. chem. Soc. **84**, 4240 (1962).

TAVERNETTI, R. R., CHRISTY, N. P., KELLY, W. G., ROGINSKY, M. S.: Abnormal protein binding of estradiol in cirrhosis. 46th Meeting Endocrine Soc. Program p. 157 (Abstract), No. 270 (1964).

— ROSENBAUM, W., KELLY, W. G., CHRISTY, N. P., ROGINSKY, M. S.: Evidence for the presence in human plasma of an estrogen-binding factor other than albumin: abnormal binding of estradiol in men with hepatic cirrhosis. J. clin. Endocr. **27**, 920 (1967).

TAYLOR, J. F.: The isolation of proteins. In: The proteins, vol. I, part A. Eds.: H. NEURATH and K. BAILEY. New York: Academic Press 1953, p. 1.

THEORELL, H., TANIGUCHI, S., ÅKESON, Å., SKURSKY, L.: Crystallization of a separate steroid-active liver alcohol dehydrogenase. Biochem. biophys. Res. Commun. **24**, 603 (1966).

THERRIAULT, D. G., TAYLOR, J. F.: Dimerization of serum albumin on extraction with an organic solvent. Biochem. biophys. Res. Commun. **3**, 560 (1960).

THERRIAULT, D. G., TAYLOR, J. F.: Lipid-protein interactions. J. Amer. Oil Chemists' Soc. **41**, 490 (1964).

THORNEYCROFT, I. H., TILLSON, S. A., SCARAMUZZI, R., CALDWELL, B. V.: Preparation and purification of antibodies to steroids. In: Immunological methods in steroid determination. Eds.: F. G. PÉRON and B. V. CALDWELL. New York: Appleton-Century Crofts (1970) (in press).

THORNTON, M., HUGHES, E. R., KELLEY, V. C., ELY, R. S.: Influence of age on plasma adrenocorticosteroids in rats, rabbits and guinea pigs. Amer. J. Physiol. **202**, 392 (1962).

THORP, J. M.: The influence of plasma proteins on the action of drugs. In: Absorption and distribution of drugs. Ed.: T. B. BINNS. Edinburgh: Livingstone Ltd. 1964, p. 64.

TILLSON, S. A., THORNEYCROFT, I. H., ABRAHAM, G. E., CALDWELL, B. V.: Solid phase radioimmunoassay of steroids. In: Immunological methods in steroid determination. Eds.: F. G. PÉRON and B. V. CALDWELL. New York: Appleton-Century Crofts (1970) (in press).

TISELIUS, A.: Electrophoresis. Methods. Enzymol. **4**, 3 (1957).

TOKITA, K., BURKE, J. F., YOSHIZAKI, H., FISCHER, S., SCHMID, K.: The constancy of the α_1-acid glycoprotein variants of normal adults under conditions of severe stress. J. clin. Invest. **45**, 1624 (1966).

— SCHMID, K.: The α_1-acid glycoprotein variants of certain Caucasian patients. Clin. Chim. Acta **17**, 39 (1967).

TOLBERT, B. M., ADAMS, P. T., BENNETT, E. L., HUGHES, A. M., KIRK, M. R., LEMMON, R. M., NOLLER, R. M., OSTWALD, R., CALVIN, M.: Observations on the radiation decomposition of some C^{14}-labelled compounds. J. Amer. chem. Soc. **75**, 1867 (1953).

TOMKINS, G. M., YIELDING, K. L.: Action of steroid hormones on glutamic dehydrogenase. In: Actions of hormones on molecular processes. Eds.: G. LITWACK and D. KRITCHEVSKY. New York-London: Academic Press 1964, p. 209.

— — CURRAN, J. F., SUMMERS, M. R., BITENSKY, M. W.: The dependence of the substrate specificity on the conformation of crystalline glutamate dehydrogenase. J. biol. Chem. **240**, 3793 (1965).

TORIBARA, T. Y.: Centrifuge type of ultrafiltration apparatus. Analyt. Chem. **25**, 1286 (1953).

— TEREPKA, A. R., DEWEY, P. A.: The ultrafiltrable calcium of human serum. I. Ultrafiltration methods and normal values. J. clin. Invest. **36**, 738 (1957).

TOUCHSTONE, J. C., CHANG, E.: Protein binding of estriol conjugates. Proc. Soc. exp. Biol. (N. Y.) **112**, 694 (1963).

TRAPP, G. A., WEST, C. D.: Determination of corticosteroid-binding proteins by an adsorption method. J. Lab. clin. Med. **73**, 861 (1969).

TRIA, E., SCANU, A. M.: Structural and functional aspects of lipoproteins in living systems. New York-London: Academic Press 1969.

TROENSEGAARD, N., KOUDAHL, B.: Cholesterin als prosthetische Gruppe im Serum-globulin. VIII. Mitteilung. Z. Physiol. Chem. **153**, 111 (1926).

TROOP, R. C.: Influence of gonadal hormones on the metabolism of cortisone. Endocrinology **64**, 671 (1959).

— BIGGS, J. T., JR.: Effect of heparin on A ring reduction of adrenal steroids by liver homogenates. Metabolism **14**, 867 (1965).

TS'O, P. O. P., LU, P.: Interaction of nucleic acids. I. Physical binding of thymine, adenine, steroids, and aromatic hydrocarbons to nucleic acids. Proc. natl. Acad. Sci. (Wash.) **51**, 17 (1964).

TURNER, R. B.: Physical and chemical properties of the steroids related to protein binding. In: Chemical specificity in biological interactions. Ed.: F. R. N. GURD. New York: Academic Press 1954, p. 29.

ULRICH, F., LONG, C. N. H.: Effects of stress on serum C^{14} levels in rats following administration of hydrocortisone-4-C^{14} and corticosterone-4-C^{14}. Endocrinology **59**, 170 (1956).

ULRICH, R., SLUSHER, M. A.: Blood levels of ACTH in individual adrenalectomized rats. Endocrinology **75**, 483 (1964).

UNGAR, F., HALBERG, F.: *In vitro* exploration of a circadian rhythm in adrenocorticotropic activity of C mouse hypophysis. Experientia **19**, 158 (1963).

UPTON, G. V., BONDY, P. K.: The binding of cortisol by plasma protein. Arch. Biochem. **78**, 197 (1958).

VAN BAELEN, H., DE MOOR, P.: The binding of progesterone by the α_1-acid glycoprotein. Communication at the 4th Meeting of the International Study Group for Steroid Hormones, Rome, Italy, Dec. 1969. Research on Steroids 4 (1970), in press.

— HEYNS, W., DE MOOR, P.: Measurement of urinary estrogens using adsorption on Sephadex. J. Clin. Endocr. **27**, 1056 (1967 a).

— — — The separation of estrogens on Sephadex. J. Chromatog. **30**, 226 (1967 b).

— — — Microheterogeneity of the testosterone binding globulin of human pregnancy serum demonstrated by isoelectric focusing. Ann. Endocr. (Paris) 30, 199 (1969).

— — SCHONNE, E., DE MOOR, P.: An estradiol binding globulin in human serum: partial purification. Ann. Endocr. (Paris) **29**, 153 (1968).

VANDENHEUVEL, F. A.: The origin, metabolism and structure of normal human serum lipoproteins. Canad. J. Biochem. **40**, 1299 (1962).

VANDE WIELE, R., FERIN, M.: Physiological studies. In: Immunological methods in steroid determination. Eds.: F. G. PÉRON and B. V. CALDWELL. New York: Appleton-Century Crofts 1970 (in press).

— — RAZIANO, J., DYRENFURTH, I.: Immunological aspects of steroid hormones. Transactions of the 4th Meeting of the International Study Group for Steroid Hormones. Rome, Dec. 1969. Research on Steroids 4 (1970), in press.

VAN REES, G. P., NOACH, E. L., VAN DIETEN, J. A. M. J.: Influence of testosterone on the secretion of thyrotrophin in the rat. Acta endocr. (Kbh.) **50**, 155 (1965).

VELLUZ, L., PETIT, A., PESEZ, M., BERRET, R.: Saponification de l'acetate de desoxycorticostérone. Bull. Soc. chim. Fr. 123 (1947).

VENDT, V. P.: Protein-sterol complexes. Vitaminy, Kiev. Akad. Nauk. Ukr. S. S. R., Sbornik **2,** 30 (1956); see Chem. Abstracts **51,** 18037 f. (1957).

VENNING, E. H.: Adrenal function in pregnancy. Endocrinology **39,** 203 (1946).

— The secretion of various hormones and the activity of the adrenal cortex in pregnancy. Macy Found. Conferences on Gestation **3,** 71 (1957).

VERMEULEN, A., FERIN, J.: The influence of 17α-methyl-19-nortestosterone (M.N.T.) on the metabolism of cortisol. Acta endocr. (Kbh.) **39,** 22 (1962).

— VERDONCK, L.: Studies on the binding of testosterone to human plasma. Steroids **11,** 609 (1968 a).

— — Studies on the binding of testosterone to plasma. Ann. Endocr. (Paris) **29,** Suppl., 149 (1968 b).

— — Testosterone assays by protein binding. Transactions of 2nd Karolinska Symposium on Research Methods in Reproductive Endocrinology. Steroid assay by protein binding, Geneva, March, 1970. Acta endocr. (Kbh.) Suppl. No. **147,** 239 (1970).

— — VAN DER STRAETEN, M., ORIE, N.: Capacity of the testosterone-binding globulin in human plasma and influence of specific binding of testosterone on its metabolic clearance rate. J. clin. Endocr. **29,** 1470 (1969).

VILLEE, C. A.: Some interactions of steroids, pyridine nucleotides and enzymes. Acta endocr. (Kbh.) **42,** 233 (1963).

— ENGEL, L. L. (Eds.): Mechanism of action of steroid hormones. New York-Oxford-London-Paris: Pergamon Press 1961.

VOGT, M.: Inhibition by hexoestrol of adrenocortical secretion in the rat. J. Physiol. **130,** 601 (1955).

VON EICHLER, O., FARAH, A.: The adrenal cortical hormones: Their origin, chemistry, physiology and pharmacology, part I. Berlin: Lange and Springer 1962.

VON HIPPEL, P., WONG, K. Y.: Neutral salts: the generality of their effects on the stability of macromolecular conformations. Science **145,** 577 (1964).

— — On the conformational stability of globular proteins. The effects of various electrolytes and nonelectrolytes on the thermal ribonuclease transition. J. biol. Chem. **240,** 3909 (1965).

VON MURALT, A. L.: The titration constants of multivalent substances. J. Amer. chem. Soc. **52,** 3518 (1930).

VOOGT, J. L., SAR, M., MEITES, J.: Influence of cycling, pregnancy, labor, and suckling on corticosterone-ACTH levels. Amer. J. Physiol. **216,** 655 (1969).

WAGH, P. V., BORNSTEIN, I., WINZLER, R. J.: The structure of a glycopeptide from human orosomucoid (α_1-acid glycoprotein). J. biol. Chem. **244,** 658 (1969).

WALBORG, E. F., JR., WARD, D. N.: The amino acid and carbohydrate components of human α_1-acid glycoprotein. Texas Rep. Biol. Med. **21,** 601 (1963).

WALLACE, E. Z., CARTER, A. C.: Studies on the mechanism of the plasma 17-hydroxycorticosteroid elevation induced in man by estrogens. J. clin. Invest. **39**, 601 (1960).

— SILVERBERG, H. I., CARTER, A.: Effect of ethinylestradiol on plasma 17-hydroxycorticosteroids, ACTH responsiveness and hydrocortisone clearance in man. Proc. Soc. exp. Biol. (N. Y.) **95**, 805 (1957).

WANG, D. Y., BULBROOK, R. D.: Binding of the sulphate esters of dehydroisoandrosterone, testosterone, 17-acetoxypregnenolone and pregnenolone in the plasma of man, rabbit and rat. J. Endocr. **39**, 405 (1967).

WANG, S. F., KAWAHARA, F. S., TALALAY, P.: The mechanism of the Δ^5-3-ketosteroid isomerase reaction: absorption and fluorescence spectra of enzyme-steroid complexes. J. biol. Chem. **238**, 576 (1963).

WARNER, D. T.: A proposed water-protein interaction and its application to the structure of the tobacco mosaic virus particle. Ann. N. Y. Acad. Sci. **125**, 605 (1965).

WARREN, J. C., CHEATUM, S. G.: Effect of neutral salts on enzyme activity and structure. Biochemistry **5**, 1702 (1966).

— — Stereospecificity of hydrogen transfer by bovine adrenal 3β-hydroxysteroid dehydrogenase. Biochim. Biophys. Acta (Amst.) **159**, 540 (1968).

— SALHANICK, H. A.: Measurement of cortisol-binding capacity in human plasma. Proc. Soc. exp. Biol. (N. Y.) **105**, 624 (1960).

— STOWRING, L., MORALES, M. F.: The effect of structure-disrupting ions on the activity of myosin and other enzymes. J. biol. Chem. **241**, 309 (1966).

WARREN, L.: The thiobarbituric acid assay of sialic acids. J. biol. Chem. **234**, 1971 (1959).

WAUD, D. R.: Pharmacological receptors. Pharmacol. Rev. **20**, 49 (1968).

WAUGH, D. F.: Protein-protein interactions. Advanc. Protein Chem. **9**, 325 (1954).

WEBER, G.: The binding of small molecules to proteins. In: Molecular biophysics. Eds.: B. PULLMAN and M. WEISSBLUTH. New York: Academic Press 1965, p. 369.

WEIMER, H. E., MEHL, J. W., WINZLER, R. J.: Studies on the mucoproteins of human plasma. V. Isolation and characterization of a homogeneous mucoprotein. J. biol. Chem. **185**, 561 (1950).

— WINZLER, R. J.: Comparative study of orosomucoid preparations from sera of six species of mammals. Proc. Soc. exp. Biol. (N. Y.) **90**, 458 (1955).

WEISS, J.: The chemical action of X-rays on some steroids in aqueous systems. Ciba Found. Colloq. Endocrinol. **7**, 142 (1953).

WESTPHAL, U.: Erniedrigung der Azorubinbindefähigkeit von menschlichem Serum in vitro. Hoppe Seylers Z. physiol. Chem. **293**, 141 (1953).

— Interaction between cortisol-4-C^{14} or progesterone-4-C^{14} and serum albumin as demonstrated by ultracentrifugation and electrophoresis. Endocrinology **57**, 456 (1955).

— Spectrophotometric demonstration of interaction between proteins and Δ^4-3-ketosteroid hormones. Fed. Proc. **15**, 382 (1956).

WESTPHAL, U.: Unpublished results (1957 a).
— Steroid-protein interactions. III. Spectrophotometric demonstration of interaction between proteins and progesterone, desoxycorticosterone and cortisol. Arch. Biochem. **66**, 71 (1957 b).
— Diffusion of progesterone and desoxycorticosterone into nitrocellulose (Lusteroid): A potential source of error in ultracentrifugation. J. Lab. clin. Med. **51**, 473 (1958).
— Interactions between steroids and proteins. In: Mechanism of action of steroid hormones. Eds.: C. A. VILLEE and L. L. ENGEL. New York: Pergamon Press 1961, p. 33.
— Binding of steroids to proteins. J. Amer. Oil Chemists' Soc. **41**, 481 (1964).
— Steroid-protein interactions. XII. Distribution of progesterone and corticosteroid hormones among serum proteins. Hoppe-Seylers Z. physiol. Chem. **346**, 243 (1966).
— Steroid-protein interactions. XIII. Concentrations and binding affinities of corticosteroid-binding globulins in sera of man, monkey, rat, rabbit, and guinea pig. Arch. Biochem. **118**, 556 (1967).
— Assay and properties of corticosteroid-binding globulin and other steroid-binding serum proteins. Methods Enzymol. **15**, 761 (1969).
— Corticosteroid-binding globulin and other steroid hormone carriers in the blood stream. J. Reprod. Fertil., Suppl. **10**, 15 (1970 a).
— Binding of hormones to serum proteins. In: Biochemical actions of hormones, vol. I, chapter 6. Ed.: G. LITWACK. New York: Academic Press 1970 b.
— General aspects of steroid-protein interaction. Transactions of the 4th Meeting of the International Study Group for Steroid Hormones, Rome, Dec. 1969. Research on Steroids **4** (1970 c), in press.
— Preparation and characteristics of corticosteroid-binding globulin (CBG, transcortin). Transactions of the 2nd Karolinska Symposium on Research Methods in Reproductive Endocrinology. Steroid assay by protein binding, Geneva, March, 1970. Acta endocr. (Kbh.), Suppl. No. **147**, 122 (1970 d).
— ALLEN, R. G.: Unpublished results.
— ASHLEY, B. D.: Steroid-protein interactions. IV. Influence of functional groups in Δ^4-3-ketosteroids on interaction with serum albumin and β-lactoglobulin. J. biol. Chem. **233**, 57 (1958 a).
— — Spectrometric determination of interaction between Δ^4-3-ketosteroids and proteins or other nitrogenous compounds. Proc. IVth Internatl. Congr. Biochem., Section 9—117, p. 118. Pergamon 1958 b.
— — Steroid-protein interactions. VI. Stereochemical aspects of interaction between Δ^4-3-ketosteroids and human serum albumin. J. biol. Chem. **234**, 2847 (1959).
— — Steroid protein interactions. VIII. Structure of Δ^4-3-ketosteroids in relation to spectrophotometric interaction with human serum albumin. J. biol. Chem. **237**, 2763 (1962).

WESTPHAL, U., ASHLEY, B. D., SELDEN, G. L.: Steroid-protein interactions. V. Comparison of spectrophotometric and equilibrium-dialysis procedures for determination of binding constants. J. Amer. chem. Soc. **80**, 5135 (1958).

— — — Steroid-protein interactions. VII. Interactions of progesterone and corticosteroids with human plasma proteins determined by multiple equilibrium dialysis. Arch. Biochem. **92**, 441 (1961).

— CARNIGHAN, R. H.: Interaction between progesterone and human α_1-acid glycoprotein (orosomucoid). Proc. 6th Intern. Congr. Biochem., Abstr. II-201, p. 187, Washington 1964.

— CHADER, G. J., HARDING, G. B.: Decomposition of cortisol-4-^{14}C to 11β-hydroxyandrost-4-ene-3,17-dione-4-^{14}C. Steroids **10**, 155 (1967).

— DE VENUTO, F.: Steroid-protein interactions. XI. Electrophoretic characterization of corticosteroid-binding proteins in serum of rat, man and other species. Biochim. Biophys. Acta (Amst.) **115**, 187 (1966).

— FIRSCHEIN, H. E., PEARCE, E. M.: Binding of hydrocortisone-4-C^{14} and progesterone-4-C^{14} to serum albumin, demonstrated by paper electrophoresis. Science **121**, 601 (1955).

— — — Transport of cortisol-4-C^{14} and progesterone-4-C^{14} by serum proteins as demonstrated by paper electrophoresis and paper chromatography. Amer. J. Physiol. **185**, 54 (1956).

— FORBES, T. R.: Biological inactivation of progesterone by binding to alpha-1 acid glycoprotein (orosomucoid). Endocrinology **73**, 504 (1963).

— GANGULY, M., KERKAY, J.: Effects of solvent environment on binding affinity between α_1-acid glycoprotein and progesterone. Fed. Proc. **25**, 799 (1966).

— GEDIGK, P.: Azorubin-binding capacity of normal and pathological sera. Proc. Soc. exp. Biol. (N. Y.) **76**, 838 (1951).

— — MEYER, F.: Über eine chromatographische Methode zur Charakterisierung von Serumeiweiß. Hoppe Seylers Z. physiol. Chem. **285**, 36 (1950).

— HARDING, G. B.: Unpublished results.

— PRIEST, S. G.: Increased electrophoretic mobility of albumin in the serum of rats subjected to carbon tetrachloride treatment or to tourniquet shock. J. clin. Invest. **34**, 1462 (1955).

— SCHMIDT-THOMÉ, J.: Über Δ^5-Pregnendion-(3,20), ein Isomeres des Progesterons. Ber. **69**, 889 (1936).

— STETS, J. F., PRIEST, S. G.: Influence of fatty acids and related anions on the azorubin-binding capacity of serum albumin. Arch. Biochem. **43**, 463 (1953).

— WILLIAMS, W. C., JR., ASHLEY, B. D.: Increased corticosteroid binding ("transcortin") activity in adrenalectomized and hypophysectomized rats. Proc. Soc. exp. Biol. (N. Y.) **109**, 926 (1962).

— — — DE VENUTO, F.: Proteinbindung der Corticosteroide im Serum adrenalektomierter und hypophysektomierter Ratten. Steroid-protein interactions. X. Hoppe Seylers Z. physiol. Chem. **332**, 54 (1963).

WHITE, F. H., JR., ANFINSEN, C. B.: Some relationships of structure to function in ribonuclease. Ann. N. Y. Acad. Sci. **81**, 515 (1959).

WIEST, W. G.: Purification of ovarian 20α-hydroxysteroid dehydrogenase. Methods Enzymol. **15**, 638 (1969).
— On the function of 20α-hydroxypregn-4-en-3-one during parturition of the rat. Endocrinology **83**, 1181 (1968).
WILLIAMS, E. J., FOSTER, J. F.: Investigation of bovine plasma albumin by differential ultraviolet spectroscopy. J. Amer. chem. Soc. **81**, 865 (1959).
WILLIAMS-ASHMAN, H. G., LIAO, S.: Sex hormones and hydrogen transport by isolated enzyme systems. In: Actions of hormones on molecular processes. Eds.: G. LITWACK and D. KRITCHEVSKY. New York-London: Academic Press 1964, p. 482.
WILSON, H., BORRIS, J. J., BAHN, R. C.: Steroids in the blood and urine of female mice bearing an ACTH-producing pituitary tumor. Endocrinology **62**, 135 (1958).
WINZLER, R. J.: Glycoproteins. In: Plasma proteins, vol. I. Ed.: F. W. PUTNAM. New York-London: Academic Press 1960, p. 309.
WISHNIA, A.: The solubility of hydrocarbon gases in protein solutions. Proc. natl. Acad. Sci. (Wash.) **48**, 2200 (1962).
— Substrate specificity at the alkane binding sites of hemoglobin and myoglobin. Biochemistry **8**, 5064 (1969 a).
— On the thermodynamic basis of induced fit. Specific alkane binding to proteins. Biochemistry **8**, 5070 (1969 b).
— PINDER, T. W., JR.: Hydrophobic interactions in proteins: Conformation changes in bovine serum albumin below pH 5. Biochemistry **3**, 1377 (1964).
— — Hydrophobic interactions in proteins. The alkane binding site of β-lactoglobulins A and B. Biochemistry **5**, 1534 (1966).
WONG, K. P., FOSTER, J. F.: The microheterogeneity of plasma albumin. VI. Membrane equilibrium salting-out as a method of demonstrating microheterogeneity of proteins. Biochemistry **8**, 4096 (1969 a).
— — The microheterogeneity of plasma albumin. VII. An investigation by the equilibrium salting-out method of the origins of microheterogeneity. Biochemistry **8**, 4104 (1969 b).
YAMADA, T., TAKEMURA, Y., KOBAYASHI, I., SHICHIJO, K.: Re-evaluation of the effect of estrogen on thyroid activity in the rat and its mechanism. Endocrinology **79**, 849 (1966).
YAMAGAMI, K., LABAT, J., PANDEY, R. S., SCHMID, K.: The free and buried tyrosyl residues of α_1-acid glycoprotein. Biochemistry **7**, 2873 (1968).
— SCHMID, K.: Conformational transitions associated with the release of the buried tryptophan residues of α_1-acid glycoprotein. J. biol. Chem. **242**, 4176 (1967).
YAMAUCHI, T., YAMASHINA, I.: Isolation and analysis of glycopeptides from polymorphic variants of α_1-acid glycoprotein of human plasma. J. Biochem. **66**, 213 (1969).
YANG, J. T.: The viscosity of macromolecules in relation to molecular conformation. Advanc. Protein Chem. **16**, 323 (1961).
YANKEELOV, J. A., JR.: Split-compartment mixing cells for difference spectroscopy. Analyt. Biochem. **6**, 287 (1963).

YANKEELOV, J. A., JR.: Modification of arginine in proteins by oligomers of 2,3-butanedione. Biochemistry **9**, 2433 (1970).

— KOSHLAND, D. E., JR.: Evidence for conformation changes induced by substrates of phosphoglucomutase. J. biol. Chem. **240**, 1593 (1965).

— MITCHELL, C. D., CRAWFORD, T. H.: A simple trimerization of 2,3-butanedione yielding a selective reagent for the modification of arginine in proteins. J. Amer. chem. Soc. **90**, 1664 (1968).

YANNONE, M. E., MUELLER, J. R., OSBORN, R. H.: Protein binding of progesterone in the peripheral plasma during pregnancy and labor. Steroids **13**, 773 (1969).

YATES, F. E., KELLER, N., RICHARDSON, U. I.: The biological activity of protein-bound corticosteroids. Fed. Proc. **22**, 166 (1963).

— URQUHART, J.: Control of plasma concentrations of adrenocortical hormones. Physiol. Rev. **42**, 359 (1962).

YIELDING, K. L., TOMKINS, G. M.: Inhibition of the enzymic oxidation of DPNH by steroid hormones. Proc. natl. Acad. Sci. (Wash.) **45**, 1730 (1959).

— — Structural alterations in crystalline glutamic dehydrogenase induced by steroid hormones. Proc. natl. Acad. Sci. (Wash.) **46**, 1483 (1960).

— — Studies on the interaction of steroid hormones with glutamic dehydrogenase. Rec. Progr. Hormone Res. **18**, 467 (1962).

YOSHIMI, T., LIPSETT, M. B.: The measurement of plasma progesterone. Steroids **11**, 527 (1968).

YOSHIZAKI, H., HUNZIKER, K., SCHMID, K.: The constancy of the types of α_1-acid glycoprotein variants in patients with uterectomy and irradiation. Clin. chim. Acta **23**, 147 (1969).

YOUNG, W. C.: The mammalian ovary. In: Sex and internal secretions, 3rd ed., vol. I. Ed.: W. C. YOUNG. Baltimore: Williams and Wilkins 1961 a, p. 449.

— Sex and internal secretions, 3rd ed. Baltimore: Williams and Wilkins 1961 b.

YUDAEV, N. A., ROZEN, V. B., MIKOSHA, A. S.: Plasma binding of hydrocortisone in estrogen-treated guinea pigs. Probl. Endokrinol. i. Gormonoterap. **10**, 73 (1964). Translated in Fed. Proc. **23**, Translation Suppl. T-1264 (1964).

ZANDER, J., FORBES, T. R., NEHER, R., DESAULLES, P.: Über biologisch-aktive Progesteronmetaboliten im menschlichen Organismus. Klin. Wschr. **35**, 143 (1957).

— — VON MÜNSTERMANN, A. M., NEHER, R.: Δ^4-3-Ketopregnen-20α-ol and Δ^4-3-ketopregnen-20β-ol, two naturally occurring metabolites of progesterone. Isolation, identification, biologic activity and concentration in human tissues. J. clin. Endocr. **18**, 337 (1958).

ZARROW, M. X.: Gestation. In: Sex and internal secretions, 3rd ed., vol. II. Baltimore: Williams and Wilkins 1961, p. 958.

— DENENBERG, V. H., HALTMEYER, G. C., THATCHER BRUMAGHIN, J.: Plasma and adrenal corticosterone levels following exposure of the two-day old rat to various stressors. Proc. Soc. exp. Biol. (N. Y.) **125**, 113 (1967).

ZARROW, M. X., YOCHIM, J. M., McCARTHY, J. L.: Experimental endocrinology. New York-London: Academic Press 1964, p. 273.

ZIMMERING, P. E., BEISER, S. M., ERLANGER, B. F.: Purification and some properties of anti-testosterone antibodies. J. Immunol. **95,** 262 (1965).

— LIEBERMAN, S., ERLANGER, B. F.: Binding of steroids to steroid-specific antibodies. Biochemistry **6,** 154 (1967).

ZIMMERMANN, E., CRITCHLOW, V.: Effects of diurnal variation in plasma corticosterone levels on adrenocortical response to stress. Proc. Soc. exp. Biol. (N. Y.) **125,** 658 (1967).

ZINNEMAN, H. H., JOHNSON, J. J., SEAL, U. S.: Effect of short-term therapy with cortisol on the urinary excretion of free amino acids. J. clin. Endocr. **23,** 996 (1963).

— MUSA, B. M., DOE, R. P.: Changes in plasma and urinary amino acids following estrogen administration to males. Metabolism **14,** 1214 (1965).

— SEAL, U. S., DOE, R. P.: Urinary amino acids in pregnancy, following progesterone, and estrogen-progesterone. J. clin. Endocr. **27,** 397 (1967).

ZUCKERMAN, S.: The ovary, vol. II. New York: Academic Press 1962.

Subject Index

For abbreviations, see Appendix A, p. 481

⟷ indicates "binding to"
→ indicates "influence on"
← indicates "influenced by"

Monographs on Endocrinology

Already Published:

Vol. 1: OHNO, S., Sex Chromosomes and Sex-linked Genes. With 33 figures. X, 192 pages. 1967. DM 38,—; US $ 9.50.

Vol. 2: EIK-NES, K. B., and E. C. HORNING, Gas Phase Chromatography of Steroids. With 85 figures. XV, 382 pages. 1968. DM 38,—; US $ 9.50.

Vol. 3: SULMAN, F. G., Hypothalamic Control of Lactation. With 58 figures. XII, 235 pages. 1970. DM 52,—; US $ 14.30.

Vol. 4: WESTPHAL, U., Steroid-Protein Interactions. With 144 figures. XIII, 567 pages. 1971. DM 86,—; US $ 23.70.

Vol. 5: MÜLLER, J., Regulation of Aldosterone Biosynthesis. With 19 figures. VII, 139 pages. 1971. DM 36,—; US $ 9.90.

In Preparation:

BAULIEU, E. E., Bicêtre: Current Problems in the Metabolism of Steroid Hormones.

BERSON, S. A./YALOW, R. S., New York: Immunoassay of Peptide Hormones.

BORTH, R., Toronto: Clinical Hormone Assays.

BREUER, H./RAO, G. S., Bonn: Metabolism of Estrogens.

CALDEYRO-BARCIA, R., Montevideo: Oxytocin.

COPP, D. H., Vancouver: Calcitonin.

EDELMAN, I. S., San Francisco: Aldosterone: Gene Action and Sodium Transport.

FEDERLIN, K., Ulm: Immunopathology of Insulin. Clinical and Experimental Studies.

GREGORY, R. A., Liverpool/GROSSMAN, M. I., Los Angeles: The Gastrins.

GURPIDE, E., Minneapolis: Tracer Methods in Hormone Research.

HORTON, W. E., London: Prostaglandins.

JENSEN, E./DESOMBRE, E., Chicago: Receptors and Action of Estradiol.

LUNENFELD, B., Tel Hashomer: Gonadotropins.

McKENZIE, J. M., Montreal: The Pathogenesis of Graves' Disease.

NEUMANN/STEINBECK/ELGER, Berlin: Hormones in Sexual Differentiation.

RENOLD, A. E./STAUFFACHER, W., Geneva: Pathophysiology of Diabetes.

ROBERTS, S., Los Angeles: Subcellular Mechanisms in the Regulation of Corticosteroidgenesis.

SHORT, R. V., Cambridge: The Ovary and its Hormones.

GPSR Compliance
The European Union's (EU) General Product Safety Regulation (GPSR) is a set of rules that requires consumer products to be safe and our obligations to ensure this.

If you have any concerns about our products, you can contact us on

ProductSafety@springernature.com

In case Publisher is established outside the EU, the EU authorized representative is:

Springer Nature Customer Service Center GmbH
Europaplatz 3
69115 Heidelberg, Germany

www.ingramcontent.com/pod-product-compliance
Lightning Source LLC
Chambersburg PA
CBHW071406230926
51846CB00001B/24
9783642462641